Springer Series in Statistics

Springer Series in Statistics

L. A. Goodman and W. H. Kruskal, Measures of Association for Cross Classifications. x, 146 pages, 1979.

J. O. Berger, Statistical Decision Theory: Foundations, Concepts, and Methods. xiv, 425 pages, 1980.

R. G. Miller, Jr., Simultaneous Statistical Inference, 2nd edition. xvi, 299 pages, 1981.

P. Brémaud, Point Processes and Queues: Martingale Dynamics. xviii, 354 pages, 1981.

E. Seneta, Non-Negative Matrices and Markov Chains. xv, 279 pages, 1981.

F. J. Anscombe, Computing in Statistical Science through APL. xvi, 426 pages, 1981.

J. W. Pratt and J. D. Gibbons, Concepts of Nonparametric Theory. xvi, 462 pages, 1981.

V. Vapnik, Estimation of Dependences based on Empirical Data. xvi, 399 pages, 1982.

H. Heyer, Theory of Statistical Experiments. x, 289 pages, 1982.

L. Sachs, Applied Statistics: A Handbook of Techniques. xxviii, 706 pages, 1982.

M. R. Leadbetter, G. Lindgren and H. Rootzen, Extremes and Related Properties of Random Sequences and Processes. xii, 336 pages, 1983.

H. Kres, Statistical Tables for Multivariate Analysis. xxii, 504 pages, 1983.

J. A. Hartigan, Bayes Theory. xii, 145 pages, 1983.

F. Mosteller, D.S. Wallace, Applied Bayesian and Classical Inference: The Case of *The Federalist* Papers. xxxv, 301 pages, 1984.

D. Pollard, Convergence of Stochastic Processes. xiv, 215 pages, 1984.

R.H. Farrell, Multivariate Calculation. xvi, 367 pages, 1985.

D. F. Andrews and A. M. Herzberg, Data: A Collection of Problems from Many Fields for the Student and Research Worker. xx, 442 pages, 1985.

D. F. Andrews
A. M. Herzberg

Data

A Collection of Problems from Many Fields for the Student and Research Worker

With 100 Tables and 11 Figures

Springer-Verlag
New York Berlin Heidelberg Tokyo

D. F. Andrews
Department of Statistics
University of Toronto
Toronto, Ontario
M5S 1A1
Canada

A. M. Herzberg
Department of Mathematics
Imperial College of Science
and Technology
London SW7 2BZ
United Kingdom

AMS Subject Classification: 62-07

Library of Congress Cataloging in Publication Data
Andrews, D. F. (David F.)
Data: a collection of problems from many fields for the student and research worker.
(Springer series in statistics)
Bibliography: p.
Includes index.
1. Mathematical statistics—Problems, exercises, etc.
I. Herzberg, A.M. II. Title. III. Series.
QA276.2.A53 1985 001.4'22 85-4637

Softcover reprint of the hardcover 1st edition 1985

This book was typeset by the authors, using a phototypesetter driven by a PDP-11/70 running a UNIX operating system.

9 8 7 6 5 4 3 2 1

ISBN-13: 978-1-4612-9563-1 e-ISBN-13: 978-1-4612-5098-2
DOI: 10.1007/978-1-4612-5098-2

Preface

Statistics provides tools and strategies for the analysis of data. While much has been written about the methodology, sometimes without reference to data, little has been said about the data. In this volume we present sets of data obtained from many situations without any direct reference to a particular type of analysis. Our view of the usefulness of bringing together a broad collection of sets of data has been shared by many friends and contributors.

Students of statistics need to gain facility with their art by applying their knowledge to many sets of data. Textbook examples tend to be small and selected primarily to illustrate a particular technique, thus failing to demonstrate the questioning, iterative nature of statistical analysis. The situations which gave rise to the more extensive sets of data given in this volume are colourful and interesting, and can be readily understood by laymen, students and research workers with diverse interests. These sets were often chosen for their perverse reluctance to yield under the naive application of standard procedures. They do not have correct solutions. They describe situations where the statistician can develop skills and learn the limitations of statistical methods. Research workers in statistics developing new statistical techniques need to try these on sets of data, preferably real rather than simulated, with all the warts and blemishes to which real data fall err. Professor S. M. Stigler in discussing the use of simulated data said "The principal disadvantage of such simulation is that no matter how clever the investigator is in his choice of specifications for sampling distributions, there is no guarantee that the pseudo-samples he generates are actually representative of real data. ...real data can exhibit many characteristics not allowed for in most simulation-based robustness studies."[1] For example, simulations typically do not allow for bias, serial correlation, time trend, outliers and missing values.

Professor G. E. P. Box and Dr. F. Yates expressed reservations about the encouragement of unthinking manipulation of numbers. We share their view that statistical methods should not be applied to numbers but rather to the situations giving rise to the data. These sets of data are not a substitute for the collaboration of the statistician with an ongoing project in which planning and long term goals are developed. Professor M. J. R. Healy has stated "Ultimately, the only way I know to learn to handle data is to practice under the eye of an expert in the craft."[2]

Some of the sets of data have been discussed frequently; these are included for convenience, and in some cases in more complete form than has usually

1. Stigler, S.M. (1977). Do robust estimators work with real data? (with discussion). *Ann. Statist.* **5**, 1055-1098.
2. Healy, M. J. R. (1979). Does medical statistics exist? *BIAS* **6**, 137-182.

been available. The remainder of the sets of data are relatively recent. Although many may have been published before, they are widely scattered and difficult to obtain.

As we wanted to prepare a collection of data illustrating statistical situations rather than statistical methods, we contacted those involved with the collection or the analysis of data. We asked these persons to write the descriptions accompanying each set of data or to check what we had written in order that the description of the data would convey some of the excitement the investigators had felt. The final collection is varied both in the field of study and complexity of analysis required. Of the topics covered, many readers will have sufficient understanding to use judgement rather than blind calculation. We have also tried to avoid associating situations with particular statistical methods and have included no references to methods in the index.

We sent requests for data to over 200 individuals. Each was sent a copy of the Galápagos Islands data (page 291) as an illustration. We stressed in our original letter that sets of data involving 2,000 entries would be considered large but not too great. The collection includes sets of over 10,000 entries. Only excerpts from larger sets of data have been included because the number of pages required to print them would have been excessive; these excerpts provide an indication of the nature of the data. All the data are available in machine readable form by contacting the authors. It should be noted that the tables have been set primarily so that they may be used directly by most common systems for statistical analysis. In some cases reformatting the tables will enhance their usefulness for visual interpretation. A preliminary volume comprising a few of the sets of data was prepared and distributed for comment.

We are indebted to Professor J.W. Tukey who gave us the idea for this collection. We thank all the contributors and those who introduced us to many of them. We should also like to thank in particular Mrs. L. Lapczak, Mrs. S. Gamble and Mr. D. Szechy who helped to keep the records and data straight on both sides of the Atlantic. Miss C. Siu and Mrs. C. Stagg assisted in the preparation of the manuscript and the data. Mr. Szechy was largely responsible for the production of the final version. To him we are deeply indebted. Miss J. Pindelska and Miss M. P. Thompson found not easily obtainable manuscripts. Mr. I. Clark drew a large proportion of the figures. The constant encouragement of many helped to sustain us.

This project was supported in part by the National Research Council of Canada, and more recently by the Natural Sciences and Engineering Research Council, and the Canada Council.

September 9, 1984 D.F.A.
Toronto, Canada A.M.H.

Acknowledgments

We are grateful to many institutions, journals and individuals for the use of the sets of data. These are noted under "Source" or "Contributor". We appreciate, in addition, the assistance as follows: of Cambridge University Press for the Iris Data; of the U. S. Department of Agriculture for the data by G. A. Wiebe on uniformity trials; of the Director of the Zürich Observatory for the Monthly Mean Sunspot Numbers; of the Astronomical Journal for Motions and Distances of Planetary Nebulae and Motions of Stars in a Star Cluster; the Lengths of Remissions of Children with Acute Leukemia is included with permission of the publishers Grune and Stratton, New York; of the Society of American Archaeology for the Lamoka Lake Site Determinations; and of the Elsevier Scientific Publishing Company for the Soil Data from the Province of Murcia, Spain.

Contents

Tables

The page numbers correspond to the first page of the chapter containing the table.

Introduction

The purpose of statistical analysis is to extract and assess the information contained in data which has arisen in many diverse situations. Thus, the methodology of statistics has been developed to address a wide variety of problems, while the theory of statistics organizes and studies this methodology. The evolution of both the theory and the methods is influenced by new situations and new forms of measurement.

Since the data are the raw materials to be refined by the statistician's tools, the availability of a broad collection of sets of data obtained from a large number of situations serves to illustrate many types of analyses and point out gaps in current statistical methodolgy. This collection of sets of data presents such problems. It may be considered as having two parts. The first part contains sets of data which have been used frequently to demonstrate statistical techniques; the second part contains situations showing the complexity and variety of issues facing statisticians. All the problems can be readily understood by nonspecialists. While some of the sets of data presented in the second part may have been published previously, most of them are not easily accessible. In many cases, different forms of statistical analysis lead to different conclusions; the reader, therefore, faces the challenge of finding an appropriate analysis.

Part one begins with the so-called "Fisher Iris Data", perhaps the most famous problem for the illustration of clustering and discrimination techniques. These data are measurements of sepal length and width and petal length and width for three types of iris. The data were collected by Dr. E. Anderson and first appeared in an article by Sir Ronald A. Fisher. Fisher, in his book *Design of Experiments* discussed a small set of data on the height of *Zea Mays* plants which had been either self-fertilized or cross-fertilized. The data were originally presented by Charles Darwin and are reproduced here with some relevant excerpts from Darwin's book.

The data on the number of pelts of the Canadian lynx sold by trappers to the Hudson Bay Company are often studied. Many theories prevail associating apparent cycles in these data with other apparent cycles including those of sunspot numbers. The numbers of pelts are given together with the price for pelts. The relation between price and trapping success may be studied.

A commonly used illustration of the Poisson distribution is based on data obtained by L. von Bortkewitsch on the number of deaths by horsekicks in the Prussian Army. As the discussion shows, the data are rather more complex than the usual description admits.

In 1919, Fisher went to work at the Rothamsted Experimental Station some 25 miles from London. He was hired to bring "modern statistical methods" to the task of analyzing a large amount of data collected on agricultural field trials over many years. Data had been collected on the Broadbalk field at

Rothamsted since 1844. The data for yields of grain and straw for 1852-1925 are given with a full description of the kinds of fertilizer used.

In order to discover the effects of natural soil variation, agriculturalists conduct uniformity trials. These assign the same treatment, for example, plant or fertilizer, to all parts of the land. The complete data for two trials, that of Mercer and Hall and Wiebe which are very much discussed and re-analyzed, are presented.

The Stanford Heart Transplantation Program began in October, 1967. The data are of medical importance and have been analyzed by many. The complete data on the 184 patients who had received transplants by February 1980 are given. Another well-examined problem concerns the frequency of disasters in coal mines. The data, intervals between coal-mining disasters, were given originally by Maguire, Pearson and Wynn. The data presented have been corrected and relate to a longer period.

Two other sets of historical interest complete part one.

Part two presents diverse problems from many areas. The problems typically resist simple techniques. The nature of the data requires special consideration in each case.

Data from astronomy and the earth sciences exhibit this diversity. Here are data on sunspots, rainfall, stream-flow, cloud seeding, stellar motions and emissions. These data present difficult problems for anaylsis. For example, rainfall is measured as a continuous variable with 0 as an extra-ordinary mode; soil measurements might be expected to vary slowly within a geological stratum and change abruptly at the edge.

Scientific calibration problems give rise to variation from many sources: laboratory, technician, measurement method, reference standard, time of measurement and many other factors. These sources may contribute different forms of variation: systematic, biased, stochastic, large and small. Problems from medical laboratories, chemical determinations, time co-ordination and other areas are presented.

Disease studies need additional considerations. Repeated measurements are often made on the same subjects. Frequently the number of cases is small but the amount of information is large. Typically the subject to subject variation is very much larger than the effect to be studied. In these cases the type of analysis and the form of the data used in the analysis can greatly influence the sensitivity of the investigation. Many times the challenge is to assess the influence of one factor in the presence of many other factors which cannot be controlled. Here examples of studies relating sunspots and melanoma, cancer and vitamin C, byssinosis and occupational dust, Down's syndrome and age, cancer and diet, cancer and race, and many others are presented.

Plants and animals can spread or move. The study of species and their locations and motions give rise to many nonstandard problems. Some of these require the statistical modelling of directions of motion and the incorporation of distance. Examples are given of plant species on the Galápagos Islands, the spread of rabies in western Europe, the movement of earthworms, the congregation of trees in a woodland and the proliferation of parasites.

Social institutions affect everyone. Communities, economies and organizations are measured regularly over an extended period of time. These systems change slowly aside from occasional dislocations. The study of populations and their economies is represented here with examples from agriculture, unemployment, family size studies, insurance and community organizations.

In additon to these broad classes of examples, there are many others including authorship, pattern recognition and the social behaviour of otters.

These data illustrate some of the types of problems handled by statistical analysis. Typically, several techniques may be appropriately applied.

1. Iris Data

Source Fisher, R.A. (1936). The use of multiple measurements in taxonomic problems. *Ann. Eugenics* 7, Pt. II, 179-188.

The data given in Table 1.1, taken from Fisher (1936), are the measurements of the sepal length and width and petal length and width in centimetres of fifty plants for each of three types of iris; *Iris setosa, Iris versicolor* and *Iris virginica*. These data are commonly referred to as the "Fisher Iris Data". Although the data were collected by Dr. Edgar Anderson, R. A. Fisher published the data on *Iris setosa* and *Iris versicolor* to demonstrate the use of discriminant functions. The *Iris virginica* data are used to extend Fisher's technique and to test Randolph's (1934) hypothesis that *Iris versicolor* is a polyploid hybrid of the two other species which is related to the fact that *Iris setosa* is a diploid species with 38 chromosomes, *Iris virginica* a tetraploid and *Iris versicolor* having 108 chromosomes is a hexaploid.

Although Dr. Anderson never published these data, he described [Anderson (1935)] how he collected information on irises:

> For some years I have been studying variation in irises but never before have I had the good fortune to meet such quantities of material for observation. On the simple assumption that if current theories are true, one should be able to find evidence of continuing evolution in any group of plants, I have been going around the world looking as sharply as possible at variation in irises. On any theory of evolution the differences between individuals get somehow built up, in time, into the differences between species. That is to say that by one process or another the differences which exist between one plant of *Iris versicolor* and its neighbor are compounded into the greater difference which distinguishes *Iris versicolor* from *Iris setosa canadensis*. It is a convenient theory and if it is true, we should be able to find the beginnings of such a compounding going on in our present day species. For that reason I have studied such irises as I could get to see, in as great detail as possible, measuring iris standard after iris standard and iris fall after iris fall, sitting squat-legged with record book and ruler in mountain meadows, in cypress swamps, on lake beaches, and in English parks. The result is still merely a ten year's harvest of dry statistics, only partially winnowed and just beginning to shape itself into generalizations which permit of summarization and the building of a few new theories to test by other means.
>
> I have found no other opportunity quite like the field from Ile Verte to Trois Pistoles. There for mile after mile one could gather irises at will and assemble for comparison one hundred full-blown flowers of *Iris versicolor* and of *Iris setosa canadensis*, each from a different plant, but all from the same pasture, and picked on the same day and measured at the same time by the same person with the same apparatus. The result is, to ordinary eyes, a few pages of singularly dry statistics, but to the biomathematician a juicy morsel

quite worth looking ten years to find.

After which rhapsody on the beauty of variation it must immediately be emphasized that *Iris setosa canadensis* varies but little in comparison with our other native blue flags. *Iris versicolor* in any New England pastures may produce ground colors all the way from mauve to blue and with hafts white or greenish or even sometimes quite a bright yellow at the juncture with the blade. *Iris setosa canadensis* by contrast is prevailingly uniform, its customary blue grey occasionally becoming a little lighter or a little darker or even a little more towards the purple, and its tiny petals producing odd variants in form and pattern, but presenting on the whole only a fraction of the variability of *Iris versicolor* from the same pasture.

The reasons for this uniformity are not far to seek. Its lower chromosome number is one, but a discussion of that and its bearings on the whole problem would be a treatise in itself. More important probably is the fact that by geological and biological evidence, *Iris setosa canadensis* is most certainly a remnant, a relict *[sic]* of what was before the glacial period a species widely spread in northern North America.

If we take a map and plot thereon all known occurrences of *Iris setosa* and *Iris setosa canadensis,* we shall find the former growing over a large area at the northwest corner of the continent, and the latter clustering in a fairly restricted circle about the Gulf of St. Lawrence, while in the great intervening stretch of territory, none of these irises has been collected. This is a characteristic distribution for plants which were almost exterminated from eastern North America by the continental ice sheet, but while *[sic]* managed to persist in the unglaciated areas about the Gulf of St. Lawrence from which center they have later spread. In Alaska the species itself, *Iris setosa,* is apparently quite as variable as our other American irises.

In Anderson (1928, 1933, 1936), the distribution in North America of *Iris versicolor* and *Iris virginica* is discussed, the 1936 paper describing the distribution of *Iris setosa* also. Some data are given in Anderson (1928) but not that presented by Fisher.

Anderson (1928, 1936) develops innovative and useful pictorial methods which he calls ideographs for representing and comparing multidimensional data.

References

Anderson, E. (1928). The problem of species in the northern blue flags, *Iris versicolor L.* and *Iris virginica L. Ann. Mo. Bot. Gard*. **15**, 241-332.

Anderson, E. (1933). The distribution of *Iris versicolor* in relation to the post-glacial Great Lakes. *Rhodora* **35**, 154-160.

Anderson, E. (1935). The irises of the Gaspé Peninsula. *Bull. Amer. Iris Soc.* **59**, 2-5.

Anderson, E. (1936). The species problem in *Iris. Ann. Mo. Bot. Gard*. **23**, 457-509.

Randolph, L. F. (1934). Chromosome numbers in native American and introduced species and cultivated varieties of Iris. *Bull. Amer. Iris Soc.* **52**, 61-66.

Table 1.1
Sepal Length and Width and Petal Length and Width in Centimetres of
Iris setosa, Iris versicolor and Iris virginica

Iris setosa				*Iris versicolor*				*Iris virginica*			
Sepal length	Sepal width	Petal length	Petal width	Sepal length	Sepal width	Petal length	Petal width	Sepal length	Sepal width	Petal length	Petal width
5.1	3.5	1.4	0.2	7.0	3.2	4.7	1.4	6.3	3.3	6.0	2.5
4.9	3.0	1.4	0.2	6.4	3.2	4.5	1.5	5.8	2.7	5.1	1.9
4.7	3.2	1.3	0.2	6.9	3.1	4.9	1.5	7.1	3.0	5.9	2.1
4.6	3.1	1.5	0.2	5.5	2.3	4.0	1.3	6.3	2.9	5.6	1.8
5.0	3.6	1.4	0.2	6.5	2.8	4.6	1.5	6.5	3.0	5.8	2.2
5.4	3.9	1.7	0.4	5.7	2.8	4.5	1.3	7.6	3.0	6.6	2.1
4.6	3.4	1.4	0.3	6.3	3.3	4.7	1.6	4.9	2.5	4.5	1.7
5.0	3.4	1.5	0.2	4.9	2.4	3.3	1.0	7.3	2.9	6.3	1.8
4.4	2.9	1.4	0.2	6.6	2.9	4.6	1.3	6.7	2.5	5.8	1.8
4.9	3.1	1.5	0.1	5.2	2.7	3.9	1.4	7.2	3.6	6.1	2.5
5.4	3.7	1.5	0.2	5.0	2.0	3.5	1.0	6.5	3.2	5.1	2.0
4.8	3.4	1.6	0.2	5.9	3.0	4.2	1.5	6.4	2.7	5.3	1.9
4.8	3.0	1.4	0.1	6.0	2.2	4.0	1.0	6.8	3.0	5.5	2.1
4.3	3.0	1.1	0.1	6.1	2.9	4.7	1.4	5.7	2.5	5.0	2.0
5.8	4.0	1.2	0.2	5.6	2.9	3.6	1.3	5.8	2.8	5.1	2.4
5.7	4.4	1.5	0.4	6.7	3.1	4.4	1.4	6.4	3.2	5.3	2.3
5.4	3.9	1.3	0.4	5.6	3.0	4.5	1.5	6.5	3.0	5.5	1.8
5.1	3.5	1.4	0.3	5.8	2.7	4.1	1.0	7.7	3.8	6.7	2.2
5.7	3.8	1.7	0.3	6.2	2.2	4.5	1.5	7.7	2.6	6.9	2.3
5.1	3.8	1.5	0.3	5.6	2.5	3.9	1.1	6.0	2.2	5.0	1.5
5.4	3.4	1.7	0.2	5.9	3.2	4.8	1.8	6.9	3.2	5.7	2.3
5.1	3.7	1.5	0.4	6.1	2.8	4.0	1.3	5.6	2.8	4.9	2.0
4.6	3.6	1.0	0.2	6.3	2.5	4.9	1.5	7.7	2.8	6.7	2.0
5.1	3.3	1.7	0.5	6.1	2.8	4.7	1.2	6.3	2.7	4.9	1.8
4.8	3.4	1.9	0.2	6.4	2.9	4.3	1.3	6.7	3.3	5.7	2.1
5.0	3.0	1.6	0.2	6.6	3.0	4.4	1.4	7.2	3.2	6.0	1.8
5.0	3.4	1.6	0.4	6.8	2.8	4.8	1.4	6.2	2.8	4.8	1.8
5.2	3.5	1.5	0.2	6.7	3.0	5.0	1.7	6.1	3.0	4.9	1.8
5.2	3.4	1.4	0.2	6.0	2.9	4.5	1.5	6.4	2.8	5.6	2.1
4.7	3.2	1.6	0.2	5.7	2.6	3.5	1.0	7.2	3.0	5.8	1.6
4.8	3.1	1.6	0.2	5.5	2.4	3.8	1.1	7.4	2.8	6.1	1.9
5.4	3.4	1.5	0.4	5.5	2.4	3.7	1.0	7.9	3.8	6.4	2.0
5.2	4.1	1.5	0.1	5.8	2.7	3.9	1.2	6.4	2.8	5.6	2.2
5.5	4.2	1.4	0.2	6.0	2.7	5.1	1.6	6.3	2.8	5.1	1.5
4.9	3.1	1.5	0.2	5.4	3.0	4.5	1.5	6.1	2.6	5.6	1.4
5.0	3.2	1.2	0.2	6.0	3.4	4.5	1.6	7.7	3.0	6.1	2.3
5.5	3.5	1.3	0.2	6.7	3.1	4.7	1.5	6.3	3.4	5.6	2.4
4.9	3.6	1.4	0.1	6.3	2.3	4.4	1.3	6.4	3.1	5.5	1.8
4.4	3.0	1.3	0.2	5.6	3.0	4.1	1.3	6.0	3.0	4.8	1.8
5.1	3.4	1.5	0.2	5.5	2.5	4.0	1.3	6.9	3.1	5.4	2.1
5.0	3.5	1.3	0.3	5.5	2.6	4.4	1.2	6.7	3.1	5.6	2.4
4.5	2.3	1.3	0.3	6.1	3.0	4.6	1.4	6.9	3.1	5.1	2.3
4.4	3.2	1.3	0.2	5.8	2.6	4.0	1.2	5.8	2.7	5.1	1.9
5.0	3.5	1.6	0.6	5.0	2.3	3.3	1.0	6.8	3.2	5.9	2.3
5.1	3.8	1.9	0.4	5.6	2.7	4.2	1.3	6.7	3.3	5.7	2.5
4.8	3.0	1.4	0.3	5.7	3.0	4.2	1.2	6.7	3.0	5.2	2.3
5.1	3.8	1.6	0.2	5.7	2.9	4.2	1.3	6.3	2.5	5.0	1.9
4.6	3.2	1.4	0.2	6.2	2.9	4.3	1.3	6.5	3.0	5.2	2.0
5.3	3.7	1.5	0.2	5.1	2.5	3.0	1.1	6.2	3.4	5.4	2.3
5.0	3.3	1.4	0.2	5.7	2.8	4.1	1.3	5.9	3.0	5.1	1.8

2. Darwin's Data on Growth Rates of Plants

Source Darwin, C. (1878). *The Effects of Cross and Self Fertilisation in the Vegetable Kingdom,* 2nd. edition. London: John Murray.
Contributor G.A. Barnard Brightlingsea U.K.

Since their use by R.A. Fisher (Fisher, 1966, Chapter 3), this small set of data, chosen from more than eighty such sets discussed in Charles Darwin's book (Darwin, 1878), has perhaps been subjected to more analysis by statisticians than any other such set. While Fisher quotes at length from Darwin's book the amount of background detail he gives is less than proportionate to the detailed statistical discussion that has been given to the data. Darwin was interested in the question of cross-fertilization *versus* the self-fertilization of plants. Darwin stated:[1]

> It often occurred to me that it would be advisable to try whether seedlings from cross-fertilised flowers were in any way superior to those from self-fertilised flowers. But as no instance was known with animals of any evil appearing in a single generation from the closest possible interbreeding, that is between brothers and sisters, I thought that the same rule would hold good with plants; and that it would be necessary at the sacrifice of too much time to self-fertilise and inter-cross plants during several successive generations, in order to arrive at any result. I ought to have reflected that such elaborate provisions favouring cross-fertilisation, as we see in innumerable plants, would not have been acquired for the sake of gaining a distant and slight advantage, or of avoiding a distant and slight evil. Moreover, the fertilisation of a flower by its own pollen corresponds to a closer form of inter-breeding than is possible with ordinary bi-sexual animals; so that an earlier result might have been expected. (8)
>
> I therefore determined to begin a long series of experiments with various plants, and these were continued for the following eleven years; and we shall see that in a large majority of cases the crossed beat the self-fertilised plants. Several of the exceptional cases, moreover, in which the crossed plants were not victorious, can be explained.
>
> My experiments were tried in the following manner. A single plant, if it produced a sufficiency of flowers, or two or three plants were placed under a net stretched on a frame, and large enough to cover the plant (together with the pot, when one was used) without touching it. This latter point is important, for if the flowers touch the net they may be cross-fertilised by bees, as I have known to happen; and when the net is wet the pollen may be injured. I used at first "white cotton net," with very fine meshes, but afterwards a kind of net with meshes one-tenth of an inch in diameter; and this I found by experience effectually excluded all insects excepting Thrips, which no net will exclude. On the plants thus protected several flowers were marked, and were

1. The numbers in brackets following the extracts refer to the pages in Darwin (1878).

fertilised with their own pollen; and an equal number on the same plants, marked in a different manner, were at the same time crossed with pollen from a distinct plant. The crossed flowers were never castrated, in order to make the experiments as like as possible to what occurs under nature with plants fertilised by the aid of insects. Therefore, some of the flowers which were crossed may have failed to be thus fertilised, and afterwards have been self-fertilised. But this and some other sources of error will presently be discussed. In some few cases of spontaneously self-fertile species, the flowers were allowed to fertilise themselves under the net; and in still fewer cases uncovered plants were allowed to be freely crossed by the insects which incessantly visited them. There are some great advantages and some disadvantages in my having occasionally varied my method of proceeding; but when there was any difference in the treatment, it is always so stated under the head of each species.

Care was taken that the seeds were thoroughly ripened before being gathered. Afterwards the crossed and self-fertilised seeds were in most cases placed on damp sand on opposite sides of a glass tumbler covered by a glass plate, with a partition between the two lots; and the glass was placed on the chimney-piece in a warm room. I could thus observe the germination of the seeds. Sometimes a few would germinate on one side before any on the other, and these were thrown away. But as often as a pair germinated at the same time, they were planted on opposite sides of a pot, with a superficial partition between the two; and I thus proceeded until from half-a-dozen to a score or more seedlings of exactly the same age were planted on the opposite sides of several pots. If one of the young seedlings became sickly or was in any way injured, it was pulled up and thrown away, as well as its antagonist on the opposite side of the same pot. (10-12)

The soil in the pots in which the seedlings were planted, or the seeds sown, was well mixed, so as to be uniform in composition. The plants on the two sides were always watered at the same time and as equally as possible; and even if this had not been done, the water would have spread almost equally to both sides, as the pots were not large. The crossed and self-fertilised plants were separated by a superficial partition, which was always kept directed towards the chief source of the light, so that the plants on both sides were equally illuminated. I do not believe it possible that two sets of plants could have been subjected to more closely similar conditions, than were my crossed and self-fertilised seedlings, as grown in the above described manner.

In comparing the two sets, the eye alone was never trusted. Generally the height of every plant on both sides was carefully measured, often more than once, viz., whilst young, sometimes again when older, and finally when fully or almost fully grown. (13)

The passages from Darwin quoted by Fisher are on pages 15-19. In fact there seems to be no clear evidence that Fisher himself went beyond Darwin's first chapter; since his interest in heterostylism in *Primula,* mentioned later in Darwin's book, may well have arisen from study of another book by Darwin, *The Different Forms of Flowers on Plants of the Same Species.* Had Fisher read even the section quoted here on *Zea Mays,* he might have been expected to refer to the comments made by Darwin in the third paragraph of that section.

This plant is monoecious, and was selected for trial on this account, no other such plant having been experimented on. It is also anemophilous, or is fertilised by the wind; and of such plants only the common beet had been tried. Some plants were raised in the greenhouse, and were crossed with pollen taken from a distinct plant; and a single plant, growing quite separately in a different part of the house, was allowed to fertilise itself spontaneously. The seeds thus obtained were placed on damp sand, and as they germinated in pairs of equal age were planted on the opposite sides of four very large pots; nevertheless they were considerably crowded. The pots were kept in the hothouse. The plants were first measured to the tips of their leaves when only between 1 and 2 feet in height, as shown in [Table 2.1].

The fifteen crossed plants here average 20.19, and the fifteen self-fertilised plants 17.57 inches in height; or as 100 to 87. Mr. Galton made a graphical representation, in accordance with the method described in the introductory chapter, of the above measurements, and adds the words "very good" to the curves thus formed.

Shortly afterwards one of the crossed plants in Pot I. died; another became much diseased and stunted; and the third never grew to its full height. They seemed to have been all injured, probably by some larva gnawing their roots. Therefore all the plants on both sides of this pot were rejected in the subsequent measurements. When the plants were fully grown they were again measured to the tips of the highest leaves, and the eleven crossed plants now averaged 68.1, and the eleven self-fertilised plants 62.34 inches in height; or as 100 to 91. In all four pots a crossed plant flowered before any one of the self-fertilised; but three of the plants did not flower at all. Those that flowered were also measured to the summits of the male flowers: the ten crossed plants average 66.51, and the nine self-fertilised plant 61.59 inches in height, or as 100 to 93.

A large number of the same cross and self-fertilised seeds were sown in the middle of the summer in the open ground in two long rows. Very much fewer of the self-fertilised than of the crossed plants produced flowers; but those that did flower, flowered almost simultaneously. When fully grown the ten tallest plants in each row were selected and measured to the summits of their male flowers. The crossed averaged to the tips of their leaves 54 inches in height, and the self-fertilised 44.65, or as 100 to 83; and the summits of their male flowers, 53.96 and 43.45 inches; or as 100 to 80. (233-235)

The data on *Zea Mays* have also been discussed by Box and Tiao (1973, pp. 153-), by Sprott (1978) and others. It is of interest to consider the following: (i) If, in the light of Darwin's comments (his page 234), the data from Pot 1 of the *Zea Mays* are discarded, do the remaining differences form a robust sample in the sense that alternative suppositions concerning their distribution have little effect on the conclusion to be drawn; (ii) Could the two anomalous differences, in Pots 1 and 4 of Table 2.1 arisen through the interchange of crossed-fertilized and self-fertilized seeds on the way from the chimney piece in the warm room to the pot. In the light of all the data, not only that presented in Table 1, but also on more than eighty other species, is the suggestion worth considering? How strong is the evidence in support of the superiority of crossed over selfed plants? Are selfed plants more variable than crossed? Did Darwin's careful pairing in fact increase the precision of his comparisons?

References

Box, G. E. P. and Tiao, G. C. (1973). *Bayesian Inference in Statistical Analysis.* Reading, Mass.: Addison-Wesley.

Fisher R. A. (1966). *Design of Experiments,* 8th edition. Edinburgh: Oliver and Boyd.

Sprott, D. A. (1978). Robustness and non-parametric procedures are not the only or the safe alternatives to normality. *Canadian J. Psych,* **32,** 180-185.

Table 2.1
Heights of *Zea Mays* *

Pot number	Crossed plants	Self-fertilized plants
1	23.5	17.375
1	12	20.375
1	21	20
2	22	20
2	19.125	18.375
2	21.5	18.625
3	22.125	18.625
3	20.375	15.25
3	18.25	16.5
3	21.625	18
3	23.25	16.25
4	21	18
4	22.125	12.75
4	23	15.5
4	12	18

* The data are in inches. Darwin gave the data in terms of 1/8 th inches rather than decimals.

3. Canadian Lynx Trappings

Source Jones, J.W. (1914). *Fur Farming in Canada.* Commission of Conservation.

Elton, C. and Nicholson, M. (1942). The ten-year cycle in numbers of the lynx in Canada. *J. Animal Ecology* **11**, 215-244.

The numbers of Canadian lynx trapped in the MacKenzie River District of Northwest Canada have been used to illustrate a number of methods associated with time series analysis. Many of these methods are associated with autoregressive or periodic models. Table 3.1 presents these data for the years 1821 to 1934. Table 3.2 presents lynx pelt sales of the Hudson's Bay Company together with the price paid for the period 1857 to 1911. These large numbers reflect trappings over a wider area of North America. Some of the data in this table appear to be in error. It is of interest whether variation in MacKenzie River District trappings is shared by all the areas contributing to Table 3.2. Is there a relation between price of pelts and size of catch? Any relation would affect the interpretation of an autoregressive model. Note that there may be a delay between the trapping and sale of a pelt.

Table 3.1
Annual Number of Lynx Trappings in the MacKenzie River District for the Period 1821 to 1934

Year	No. Lynx trapped	Year	No. Lynx trapped	Year	No. Lynx trapped
1821	269	1859	684	1897	587
1822	321	1860	299	1898	105
1823	585	1861	236	1899	153
1824	871	1862	245	1900	387
1825	1475	1863	552	1901	758
1826	2821	1864	1623	1902	1307
1827	3928	1865	3311	1903	3465
1828	5943	1866	6721	1904	6991
1829	4950	1867	4245	1905	6313
1830	2577	1868	687	1906	3794
1831	523	1869	255	1907	1836
1832	98	1870	473	1908	345
1833	184	1871	358	1909	382
1834	279	1872	784	1910	808
1835	409	1873	1594	1911	1388
1836	2285	1874	1676	1912	2713
1837	2685	1875	2251	1913	3800
1838	3409	1876	1426	1914	3091
1839	1824	1877	756	1915	2985
1840	409	1878	299	1916	3790
1841	151	1879	201	1917	674
1842	45	1880	229	1918	81
1843	68	1881	469	1919	80
1844	213	1882	736	1920	108
1845	546	1883	2042	1921	229
1846	1033	1884	2811	1922	399
1847	2129	1885	4431	1923	1132
1848	2536	1886	2511	1924	2432
1849	957	1887	389	1925	3574
1850	361	1888	73	1926	2935
1851	377	1889	39	1927	1537
1852	225	1890	49	1928	529
1853	360	1891	59	1929	485
1854	731	1892	188	1930	662
1855	1638	1893	377	1931	1000
1856	2725	1894	1292	1932	1590
1857	2871	1895	4031	1933	2657
1858	2119	1896	3495	1934	3396

Table 3.2
Number of Lynx Pelts and Unit Price Paid to the Hudson's Bay Company for the Years 1857-1911

Year	Number of pelts	Price *	Year	Number of pelts	Price *
1857	23362	12/8	1885	27187	11/8
1858	31642	8/1	1886	51511	18/9
1859	33757	9/6	1887	74050	9/9
1860	23226	10/ -	1888	78773	9/4
1861	15178	8/1	1889	33899	19/1
1862	7272	8/6	1890	18886	13/79(*sic*)
1863	4448	12/3	1891	11520	13/7
1864	4926	14/6	1892	8352	19/4
1865	5437	12/7	1893	8660	16/7
1866	16498	11/7	1894	12902	11/6
1867	35971	8/2	1895	20331	12/1
1868	76556	6/10	1896	36853	7/9
1869	68392	6/1	1897	56407	6/2
1870	37447	5/6	1898	39437	7/1
1871	15686	6/6	1899	26761	10/1
1872	7942	11/10	1900	15185	27/7
1873	5123	19/10	1901	4473	17/3
1874	7106	14/4	1902	5781	29/6
1875	11250	14/2	1903	9117	45/8
1876	18774	13/ -	1904	19267	24/4
1877	30508	8/3	1905	36116	27/ -
1878	42834	7/5	1906	58850	26/10
1879	27345	8/7	1907	61478	27/3
1880	17834	10/1	1908	36300	38/501 (*sic*)
1881	15386	12/7	1909	9704	87/4
1882	9443	14/8	1910	3410	123/1
1883	7599	16/9	1911	3774	102/4
1884	8061	19/2			

* Price in shillings and pence.

4. The Number of Deaths by Horsekicks in the Prussian Army

Source Bortkewitsch, L. von (1898). *Das Gesetz der kleinen Zahlen*. Leipzig: Teubner.

Suggested by P. Armitage University of Oxford

The number of men killed by horsekicks in the Prussian Army introduces the Poisson distribution to many students. The illustration was initially given by Bortkewitsch (1898). Winsor (1947) has reproduced the original tables with some discussion. Bishop, Fienberg and Holland (1975) also discuss the data.

The original data are given in Table 4.1. There are 14 corps with number of deaths recorded over twenty years during the period from 1875-1894.

References

Bishop, Y.M.M., Fienberg, S.E. and Holland, P.W. (1975). *Discrete Multivariate Analysis: Theory and Practice.* Cambridge, Mass.: The M.I.T. Press.

Winsor, C. P. (1947). Quotations "Das Gesetz der kleinen Zahlen". *Human Biology* **19**, 154-161.

Table 4.1
Number of Deaths by Horsekicks in the Prussian Army from 1875-1894 for 14 Corps

Year	G*	I*	II	III	IV	V	VI*	VII	VIII	IX	X	XI*	XIV	XV	Total
1875								1	1				1		3
1876	2				1								1	1	5
1877	2						1	1			1		2		7
1878	1	2	2	1	1						1		1		9
1879				1	1	2	2		1			2	1		10
1880		3	2	1	1	1				2	1	4	3		18
1881	1			2	1			1		1					6
1882	1	2					1		1	1	2	1	4	1	14
1883			1	2		1	2	1		1		3			11
1884	3		1					1			2		1	1	9
1885							1			2		1		1	5
1886	2	1			1	1	1			1		1	3		11
1887	1	1	2	1			3	2	1	1		1	2		15
1888		1	1			1	1					1	1		6
1889			1	1		1	1			1	2	2		2	11
1890	1	2		2		1	1	2		2	1	1	2	2	17
1891				1	1	1		1	1		3	3	1		12
1892	1	3	2		1	1	3		1	1		1	1		15
1893		1				1		2			1	3			8
1894	1								1		1	1			4
Total	16	16	12	12	8	11	17	12	7	13	15	25	24	8	196

* G indicates Guard Corps
G,I,VI and XI Corps' organization differ from the others

5. The Yields of Wheat on the Broadbalk Fields at Rothamsted

Source Rothamsted Experimental Station

Contributor G. Dyke Rothamsted Experimental Station

In 1919, Sir Ronald A. Fisher went to Rothamsted Experimental Station. He was hired to bring "modern statistical methods" to the task of analyzing a large amount of data collected on agricultural field trials over many years. Data had been collected on the Broadbalk field since 1844. The first experimental crop had been sown in the autumn of 1843 and harvested in 1844. Each year since, wheat has been sown and harvested on all or part of the field. The data for the yields of grain and straw for the "classical" period 1852-1925 are given in Table 5.1. From 1843-1851, many plot-treatments were varied from year to year; after 1925, parts of all plots were fallowed each year. The ten-year yields were given by Garner and Dyke (1969). The annual yields given in Table 5.1 were kindly supplied by G.V. Dyke and the late J.H.A. Dunwoody. Fisher (1921) discussed and analyzed the yield of grain for these data and Fisher (1924) discussed the influence of the rainfall. Figure 5.1 shows a recent plan of the Broadbalk field. Exhibit 5.1 gives the organic manures and inorganic fertilizers that were applied in various combinations to the plots.

References

Fisher, R.A. (1921). Studies in crop variation, I. An examination of the yield of dressed grain from Broadbalk. *J. Agric. Sci.* **11**, 107-135.

Fisher, R.A. (1924). The influence of rainfall on the yield of wheat at Rothamsted. *Phil. Trans. Roy. Soc., London B* **213**, 89-142.

Garner, H.V. and Dyke, G.V. (1969). The Broadbalk yields. In *Rothamsted Experimental Station, Report for 1968, Part 2.* Harpenden, U.K.: Lawes Agricultural Trust, pp. 26-49.

Grey, E. (1922). *Rothamsted Experimental Station; Reminiscences, Tales and Anecdotes of the Laboratories, Staff and Experimental Fields, 1872 - 1922.* London: Dangerfield.

Johnston, A.E. and Garner, H.V. (1969). Historical introduction. In *Rothamsted Experimental Station, Report for 1968, Part 1.* Harpenden, U.K.: Lawes Agricultural Trust, pp. 12-25.

Moffatt, J.R. (1939). Agricultural methods adopted in the Rothamsted classical and modern field experiments. *Emp. J. Exp. Agric.* 7, 251-260.

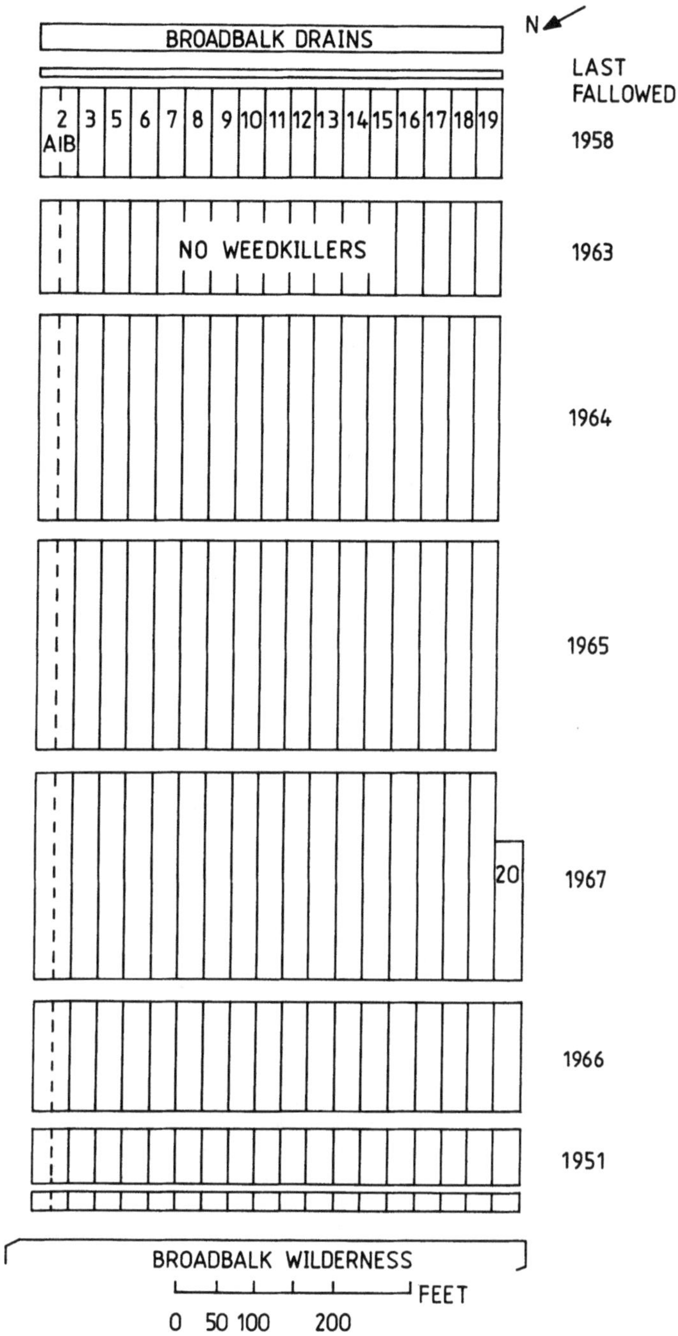

Figure 5.1
Plan of Broadbalk Showing the Arrangement of the Plots and Some Recent Modifications

Table 5.1: Yearly Yields of Grain and Straw for

	Plot							
Year	2B	3	5	6	7	8	9	10
							Grain, tonnes	
1852	1.92	0.97	1.18	1.45	1.83	1.88	1.74	1.49
1853	1.26	0.45	0.67	1.15	1.53	1.55	0.75	0.86
1854	3.00	1.61	1.74	2.48	3.30	3.52	2.77	2.66
1855	2.51	1.26	1.29	2.00	2.37	2.28	1.98	1.73
1856	2.55	1.37	1.35	1.95	2.55	2.75	2.07	1.81
1857	2.90	1.43	1.64	2.47	3.19	3.47	2.79	2.24
1858	2.82	1.32	1.35	2.06	2.79	3.00	2.16	1.80
1859	2.54	1.23	1.43	2.05	2.35	2.28	1.80	1.52
1860	2.09	0.84	1.03	1.42	1.80	1.99	1.69	1.10
1861	2.47	0.89	1.19	2.00	2.49	2.51	1.72	1.06
1862	2.74	1.18	1.24	1.97	2.50	2.76	2.43	1.71
1863	3.23	1.33	1.45	2.83	3.91	4.05	3.53	3.05
1864	2.91	1.21	1.22	2.24	3.36	3.69	3.09	2.52
1865	2.67	0.98	1.03	1.80	2.89	3.18	2.74	2.01
1866	2.32	0.91	0.94	1.45	2.12	2.30	2.26	1.98
1867	1.97	0.61	0.65	1.12	1.59	2.19	1.82	1.32
1868	2.92	1.23	1.27	2.06	2.77	3.29	2.67	1.95
1869	2.53	0.97	1.06	1.47	1.97	2.35	2.14	1.34
1870	2.64	1.10	1.35	2.22	2.94	3.32	2.62	1.63
1871	2.80	0.74	0.87	1.22	1.69	2.07	1.86	0.75
1872	2.29	0.80	0.94	1.46	2.17	2.60	2.24	1.37
1873	1.82	0.81	0.85	1.08	1.48	1.84	1.94	1.36
1874	2.72	0.80	0.89	1.74	2.74	2.86	2.06	1.77
1875	2.12	0.60	0.66	1.19	1.89	2.17	1.69	0.96
1876	1.73	0.59	0.72	1.14	1.77	2.21	1.69	0.88
1877	1.66	0.63	0.77	0.98	1.36	1.69	2.26	1.20
1878	2.12	0.87	1.01	1.62	2.31	2.77	2.10	2.00
1879	1.19	0.36	0.43	0.77	1.18	1.51	1.01	0.36
1880	2.66	0.81	1.21	1.87	2.41	2.47	1.56	0.80
1881	2.14	0.91	0.90	1.52	1.92	2.20	2.07	1.36
1882	2.25	0.77	0.87	1.64	2.52	2.60	1.98	1.81
1883	2.52	0.96	1.13	2.00	2.67	3.12	2.69	1.57
1884	2.36	0.93	1.10	1.85	2.78	3.19	2.53	1.98
1885	2.82	0.95	1.02	1.52	2.21	2.60	1.88	1.67
1886	2.61	0.65	0.80	1.56	2.50	3.02	1.68	1.15
1887	2.51	0.98	1.02	1.65	2.10	2.50	1.91	1.55
1888	2.61	0.70	0.84	1.63	2.46	2.48	1.47	0.85
1889	2.75	0.86	1.05	1.59	2.12	2.41	1.32	0.91
1890	3.49	1.32	1.32	2.45	3.06	3.13	2.15	1.87

Eighteen Plots in Broadbalk, Rothamsted, 1852-1925 *

	Plot								
Year	11	12	13	14	15	16	17	18	19
per hectare									
1852	1.60	1.68	1.66	1.70	1.67	1.96	1.74	1.00	1.77
1853	1.17	1.49	1.53	1.50	1.40	1.70	0.59	1.33	1.30
1854	3.15	3.31	3.23	3.26	3.14	3.66	3.18	1.70	2.99
1855	1.56	2.30	2.26	2.30	2.37	2.36	1.31	2.41	2.20
1856	2.21	2.34	2.27	2.44	2.22	2.73	2.20	1.27	2.31
1857	2.71	3.07	3.06	3.07	3.05	3.54	1.81	2.85	2.91
1858	2.28	2.70	2.68	2.72	2.65	3.04	2.43	1.57	2.44
1859	1.86	2.34	2.31	2.31	2.32	2.26	1.35	2.22	2.13
1860	1.44	1.81	1.78	1.76	1.75	2.10	1.66	1.06	1.61
1861	1.70	2.33	2.51	2.41	2.49	2.67	1.48	2.34	2.36
1862	1.93	2.31	2.23	2.18	2.19	2.50	1.92	1.32	1.68
1863	3.38	3.92	3.86	3.91	3.50	4.10	1.55	3.37	3.42
1864	2.58	3.23	3.18	3.07	2.73	3.76	2.63	1.26	2.73
1865	1.95	2.52	2.71	2.65	2.64	2.38	1.23	2.23	2.33
1866	2.01	2.03	1.81	2.01	1.87	1.25	1.90	0.91	1.90
1867	1.56	1.73	1.68	1.60	1.60	1.01	0.75	1.59	1.59
1868	2.48	2.93	2.82	3.07	3.11	1.64	2.73	1.35	2.71
1869	1.51	1.89	1.94	1.95	1.87	1.12	1.13	1.60	1.62
1870	1.87	2.56	2.73	2.65	2.83	1.35	2.52	1.41	2.40
1871	0.82	1.52	2.16	1.80	2.24	0.94	1.09	1.99	1.58
1872	1.97	2.10	2.17	2.18	2.27	0.93	1.88	0.94	2.01
1873	1.29	1.55	1.59	1.61	2.19	0.86	0.80	1.37	1.36
1874	2.28	2.72	2.58	2.51	2.03	0.83	2.31	0.95	2.57
1875	1.26	1.77	1.93	1.86	1.92	0.72	0.82	1.77	1.56
1876	1.01	1.38	1.82	1.56	1.86	0.74	1.86	0.70	1.29
1877	1.25	1.26	1.27	1.25	2.25	0.66	0.70	0.85	1.30
1878	2.13	2.17	2.13	2.29	1.59	0.95	2.10	1.08	1.93
1879	0.82	1.05	1.21	1.23	0.41	0.38	0.25	1.46	0.46
1880	1.79	2.05	2.30	2.14	2.56	0.99	2.30	1.03	2.31
1881	1.53	1.61	1.92	1.86	1.70	0.90	0.89	2.13	1.72
1882	2.22	2.49	2.33	2.45	2.07	0.76	2.23	1.11	2.18
1883	1.94	2.29	2.52	2.48	2.45	1.15	1.14	2.74	2.23
1884	2.31	2.60	2.43	2.62	2.51	2.63	2.47	0.93	2.29
1885	1.58	1.91	1.91	1.89	1.77	2.63	0.86	2.29	2.06
1886	1.17	1.84	2.63	2.15	1.82	3.08	2.63	0.95	2.00
1887	1.53	2.12	1.91	2.05	2.42	2.87	0.75	2.24	1.93
1888	0.80	1.62	2.27	1.76	1.98	2.27	2.19	0.88	1.95
1889	1.12	1.66	1.82	1.70	1.89	2.00	0.70	1.62	1.50
1890	2.34	2.87	3.17	2.72	3.30	2.91	2.89	1.60	2.55

Table 5.1

	Plot							
Year	2B	3	5	6	7	8	9	10
1891	3.22	0.88	0.79	1.76	2.72	2.65	1.94	1.60
1892	2.37	0.68	0.73	1.58	2.26	2.70	1.27	0.84
1893	2.52	0.78	1.02	1.42	1.48	1.58	1.00	0.61
1894	3.23	1.25	1.55	2.75	3.49	3.54	3.04	2.18
1895	3.17	0.79	1.11	1.56	2.31	2.92	1.81	0.71
1896	3.26	1.19	1.45	2.15	2.76	3.33	2.73	1.65
1897	2.72	0.69	0.93	1.43	2.09	2.73	1.78	1.24
1898	3.03	0.92	1.03	1.55	2.18	2.28	1.89	1.60
1899	3.02	0.86	0.94	1.34	2.22	2.78	1.86	1.66
1900	2.36	0.86	0.92	1.37	2.09	3.08	1.66	1.37
1901	2.83	0.85	1.06	1.61	2.12	3.00	1.80	1.47
1902	2.76	0.90	1.07	1.75	2.53	2.94	2.25	1.64
1903	2.07	0.54	0.58	1.23	1.83	2.37	1.57	0.91
1904	1.63	0.32	0.51	0.84	1.43	1.80	1.30	0.74
1905	3.02	1.40	1.84	2.47	3.15	3.19	2.80	1.33
1906	3.27	1.14	1.30	2.14	2.92	3.66	2.57	1.77
1907	2.75	0.69	0.88	1.74	2.54	2.67	2.19	2.04
1908	2.97	0.91	1.21	1.65	2.49	3.47	2.38	1.66
1909	2.78	0.68	0.78	1.30	2.08	2.59	1.82	0.87
1910	2.19	0.57	0.77	1.32	1.96	2.16	1.80	1.17
1911	2.84	0.94	1.12	1.32	1.98	2.83	2.32	1.76
1912	1.39	0.35	0.44	0.21	0.57	0.81	0.76	0.24
1913	1.70	0.39	0.52	0.86	1.27	2.00	1.21	0.86
1914	2.26	0.45	0.74	1.31	1.85	2.20	1.43	0.98
1915	2.78	0.99	1.27	2.11	2.68	2.98	2.47	1.63
1916	2.01	0.84	1.24	1.38	1.85	1.76	1.65	1.33
1917	1.23	0.62	0.69	1.42	1.81	2.13	1.57	0.93
1918	2.87	0.85	0.75	1.30	2.09	2.39	1.90	1.33
1919	2.12	0.73	0.73	1.38	2.24	2.60	1.66	1.23
1920	2.39	0.66	0.63	1.12	1.95	2.20	1.85	1.11
1921	2.23	0.76	0.65	1.14	1.57	1.70	1.23	1.23
1922	2.73	0.63	0.75	1.14	1.84	1.86	1.45	0.79
1923	1.51	0.32	0.31	0.57	1.15	1.36	0.74	0.69
1924	1.01	0.17	0.30	0.71	1.64	1.76	0.95	0.43
1925	1.34	0.46	0.52	0.76	1.38	1.43	1.11	1.08

(cont.)

	Plot								
Year	11	12	13	14	15	16	17	18	19
1891	1.68	2.41	2.59	2.47	2.68	2.83	1.00	2.15	1.93
1892	1.11	1.70	2.11	1.71	2.04	2.27	2.07	0.87	1.75
1893	0.53	0.81	1.20	0.92	1.34	1.41	0.90	1.50	1.27
1894	2.87	3.39	3.39	3.19	2.93	3.38	2.65	1.93	2.71
1895	1.09	1.63	2.07	1.54	1.81	2.37	0.90	2.07	1.62
1896	1.78	2.52	2.48	2.18	2.23	2.79	2.63	1.23	2.60
1897	1.17	1.58	2.00	1.41	1.49	2.01	0.82	2.24	1.64
1898	1.35	1.90	2.01	1.78	2.19	1.80	1.96	1.12	2.20
1899	1.53	2.04	1.87	1.99	1.89	2.66	0.98	1.89	2.02
1900	1.26	1.72	2.00	1.59	1.43	2.47	2.05	0.84	1.67
1901	1.43	1.89	2.11	1.60	2.12	2.10	1.28	2.06	1.99
1902	1.56	2.25	2.59	2.20	2.68	2.22	2.47	1.40	2.34
1903	1.05	1.18	1.86	1.16	1.43	1.86	0.41	1.77	1.26
1904	0.89	1.13	1.52	1.22	1.69	1.76	1.52	0.25	1.31
1905	1.69	2.41	3.05	2.11	2.89	2.76	1.94	2.40	1.80
1906	1.82	2.43	3.18	2.47	3.31	3.30	3.27	2.18	2.82
1907	2.50	3.05	2.55	2.74	2.48	2.66	0.89	2.26	2.15
1908	1.67	2.48	2.64	1.99	2.37	2.85	2.44	1.08	2.07
1909	0.63	1.54	2.20	1.34	2.12	2.24	0.72	2.29	1.62
1910	1.70	2.08	1.93	1.90	1.59	1.95	1.95	0.74	1.47
1911	1.49	1.98	2.27	1.80	1.86	3.16	1.05	2.11	2.19
1912	0.23	0.47	0.52	0.29	0.42	0.87	0.51	0.47	0.59
1913	0.95	1.41	1.44	1.26	1.53	1.70	0.55	1.50	1.48
1914	1.23	1.79	1.49	1.20	1.53	1.53	1.25	0.42	0.77
1915	2.34	2.70	2.60	2.46	1.51	2.61	1.27	1.90	2.04
1916	1.08	1.44	1.61	1.32	1.32	1.57	1.49	0.91	1.32
1917	1.10	1.35	1.97	1.41	1.95	2.00	0.79	1.63	0.89
1918	1.60	2.02	1.90	2.04	2.07	2.43	1.69	1.21	1.35
1919	1.11	1.47	2.06	1.77	1.58	2.48	0.95	2.20	0.95
1920	1.01	1.48	2.02	1.36	1.29	1.90	1.44	0.49	0.94
1921	0.81	1.21	1.35	1.34	1.60	1.79	0.78	1.82	1.39
1922	0.46	0.73	1.62	0.78	1.08	2.01	1.57	0.96	1.19
1923	0.71	0.87	1.05	1.02	1.26	1.25	0.34	1.06	1.18
1924	0.45	0.75	0.99	0.77	0.50	1.55	0.57	0.58	0.46
1925	1.39	1.46	1.63	1.47	1.30	1.57	0.75	1.17	0.70

Table 5.1

	Plot							
Year	2B	3	5	6	7	8	9	10
							Straw, tonnes	
1852	3.87	1.78	2.21	2.91	4.27	4.25	3.97	3.14
1853	3.78	1.73	2.29	3.12	4.19	4.42	2.61	2.65
1854	4.99	2.51	2.82	4.43	6.22	6.87	4.70	4.52
1855	4.31	2.03	2.03	3.29	4.52	4.58	4.01	3.24
1856	4.84	2.37	2.32	3.44	5.02	5.76	3.94	3.35
1857	3.72	1.75	1.88	3.09	4.24	4.90	3.84	2.95
1858	4.30	1.89	1.78	2.96	4.52	5.23	3.93	2.66
1859	5.39	2.47	2.64	4.26	5.31	6.14	4.97	3.45
1860	3.86	1.67	1.82	2.56	3.44	4.31	4.43	2.56
1861	3.48	1.49	1.74	2.95	3.95	4.32	3.71	2.29
1862	4.70	1.96	2.07	3.33	4.38	5.24	5.22	3.05
1863	4.80	1.81	1.94	4.16	6.57	7.40	5.89	4.22
1864	4.36	1.51	1.54	3.79	5.57	6.26	5.20	3.41
1865	3.47	1.22	1.32	2.26	4.11	5.16	4.38	2.84
1866	4.55	1.48	1.64	2.52	4.35	5.95	5.60	3.27
1867	3.51	1.09	1.16	1.93	3.10	4.70	4.58	2.33
1868	4.70	1.24	1.51	2.88	4.31	5.39	4.44	2.54
1869	4.41	1.52	1.79	2.54	3.60	4.39	4.71	2.49
1870	3.06	1.21	1.52	2.64	3.60	4.11	3.36	1.89
1871	5.05	1.33	1.61	2.57	3.46	4.41	4.11	1.45
1872	4.22	1.33	1.49	2.88	4.29	5.68	5.46	2.89
1873	2.76	1.06	1.17	1.70	2.27	2.99	3.53	1.84
1874	4.98	1.11	0.99	2.49	5.21	6.80	4.02	2.45
1875	4.17	1.07	1.26	1.80	3.83	4.84	3.98	1.88
1876	2.40	0.76	0.88	1.45	2.48	3.29	2.69	1.23
1877	2.54	0.87	0.99	1.32	2.06	2.51	3.58	1.58
1878	4.53	1.16	1.48	3.30	5.55	6.92	4.78	3.30
1879	2.51	0.85	0.96	1.78	3.38	4.68	3.04	1.14
1880	4.37	1.35	1.94	3.26	4.49	5.02	3.26	1.61
1881	2.65	1.11	1.01	1.83	2.47	3.26	3.06	1.51
1882	4.48	1.18	1.44	3.27	6.42	7.97	5.54	3.25
1883	3.28	1.11	1.28	2.92	4.07	4.81	4.24	1.84
1884	3.81	0.99	1.31	2.81	4.53	5.84	4.29	2.77
1885	4.38	1.17	1.36	2.22	3.70	5.18	3.17	2.35
1886	3.09	0.67	0.85	1.77	3.45	4.40	2.16	1.17
1887	3.88	0.97	1.13	2.08	3.02	4.05	2.66	2.04
1888	4.52	1.00	1.24	2.44	4.12	4.70	2.79	1.58
1889	4.86	1.10	1.31	2.15	3.65	4.58	2.36	1.55
1890	6.11	1.52	1.56	3.77	5.39	5.76	3.58	2.48

(cont.)

	Plot								
Year	11	12	13	14	15	16	17	18	19
per hectare									
1852	3.36	3.64	3.65	4.01	3.72	5.21	4.04	1.93	3.81
1853	2.93	4.13	4.27	4.05	3.99	5.56	2.28	4.02	3.60
1854	5.70	6.15	6.13	6.15	5.70	7.47	5.70	2.69	5.24
1855	3.16	4.23	4.13	4.25	4.51	5.34	2.12	4.61	4.32
1856	3.90	4.24	4.14	4.66	4.08	6.16	3.94	2.21	3.99
1857	3.29	4.05	4.12	4.10	4.18	5.26	2.24	3.82	3.58
1858	3.26	4.11	4.13	4.17	4.04	5.39	3.63	2.23	3.57
1859	3.98	5.21	5.35	5.30	5.47	6.57	2.72	5.15	4.51
1860	2.91	3.54	3.38	3.45	3.34	4.66	3.18	1.94	3.07
1861	2.93	3.66	3.82	3.74	3.81	4.91	2.10	3.60	3.63
1862	3.20	4.07	4.07	3.89	4.01	5.06	3.45	2.32	2.97
1863	4.87	6.06	6.47	6.03	5.76	7.70	2.14	5.39	5.07
1864	4.10	4.86	5.10	4.55	4.49	6.72	4.25	1.71	3.77
1865	2.84	3.48	3.79	3.54	3.65	3.24	1.65	3.17	3.35
1866	3.58	3.93	3.91	3.88	3.56	2.20	3.89	1.70	3.75
1867	2.59	2.98	3.07	2.88	3.03	1.81	1.37	2.93	2.94
1868	3.16	3.96	4.51	4.12	4.98	2.29	4.09	1.80	3.54
1869	2.75	3.16	3.41	3.40	3.34	1.84	1.90	2.74	2.88
1870	2.14	2.85	3.25	2.90	3.47	1.51	2.99	1.54	2.51
1871	1.53	2.88	4.24	3.36	4.15	1.73	2.04	3.74	3.01
1872	3.83	4.10	4.37	4.25	4.56	1.74	3.70	1.87	3.70
1873	1.77	2.23	2.32	2.41	3.45	1.29	1.23	2.14	2.08
1874	3.50	4.36	4.46	4.03	3.11	1.29	3.91	1.15	3.51
1875	2.58	3.41	3.66	3.50	3.59	1.33	1.51	3.37	2.80
1876	1.42	2.45	2.73	2.18	2.72	0.98	2.85	1.00	1.76
1877	1.57	1.67	1.76	1.76	3.12	0.84	0.87	1.19	1.63
1878	4.81	5.21	5.25	5.50	3.44	1.49	5.19	1.74	3.53
1879	2.26	2.77	3.42	3.25	1.01	0.91	0.70	3.75	1.50
1880	3.20	3.66	4.21	3.73	4.51	1.68	3.73	1.51	3.65
1881	1.79	2.11	2.54	2.45	2.21	1.04	1.01	2.86	2.03
1882	4.33	5.37	5.48	5.33	4.59	1.40	4.66	1.73	3.88
1883	2.59	3.09	3.77	3.40	3.65	1.24	1.30	4.31	3.09
1884	3.53	4.11	4.09	4.04	4.97	5.11	4.30	1.29	3.60
1885	2.53	3.20	3.66	3.23	3.09	5.45	1.18	3.59	2.97
1886	1.42	2.19	3.59	2.71	2.32	4.81	3.67	1.04	2.29
1887	2.23	2.94	2.68	2.90	3.38	4.55	1.00	3.15	2.63
1888	1.55	2.75	3.62	2.96	3.31	4.46	3.44	1.35	3.20
1889	1.82	2.81	3.86	3.39	3.07	4.56	1.05	3.48	3.40
1890	3.73	4.58	5.61	4.33	5.82	5.74	4.43	2.03	3.70

Table 5.1

	Plot							
Year	2B	3	5	6	7	8	9	10
1891	6.60	1.31	1.24	3.25	5.80	7.44	3.62	2.36
1892	3.81	0.93	1.03	2.36	3.61	4.48	2.19	1.51
1893	2.53	0.71	0.88	1.29	1.48	1.74	1.08	0.73
1894	6.34	1.64	2.15	4.49	6.76	8.13	5.48	3.26
1895	3.91	0.81	1.16	1.71	2.76	3.77	2.23	0.93
1896	5.56	1.48	1.88	3.12	4.19	5.67	3.58	2.36
1897	4.28	1.01	1.31	2.14	3.55	4.90	3.10	2.05
1898	6.99	1.54	2.01	3.28	5.58	6.86	3.93	3.10
1899	6.59	1.18	1.67	2.40	5.06	7.52	3.90	2.81
1900	4.22	1.13	1.35	2.11	3.28	4.98	2.72	1.91
1901	4.38	0.98	1.24	2.20	2.74	4.34	2.36	1.71
1902	5.88	1.17	1.43	2.60	5.01	6.04	3.59	2.12
1903	3.81	0.66	0.93	2.33	3.57	5.33	2.95	1.44
1904	2.88	0.76	0.87	1.63	2.89	3.57	2.81	1.56
1905	6.44	2.48	3.08	4.80	5.86	6.44	5.19	2.60
1906	4.83	1.28	1.57	2.85	4.07	5.28	3.38	2.09
1907	7.33	1.23	1.89	3.94	7.01	8.99	5.77	4.15
1908	4.28	0.96	1.37	2.38	3.79	5.52	3.35	1.92
1909	6.37	1.15	1.53	2.78	4.50	5.90	3.76	2.18
1910	4.84	1.17	1.31	2.59	4.35	5.65	3.82	2.40
1911	4.69	1.23	1.61	2.24	3.46	4.48	3.64	2.16
1912	2.44	0.71	0.73	0.49	1.06	1.69	1.40	0.43
1913	3.74	0.55	0.79	1.62	3.14	3.81	2.45	1.64
1914	4.94	0.65	0.97	2.00	3.42	5.27	2.49	1.58
1915	5.14	1.56	1.98	3.45	4.33	5.14	4.04	2.56
1916	4.41	1.45	2.05	2.54	4.34	4.64	3.69	3.02
1917	1.93	0.70	0.88	1.82	2.35	3.21	2.31	1.23
1918	5.21	1.19	1.26	2.44	4.33	5.64	3.55	2.24
1919	2.71	0.90	0.92	1.86	2.90	3.70	2.28	1.68
1920	5.55	1.04	1.02	1.95	3.90	5.50	3.33	2.32
1921	4.70	0.98	0.83	1.95	3.28	4.01	2.42	2.02
1922	4.46	0.96	1.21	1.83	3.38	4.20	2.37	1.69
1923	4.43	0.46	0.52	1.23	3.34	4.32	2.36	2.13
1924	2.62	0.41	0.48	1.18	3.37	4.32	2.01	1.24
1925	2.76	0.77	0.68	1.09	2.38	2.94	2.10	1.48

* Yields at threshing, about 85% dry matter.
One ton = 1.016 tonne.

(cont.)

	Plot								
Year	11	12	13	14	15	16	17	18	19
1891	3.41	4.46	5.07	4.54	4.76	6.68	1.51	3.98	3.46
1892	1.96	2.48	3.26	2.65	3.22	4.27	3.11	1.30	2.73
1893	0.78	0.95	1.23	1.01	1.33	1.45	0.85	1.41	1.29
1894	4.83	5.32	6.26	5.44	5.67	7.08	4.76	2.88	4.90
1895	1.59	2.05	2.58	1.99	2.13	3.20	0.96	2.62	1.99
1896	2.87	3.71	3.82	3.22	3.34	4.42	3.94	1.67	4.05
1897	2.27	2.68	3.41	2.52	2.41	4.13	1.27	3.67	2.64
1898	3.12	4.54	5.09	4.33	5.70	4.25	4.93	1.95	4.94
1899	2.59	3.73	4.49	3.85	4.03	5.62	1.60	4.36	4.19
1900	1.83	2.29	2.87	2.26	5.60	4.35	2.96	1.24	2.68
1901	1.73	2.31	3.00	2.05	3.01	2.79	1.52	2.66	2.44
1902	2.48	4.04	4.72	3.42	4.67	4.45	5.04	2.11	4.37
1903	2.02	1.99	3.33	2.38	2.18	3.83	0.65	3.69	2.58
1904	2.16	2.51	3.30	2.17	3.11	3.56	3.27	0.67	2.60
1905	3.47	4.61	5.57	3.78	5.27	5.85	3.40	4.50	3.78
1906	2.22	3.03	4.52	3.15	4.52	4.83	3.20	2.74	3.72
1907	5.09	6.44	7.19	5.89	6.10	8.25	1.63	5.56	4.93
1908	2.44	3.03	3.72	2.68	3.23	4.50	3.62	1.28	2.66
1909	2.07	3.65	4.98	3.31	4.84	5.37	1.43	4.46	3.71
1910	3.47	4.10	4.27	3.83	3.49	5.76	3.83	1.42	2.95
1911	1.91	2.59	3.44	2.38	2.80	5.32	1.47	3.09	3.10
1912	0.55	1.00	1.19	0.73	0.89	1.89	0.94	0.90	1.09
1913	1.73	2.69	3.64	2.54	2.91	4.12	0.88	3.44	2.74
1914	2.03	2.76	2.72	2.13	2.83	3.29	2.46	0.56	1.50
1915	3.45	4.06	4.77	3.89	2.96	5.33	2.36	3.51	3.42
1916	3.00	3.64	4.02	3.46	2.79	4.04	3.76	1.61	2.47
1917	1.43	1.71	2.55	1.79	2.64	3.00	0.95	1.99	1.21
1918	2.68	3.34	3.69	3.54	4.04	5.23	3.08	1.43	1.91
1919	2.00	2.27	3.18	2.68	2.14	3.72	1.10	2.96	1.30
1920	2.55	3.51	4.31	3.44	2.97	4.13	2.55	0.92	1.58
1921	2.14	2.72	3.25	2.78	3.21	3.99	1.09	3.22	2.48
1922	2.02	2.39	3.02	2.25	2.76	4.03	2.90	1.72	2.64
1923	2.67	2.70	3.27	3.02	2.87	3.74	0.87	2.88	2.82
1924	1.84	2.05	2.43	2.05	1.56	3.67	1.30	0.91	1.64
1925	1.94	2.33	2.88	2.65	1.91	2.96	0.95	1.97	1.48

Exhibit 5.1
Organic Manures and Inorganic Fertilizers Applied to the Eighteen Plots *

TABLE 2·1

MANURIAL HISTORY OF THE BROADBALK WINTER WHEAT PLOTS
1852–1967

(For treatments 1844–51 see over)

Plot		Treatment from 1852
2A	FYM since 1885	In 1885 this new plot was made from two half plots, the one on the south had been unmanured since 1844 and the other was half of the original plot 1 which had KNaMg between 1844 and 1883 and was fallowed in 1884.
2B	FYM	
3	None	Originally 2 half plots 3 and 4 3, unmanured 4, 1844–51, NP‡; since 1852 unmanured..
5	P K Na Mg	
6	N_1 P K Na Mg	
7	N_2 P K Na Mg	
8	N_3 P K Na Mg	
9	N_1* P K Na Mg	Since 1894; previously 9a 1852–54 N_1*, 1855–84 N_2*P K Na Mg, 1885–93 N_1*P K Na Mg. 9b 1852–54 N_2*, 1855–84 N_2*, 1885–93 N_1*.
10	N_2	
11	N_2 P	
12	N_2 P Na§	
13	N_2 P K	
14	N_2 P Mg§	
15	N_2† P K Na Mg	Since 1873 all N in autumn. See note in Table 2·2 about times N was applied. 15a 1852–72 N_2 P K Na Mg‡ 15b 1852–72 $N_{1\cdot 5}$ P K Na Mg‡ + 500 lb rape cake
16	N_2* P K Na Mg	Since 1884; previously 1852–64 N_4PKNaMg, 1865–83 unmanured.
17	N_2	applied in alternate years: Plot 17, N in even years, Plot 18, N in odd years
18	P K Na Mg	
19	Castor meal (R)	The size of the present plot only since 1904. The original plot 19 was a half width plot, 1852–78 $N_{1\cdot 5}$ + P‡ + 500 lb rape cake, 1879–1904, rape cake. In 1894 the original plot 20 unmanured, a half width plot 500 lks long at the west end of the plots was taken into plot 19. In 1904 the total length of the plot was made full width by taking in land from the headland.
20	N_2 K Na Mg	Since 1906. The original plot 20 was unmanured after 1846 and was taken in to plot 19 in 1894. A new plot 20, width as plots 3–18 was made on ground previously known as Knott Wood Butts and unmanured. 1894–1905 unmanured.

N Nitrogen as ammonium salts.
N* Nitrogen as sodium nitrate.
N† All nitrogen as ammonium sulphate in autumn. See note Table 2·2.
‡ P made with hydrochloric acid, all N as ammonium sulphate.
§ See Table 2·2 for amount, Na and Mg are applied in chemically equivalent amounts.

* Exhibit 5.1 is taken from Johnston and Garner (1969).

Exhibit 5.1 *(cont.)*

TABLE 2·2

DETAILS OF MANURES, DRAINING, CULTIVATIONS AND CROPPING, BROADBALK WINTER WHEAT, 1844–1967

Fertilisers and FYM applied annually in autumn before sowing, except N, see below.

Amounts per acre†

NITROGEN

Ammonium sulphate at 210, 420, 630 lb, (N_1, N_2, N_3) supplying 43, 86, 129 lb N. 172 lb N (N_4) was tested until 1864.

Before 1917 a mixture of equal weights of ammonium sulphate and ammonium chloride was used (except 1887 ammonium sulphate only). In 1901 the mixed salts applied in spring were compared with ammonium bicarbonate on quarter plots in the west half of the experiment. From 1844–51 various amounts of ammonium sulphate, ammonium chloride and the mixed salts were used.

Sodium nitrate at 275, 550 lb from 1885, (N_1*, N_2*) supplying 43 and 86 lb N; previously N_2* only from 1855.

PHOSPHORUS

Superphosphate containing 29–30 lb P, 66–69 lb P_2O_5 (366 lb 18·5% P_2O_5 superphosphate).

Before 1889 superphosphate was made on the Farm from calcined bone dust and acid. From 1844 to 1847 various proportions of bone dust and sulphuric acid or hydrochloric acid were used. From 1848 to 1888 200 lb calcined bone dust was treated with 150 lb sulphuric acid (sp. gr. 1·7). From 1889 superphosphate was supplied ready made and the weight adjusted to give the same amount of phosphorus as in the period 1848–88. 1898–1902 basic slag (400 lb) used instead of superphosphate.

Phosphorus dressings omitted 1915.

POTASSIUM

Potassium sulphate *1852–1858* 300 lb containing 120 lb K (145 lb K_2O)
1859–1967 200 lb containing 80 lb K (96 lb K_2O)

Potassium dressings omitted 1915, 1917, 1918, 1919.

From 1843 to 1851 various weights of pearl ash (45% K) and potassium sulphate were tested.

SODIUM

Sodium sulphate *1852–1858* 200 lb containing 28 lb Na
1859–1967 100 lb containing 14 lb Na

Sodium dressings omitted 1915

exception plot 12 *1852–1858* 550 lb containing 77 lb Na
1859–1967 336½ lb containing 51 lb Na

Sodium dressings omitted 1915, 1917, 1918, 1919.

From 1843 to 1851 various amounts of soda ash (34% Na) and sodium chloride were tested.

MAGNESIUM

Magnesium sulphate *1852–1967* 100 lb containing 10 lb Mg

Magnesium dressings omitted 1915

exception plot 14 *1852–1858* 420 lb containing 42 lb Mg
1859–1967 280 lb containing 28 lb Mg

Magnesium dressings omitted 1915, 1917, 1918, 1919.

From 1843 to 1851 various amounts and forms of magnesium were tested.

† Plot areas. The sizes of the whole plots were decreased in autumn 1893 when the 'a' and 'b' halves of the plots were ploughed together leaving a path on either side of the plot; to maintain the amounts of nutrients per acre the manures applied per plot were correspondingly smaller.

1 lb. per acre = 1.121 kg. per hectare.
1 ton per acre = 2.50171 tons per hectare.

Exhibit 5.1 *(cont.)*

TABLE 2·2—continued

SILICATE TEST

1862–1863. A strip 16·5 ft wide at the eastern end of all plots was dressed with a mixture of 200 lb sodium silicate + 200 lb calcium silicate. Another 16·5 ft wide strip received a mixture, at 400 lb per acre, of 8 cwt soluble silica rock, 1 cwt fresh burnt lime and 20 lb common washing soda mixed together with hot water to a thick paste.

1864–1866. 288 lb of the above silicate rock mixture tested on the 'a' halves of plots 5, 6, 7, 8, 9, 16, 18.

1868–1879 on the 'a' halves of plots 5, 6, 7, 8, 11, 12, 13, 14 and of 17 or 18, whichever received PKNaMg, the total straw grown on that plot the previous season was chaffed, spread and ploughed in. The straw was used as a source of silica and continued the tests of silica made in 1862–66.

FARMYARD MANURE

14 tons The analyses Lawes and Gilbert had in the 1840s showed that 14 tons 'average' FYM contained 200 lb total N. In recent years the FYM has been made by fattening bullocks treading straw in covered yards during the winter.

CASTOR MEAL

1852–1878	500 lb rape cake tested with N and P
1879–1882	1700 lb rape cake (probably supplied 86 lb N)
1883–1940	1889 lb rape cake (probably supplied 89 lb N) dressings omitted 1917–20
1941–1954	1889 lb castor bean meal
1955–1967	weight per acre of castor bean meal adjusted for analysis so that total nitrogen applied equalled 86 lb N.

LIMING

For an account of the chalking before the experiment started see Avery and Bullock page 67.

To correct acidity that had developed on parts of some plots, especially at their eastern ends, ground chalk (5 tons $CaCO_3$/acre) was applied to all of section VB in autumn 1954 and to all of section VA and plot 19 section IV in autumn 1963. To prevent further acidity developing, a scheme was introduced whereby chalk was applied annually to the stubble before ploughing, except for the section to be fallowed which also received no fertilisers. The amount was based on the acidifying effect of the ammonium sulphate (100 lb $CaCO_3$ for every 14 lb N as ammonium sulphate) and the castor meal (50 lb $CaCO_3$ for every 14 lb N as castor meal). The first dressing in autumn 1954 was double this amount to assist the recovery of some of the acid plots. A survey of the surface soil pH values in 1967 showed that some small areas required further chalking and various amounts of chalk (23–69 cwt $CaCO_3$) were applied in autumn 1967.

DRAINING

Although Lawes described the soil of Broadbalk as having a good natural drainage, because it was an experiment it was necessary to get on the land for longer periods than under normal agricultural practice and it was decided to improve the drainage. Tile drains were put under the centre furrow separating the 'a' and 'b' halves of each plot (except the present plot 20) in autumn 1849. Plot 2A (FYM since 1885) was drained in autumn 1884 before the first application of FYM was given. Tiles of 'horseshoe and sole' type (2 in. in diameter) were laid between 2 and 2·5 ft deep except under the west side of section III where they were about 3 ft deep. The fall of the drains was given by the general, but not uniform, slope of the field from west to east, about 12 ft on plot 19 and 16 ft on plot 2. The tile drains discharged into a 4-in. cross main at the east side of the field; the cross main ran into a well in the chalk, outside the field at the north-east corner. To make drainage better on the western half of the field about twelve cross drains were put in under the parallel plot drains. These secondary drains were 3 to 5 ft deep and were led into the nearest dell. Records suggest they discharged into some sort of soakaway dug down to the chalk at the centre of the dell.

The system was originally intended only to drain the land but, because there was one drain under each plot, Lawes and Gilbert realised that they could be used to measure losses of plant nutrients in drainage water from the different fertiliser treatments. Small pits were dug at the intersection of each plot drain and the cross main in December 1866 to sample the runnings from plots 2 to 16. This was not ideal as there was always risk of the sample being contaminated. The drains from plots 17, 18 and 19 were opened

Exhibit 5.1 *(cont.)*

TABLE 2·2—continued

in November 1878 and in spring 1879 the arrangements for collecting the runnings were improved. The drain from each plot discharged into its own pit which overflowed into a separate, deepened main drain which was kept open. In 1897–98 this drain was further enlarged, the base concreted and the sides bricked

CULTIVATIONS

Details of cultivations were discussed by Moffatt (1939), and in general terms by Grey (1922) for many years Field Superintendent.

PLOUGHING AND FERTILISER APPLICATIONS. 1844–1914. Previous crop stubble was either shallow ploughed or harrowed and when there was much weed and plant debris this was carted off the plots. Fertilisers and FYM were then applied and the plots reploughed, except 1891–99 when the plots were ploughed immediately after harvest and again just before the fertilisers were applied; the fertilisers were then worked in by harrowing. Seedbed preparation and drilling was always done soon after the second ploughing.

Since 1915 the plots have been ploughed once, after the FYM was applied. The fertilisers were then applied by manure distributor on the partly prepared seedbed, final cultivations were done and the seed drilled. Before about 1880 the seed was usually drilled in November, 1880–1945 usually in October.

N fertilisers have been applied at various times.

1. Ammonium salts, after 1917 ammonium sulphate	*1843–1872*	all in autumn
	1873–1877	all in autumn except plot 15 all in spring
	1878–1883	all in spring except plot 15 all in autumn
	1884–1967	two dressings, 21·5 lb N applied in autumn, balance applied in spring except plot 15 all in autumn.
2. Sodium nitrate	*1867–1967*	always applied all in spring; from 1899 the dressing supplying 86 lb N in two equal amounts, the dates of application differing from 6 days to 6 weeks.

After the drains were put under each plot in the autumn of 1849 (plot 2A was drained in 1884) the field was always ploughed to leave a wide furrow over the drain, the ridges on the outer edges of the plots being spread back on to the plots by hand or by harrowing. The furrows were not cropped. Since 1894 the 'a' and 'b' halves of the plots were cropped and harvested as one plot. Tractors were first used for cultivation in 1920 and for ploughing from 1925

WEED CONTROL AND FALLOWING. Up to 1914 the plots were hand hoed in most years, wild oats were pulled by hand. In 1889 only alternate rows were sown on the west half of the field to make hoeing easier and in 1890 alternate rows were sown on the east half of the field. In 1904 and 1905 north or south halves of the plots only were cropped and the fallow halves were worked to kill weeds. In 1914 the west halves of all plots were fallowed and in 1915 the east halves.

Exhibit 5.1 *(cont.)*

TABLE 2·2—continued

HARVESTING. Plots were originally cut by hand; they were first cut by a self binder in 1902, but hand cutting with scythes was often necessary when the ground was very wet or the crop badly laid. Sheaves from each plot were stooked on that plot, carted, stored and then threshed during the winter.

CROPPING

Some part of every plot has grown winter wheat each year.

Seed rates ranged from 110–190 lb/acre.

Row spacing
1844–1862 rows at 12·5 in. (12 rows on a land)
1863–1893 rows at 7·8 in. (18 rows on a land, i.e. each 'a' and 'b' half plot except 1889 and 1890, see section on weed control)
1894–1905 rows at 8·9 in. (28 rows on the 'new' wide plots)
1906–1927 rows at 12 in. (20 rows/plot except 1913 rows at 7·6 in.)

Varieties
1844–1848 Old Red Lammas
1849–1852 Old Red Cluster
1853–1881 Red Rostock
1882–1899 Red Club
1900–1916 Squarehead's Master, except 1905 Giant Red, 1910 Browick Red, 1911–12 Little Joss
1917–1945 Red Standard except 1929, 1940, 1941, 1943 Squarehead's Master; 1942, Stand up.

Seed dressings. In 1912 the seed was dressed with copper sulphate and in 1923 and 1924 it was treated with formalin.

6. Uniformity Trials: Variation and Correlation in the Yield of Wheat

Source Mercer, W.B. and Hall, A.D. (1911). The experimental error of field trials. *J. Agric. Sci.* **4**, 107-132.

Wiebe, G.A. (1935). Variation and correlation among 1500 wheat nursery plots. *J. Agric. Research* **50**, 331-357.

Contributor G.N. Wilkinson Waite Agricultural Research Institute University of Adelaide.

There has been much interest in examining the effects of possibly correlated errors among plots in agricultural field trials. Many uniformity trials have been conducted to investigate this aspect. In these trials, every plot receives the same treatment. Two well-known experiments of this kind are those of Mercer and Hall (1911) and Wiebe (1935). Their data have been used extensively in other investigations; see, for example Barbacki and Fisher (1936), Besag (1974) and Wilkinson, Eckert, Hancock and Mayo (1983). Further, Fairfield Smith (1938) discussed these and other uniformity trials and determined an empirical relationship between variance of yield as a function of plot size of field experiments and plot to plot correlation. Papadakis (1937), Bartlett (1938, 1978, 1981) and Wilkinson *et al.* (1983) discuss methods for adjusting the analysis of field experiments by using the yields of neighbouring plots.

Mercer and Hall stressed that "the magnitude of the experimental error attached to field plots was very important in Agricultural Science because in its proper recognition depends the degree of confidence which may be attached to the results obtained in field work. A very cursory examination of the results of any set of field trials will serve to show that a pair of plots similarly treated may be expected to yield considerably different results, even when the soil appears to be uniform and the conditions under which the experiment is conducted are carefully designed to reduce errors in weighing and measurement". Mercer and Hall describe the trials in the following way:

> In order to shed light on the question, during the year 1910 an attempt was made at Rothamsted to estimate the variations in the yield of various sized plots of ordinary field crops which had been subjected to no special treatment and appeared to the eye sensibly uniform. The fields were selected, one of wheat which promised to be a fair crop for the season and was generally standing up well, the other of mangolds* which looked a uniformly and fairly heavy crop for the season and soil. In the wheat field a very

* The mangold experiment will not be discussed here but is documented in, for example, Daniel (1976).

> uniform area was selected, one acre of which was harvested in several plots, each one five-hundredth of an acre in area. The small sheaves which each plot yielded were then stored and eventually threshed out by a hand machine, corn and straw being separately weighed. In measuring the plot a fixed number of rows of the drill were taken, eleven in this case, and a fixed length of 10.82 feet measured along the rows for cutting. Thus if there were any variation in the breadth of the drills the assumed one five-hundredth of an acre would not represent the actual land area of each plot. These variations were small: moreover in all experimental crop work it is necessary to see that comparative plots do contain exactly the same number of drills, so as to make the measured area one of crop rather than land.

Table 6.1 gives the yields of grain and straw per plot in pounds, the table being arranged as the plots occurred in the field.

The data given in Table 6.2, are the yields of wheat on 1500 nursery plots in an agricultural uniformity trial discussed by Wiebe (1935). Dr. Wiebe describes the experiment as follows:

> The material for this study consisted of 1,500 rows of Federation wheat (C.I. no. 4734),* which is well adapted to the region where it was grown. The seed was pure. The crop was grown in the summer of 1927 on the west end of series 100 on the Aberdeen Substation, Aberdeen, Idaho. The ground had been uniformly cropped the previous year to field peas. The plot was seeded on April 18 with a grain drill that sowed eight rows at a time. The rows were 12 inches apart and 15 feet long. Because the land sloped gently to the south and west, levees were necessary at 15- and 30-foot intervals. Figure 1 [Figure 6.1] shows the general arrangement and the system of identifying individual series and rows. During the growing season the crop was irrigated four times. The average annual rainfall at Aberdeen is about 7 inches. The individual rows were harvested in August and threshed with a small nursery thresher. The grain yields, recorded in grains per row, are shown in table 1 [Table 6.2]. A contour map of the yields obtained in the plot are shown in figure 2 [Figure 6.2].
>
> The wheat yielded 63 bushels per acre, which is above average for that region. No disease or insect damage was noticed. Although the height of plants and the weights of straw per row were recorded, only the yield of grain is discussed in this paper.

References

Barbacki, S. and Fisher, R.A. (1936). A test of the supposed precision of systematic arrangements. *Ann. Eugen.* 7, 189-193.

Bartlett, M.S. (1938). The approximate recovery of information from field experiments with large blocks. *J. Agric. Sci.* **28**, 418-427.

Bartlett, M.S. (1978). Nearest neighbour models in the analysis of field experiments (with discussion). *J.R. Statist. Soc. B* **40**, 147-174.

* Accession number of the Division of Cereal Crops and Diseases, Bureau of Plant Industry, U.S. Department of Agriculture.

Bartlett, M.S. (1981). A further note on the use of neighbouring plot values in the analysis of field experiments. *J.R. Statist. Soc. B* **43**, 100-102.

Besag, J. (1974). Spatial interaction and the statistical analysis of lattice systems. *J.R. Statist. Soc. B* **36**, 192-236.

Daniel, C. (1976). *Applications of Statistics to Industrial Experimentation.* New York: Wiley.

Fairfield Smith, H. (1938). An empirical law describing heterogeneity in the yields of agricultural crops. *J. Agric. Sci.* **28**, 1-23.

Papadakis, J.S. (1937). Méthode statistique pour des expériences sur champ. *Bull. Inst. Amél. Plantes à Solonique* **23**.

Wilkinson, G.N., Eckert, S.R. Hancock, T.W. and Mayo, O. (1983). Nearest neighbour (NN) analysis of field experiments (with discussion). *J. R. Statist. Soc. B* **45**, 151-211.

Table 6.1: The Plan of
of Grain and Straw

Code		Plots												
1	0	3.63	4.15	4.06	5.13	3.04	4.48	4.75	4.04	4.14	4.00	4.37	4.02	4.58
1	1	6.37	6.85	7.19	7.99	4.71	6.08	7.31	6.08	6.98	5.87	6.75	6.10	7.23
2	0	4.07	4.21	4.15	4.64	4.03	3.74	4.56	4.27	4.03	4.50	3.97	4.19	4.05
2	1	6.24	7.29	7.41	7.80	6.34	6.63	7.88	6.35	6.91	6.50	6.09	6.43	6.57
3	0	4.51	4.29	4.40	4.69	3.77	4.46	4.76	3.76	3.30	3.67	3.94	4.07	3.73
3	1	7.05	7.71	7.35	7.50	6.17	6.98	8.18	5.93	5.95	6.20	6.18	6.37	6.02
4	0	3.90	4.64	4.05	4.04	3.49	3.91	4.52	4.52	3.05	4.59	4.01	3.34	4.06
4	1	6.91	8.23	7.89	6.66	5.70	6.46	7.60	7.29	5.82	5.41	5.99	5.60	6.19
5	0	3.63	4.27	4.92	4.64	3.76	4.10	4.40	4.17	3.67	5.07	3.83	3.63	3.74
5	1	5.93	7.73	8.58	7.86	6.05	6.77	7.91	7.33	7.33	8.05	6.36	6.43	6.13
6	0	3.16	3.55	4.08	4.73	3.61	3.66	4.39	3.84	4.26	4.36	3.79	4.09	3.72
6	1	5.59	6.45	7.04	7.98	5.89	6.15	7.36	6.28	7.61	5.58	5.46	6.10	6.03
7	0	3.18	3.50	4.23	4.39	3.28	3.56	4.94	4.06	4.32	4.86	3.96	3.74	4.33
7	1	5.32	5.87	7.02	6.98	4.97	6.06	8.06	6.81	7.37	7.51	6.23	6.38	6.79
8	0	3.42	3.35	4.07	4.66	3.72	3.84	4.44	3.40	4.07	4.93	3.93	3.04	3.72
8	1	5.52	5.71	7.05	7.28	5.78	6.10	7.50	5.97	6.99	7.57	6.13	4.96	5.97
9	0	3.97	3.61	4.67	4.49	3.75	4.11	4.64	2.99	4.37	5.02	3.56	3.59	4.05
9	1	6.03	6.01	7.64	6.95	5.94	6.83	7.92	5.07	7.25	8.23	5.75	6.03	6.82
10	0	3.40	3.71	4.27	4.42	4.13	4.20	4.66	3.61	3.99	4.44	3.86	3.99	3.37
10	1	5.66	6.29	7.17	6.95	7.31	6.86	7.59	6.33	7.26	7.75	6.14	6.26	6.25
11	0	3.39	3.64	3.84	4.51	4.01	4.21	4.77	3.95	4.17	4.39	4.17	4.17	4.09
11	1	5.61	6.30	6.60	7.86	7.18	8.23	8.23	7.11	7.52	7.73	7.20	7.08	7.28
12	0	4.43	3.70	3.82	4.45	3.59	4.37	4.45	4.08	3.72	4.56	4.10	3.07	3.99
12	1	7.07	6.17	6.87	7.17	6.53	8.75	8.74	7.17	7.28	7.73	6.90	6.12	7.13
13	0	4.52	3.79	4.41	4.57	3.94	4.47	4.42	3.92	3.86	4.77	4.99	3.91	4.09
13	1	7.10	6.33	7.03	7.93	7.06	8.53	8.02	6.70	7.20	7.67	7.82	7.34	7.72
14	0	4.46	4.09	4.39	4.31	4.29	4.47	4.37	3.44	3.82	4.63	4.36	3.79	3.56
14	1	7.16	7.22	7.73	7.31	7.08	8.15	7.69	6.62	7.05	7.87	7.39	6.33	6.69
15	0	3.46	4.42	4.29	4.08	3.96	3.96	3.89	4.11	3.73	4.03	4.09	3.82	3.57
15	1	8.85	5.20	7.52	6.67	6.54	7.10	6.86	7.58	6.89	7.16	7.03	7.30	6.55
16	0	5.13	3.89	4.26	4.32	3.78	3.54	4.27	4.12	4.13	4.47	3.41	3.55	3.16
16	1	8.37	7.05	6.99	6.93	6.72	6.46	7.79	7.32	7.24	7.84	5.96	6.70	5.84
17	0	4.23	3.87	4.23	4.58	3.19	3.49	3.91	4.41	4.21	4.61	4.27	4.06	3.75
17	1	6.89	6.82	7.14	7.73	6.06	6.63	7.34	7.53	7.41	7.51	7.17	7.00	6.31
18	0	4.38	4.12	4.39	3.92	4.84	3.94	4.38	4.24	3.96	4.29	4.52	4.19	4.49
18	1	6.72	7.38	7.55	6.70	8.85	6.75	7.43	7.32	7.04	6.96	7.73	7.30	7.57
19	0	3.85	4.28	4.69	5.16	4.46	4.41	4.68	4.37	4.15	4.91	4.68	5.13	4.19
19	1	6.59	7.03	8.06	8.78	7.54	8.15	7.51	7.19	7.47	7.96	8.07	8.31	6.93
20	0	3.61	4.22	4.42	5.09	3.66	4.22	4.06	3.97	3.89	4.46	4.44	4.52	3.70
20	1	6.20	7.65	8.45	8.72	7.09	7.72	7.06	7.53	7.36	6.91	6.87	8.17	6.80

* The upper figure is grain, the lower figure is straw.

the Wheat Field and the Yield of Five Hundred Wheat Plots *

Code		Plots											
1	0	3.92	3.64	3.66	3.57	3.51	4.27	3.72	3.36	3.17	2.97	4.23	4.53
1	1	6.33	5.11	5.96	5.12	5.05	6.54	5.47	4.76	4.95	4.53	6.08	6.78
2	0	3.97	3.61	3.82	3.44	3.92	4.26	4.36	3.69	3.53	3.14	4.09	3.94
2	1	6.03	5.58	5.80	5.00	5.83	8.61	6.14	5.56	5.09	5.11	5.91	5.68
3	0	4.58	3.64	4.07	3.44	3.53	4.20	4.31	4.03	3.66	3.59	3.97	4.38
3	1	7.23	5.86	6.74	5.56	4.91	6.55	6.44	6.17	6.15	5.41	6.28	7.49
4	0	3.19	3.75	4.54	3.97	3.77	4.30	4.10	3.81	3.89	3.32	3.46	3.64
4	1	6.56	4.62	7.08	6.03	5.79	5.95	5.96	6.13	5.92	4.62	5.41	6.55
5	0	4.14	3.70	3.92	3.79	4.29	4.22	3.74	3.55	3.67	3.57	3.96	4.31
5	1	5.98	7.67	6.14	5.33	5.58	6.15	5.76	5.89	5.45	5.24	5.60	6.56
6	0	3.76	3.37	4.01	3.87	4.35	4.24	3.58	4.20	3.94	4.24	3.75	4.29
6	1	5.49	5.00	5.99	5.57	6.09	5.88	5.61	5.92	5.87	5.82	5.50	6.15
7	0	3.77	3.71	4.59	3.97	4.38	3.81	4.06	3.42	3.05	3.44	2.78	3.44
7	1	5.48	5.66	7.28	6.03	6.24	5.69	6.25	5.45	4.57	4.56	4.28	5.68
8	0	3.93	3.71	4.76	3.83	3.71	3.54	3.66	3.95	3.84	3.76	3.47	4.24
8	1	6.07	5.79	6.49	6.29	5.91	5.21	5.78	5.92	5.66	5.24	5.59	7.26
9	0	3.96	3.75	4.73	4.24	4.21	3.85	4.41	4.21	3.63	4.17	3.44	4.55
9	1	6.35	5.12	8.64	6.45	6.29	6.15	6.15	6.04	5.81	5.58	4.81	6.32
10	0	3.47	3.09	4.20	4.09	4.07	4.09	3.95	4.08	4.03	3.97	2.84	3.91
10	1	5.78	5.47	6.49	6.16	6.18	5.47	6.11	7.00	5.72	5.65	4.10	5.96
11	0	3.29	3.37	3.74	3.41	3.86	4.36	4.54	4.24	4.08	3.89	3.47	3.29
11	1	5.71	6.44	8.63	5.78	6.14	7.39	7.46	7.20	6.54	5.98	5.84	5.65
12	0	3.14	4.86	4.36	3.51	3.47	3.94	4.47	4.11	3.97	4.07	3.56	3.83
12	1	5.05	6.39	7.26	6.11	5.90	6.68	7.84	6.95	6.47	5.80	6.38	6.29
13	0	3.05	3.39	3.60	4.13	3.89	3.67	4.54	4.11	4.58	4.02	3.93	4.33
13	1	5.70	5.86	6.27	6.87	6.23	6.20	7.33	6.64	6.79	6.35	5.69	7.11
14	0	3.29	3.64	3.60	3.19	3.80	3.72	3.91	3.35	4.11	4.39	3.47	3.93
14	1	5.71	6.36	5.84	5.87	6.14	6.34	6.96	6.27	6.64	6.11	5.78	6.07
15	0	3.43	3.73	3.39	3.08	3.48	3.05	3.65	3.71	3.25	3.69	3.43	3.38
15	1	5.38	8.58	6.42	5.42	5.52	5.20	6.60	6.29	6.37	5.18	5.82	5.68
16	0	3.47	3.30	3.39	2.92	3.23	3.25	3.86	3.22	3.69	3.80	3.79	3.63
16	1	5.84	5.70	5.80	4.95	5.33	5.25	6.64	5.40	5.93	5.70	6.21	5.99
17	0	3.91	3.51	3.45	3.05	3.68	3.52	3.91	3.87	3.87	4.21	3.68	4.06
17	1	6.21	5.99	6.05	7.64	5.82	5.85	6.71	6.13	7.50	5.48	6.01	6.88
18	0	3.82	3.60	3.14	2.73	3.09	3.66	3.77	3.48	3.76	3.69	3.84	3.67
18	1	6.37	6.34	5.48	4.77	5.41	5.84	6.98	6.14	6.11	5.43	6.35	6.33
19	0	4.41	3.54	3.01	2.85	3.36	3.85	4.15	3.93	3.91	4.33	4.21	4.19
19	1	6.78	5.58	5.68	4.96	6.14	6.15	6.85	6.57	6.09	6.04	6.98	6.93
20	0	4.28	3.24	3.29	3.48	3.49	3.68	3.36	3.71	3.54	3.59	3.76	3.36
20	1	6.97	5.95	5.58	5.52	5.82	6.76	6.08	6.35	6.21	4.66	6.36	6.33

Table 6.2
Yield of Grain from Each of 1,500 Fifteen-foot Rows of Wheat, 12 Inches Apart, Grouped in 12 Series of 125 Rows Each

Series											
1	2	3	4	5	6	7	8	9	10	11	12
715	595	580	580	615	610	540	515	557	665	560	612
770	710	655	675	700	690	565	585	550	574	511	618
760	715	690	690	655	725	665	640	665	705	644	705
665	615	685	555	585	630	550	520	553	616	573	570
755	730	670	580	545	620	580	525	495	565	599	612
745	670	585	560	550	710	590	545	538	587	600	664
645	690	550	520	450	630	535	505	530	536	611	578
585	495	455	470	445	555	500	450	420	461	531	559
560	540	450	500	505	595	545	530	498	538	453	600
685	730	610	500	555	645	605	535	534	593	616	638
755	810	665	570	525	715	650	550	613	607	742	657
640	635	585	465	430	615	550	515	369	493	635	567
725	655	530	455	465	560	550	460	455	503	519	555
715	775	615	545	530	685	595	510	507	561	581	537
700	705	555	440	455	630	555	425	476	648	532	552
640	655	495	435	485	585	475	405	422	516	458	559
620	635	495	445	455	620	505	465	419	591	545	537
750	630	555	455	510	575	530	470	427	545	562	582
760	675	610	540	530	675	580	460	513	599	595	675
630	645	490	445	440	585	490	420	460	542	474	563
625	540	495	415	445	530	440	430	455	485	461	533
670	670	560	485	445	590	555	450	516	570	553	623
655	630	540	465	460	560	465	435	466	454	458	503
570	535	525	455	415	510	460	420	393	482	426	446
575	495	540	480	395	555	440	450	437	541	525	514
690	645	595	515	470	550	515	445	508	508	517	533
730	730	710	565	465	640	565	550	494	591	594	597
620	635	595	485	435	530	490	470	465	521	454	537
620	600	530	455	425	545	395	455	482	484	505	578
665	685	685	520	480	525	550	465	497	596	563	634
620	605	530	440	450	500	470	445	510	531	478	514
615	525	610	515	395	510	455	530	485	554	448	503
595	625	600	440	435	510	515	525	500	522	496	514
640	580	630	485	480	610	525	515	574	525	458	518
695	650	695	550	515	685	570	525	613	560	532	525
630	610	545	465	460	555	500	575	438	477	480	507
595	555	580	505	465	540	425	490	467	479	444	507
645	635	660	550	525	645	515	520	477	510	465	533
705	660	595	495	465	595	450	505	417	489	399	533
575	530	600	460	410	500	460	505	459	431	438	578
530	555	570	450	425	490	450	495	426	443	449	469

Table 6.2 *(cont.)*

Series											
1	2	3	4	5	6	7	8	9	10	11	12
715	680	685	505	535	600	495	480	474	401	495	533
795	715	740	590	585	585	535	560	518	513	541	548
660	635	620	495	510	585	520	475	444	405	509	469
615	580	505	490	505	555	450	470	468	442	474	484
690	645	630	600	565	645	530	550	454	474	592	548
675	615	640	505	550	615	535	500	465	456	513	563
770	625	685	570	575	685	540	475	534	456	467	548
695	700	770	540	515	685	530	495	508	454	422	600
660	660	705	570	545	645	500	540	494	500	532	687
720	730	825	665	605	665	520	550	584	546	537	630
670	640	710	500	535	575	495	455	512	497	465	600
655	590	675	575	515	570	490	470	460	427	472	589
780	790	855	625	575	655	530	540	575	502	503	582
750	705	770	575	560	565	510	435	486	470	502	570
585	620	685	620	550	520	435	460	429	485	534	619
665	605	680	570	560	560	445	465	461	463	533	627
735	760	790	645	605	595	560	485	466	487	476	645
795	795	850	700	655	635	540	470	521	582	584	750
645	685	720	610	520	550	460	440	425	485	462	698
630	695	665	630	510	575	445	430	433	423	450	627
685	840	875	640	575	625	570	490	514	539	486	717
695	780	790	575	520	600	545	470	455	486	488	619
705	745	840	710	640	610	500	530	498	479	434	574
655	735	840	640	610	585	560	450	469	562	450	570
640	880	795	660	580	615	585	460	461	472	504	600
790	765	890	770	660	690	575	490	485	475	509	582
670	690	785	645	510	560	535	405	474	455	458	555
685	790	770	665	600	545	555	390	389	461	439	615
660	825	960	750	660	680	620	510	501	490	534	694
705	805	860	635	540	650	555	435	479	457	451	649
610	720	705	615	540	625	595	380	379	483	502	608
640	735	805	665	535	605	580	430	427	441	498	638
690	855	905	700	615	650	615	495	531	448	453	645
715	765	945	820	695	750	685	530	511	580	510	762
685	750	825	715	580	660	655	505	414	451	542	642
685	790	790	695	595	625	510	525	425	471	478	514
775	845	995	820	655	675	640	545	474	513	599	597
645	870	890	740	615	655	555	490	433	558	590	567
725	825	920	640	600	690	575	535	451	484	510	567
650	865	935	690	645	640	530	520	476	507	525	563
695	860	960	725	615	710	660	525	438	539	646	623
735	910	975	775	680	700	770	635	506	526	514	638
645	745	815	700	605	615	605	550	455	504	503	589

Table 6.2 *(cont.)*

Series											
1	2	3	4	5	6	7	8	9	10	11	12
630	810	730	635	535	650	735	535	458	554	535	615
680	745	840	730	645	650	775	590	557	561	565	619
620	730	775	680	610	610	680	515	537	535	553	570
620	745	660	565	520	635	610	450	476	550	562	575
560	675	690	635	525	605	635	515	494	516	558	559
625	700	725	645	645	580	615	640	516	618	615	604
700	765	725	615	640	705	710	590	619	627	620	645
685	785	655	600	570	615	670	575	506	602	530	600
625	550	590	590	605	505	560	595	539	629	571	698
745	790	675	600	625	685	725	695	686	608	568	672
680	670	630	640	645	650	695	610	627	579	537	675
655	730	615	650	640	645	655	620	611	642	685	630
625	700	675	720	695	680	720	615	595	556	648	578
670	735	645	620	705	695	650	580	623	541	580	630
670	820	685	665	715	715	790	660	658	689	725	675
555	670	590	580	550	665	640	605	553	580	605	638
650	685	505	525	595	590	560	575	592	559	584	518
625	760	625	545	635	635	670	625	672	656	577	645
620	740	575	565	595	610	650	595	515	641	533	623
580	650	570	575	450	445	670	645	587	634	534	574
615	705	550	515	530	485	610	570	565	652	518	623
675	690	615	610	560	625	745	540	584	616	505	668
720	820	720	650	590	615	770	645	615	755	625	713
640	780	640	660	605	540	695	580	615	679	540	570
630	800	715	710	605	575	490	485	564	665	665	582
625	830	795	765	610	630	775	630	614	760	730	619
470	685	680	735	560	580	670	610	556	539	488	608
445	555	605	685	540	615	610	525	640	639	594	739
400	570	555	655	475	615	715	595	595	740	671	795
480	495	650	690	535	600	760	655	488	621	578	615
555	660	715	690	635	655	775	665	652	720	689	728
520	530	650	605	480	545	585	610	561	670	622	623
490	475	595	660	475	545	635	550	524	538	626	552
500	610	590	680	555	625	670	660	581	613	583	585
425	560	570	570	455	605	605	530	544	645	612	537
535	550	540	630	480	565	745	515	467	606	563	510
560	500	560	575	500	475	610	575	492	591	532	499
530	610	600	620	480	585	650	570	577	612	654	630
570	585	635	765	550	675	765	620	608	705	677	660
505	500	580	655	470	565	570	555	537	585	589	619
465	430	510	680	460	600	670	615	620	594	616	784

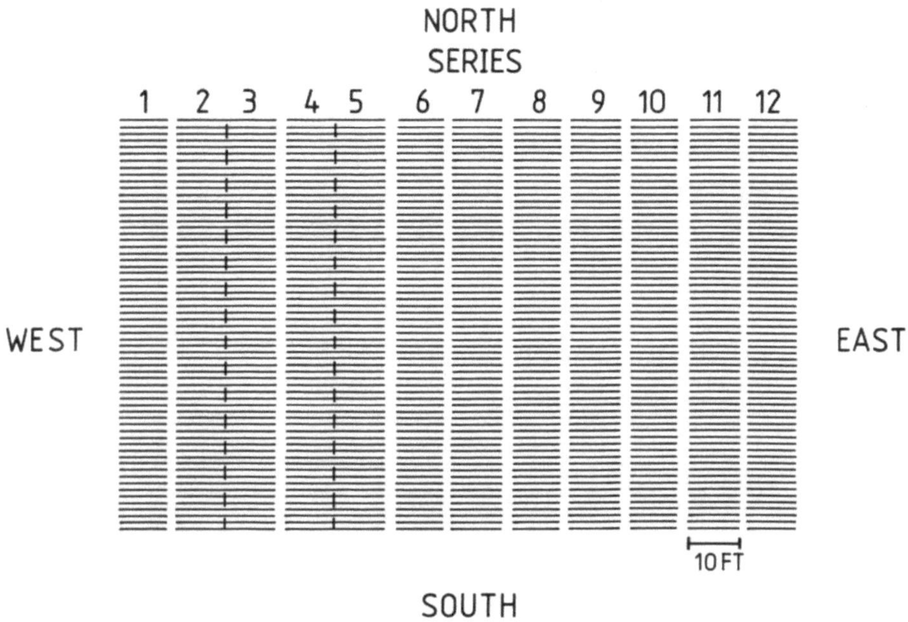

Figure 6.1
Field Plan of Uniformity Trial Showing the Arrangement of Rows
Series 2 and 3, 4 and 5 were grown as 30-foot rows
but were cut in 15-foot lengths at harvest time.

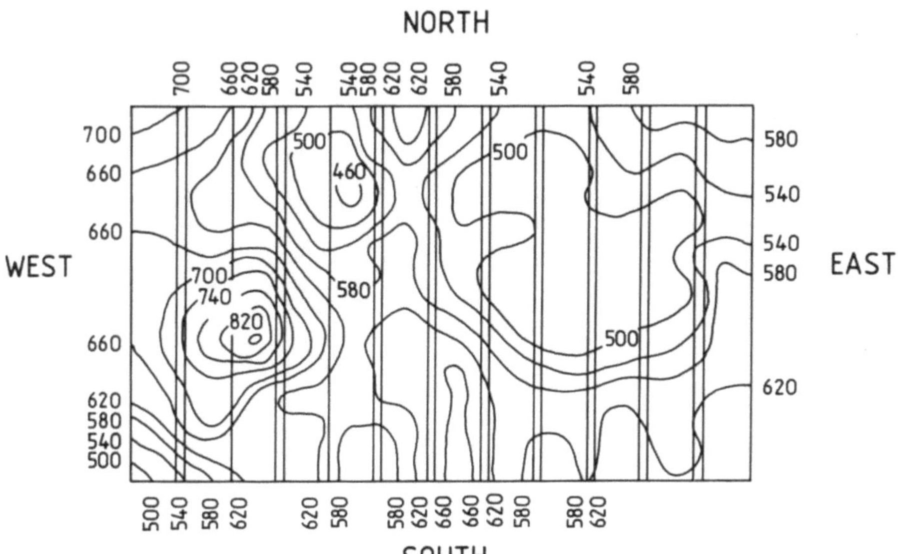

Figure 6.2
Contour Map of the Grain Yields

7. Stanford Heart Transplant Data

Source Miller, R.G. and Halpern, J.W. (1982). Regression with censored data. *Biometrika* **69**, 521-531.

Contributors Rupert G. Miller and Jerry W. Halpern Stanford University

The Stanford Heart Transplantation Program began in October 1967. Patients are admitted to the program after review by a committee, and then they wait for donor hearts to become available. While waiting, some may die or be transferred out of the program, but most receive a transplant. The cut-off date for the data presented in Table 7.1 was in February 1980, and by that time 184 patients had received a transplant.

The data displayed in Table 7.1 summarize survival time in days after transplant, an indicator on whether the patient is dead or alive, the age in years at time of transplant, and a mismatch score which measures the dissimilarity between donor and recipient tissue with respect to HLA antigens. Several patients had multiple transplants; their overall survival times are the values recorded.

An earlier version of these data, with cut-off date in April 1974, was analyzed in Miller (1976), Crowley and Hu (1977), Buckley and James (1979), Koul, Susarla and Van Ryzin (1981). Miller and Halpern (1982) compared various techniques for handling regression with censored data on this updated set of data. Pettitt (1983) also used these data to illustrate his regression methods based on ranks.

References

Buckley, J. and James, I. (1979). Linear regression with censored data. *Biometrika* **66**, 429-436.

Crowley, J. and Hu, M. (1977). Covariance analysis of heart transplant survival data. *J. Amer. Statist. Ass.* **72**, 27-36.

Koul, H., Susarla, V. and Van Ryzin, J. (1981). Regression analysis with randomly right-censored data. *Ann. Statist.* **9**, 1276-1288.

Miller, R.G. (1976). Least squares regression with censored data. *Biometrika* **63**, 449-464.

Pettitt, A.N. (1983). Approximate methods using ranks for regression with censored data. *Biometrika* **70**. 121-132.

Table 7.1
Stanford Heart
Transplant Data, February 1980

Patient no.	Survival time, days	Status *	Age at transplant	Mismatch score **
1	15	1	54	1.11
2	3	1	40	1.66
3	46	1	42	0.61
4	623	1	51	1.32
5	126	1	48	0.36
6	64	1	54	1.89
7	1350	1	54	0.87
8	23	1	56	2.05
9	279	1	49	1.12
10	1024	1	43	1.13
11	10	1	56	2.76
12	39	1	42	1.38
13	730	1	58	0.96
14	1961	1	33	1.06
15	136	1	52	1.62
16	1	1	54	0.47
17	836	1	44	1.58
18	60	1	64	0.69
19	3695	0	40	0.38
20	1996	1	49	0.91
21	0	1	41	0.87
22	47	1	62	0.87
23	54	1	49	2.09
24	51	1	50	-
25	2878	1	49	0.75
26	3410	0	45	0.98
27	44	1	36	0.0
28	994	1	48	0.81
29	51	1	47	1.38
30	1478	1	36	1.35
31	254	1	48	1.08
32	897	1	46	-
33	148	1	47	-
34	51	1	52	1.51
35	323	1	48	1.82
36	3021	0	38	0.98
37	66	1	49	0.66
38	2984	0	32	0.19
39	2723	1	32	1.93
40	550	1	48	0.12
41	66	1	51	1.12
42	65	1	45	1.68
43	227	1	19	1.02
44	2805	0	48	1.20
45	25	1	53	1.68

Table 7.1 *(cont.)*

Patient no.	Survival time, days	Status *	Age at transplant	Mismatch score **
46	631	1	26	1.46
47	2734	0	47	0.97
48	12	1	29	0.61
49	63	1	56	2.16
50	2474	1	52	1.70
51	1384	1	46	1.41
52	544	1	52	1.94
53	29	1	53	1.08
54	48	1	53	3.05
55	297	1	42	0.60
56	1318	1	48	1.44
57	1352	1	54	0.68
58	50	1	46	2.25
59	547	1	49	0.81
60	431	1	47	0.33
61	68	1	51	1.33
62	26	1	52	0.82
63	161	1	43	1.20
64	14	1	40	-
65	2313	0	26	0.46
66	1634	1	23	1.78
67	146	1	45	0.16
68	48	1	28	0.77
69	2127	1	35	0.67
70	263	1	49	0.48
71	2106	0	40	0.86
72	293	1	43	0.70
73	2025	0	30	1.44
74	2006	0	15	1.26
75	2000	0	45	1.46
76	1995	0	47	1.65
77	1945	0	38	1.28
78	65	1	55	0.69
79	731	1	38	0.42
80	1866	0	49	0.51
81	538	1	49	2.76
82	1846	0	44	0.83
83	68	1	35	0.85
84	1778	0	27	0.70
85	1722	0	40	0.95
86	928	1	50	1.12
87	1718	0	39	1.77
88	22	1	27	1.64
89	40	1	42	1.59
90	7	1	28	1.00
91	1638	0	48	0.43
92	1612	0	51	1.25

Table 7.1 *(cont.)*

Patient no.	Survival time, days	Status *	Age at transplant	Mismatch score **
93	25	1	52	0.53
94	1534	1	44	1.71
95	1547	0	50	0.18
96	1271	1	32	1.05
97	44	1	46	1.71
98	1247	1	41	0.43
99	1232	1	18	0.70
100	191	1	42	1.74
101	1393	0	46	0.95
102	1202	1	38	-
103	1378	0	41	1.65
104	1373	0	41	1.38
105	274	1	31	0.58
106	31	1	33	0.36
107	1341	0	50	1.13
108	42	1	19	0.63
109	381	1	45	0.98
110	1264	0	52	0.64
111	1262	0	34	1.68
112	1261	0	47	0.82
113	47	1	36	0.16
114	1193	0	24	1.15
115	626	1	53	1.74
116	48	1	51	0.99
117	1150	1	32	2.25
118	45	1	48	0.65
119	1116	0	14	0.54
120	1107	0	18	0.25
121	1102	0	39	1.35
122	195	1	39	0.73
123	30	1	34	0.84
124	1040	0	43	0.50
125	993	0	30	0.95
126	950	0	46	-
127	729	1	49	1.10
128	121	1	45	-
129	202	1	48	1.24
130	841	0	48	0.86
131	834	0	49	-
132	265	1	49	1.22
133	1	1	21	0.47
134	793	0	19	1.98
135	328	1	34	1.02
136	781	0	20	1.12
137	752	0	43	1.50
138	738	0	41	0.53
139	86	1	12	1.26

Table 7.1 *(cont.)*

Patient no.	Survival time, days	Status *	Age at transplant	Mismatch score **
140	132	1	46	1.09
141	663	0	36	0.47
142	660	0	42	0.75
143	221	1	35	1.04
144	90	1	38	1.00
145	619	0	47	0.90
146	618	0	50	0.82
147	576	0	53	2.25
148	563	0	41	-
149	36	1	45	0.20
150	549	0	40	2.53
151	548	0	30	0.47
152	541	0	47	0.43
153	534	0	20	-
154	169	1	51	1.89
155	122	1	51	1.33
156	382	1	36	-
157	468	0	24	1.39
158	464	0	38	2.07
159	10	1	13	1.49
160	5	1	20	-
161	136	1	55	-
162	406	0	39	1.18
163	391	0	27	1.17
164	374	0	47	-
165	50	1	50	0.50
166	139	1	51	0.96
167	322	0	36	1.73
168	292	0	43	1.40
169	278	0	41	0.98
170	22	1	45	-
171	231	0	52	-
172	145	1	50	0.96
173	188	0	52	-
174	176	0	29	1.72
175	138	1	41	-
176	149	0	21	-
177	119	0	20	-
178	107	0	46	-
179	98	0	19	-
180	89	0	27	-
181	60	0	13	-
182	56	0	27	-
183	2	0	39	-
184	1	0	27	-

* Status: dead, 1; alive,0.

** - denotes missing score.

8. Coal-Mining Disasters

Source Jarrett, R.G. (1979). A note on the intervals between coal-mining disasters. *Biometrika* **66**, 191-193.

Contributor R. G. Jarrett CSIRO, Melbourne

The data on intervals between coal-mining disasters in Britain involving 10 or more men killed (Maguire, Pearson and Wynn, 1952) has been used by a number of authors to illustrate various techniques that can be applied to point processes; see, for example, Barnard (1953), Cox and Lewis (1966) and Boneva, Kendall and Stefanov (1971). See also Maguire, Pearson and Wynn (1953).

It was noticed that the sum of the 109 intervals did not agree with the number of days between 6 December 1875 and 29 May 1951, the starting and finishing dates quoted in the original paper. The data was traced to the Colliery Year Book and Coal Trades Directory, published until 1962 by the National Coal Board in London. Earlier issues gave data back to 15 March 1851. The data given in Table 8.1 have been corrected and extended to cover 191 disasters from 15 March 1851 to 22 March 1962 inclusive, a total of 40,550 days. Those disasters included are those caused by explosions of fire-damp or coal-dust. It may be noted that two accidents occurred on the same day, 6 December 1875, the date used by Maguire, Pearson and Wynn as their starting point. Also given in the present table are the number of deaths in each accident.

Informative analyses of the data are given by Cox and Lewis (1966, p.42) who consider models in which the log of the rate of the process is a polynomial in time. With the extended set of data, the rate appears to be uniform up to about 125 events and then falls off quite considerably. Further information on man-hours worked, tons of coal produced, increased mechanization, etc. might be useful in explaining the reasons for the varying rate of the process.

References

Barnard, G. A. (1953). Time intervals between accidents - a note on Maguire, Pearson and Wynn's paper. *Biometrika* **40**, 212-213.

Boneva, L.I., Kendall, D.G. and Stefanov, I. (1971). Spline transformations: three new diagnostic aids for the statistical data-analyst (with discussion). *J. R. Statist. Soc. B* **33**, 1-71.

Cox, D. R. and Lewis, P. A. W. (1966). *The Statistical Analysis of Series of Events.* London: Methuen.

Maguire, B. A., Pearson, E. S. and Wynn, A. H. A. (1952). The time intervals between industrial accidents. *Biometrika* **39**, 168-180.

Maguire, B.A., Pearson, E.S. and Wynn, A.H.A. (1953). Further notes on the analysis of accident data. *Biometrika* **40**, 213-216.

Table 8.1
Number of Coal-Mining Disasters
between March 15, 1851 and March 22, 1962

Day	Date	Month	Year	Day of year	Interval, days	Number of deaths
Sat	15	Mar	1851	74	****	61
Tue	19	Aug	1851	231	157	35
Sat	20	Dec	1851	354	123	52
Mon	22	Dec	1851	356	2	13
Sat	24	April	1852	115	124	12
Thu	6	May	1852	127	12	22
Mon	10	May	1852	131	4	65
Thu	20	May	1852	141	10	36
Wed	22	Dec	1852	357	216	10
Sat	12	Mar	1853	71	80	10
Thu	24	Mar	1853	83	12	58
Tue	26	April	1853	116	33	11
Fri	1	July	1853	182	66	20
Sat	18	Feb	1854	49	232	89
Sat	24	May	1856	145	826	12
Thu	3	July	1856	185	40	11
Tue	15	July	1856	197	12	114
Wed	13	Aug	1856	226	29	11
Thu	19	Feb	1857	50	190	189
Wed	27	May	1857	147	97	13
Fri	31	July	1857	212	65	40
Tue	2	Feb	1858	33	186	53
Thu	25	Feb	1858	56	23	19
Fri	28	May	1858	148	92	12
Sat	11	Dec	1858	345	197	25
Wed	15	Feb	1860	46	431	13
Fri	2	Mar	1860	62	16	76
Fri	3	Aug	1860	216	154	13
Tue	6	Nov	1860	311	95	12
Sat	1	Dec	1860	336	25	142
Thu	20	Dec	1860	355	19	22
Fri	8	Mar	1861	67	78	13
Thu	26	Sept	1861	269	202	10
Fri	1	Nov	1861	305	36	13
Wed	19	Feb	1862	50	110	47
Sat	22	Nov	1862	326	276	16
Mon	8	Dec	1862	342	16	59
Fri	6	Mar	1863	65	88	26
Sat	17	Oct	1863	290	225	39
Wed	9	Dec	1863	343	53	13
Sat	26	Dec	1863	360	17	15
Fri	16	June	1865	167	538	26
Wed	20	Dec	1865	354	187	34
Tue	23	Jan	1866	23	34	30
Fri	4	May	1866	124	101	12
Thu	14	June	1866	165	41	38
Wed	31	Oct	1866	304	139	24

Table 8.1 *(cont.)*

Day	Date	Month	Year	Day of year	Interval, days	Number of deaths
Wed	12	Dec	1866	346	42	361
Thu	13	Dec	1866	347	1	91
Tue	20	Aug	1867	232	250	14
Fri	8	Nov	1867	312	80	178
Mon	11	Nov	1867	315	3	12
Wed	30	Sept	1868	274	324	10
Wed	25	Nov	1868	330	56	62
Sat	26	Dec	1868	361	31	26
Thu	1	April	1869	91	96	37
Thu	10	June	1869	161	70	53
Wed	21	July	1869	202	41	59
Fri	22	Oct	1869	295	93	11
Mon	15	Nov	1869	319	24	27
Mon	14	Feb	1870	45	91	30
Thu	7	July	1870	188	143	19
Sat	23	July	1870	204	16	19
Fri	19	Aug	1870	231	27	20
Tue	10	Jan	1871	10	144	26
Fri	24	Feb	1871	55	45	38
Thu	2	Mar	1871	61	6	19
Tue	26	Sept	1871	269	208	70
Wed	25	Oct	1871	298	29	26
Wed	14	Feb	1872	45	112	11
Thu	28	Mar	1872	88	43	27
Mon	7	Oct	1872	281	193	34
Tue	18	Feb	1873	49	134	18
Tue	14	April	1874	104	420	54
Sat	18	July	1874	199	95	15
Fri	20	Nov	1874	324	125	23
Thu	24	Dec	1874	358	34	17
Fri	30	April	1875	120	127	43
Sat	4	Dec	1875	338	218	23
Mon	6	Dec	1875	340	2	143
Mon	6	Dec	1875	340	0	16
Mon	18	Dec	1876	353	378	23
Tue	23	Jan	1877	23	36	18
Wed	7	Feb	1877	38	15	10
Sat	10	Mar	1877	69	31	18
Thu	11	Oct	1877	284	215	36
Mon	22	Oct	1877	295	11	207
Fri	8	Mar	1878	67	137	17
Tue	12	Mar	1878	71	4	43
Wed	27	Mar	1878	86	15	23
Fri	7	June	1878	158	72	189
Wed	11	Sept	1878	254	96	268

Table 8.1 *(cont.)*

Day	Date	Month	Year	Day of year	Interval, days	Number of deaths
Mon	13	Jan	1879	13	124	63
Tue	4	Mar	1879	63	50	21
Wed	2	July	1879	183	120	28
Wed	21	Jan	1880	21	203	62
Thu	15	July	1880	197	176	120
Wed	8	Sept	1880	252	55	164
Fri	10	Dec	1880	345	93	101
Mon	7	Feb	1881	38	59	25
Mon	19	Dec	1881	353	315	48
Thu	16	Feb	1882	47	59	74
Tue	18	April	1882	108	61	37
Wed	19	April	1882	109	1	13
Tue	2	May	1882	122	13	32
Tue	7	Nov	1882	311	189	45
Thu	18	Oct	1883	291	345	20
Wed	7	Nov	1883	311	20	68
Sun	27	Jan	1884	27	81	14
Sat	8	Nov	1884	313	286	14
Mon	2	Mar	1885	61	114	42
Thu	18	June	1885	169	108	178
Wed	23	Dec	1885	357	188	81
Fri	13	Aug	1886	225	233	38
Fri	10	Sept	1886	253	28	10
Sat	2	Oct	1886	275	22	22
Thu	2	Dec	1886	336	61	28
Fri	18	Feb	1887	49	78	39
Sat	28	May	1887	148	99	73
Wed	18	April	1888	109	326	30
Fri	18	Jan	1889	18	275	23
Wed	13	Mar	1889	72	54	20
Wed	16	Oct	1889	289	217	64
Thu	6	Feb	1890	37	113	176
Mon	10	Mar	1890	69	32	87
Thu	2	April	1891	92	388	10
Mon	31	Aug	1891	243	151	10
Fri	26	Aug	1892	239	361	112
Tue	4	July	1893	185	312	139
Sat	23	June	1894	174	354	290
Fri	26	April	1895	116	307	13
Sun	26	Jan	1896	26	275	57
Mon	13	April	1896	104	78	20
Thu	30	April	1896	121	17	63
Fri	18	Aug	1899	230	1205	19
Fri	24	May	1901	144	644	81
Wed	3	Sept	1902	246	467	16
Sat	21	Jan	1905	21	871	11
Fri	10	Mar	1905	69	48	33
Tue	11	July	1905	192	123	119
Wed	10	Oct	1906	283	456	25

Table 8.1 *(cont.)*

Day	Date	Month	Year	Day of year	Interval, days	Number of deaths
Thu	20	Feb	1908	51	498	14
Thu	9	April	1908	100	49	10
Tue	18	Aug	1908	231	131	75
Tue	16	Feb	1909	47	182	168
Fri	29	Oct	1909	302	255	27
Wed	11	May	1910	131	194	136
Wed	21	Dec	1910	355	224	344
Tue	9	July	1912	191	566	88
Tue	14	Oct	1913	287	462	439
Sat	30	May	1914	150	228	12
Sun	13	Aug	1916	226	806	13
Sat	12	Jan	1918	12	517	155
Thu	13	July	1922	194	1643	12
Tue	5	Sept	1922	248	54	39
Sat	28	July	1923	209	326	27
Tue	1	Mar	1927	60	1312	52
Sun	12	Feb	1928	43	348	13
Wed	26	Feb	1930	57	745	13
Wed	1	Oct	1930	274	217	14
Thu	29	Jan	1931	29	120	27
Sat	31	Oct	1931	304	275	10
Fri	20	Nov	1931	324	20	45
Mon	25	Jan	1932	25	66	11
Sat	12	Nov	1932	317	292	27
Wed	16	Nov	1932	321	4	11
Sun	19	Nov	1933	323	368	14
Sat	22	Sept	1934	265	307	265
Sat	24	Aug	1935	236	336	10
Thu	12	Sept	1935	255	19	19
Thu	6	Aug	1936	219	329	58
Fri	2	July	1937	183	330	30
Tue	10	May	1938	130	312	79
Sat	28	Oct	1939	301	536	35
Thu	21	Mar	1940	81	145	11
Tue	4	June	1940	156	75	10
Tue	3	June	1941	154	364	12
Thu	10	July	1941	191	37	16
Tue	29	July	1941	210	19	22
Thu	1	Jan	1942	1	156	57
Tue	17	Feb	1942	48	47	12
Fri	26	June	1942	177	129	13
Thu	12	Dec	1946	346	1630	15
Fri	10	Jan	1947	10	29	15
Fri	15	Aug	1947	227	217	104
Fri	22	Aug	1947	234	7	21
Tue	9	Sept	1947	252	18	12
Tue	29	May	1951	149	1358	83
Tue	19	Nov	1957	323	2366	17
Tue	28	June	1960	180	952	45
Thu	22	Mar	1962	81	632	19

9. Recoiling of Guns

Source Lord Brouncker (1734). Experiments of the recoiling of guns. In *The History of the Royal Society of London.* MDCCXXXIV by Thomas Sprat, D. D., the late Lord Bishop of Rochester (Fourth edition, First edition published 1667), pp. 233-239.

Contributor S. M. Stigler University of Chicago

The Lord Brouncker was commanded by The Royal Society of London to make some experiments on the recoiling of guns. A gun was fixed to a triangular frame which could then be fastened at one or more places (Fig. 9.1). He writes:

> When I was commanded by this Society, to make some Experiments of the Recoiling of Guns: In order to the discovery of the cause thereof, I caused this Engine that lies here before you to be prepared, and with it (assisted by some of the most eminent of this Society) I had divers shots made in the Court of this College, near the length thereof from the mark, with a full charge (about a four-penny weight) of Powder; but without any other success, than that there was nothing regular in that way, which was by laying it upon a heavy Table, unto which it was sometimes fastened with Screws at all the four places R,L,V,B, sometimes only at R or L, having wheels affixed at L and V, or R and B, that it might the more easily recoil.
>
> This uncertainty I did then conceive might arise from one or more of these three causes, *viz.*
>
> 1. The violent trembling motion of the Gun, whence the Bullet might casually receive some literal impulse from the nose of the Piece at the parting from it.
> 2. The yielding of the Table, which was sensible.
> 3. The difficulty of aiming well by the Sight and Button so far from the Mark.
>
> Therefore to avoid all these, the Experiments I caus'd to be made before you in the Gallery of this Colledge [*sic*], you may be pleased to remember were performed, first, taking only eight grains of Powder for the charge. Secondly, laying the Engine upon the Floor; and, Thirdly, aiming by a Thread at M, a Mark about an inch and ¼ from the Mouth of the Gun (the edge of a knife being put for the Mark, the better to discern the line that was shot in) and they thus succeeded.
>
> When the Piece was fastened to the Floor both at R and L, the Bullet then did so fully hit the Mark, that it was divided by it into two parts, whose difference in weight was less than ten grains (about the thirty third part of the whole Bullet) although the lesser part was a little hollow, and

that from which the neck of Lead was a little too close pared off: But when hindred from Recoiling only at R, the Bullet mist the mark towards L or A, for the whole Bullet, less than two grains expected, went on that side: And in like manner when hindred from Recoiling at L, the Bullet mist the Mark towards R or B, the whole Bullet, less than two grains excepted, passing the knife on that side thereof.

I had the honour to make other Experiments with the same Engine, lately at White-Hall, before his Majesty and his Highness Royal within the Tilt-yard Gallery, where there is the hearth of a chimney raised a little above the Floor, about the distance of thirteen feet from the opposite wall, against which I caused a plank to be placed, and the Engine to be laid first against the middle of the Hearth, that it might not recoil at all, and that part of the board to be marked against which 'twas levelled, known by a line stretched from the Breech of the Piece unto the Board, directly over the sight and button; and the fire being given (the charge being but eight grains of Powder as before) the Bullet did fully hit the mark. Secondly, the Piece (charged and levelled in the same manner) was laid at the end of the Hearth next the Park, so that very little of the corner R rested against it, and then the Bullet miss'd the mark about an inch and a quarter towards the Park, or A. The like being done at the other end of the Hearth, the Bullet then miss'd the mark as much the other way; and afterwards with double that charge something more, as before I had found it less with a smaller charge.

Since this (at first designing only to experiment the several distances that the Bullet is carried wide of the mark with different charges of Powder) I made these Experiments following.

In the first Column whereof you have the corner stopt from recoiling.

In the second the grains of Powder with which the Piece was charged.

In the third the distance the Bullet was shot wide from the mark, in inches, tenths, and parts of tenths.

In the fourth the side on which the Bullet was carried.

In the last the distance of the mark from the muzzle of the Gun in feet.

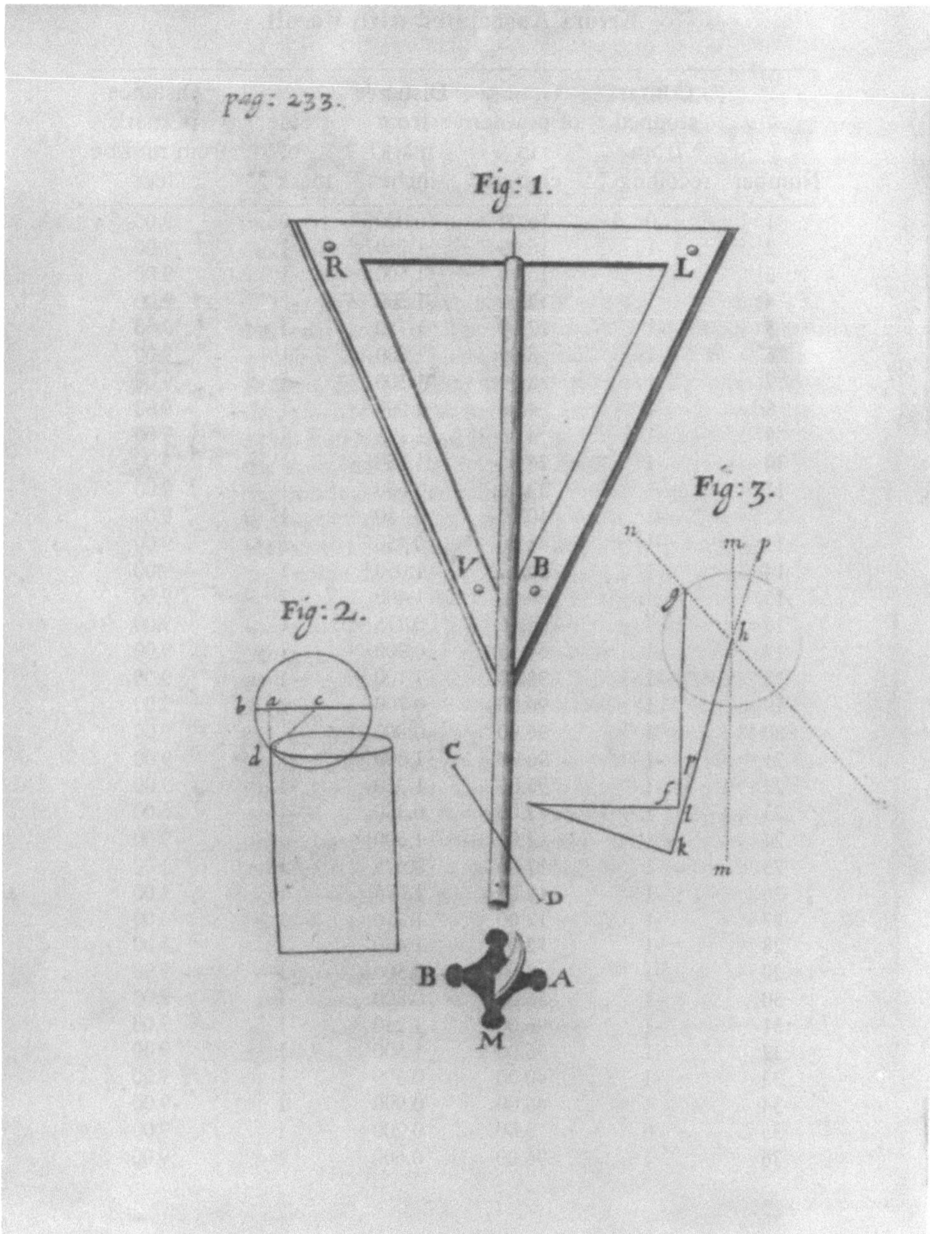

Figure 9.1
Frame used for the Recoiling of Guns Experiment
[Copy of the original, facing page 233.]

Table 9.1
Errors Associated with Recoil

Number	Corner stopped from recoiling *	Grains of powder in charge	Distance from mark, inches	Side of mark **	Distance of mark from muzzle, feet
1	0	16.00	0.000	0	9.00
2	-1	16.00	1.750	1	9.00
3	1	16.00	1.500	-1	9.00
4	1	12.00	1.500	-1	9.00
5	-1	12.00	1.750	1	9.00
6	-1	8.00	1.600	1	9.00
7	1	8.00	1.100	-1	9.00
8	1	4.00	1.000	-1	9.00
9	-1	4.00	1.125	1	9.00
10	-1	24.00	1.150	1	9.00
11	-1	32.00	0.600	1	9.00
12	-1	40.00	0.150	1	9.00
13	-1	48.00	0.450	-1	9.00
14	1	20.00	1.050	-1	9.00
15	1	20.00	1.400	-1	9.00
16	1	64.75	0.775	1	9.00
17	-1	64.75	0.900	-1	9.00
18	-1	96.00	1.100	-1	9.00
19	1	96.00	0.700	1	9.00
20	1	96.00	0.800	1	9.00
21	-1	96.00	1.300	-1	9.00
22	-1	96.00	1.350	-1	9.00
23	1	12.00	0.350	-1	6.00
24	1	12.00	1.300	-1	9.00
25	-1	12.00	0.675	1	2.00
26	-1	12.00	1.075	1	4.00
27	-1	12.00	1.250	1	6.00
28	-1	12.00	1.450	1	8.00
29	-1	48.00	0.500	-1	9.00
30	-1	56.00	0.800	-1	9.00
31	-1	96.00	1.250	-1	9.00
32	-1	96.00	1.500	-1	9.00
33	-1	40.00	0.500	-1	9.00
34	1	46.00	0.900	1	9.00
35	0	8.00	0.200	1	9.00
36	1	96.00	0.600	1	9.00

Table 9.1 *(cont.)*

Number	Corner stopped from recoiling *	Grains of powder in charge	Distance from mark, inches	Side of mark **	Distance of mark from muzzle, feet
37	-1	96.00	0.900	-1	9.00
38	-1	40.00	0.175	-1	9.00
39	-1	38.00	0.150	1	9.00
40	-1	39.00	0.050	-1	9.00
41	1	39.00	0.100	-1	9.00
42	1	12.00	0.600	-1	2.00
43	1	12.00	0.925	-1	4.00
44	1	12.00	1.200	-1	6.00
45	1	12.00	1.550	-1	8.00
46	1	12.00	1.900	-1	9.00
47	0	12.00	0.100	-1	9.00
48	1	12.00	0.300	-1	0.25
49	-1	12.00	0.350	1	0.25
50	-1	96.00	0.050	1	0.25
51	1	96.00	0.050	-1	0.25
52	1	96.00	0.200	1	2.00
53	-1	96.00	0.200	-1	
54	-1	48.00	0.150	1	0.25
55	-1	48.00	0.000	0	2.00
56	1	48.00	0.125	-1	0.25
57	1	39.00	0.350	-1	9.00
58	1	39.00	0.200	-1	9.00
59	1	40.00	0.200	-1	9.00
60	1	40.00	0.000	0	9.00
61	1	40.00	0.200	-1	9.00
62	1	96.00	0.650	1	9.00
63	-1	96.00	1.050	-1	9.00
64	-1	96.00	1.050	-1	9.00
65	1	96.00	0.750	1	9.00
66	1	96.00	1.000	1	9.00
67	1	40.00	0.800	-1	9.00
68	-1	40.00	0.500	1	9.00
69	-1	48.00	0.100	1	9.00
70	1	48.00	0.025	-1	2.00
71	-1	12.00	1.700	1	9.00
72	-1	12.00	0.225	1	0.25
73	-1	12.00	0.650	1	2.00

Table 9.1 *(cont.)*

Number	Corner stopped from recoiling *	Grains of powder in charge	Distance from mark, inches	Side of mark **	Distance of mark from muzzle, feet
74	-1	12.00	1.000	1	4.00
75	-1	12.00	1.175	1	6.00
76	-1	48.00	0.075	1	0.25
77	-1	48.00	0.150	-1	2.00
78	-1	48.00	0.250	-1	4.00
79	-1	48.00	0.550	-1	6.00
80	-1	4.00	0.150	1	0.25
81	-1	4.00	0.150	1	0.25
82	-1	8.00	0.325	1	0.25
83	-1	12.00	0.300	1	0.25
84	-1	16.00	0.225	1	0.25
85	1	48.00	0.000	0	9.00
86	1	48.00	0.100	-1	9.00
87	-1	48.00	0.025	1	9.00
88	-1	4.00	1.600	1	9.00
89	-1	4.00	1.550	1	9.00
90	1	4.00	1.600	-1	9.00
91	1	8.00	1.800	-1	9.00
92	-1	8.00	1.800	1	9.00
93	-1	12.00	2.000	1	9.00
94	1	12.00	2.100	-1	9.00
95	1	16.00	1.750	-1	9.00
96	-1	16.00	1.800	1	9.00
97	-1	20.00	1.500	1	9.00
98	-1	20.00	0.200	1	0.25
99	-1	24.00	0.275	1	0.25
100	-1	28.00	0.100	1	0.25
101	-1	32.00	0.150	1	0.25
102	-1	36.00	0.125	1	0.25
103	-1	40.00	0.100	1	0.25
104	-1	44.00	0.050	1	0.25
105	-1	48.00	0.050	1	0.25
106	-1	52.00	0.025	1	0.25
107	-1	56.00	0.025	1	0.25
108	-1	60.00	0.0125	1	0.25
109	-1	64.00	0.0125	1	0.25
110	-1	96.00	0.000	-1	0.25

* 0 denotes B; -1 denotes L; 1 denotes R.

** 0 denotes N; -1 denotes L; 1 denotes R.

10. Old Maps

Contributors J. Morrison U.S. Geological Survey and B.L. Joiner University of Wisconsin

This problem is concerned with the classification of old maps of the Great Lakes area. The data are taken from the eleven maps listed in Table 10.2. These maps are believed to be representative of the period of time commencing with the widespread knowledge that five major lakes existed in the interior of North America, and ending when relatively large scale hydrographic surveys of the lakes' shorelines were being done.

The data shown in Table 10.1 consist of the latitude, ϕ, and longitude, λ, co-ordinates, as determined for each map, for each of 39 points easily identifiable on the eleven maps. These data were obtained by placing a grid over the old maps and doing a linear interpolation. Interpolation accuracy is felt to be good except for the indicated numbers. Also included are the current co-ordinates for the 39 points.

It is conjectured that there are five key ways a map might be systematically in error. These are: a constant error in latitude, a constant error in longitude, a proportional error in latitude, a proportional error in longitude, an error resulting in a non-zero angle between true North and the map's North. In addition, particular geographic locations may be misplaced: either single locations, or groups of locations, for example, one whole lake, may be off.

The primary task is to develop a methodology for parameterizing each map with respect to these characteristics and with respect to any other characteristics that seem to be important.

Table 10.1: Latitudes and Longitudes of

	Actual		Map									
			Coronelli 1688		Del'Isle 1703		Popple 1733		Bellin 1744		Bellin 1755	
	Lat.	Long.	Lat.	Long.	Lat.	Long.	Lat.	Long.	Lat.	Long.	Lat.	Long.
1	41.68	82.85	41.62	87.67	-41.12	-83.03	-41.03	-85.10	41.28	82.42	43.13	82.25
2	41.90	82.52	42.12	87.00	41.88	82.00	41.87	85.37	42.08	81.17	42.35	81.77
3	42.25	81.92	42.07	85.97	42.13	80.78	41.98	83.78	42.22	80.20	42.62	81.37
4	42.55	80.03	42.47	84.17	42.13	79.55	42.00	83.18	42.07	80.27	42.83	79.92
5	44.05	83.03	43.42	87.25	44.23	83.60	44.13	86.85	44.60	83.50	44.62	85.50
6	43.65	83.95	43.12	87.92	43.25	84.48	43.80	87.10	43.72	83.23	43.62	85.25
7	44.25	83.45	43.63	87.62	44.43	83.83	44.20	87.00	44.78	83.83	44.82	85.75
8	45.87	84.87	45.05	89.62	-45.12	-85.08	-45.13	-87.15	45.67	85.62	45.67	87.67
9	45.93	80.90	45.38	85.58	45.57	85.37	45.15	83.90	45.75	78.73	45.90	80.78
10	46.23	83.57	45.60	87.75	45.73	84.42	45.10	87.22	46.00	84.30	45.95	86.25
11	45.80	81.58	45.38	86.57	45.17	82.55	44.90	85.08	45.82	81.33	45.78	83.33
12	45.90	83.22	45.22	87.67	45.33	84.18	44.85	86.95	45.85	83.80	45.70	86.28
13	44.25	86.33	43.53	89.87	-44.00	-87.03	-43.15	-89.10	44.25	85.92	44.20	88.17
14	43.95	86.45	43.17	89.83	43.93	86.00	43.57	88.77	44.08	85.77	43.75	87.80
15	43.07	86.25	42.20	90.13	43.12	86.18	42.97	89.10	42.87	85.13	42.80	87.17
16	42.40	86.28	41.87	90.33	42.47	86.20	42.20	89.22	41.85	84.92	41.72	87.05
17	42.12	86.48	41.75	90.65	41.88	86.22	41.87	89.58	41.63	85.25	41.55	87.38
18	41.88	87.62	41.88	93.63	42.17	88.12	41.67	90.65	41.75	86.50	41.75	88.67
19	44.53	88.00	43.58	93.92	43.82	89.15	43.22	91.53	43.42	88.33	43.33	90.33
20	45.95	86.25	45.47	90.70	-45.12	-87.03	-45.13	-89.10	45.58	87.67	45.62	89.58
21	47.93	84.85	48.75	88.00	48.08	85.28	47.90	88.13	49.25	85.75	49.25	87.70
22	46.93	84.53	47.67	88.70	47.28	85.55	47.20	88.53	46.88	85.08	46.72	87.17
23	48.18	88.42	47.92	94.28	47.75	89.62	48.63	91.47	48.12	90.63	48.08	92.55
24	47.85	89.23	47.53	95.75	47.67	90.78	48.60	92.45	47.80	92.33	47.80	94.33
25	46.75	92.10	46.20	99.22	-46.08	-93.15	47.92	94.57	46.42	94.22	46.53	96.50
26	46.87	89.33	45.85	95.95	46.13	91.12	-46.15	-93.00	46.58	90.78	46.48	92.78
27	46.75	88.45	45.60	94.42	45.97	89.57	46.15	91.77	46.00	89.17	45.83	91.08
28	47.40	87.70	46.88	93.55	46.82	89.05	47.92	91.53	47.08	89.25	46.98	91.17
29	43.27	79.03	43.00	83.42	43.20	78.58	42.80	80.50	42.93	76.43	43.50	78.75
30	44.22	76.48	44.12	80.50	44.23	79.10	45.00	77.95	44.13	72.83	44.50	77.00
31	43.63	79.47	43.55	83.92	43.85	79.15	43.72	81.05	44.08	76.72	-43.08	-79.00
32	44.20	76.97	-43.13	-80.03	-44.00	-76.07	-43.15	-78.13	44.42	74.00	-44.03	-75.08
33	44.10	77.58	-43.12	-81.00	-44.00	-77.02	-43.13	-79.08	44.53	75.08	44.38	77.85
34	42.90	78.92	42.87	83.57	42.68	78.78	42.15	80.47	42.98	76.75	43.33	78.83
35	42.37	82.92	42.53	87.33	42.48	82.05	42.28	85.28	42.55	82.00	42.88	82.83
36	42.62	82.53	42.80	86.92	42.97	81.88	42.60	85.10	42.82	81.75	43.25	82.75
37	45.27	81.65	44.30	84.92	44.52	81.57	44.32	84.42	44.80	80.37	44.83	81.50
38	45.30	86.98	44.37	91.92	45.17	86.15	44.48	90.45	44.58	87.67	44.57	89.57
39	46.48	84.63	46.10	89.45	46.13	88.02	46.07	88.63	46.20	85.70	46.10	87.57

* - denotes that interpolation accuracy is not good.

39 points as represented by 11 maps *

	Map											
	Mitchell 1755		D'Anville 1755		Lattre 1784		Arrowsmith 1802		Cary 1805		Melish 1818	
	Lat.	Long.	Lat.	Long.	Lat.	Long.	Lat.	Long.	Lat.	Long.	Lat.	Long.
1	42.05	82.88	42.15	81.63	42.03	83.53	42.02	82.90	42.00	82.90	-41.12	-82.12
2	42.23	82.35	42.38	80.78	42.25	82.45	42.12	82.42	42.08	82.37	42.00	82.08
3	42.37	81.30	42.53	80.35	42.33	81.98	42.50	81.87	42.48	81.87	42.32	81.75
4	42.37	80.03	42.87	78.73	42.33	80.65	42.68	80.15	42.67	80.18	42.55	80.08
5	44.75	83.57	44.47	81.75	44.75	84.33	44.35	82.90	44.32	82.90	44.27	83.00
6	44.03	83.47	44.03	81.72	44.03	84.22	44.07	83.60	44.03	83.57	-44.07	-83.12
7	44.95	83.72	44.62	81.97	44.92	84.47	44.50	83.15	44.47	83.13	44.63	83.13
8	45.55	85.10	45.18	84.08	-45.10	-85.03	-46.00	-84.07	-46.00	-84.08	-46.13	-84.10
9	45.88	80.03	46.00	79.93	45.63	80.68	45.83	81.05	45.80	81.05	45.72	81.38
10	45.80	84.03	45.48	81.83	45.72	84.68	46.08	82.73	46.05	82.72	45.92	83.82
11	45.70	81.97	45.57	79.52	45.67	82.60	-45.15	-81.05	-45.15	-81.05	-46.15	-81.08
12	45.67	83.68	45.33	81.35	45.55	84.25	-46.00	-82.13	-46.00	-82.13	-46.13	-83.00
13	44.45	85.38	44.15	83.55	44.45	86.00	44.22	84.87	44.20	84.92	44.08	85.25
14	44.33	85.28	43.93	83.52	44.37	85.88	44.02	84.88	44.00	84.90	43.82	85.25
15	43.42	84.85	43.47	83.97	43.43	85.55	43.33	84.90	43.30	84.93	42.92	85.12
16	42.58	84.75	42.77	84.52	42.55	85.63	42.72	85.08	42.70	85.08	42.18	85.12
17	42.43	85.12	42.53	84.82	42.25	85.77	42.50	85.22	42.50	85.18	42.00	85.18
18	42.63	86.08	42.62	85.63	42.63	86.80	42.52	86.03	42.52	86.03	42.08	86.15
19	43.83	87.15	44.00	85.67	43.82	87.85	44.28	86.35	44.30	86.33	44.18	86.88
20	45.50	86.67	45.30	84.00	45.52	87.18	45.60	85.25	45.55	85.22	45.63	85.45
21	48.33	84.92	47.70	83.47	48.22	85.63	47.95	84.97	47.92	84.93	47.92	85.00
22	46.42	84.60	46.30	82.58	46.40	85.27	47.30	84.37	-46.15	-84.02	-47.13	-84.03
23	47.37	88.57	46.68	86.35	47.42	89.17	48.28	88.63	48.27	88.68	48.27	88.73
24	47.22	89.78	46.38	87.05	47.12	90.50	48.05	89.42	48.07	89.37	48.13	89.40
25	46.10	91.53	45.65	89.47	45.97	92.23	46.80	92.17	46.78	92.18	46.73	92.17
26	46.22	88.78	45.77	87.13	46.15	89.42	46.88	89.42	46.92	89.40	46.92	89.32
27	45.72	87.62	45.45	86.05	45.72	88.38	46.83	88.63	46.83	88.67	46.82	88.73
28	46.62	87.60	46.23	85.75	46.57	88.22	45.58	88.05	47.58	88.05	47.73	88.13
29	43.42	78.25	43.52	77.48	43.35	79.00	43.27	79.07	43.23	79.07	43.27	79.00
30	44.45	75.87	44.78	74.90	44.40	76.50	44.17	76.72	44.15	76.72	44.27	76.50
31	43.80	78.70	43.80	77.52	43.85	79.55	43.67	79.52	43.63	79.53	-44.07	-79.08
32	44.57	76.27	44.80	75.35	44.58	77.00	44.12	77.25	44.13	77.23	-45.02	-77.02
33	44.65	76.72	44.70	75.63	44.78	77.48	44.05	77.77	44.05	77.78	44.08	77.57
34	42.97	78.37	43.13	77.52	42.97	79.08	42.88	78.98	42.88	78.98	42.73	80.65
35	42.88	82.62	42.77	81.27	42.80	83.25	42.62	82.77	42.63	82.70	42.68	82.63
36	43.08	82.38	42.87	80.88	43.02	83.07	42.82	82.28	42.82	82.22	43.00	82.13
37	44.97	81.25	45.08	79.55	44.92	82.02	45.40	80.83	45.42	80.83	45.08	81.65
38	44.73	86.58	44.70	84.63	44.75	87.25	45.08	85.52	45.07	85.52	45.08	86.07
39	45.90	85.03	45.77	82.93	45.88	85.72	46.42	84.33	46.40	84.32	46.45	84.88

Table 10.2
Maps used for Table 10.1

1688	Coronelli	Partie Occidentale du Canada ou de la Nouvelle France
1703	Del'Isle	Carte du Canada ou de la Nouvelle France et des Découvertes qui y ont été faites
1733	Popple	A map of the British Empire in America
1744	Bellin	Carte de Lacs du Canada
1755	Bellin	Partie Occidentale de la Nouvelle France ou Canada
1755	Mitchell	A map of the British and French Dominions in North America
1755	D'Anville	Canada Louisiane et Terres Angloises
1784	Lattre	Carte de Etats-Unis de l'Amérique
1802	Arrowsmith	A Map of the United States of North America
1805	Cary	A New Map of Part of the United States of North America exhibiting the Western Territory Kentucky, Pennsylvania, Maryland, Virginia, etc., also the Lakes Superior, Michigan, Huron, Ontario, and Erie with Upper and Lower Canada, etc., from the latest authorities.
1818	Melish	United States of America

11. Monthly Mean Sunspot Numbers

Source Waldmeir, M. (1961). *The Sunspot Activity in the Years 1610 - 1960*. Zürich: Schulthess and *International Astronomical Union Quarterly Bulletin on Solar Activity*, Tokyo.

Contributor D. R. Brillinger University of California, Berkeley

Daily relative sunspot numbers are based upon counts of spots and group entities of spots on the sun's surface at some time each day. Wolf devised a relative number with the intent of reducing the spot counts of different observers and telescopes to a common basis. The relative number is $k(f + 10g)$ where, for a given day, g is the number of groups, irrespective of the number of component spots, f is the total number of component spots which can be counted in these groups and may range from 1 to 50 or more in the case of complex groups, and k is a scale factor depending on the estimated efficiency of the observer and his telescope. The daily sunspot numbers are evaluated on the basis of more than fifty observing stations around the world and related to the observations of a reference station (which changed in 1982). The series provided, Table 11.1, is that of the simple mean of the daily values for each month.

The series is of interest in its own right as reflecting the varying behaviour of the sun. It is also of interest as an exogenous factor affecting the earth. Quite a number of statistical studies of the series have been carried out; see the brief bibliography provided.

Brief Bibliography

Bloomfield, P. (1976). *Fourier Analysis of Time Series: An Introduction.* New York: Wiley.

Brillinger, D.R. and Rosenblatt, M. (1967). Computation and interpretation of k-th order spectra. In *Spectral Analysis of Time Series* edited by B. Harris, pp.189-232. New York: Wiley.

Campbell, W. H. (1968). Correlation of sunspot numbers with the quantity of S. Chapman publications. *Trans. Amer. Geophys. Union* **49**, 609-610.

Craddock, J. M. (1967). An experiment in the analysis and prediction of time series. *The Statistician* **17**, 257-268.

Granger, C.W.J. (1957). A statistical model for sunspot activity. *Astrophys. J.* **126**, 152-158.

Gray, H.L. and Woodward, W.A. (1978). New ARMA models for Wolfer's sunspot data. *Commun. Statist. Simulat., Comput. B* 7, 97-115.

Izenman, A.J. (1983). J. R. Wolf and H. A. Wolfer: An historical note on the Zurich sunspot relative numbers. *J. R. Statist. Soc. A* **136**, 311-318.

Moran, P. A. P. (1954). Some experiments on the prediction of sunspot numbers. *J. R. Statist. Soc. B* **16**, 112-117.

Morris, J. (1977). Forecasting the sunspot cycle (with discussion). *J. R. Statist. Soc. A* **140**, 437-447.

Newton, H. W. (1958). *The Face of the Sun* . London: Penguin.

Pittock, A.B. (1978). A critical look at long-term sun-weather relationships. *Rev. Geophysics and Space Physics* **16**, 400-420.

Schuster, A. (1906). On the periodicities of sunspots. *Phil. Trans. Roy. Soc. A* **206**, 69-100.

Whittle, P. (1954). A statistical investigation of sunspot observations with special reference to H. Alfven's sunspot model. *Astrophys. J.* **120**, 251-260.

Yule, G.U. (1927). On a method of investigating periodicities in disturbed series, with special reference to Wolfer's sunspot numbers. *Phil. Trans. Roy. Soc. A* **226**, 267-298.

Table 11.1
Zürich Monthly Sunspot Numbers 1749 - 1983

Year	Jan	Feb	March	April	May	June	July	August	Sept	Oct	Nov	Dec
1749	58.0	62.6	70.0	55.7	85.0	83.5	94.8	66.3	75.9	75.5	158.6	85.2
1750	73.3	75.9	89.2	88.3	90.0	100.0	85.4	103.0	91.2	65.7	63.3	75.4
1751	70.0	43.5	45.3	56.4	60.7	50.7	66.3	59.8	23.5	23.2	28.5	44.0
1752	35.0	50.0	71.0	59.3	59.7	39.6	78.4	29.3	27.1	46.6	37.6	40.0
1753	44.0	32.0	45.7	38.0	36.0	31.7	22.2	39.0	28.0	25.0	20.0	6.7
1754	0.0	3.0	1.7	13.7	20.7	26.7	18.8	12.3	8.2	24.1	13.2	4.2
1755	10.2	11.2	6.8	6.5	0.0	0.0	8.6	3.2	17.8	23.7	6.8	20.0
1756	12.5	7.1	5.4	9.4	12.5	12.9	3.6	6.4	11.8	14.3	17.0	9.4
1757	14.1	21.2	26.2	30.0	38.1	12.8	25.0	51.3	39.7	32.5	64.7	33.5
1758	37.6	52.0	49.0	72.3	46.4	45.0	44.0	38.7	62.5	37.7	43.0	43.0
1759	48.3	44.0	46.8	47.0	49.0	50.0	51.0	71.3	77.2	59.7	46.3	57.0
1760	67.3	59.5	74.7	58.3	72.0	48.3	66.0	75.6	61.3	50.6	59.7	61.0
1761	70.0	91.0	80.7	71.7	107.2	99.3	94.1	91.1	100.7	88.7	89.7	46.0
1762	43.8	72.8	45.7	60.2	39.9	77.1	33.8	67.7	68.5	69.3	77.8	77.2
1763	56.5	31.9	34.2	32.9	32.7	35.8	54.2	26.5	68.1	46.3	60.9	61.4
1764	59.7	59.7	40.2	34.4	44.3	30.0	30.0	30.0	28.2	28.0	26.0	25.7
1765	24.0	26.0	25.0	22.0	20.2	20.0	27.0	29.7	16.0	14.0	14.0	13.0
1766	12.0	11.0	36.6	6.0	26.8	3.0	3.3	4.0	4.3	5.0	5.7	19.2
1767	27.4	30.0	43.0	32.9	29.8	33.3	21.9	40.8	42.7	44.1	54.7	53.3
1768	53.5	66.1	46.3	42.7	77.7	77.4	52.6	66.8	74.8	77.8	90.6	111.8
1769	73.9	64.2	64.3	96.7	73.6	94.4	118.6	120.3	148.8	158.2	148.1	112.0
1770	104.0	142.5	80.1	51.0	70.1	83.3	109.8	126.3	104.4	103.6	132.2	102.3
1771	36.0	46.2	46.7	64.9	152.7	119.5	67.7	58.5	101.4	90.0	99.7	95.7
1772	100.9	90.8	31.1	92.2	38.0	57.0	77.3	56.2	50.5	78.6	61.3	64.0
1773	54.6	29.0	51.2	32.9	41.1	28.4	27.7	12.7	29.3	26.3	40.9	43.2
1774	46.8	65.4	55.7	43.8	51.3	28.5	17.5	6.6	7.9	14.0	17.7	12.2
1775	4.4	0.0	11.6	11.2	3.9	12.3	1.0	7.9	3.2	5.6	15.1	7.9
1776	21.7	11.6	6.3	21.8	11.2	19.0	1.0	24.2	16.0	30.0	35.0	40.0
1777	45.0	36.5	39.0	95.5	80.3	80.7	95.0	112.0	116.2	106.5	146.0	157.3
1778	177.3	109.3	134.0	145.0	238.9	171.6	153.0	140.0	171.7	156.3	150.3	105.0
1779	114.7	165.7	118.0	145.0	140.0	113.7	143.0	112.0	111.0	124.0	114.0	110.0
1780	70.0	98.0	98.0	95.0	107.2	88.0	86.0	86.0	93.7	77.0	60.0	58.7
1781	98.7	74.7	53.0	68.3	104.7	97.7	73.5	66.0	51.0	27.3	67.0	35.2
1782	54.0	37.5	37.0	41.0	54.3	38.0	37.0	44.0	34.0	23.2	31.5	30.0
1783	28.0	38.7	26.7	28.3	23.0	25.2	32.2	20.0	18.0	8.0	15.0	10.5
1784	13.0	8.0	11.0	10.0	6.0	9.0	6.0	10.0	10.0	8.0	17.0	14.0
1785	6.5	8.0	9.0	15.7	20.7	26.3	36.3	20.0	32.0	47.2	40.2	27.3
1786	37.2	47.6	47.7	85.4	92.3	59.0	83.0	89.7	111.5	112.3	116.0	112.7
1787	134.7	106.0	87.4	127.2	134.8	99.2	128.0	137.2	157.3	157.0	141.5	174.0
1788	138.0	129.2	143.3	108.5	113.0	154.2	141.5	136.0	141.0	142.0	94.7	129.5
1789	114.0	125.3	120.0	123.3	123.5	120.0	117.0	103.0	112.0	89.7	134.0	135.5
1790	103.0	127.5	96.3	94.0	93.0	91.0	69.3	87.0	77.3	84.3	82.0	74.0
1791	72.7	62.0	74.0	77.2	73.7	64.2	71.0	43.0	66.5	61.7	67.0	66.0

Table 11.1 *(cont.)*

Year	Jan	Feb	March	April	May	June	July	August	Sept	Oct	Nov	Dec
1792	58.0	64.0	63.0	75.7	62.0	61.0	45.8	60.0	59.0	59.0	57.0	56.0
1793	56.0	55.0	55.5	53.0	52.3	51.0	50.0	29.3	24.0	47.0	44.0	45.7
1794	45.0	44.0	38.0	28.4	55.7	41.5	41.0	40.0	11.1	28.5	67.4	51.4
1795	21.4	39.9	12.6	18.6	31.0	17.1	12.9	25.7	13.5	19.5	25.0	18.0
1796	22.0	23.8	15.7	31.7	21.0	6.7	26.9	1.5	18.4	11.0	8.4	5.1
1797	14.4	4.2	4.0	4.0	7.3	11.1	4.3	6.0	5.7	6.9	5.8	3.0
1798	2.0	4.0	12.4	1.1	0.0	0.0	0.0	3.0	2.4	1.5	12.5	9.9
1799	1.6	12.6	21.7	8.4	8.2	10.6	2.1	0.0	0.0	4.6	2.7	8.6
1800	6.9	9.3	13.9	0.0	5.0	23.7	21.0	19.5	11.5	12.3	10.5	40.1
1801	27.0	29.0	30.0	31.0	32.0	31.2	35.0	38.7	33.5	32.6	39.8	48.2
1802	47.8	47.0	40.8	42.0	44.0	46.0	48.0	50.0	51.8	38.5	34.5	50.0
1803	50.0	50.8	29.5	25.0	44.3	36.0	48.3	34.1	45.3	54.3	51.0	48.0
1804	45.3	48.3	48.0	50.6	33.4	34.8	29.8	43.1	53.0	62.3	61.0	60.0
1805	61.0	44.1	51.4	37.5	39.0	40.5	37.6	42.7	44.4	29.4	41.0	38.3
1806	39.0	29.6	32.7	27.7	26.4	25.6	30.0	26.3	24.0	27.0	25.0	24.0
1807	12.0	12.2	9.6	23.8	10.0	12.0	12.7	12.0	5.7	8.0	2.6	0.0
1808	0.0	4.5	0.0	12.3	13.5	13.5	6.7	8.0	11.7	4.7	10.5	12.3
1809	7.2	9.2	0.9	2.5	2.0	7.7	0.3	0.2	0.4	0.0	0.0	0.0
1810	0.0	0.0	0.0	0.0	0.0	0.0	0.0	0.0	0.0	0.0	0.0	0.0
1811	0.0	0.0	0.0	0.0	0.0	0.0	6.6	0.0	2.4	6.1	0.8	1.1
1812	11.3	1.9	0.7	0.0	1.0	1.3	0.5	15.6	5.2	3.9	7.9	10.1
1813	0.0	10.3	1.9	16.6	5.5	11.2	18.3	8.4	15.3	27.8	16.7	14.3
1814	22.2	12.0	5.7	23.8	5.8	14.9	18.5	2.3	8.1	19.3	14.5	20.1
1815	19.2	32.2	26.2	31.6	9.8	55.9	35.5	47.2	31.5	33.5	37.2	65.0
1816	26.3	68.8	73.7	58.8	44.3	43.6	38.8	23.2	47.8	56.4	38.1	29.9
1817	36.4	57.9	96.2	26.4	21.2	40.0	50.0	45.0	36.7	25.6	28.9	28.4
1818	34.9	22.4	25.4	34.5	53.1	36.4	28.0	31.5	26.1	31.7	10.9	25.8
1819	32.5	20.7	3.7	20.2	19.6	35.0	31.4	26.1	14.9	27.5	25.1	30.6
1820	19.2	26.6	4.5	19.4	29.3	10.8	20.6	25.9	5.2	9.0	7.9	9.7
1821	21.5	4.3	5.7	9.2	1.7	1.8	2.5	4.8	4.4	18.8	4.4	0.0
1822	0.0	0.9	16.1	13.5	1.5	5.6	7.9	2.1	0.0	0.4	0.0	0.0
1823	0.0	0.0	0.6	0.0	0.0	0.0	0.5	0.0	0.0	0.0	0.0	20.4
1824	21.6	10.8	0.0	19.4	2.8	0.0	0.0	1.4	20.5	25.2	0.0	0.8
1825	5.0	15.5	22.4	3.8	15.4	15.4	30.9	25.4	15.7	15.6	11.7	22.0
1826	17.7	18.2	36.7	24.0	32.4	37.1	52.5	39.6	18.9	50.6	39.5	68.1
1827	34.6	47.4	57.8	46.0	56.3	56.7	42.9	53.7	49.6	57.2	48.2	46.1
1828	52.8	64.4	65.0	61.1	89.1	98.0	54.3	76.4	50.4	54.7	57.0	46.6
1829	43.0	49.4	72.3	95.0	67.5	73.9	90.8	78.3	52.8	57.2	67.6	56.5
1830	52.2	72.1	84.6	107.1	66.3	65.1	43.9	50.7	62.1	84.4	81.2	82.1
1831	47.5	50.1	93.4	54.6	38.1	33.4	45.2	54.9	37.9	46.2	43.5	28.9
1832	30.9	55.5	55.1	26.9	41.3	26.7	13.9	8.9	8.2	21.1	14.3	27.5
1833	11.3	14.9	11.8	2.8	12.9	1.0	7.0	5.7	11.6	7.5	5.9	9.9
1834	4.9	18.1	3.9	1.4	8.8	7.8	8.7	4.0	11.5	24.8	30.5	34.5
1835	7.5	24.5	19.7	61.5	43.6	33.2	59.8	59.0	100.8	95.2	100.0	77.5
1836	88.6	107.6	98.1	142.9	111.4	124.7	116.7	107.8	95.1	137.4	120.9	206.2
1837	188.0	175.6	134.6	138.2	111.3	158.0	162.8	134.0	96.3	123.7	107.0	129.8

Table 11.1 *(cont.)*

Year	Jan	Feb	March	April	May	June	July	August	Sept	Oct	Nov	Dec
1838	144.9	84.8	140.8	126.6	137.6	94.5	108.2	78.8	73.6	90.8	77.4	79.8
1839	107.6	102.5	77.7	61.8	53.8	54.6	84.7	131.2	132.7	90.8	68.8	63.6
1840	81.2	87.7	55.5	65.9	69.2	48.5	60.7	57.8	74.0	49.8	54.3	53.7
1841	24.0	29.9	29.7	42.6	67.4	55.7	30.8	39.3	35.1	28.5	19.8	38.8
1842	20.4	22.1	21.7	26.9	24.9	20.5	12.6	26.5	18.5	38.1	40.5	17.6
1843	13.3	3.5	8.3	8.8	21.1	10.5	9.5	11.8	4.2	5.3	19.1	12.7
1844	9.4	14.7	13.6	20.8	12.0	3.7	21.2	23.9	6.9	21.5	10.7	21.6
1845	25.7	43.6	43.3	56.9	47.8	31.1	30.6	32.3	29.6	40.7	39.4	59.7
1846	38.7	51.0	63.9	69.2	59.9	65.1	46.5	54.8	107.1	55.9	60.4	65.5
1847	62.6	44.9	85.7	44.7	75.4	85.3	52.2	140.6	161.2	180.4	138.9	109.6
1848	159.1	111.8	108.9	107.1	102.2	123.8	139.2	132.5	100.3	132.4	114.6	159.9
1849	156.7	131.7	96.5	102.5	80.6	81.2	78.0	61.3	93.7	71.5	99.7	97.0
1850	78.0	89.4	82.6	44.1	61.6	70.0	39.1	61.6	86.2	71.0	54.8	60.0
1851	75.5	105.4	64.6	56.5	62.6	63.2	36.1	57.4	67.9	62.5	50.9	71.4
1852	68.4	67.5	61.2	65.4	54.9	46.9	42.0	39.7	37.5	67.3	54.3	45.4
1853	41.1	42.9	37.7	47.6	34.7	40.0	45.9	50.4	33.5	42.3	28.8	23.4
1854	15.4	20.0	20.7	26.4	24.0	21.1	18.7	15.8	22.4	12.7	28.2	21.4
1855	12.3	11.4	17.4	4.4	9.1	5.3	0.4	3.1	0.0	9.7	4.3	3.1
1856	0.5	4.9	0.4	6.5	0.0	5.0	4.6	5.9	4.4	4.5	7.7	7.2
1857	13.7	7.4	5.2	11.1	29.2	16.0	22.2	16.9	42.4	40.6	31.4	37.2
1858	39.0	34.9	57.5	38.3	41.4	44.5	56.7	55.3	80.1	91.2	51.9	66.9
1859	83.7	87.6	90.3	85.7	91.0	87.1	95.2	106.8	105.8	114.6	97.2	81.0
1860	81.5	88.0	98.9	71.4	107.1	108.6	116.7	100.3	92.2	90.1	97.9	95.6
1861	62.3	77.8	101.0	98.5	56.8	87.8	78.0	82.5	79.9	67.2	53.7	80.5
1862	63.1	64.5	43.6	53.7	64.4	84.0	73.4	62.5	66.6	42.0	50.6	40.9
1863	48.3	56.7	66.4	40.6	53.8	40.8	32.7	48.1	22.0	39.9	37.7	41.2
1864	57.7	47.1	66.3	35.8	40.6	57.8	54.7	54.8	28.5	33.9	57.6	28.6
1865	48.7	39.3	39.5	29.4	34.5	33.6	26.8	37.8	21.6	17.1	24.6	12.8
1866	31.6	38.4	24.6	17.6	12.9	16.5	9.3	12.7	7.3	14.1	9.0	1.5
1867	0.0	0.7	9.2	5.1	2.9	1.5	5.0	4.9	9.8	13.5	9.3	25.2
1868	15.6	15.8	26.5	36.6	26.7	31.1	28.6	34.4	43.8	61.7	59.1	67.6
1869	60.9	59.3	52.7	41.0	104.0	108.4	59.2	79.6	80.6	59.4	77.4	104.3
1870	77.3	114.9	159.4	160.0	176.0	135.6	132.4	153.8	136.0	146.4	147.5	130.0
1871	88.3	125.3	143.2	162.4	145.5	91.7	103.0	110.0	80.3	89.0	105.4	90.3
1872	79.5	120.1	88.4	102.1	107.6	109.9	105.5	92.9	114.6	103.5	112.0	83.9
1873	86.7	107.0	98.3	76.2	47.9	44.8	66.9	68.2	47.5	47.4	55.4	49.2
1874	60.8	64.2	46.4	32.0	44.6	38.2	67.8	61.3	28.0	34.3	28.9	29.3
1875	14.6	22.2	33.8	29.1	11.5	23.9	12.5	14.6	2.4	12.7	17.7	9.9
1876	14.3	15.0	31.2	2.3	5.1	1.6	15.2	8.8	9.9	14.3	9.9	8.2
1877	24.4	8.7	11.7	15.8	21.2	13.4	5.9	6.3	16.4	6.7	14.5	2.3
1878	3.3	6.0	7.8	0.1	5.8	6.4	0.1	0.0	5.3	1.1	4.1	0.5
1879	0.8	0.6	0.0	6.2	2.4	4.8	7.5	10.7	6.1	12.3	12.9	7.2
1880	24.0	27.5	19.5	19.3	23.5	34.1	21.9	48.1	66.0	43.0	30.7	29.6
1881	36.4	53.2	51.5	51.7	43.5	60.5	76.9	58.0	53.2	64.0	54.8	47.3
1882	45.0	69.3	67.5	95.8	64.1	45.2	45.4	40.4	57.7	59.2	84.4	41.8
1883	60.6	46.9	42.8	82.1	32.1	76.5	80.6	46.0	52.6	83.8	84.5	75.9

Table 11.1 *(cont.)*

Year	Jan	Feb	March	April	May	June	July	August	Sept	Oct	Nov	Dec
1884	91.5	86.9	86.8	76.1	66.5	51.2	53.1	55.8	61.9	47.8	36.6	47.2
1885	42.8	71.8	49.8	55.0	73.0	83.7	66.5	50.0	39.6	38.7	33.3	21.7
1886	29.9	25.9	57.3	43.7	30.7	27.1	30.3	16.9	21.4	8.6	0.3	12.4
1887	10.3	13.2	4.2	6.9	20.0	15.7	23.3	21.4	7.4	6.6	6.9	20.7
1888	12.7	7.1	7.8	5.1	7.0	7.1	3.1	2.8	8.8	2.1	10.7	6.7
1889	0.8	8.5	7.0	4.3	2.4	6.4	9.7	20.6	6.5	2.1	0.2	6.7
1890	5.3	0.6	5.1	1.6	4.8	1.3	11.6	8.5	17.2	11.2	9.6	7.8
1891	13.5	22.2	10.4	20.5	41.1	48.3	58.8	33.2	53.8	51.5	41.9	32.3
1892	69.1	75.6	49.9	69.6	79.6	76.3	76.8	101.4	62.8	70.5	65.4	78.6
1893	75.0	73.0	65.7	88.1	84.7	88.2	88.8	129.2	77.9	79.7	75.1	93.8
1894	83.2	84.6	52.3	81.6	101.2	98.9	106.0	70.3	65.9	75.5	56.6	60.0
1895	63.3	67.2	61.0	76.9	67.5	71.5	47.8	68.9	57.7	67.9	47.2	70.7
1896	29.0	57.4	52.0	43.8	27.7	49.0	45.0	27.2	61.3	28.4	38.0	42.6
1897	40.6	29.4	29.1	31.0	20.0	11.3	27.6	21.8	48.1	14.3	8.4	33.3
1898	30.2	36.4	38.3	14.5	25.8	22.3	9.0	31.4	34.8	34.4	30.9	12.6
1899	19.5	9.2	18.1	14.2	7.7	20.5	13.5	2.9	8.4	13.0	7.8	10.5
1900	9.4	13.6	8.6	16.0	15.2	12.1	8.3	4.3	8.3	12.9	4.5	0.3
1901	0.2	2.4	4.5	0.0	10.2	5.8	0.7	1.0	0.6	3.7	3.8	0.0
1902	5.2	0.0	12.4	0.0	2.8	1.4	0.9	2.3	7.6	16.3	10.3	1.1
1903	8.3	17.0	13.5	26.1	14.6	16.3	27.9	28.8	11.1	38.9	44.5	45.6
1904	31.6	24.5	37.2	43.0	39.5	41.9	50.6	58.2	30.1	54.2	38.0	54.6
1905	54.8	85.8	56.5	39.3	48.0	49.0	73.0	58.8	55.0	78.7	107.2	55.5
1906	45.5	31.3	64.5	55.3	57.7	63.2	103.6	47.7	56.1	17.8	38.9	64.7
1907	76.4	108.2	60.7	52.6	42.9	40.4	49.7	54.3	85.0	65.4	61.5	47.3
1908	39.2	33.9	28.7	57.6	40.8	48.1	39.5	90.5	86.9	32.3	45.5	39.5
1909	56.7	46.6	66.3	32.3	36.0	22.6	35.8	23.1	38.8	58.4	55.8	54.2
1910	26.4	31.5	21.4	8.4	22.2	12.3	14.1	11.5	26.2	38.3	4.9	5.8
1911	3.4	9.0	7.8	16.5	9.0	2.2	3.5	4.0	4.0	2.6	4.2	2.2
1912	0.3	0.0	4.9	4.5	4.4	4.1	3.0	0.3	9.5	4.6	1.1	6.4
1913	2.3	2.9	0.5	0.9	0.0	0.0	1.7	0.2	1.2	3.1	0.7	3.8
1914	2.8	2.6	3.1	17.3	5.2	11.4	5.4	7.7	12.7	8.2	16.4	22.3
1915	23.0	42.3	38.8	41.3	33.0	68.8	71.6	69.6	49.5	53.5	42.5	34.5
1916	45.3	55.4	67.0	71.8	74.5	67.7	53.5	35.2	45.1	50.7	65.6	53.0
1917	74.7	71.9	94.8	74.7	114.1	114.9	119.8	154.5	129.4	72.2	96.4	129.3
1918	96.0	65.3	72.2	80.5	76.7	59.4	107.6	101.7	79.9	85.0	83.4	59.2
1919	48.1	79.5	66.5	51.8	88.1	111.2	64.7	69.0	54.7	52.8	42.0	34.9
1920	51.1	53.9	70.2	14.8	33.3	38.7	27.5	19.2	36.3	49.6	27.2	29.9
1921	31.5	28.3	26.7	32.4	22.2	33.7	41.9	22.8	17.8	18.2	17.8	20.3
1922	11.8	26.4	54.7	11.0	8.0	5.8	10.9	6.5	4.7	6.2	7.4	17.5
1923	4.5	1.5	3.3	6.1	3.2	9.1	3.5	0.5	13.2	11.6	10.0	2.8
1924	0.5	5.1	1.8	11.3	20.8	24.0	28.1	19.3	25.1	25.6	22.5	16.5
1925	5.5	23.2	18.0	31.7	42.8	47.5	38.5	37.9	60.2	69.2	58.6	98.6
1926	71.8	70.0	62.5	38.5	64.3	73.5	52.3	61.6	60.8	71.5	60.5	79.4
1927	81.6	93.0	69.6	93.5	79.1	59.1	54.9	53.8	68.4	63.1	67.2	45.2
1928	83.5	73.5	85.4	80.6	76.9	91.4	98.0	83.8	89.7	61.4	50.3	59.0
1929	68.9	64.1	50.2	52.8	58.2	71.9	70.2	65.8	34.4	54.0	81.1	108.0

Table 11.1 *(cont.)*

Year	Jan	Feb	March	April	May	June	July	August	Sept	Oct	Nov	Dec
1930	65.3	49.2	35.0	38.2	36.8	28.8	21.9	24.9	32.1	34.4	35.6	25.8
1931	14.6	43.1	30.0	31.2	24.6	15.3	17.4	13.0	19.0	10.0	18.7	17.8
1932	12.1	10.6	11.2	11.2	17.9	22.2	9.6	6.8	4.0	8.9	8.2	11.0
1933	12.3	22.2	10.1	2.9	3.2	5.2	2.8	0.2	5.1	3.0	0.6	0.3
1934	3.4	7.8	4.3	11.3	19.7	6.7	9.3	8.3	4.0	5.7	8.7	15.4
1935	18.9	20.5	23.1	12.2	27.3	45.7	33.9	30.1	42.1	53.2	64.2	61.5
1936	62.8	74.3	77.1	74.9	54.6	70.0	52.3	87.0	76.0	89.0	115.4	123.4
1937	132.5	128.5	83.9	109.3	116.7	130.3	145.1	137.7	100.7	124.9	74.4	88.8
1938	98.4	119.2	86.5	101.0	127.4	97.5	165.3	115.7	89.6	99.1	122.2	92.7
1939	80.3	77.4	64.6	109.1	118.3	101.0	97.6	105.8	112.6	88.1	68.1	42.1
1940	50.5	59.4	83.3	60.7	54.4	83.9	67.5	105.5	66.5	55.0	58.4	68.3
1941	45.6	44.5	46.4	32.8	29.5	59.8	66.9	60.0	65.9	46.3	38.3	33.7
1942	35.6	52.8	54.2	60.7	25.0	11.4	17.7	20.2	17.2	19.2	30.7	22.5
1943	12.4	28.9	27.4	26.1	14.1	7.6	13.2	19.4	10.0	7.8	10.2	18.8
1944	3.7	0.5	11.0	0.3	2.5	5.0	5.0	16.7	14.3	16.9	10.8	28.4
1945	18.5	12.7	21.5	32.0	30.6	36.2	42.6	25.9	34.9	68.8	46.0	27.4
1946	47.6	86.2	76.6	75.7	84.9	73.5	116.2	107.2	94.4	102.3	123.8	121.7
1947	115.7	113.4	129.8	149.8	201.3	163.9	157.9	188.8	169.4	163.6	128.0	116.5
1948	108.5	86.1	94.8	189.7	174.0	167.8	142.2	157.9	143.3	136.3	95.8	138.0
1949	119.1	182.3	157.5	147.0	106.2	121.7	125.8	123.8	145.3	131.6	143.5	117.6
1950	101.6	94.8	109.7	113.4	106.2	83.6	91.0	85.2	51.3	61.4	54.8	54.1
1951	59.9	59.9	59.9	92.9	108.5	100.6	61.5	61.0	83.1	51.6	52.4	45.8
1952	40.7	22.7	22.0	29.1	23.4	36.4	39.3	54.9	28.2	23.8	22.1	34.3
1953	26.5	3.9	10.0	27.8	12.5	21.8	8.6	23.5	19.3	8.2	1.6	2.5
1954	0.2	0.5	10.9	1.8	0.8	0.2	4.8	8.4	1.5	7.0	9.2	7.6
1955	23.1	20.8	4.9	11.3	28.9	31.7	26.7	40.7	42.7	58.5	89.2	76.9
1956	73.6	124.0	118.4	110.7	136.6	116.6	129.1	169.6	173.2	155.3	201.3	192.1
1957	165.0	130.2	157.4	175.2	164.6	200.7	187.2	158.0	235.8	253.8	210.9	239.4
1958	202.5	164.9	190.7	196.0	175.3	171.5	191.4	200.2	201.2	181.5	152.3	187.6
1959	217.4	143.1	185.7	163.3	172.0	168.7	149.6	199.6	145.2	111.4	124.0	125.0
1960	146.3	106.0	102.2	122.0	119.6	110.2	121.7	134.1	127.2	82.8	89.6	85.6
1961	57.9	46.1	53.0	61.4	51.0	77.4	70.2	55.9	63.6	37.7	32.6	40.0
1962	38.7	50.3	45.6	46.4	43.7	42.0	21.8	21.8	51.3	39.5	26.9	23.2
1963	19.8	24.4	17.1	29.3	43.0	35.9	19.6	33.2	38.8	35.3	23.4	14.9
1964	15.3	17.7	16.5	8.6	9.5	9.1	3.1	9.3	4.7	6.1	7.4	15.1
1965	17.5	14.2	11.7	6.8	24.1	15.9	11.9	8.9	16.8	20.1	15.8	17.0
1966	28.2	24.4	25.3	48.7	45.3	47.7	56.7	51.2	50.2	57.2	57.2	70.4
1967	110.9	93.6	111.8	69.5	86.5	67.3	91.5	107.2	76.8	88.2	94.3	126.4
1968	121.8	111.9	92.2	81.2	127.2	110.3	96.1	109.3	117.2	107.7	86.0	109.8
1969	104.4	120.5	135.8	106.8	120.0	106.0	96.8	98.0	91.3	95.7	93.5	97.9
1970	111.5	127.8	102.9	109.5	127.5	106.8	112.5	93.0	99.5	86.6	95.2	83.5
1971	91.3	79.0	60.7	71.8	57.5	49.8	81.0	61.4	50.2	51.7	63.2	82.2
1972	61.5	88.4	80.1	63.2	80.5	88.0	76.5	76.8	64.0	61.3	41.6	45.3
1973	43.4	42.9	46.0	57.7	42.4	39.5	23.1	25.6	59.3	30.7	23.9	23.3
1974	27.6	26.0	21.3	40.3	39.5	36.0	55.8	33.6	40.2	47.1	25.0	20.5
1975	18.9	11.5	11.5	5.1	9.0	11.4	28.2	39.7	13.9	9.1	19.4	7.8

Table 11.1 *(cont.)*

Year	Jan	Feb	March	April	May	June	July	August	Sept	Oct	Nov	Dec
1976	8.1	4.3	21.9	18.8	12.4	12.2	1.9	16.4	13.5	20.6	5.2	15.3
1977	16.4	23.1	8.7	12.9	18.6	38.5	21.4	30.1	44.0	43.8	29.1	43.2
1978	51.9	93.6	76.5	99.7	82.7	95.1	70.4	58.1	138.2	125.1	97.9	122.7
1979	166.6	137.5	138.0	101.5	134.4	149.5	159.4	142.2	188.4	186.2	183.3	176.3
1980	159.6	155.0	126.2	164.1	179.9	157.3	136.3	135.4	155.0	164.7	147.9	174.4
1981	114.0	141.3	135.5	156.4	127.5	90.0	143.8	158.7	167.3	162.4	137.5	150.1
1982	111.2	163.6	153.8	122.0	82.2	110.4	106.1	107.6	118.8	94.7	98.1	127.0
1983	84.3	51.0	66.5	80.7	99.2	91.1	82.2	71.8	50.3	55.8	33.3	33.4

12. Ozone Column

Source Birrer, W. (1975). Homogenisiering und Diskussien der Totalozon - Missreihe von Arosa 1926 - 1971. Laboratorium für Atmosphärenphysik

Contributor P. Bloomfield University of North Carolina, Chapel Hill

The data given in Table 12.1 are monthly mean thickness in Dobson units, one milli-centimetre ozone at standard temperature and pressure, of the ozone column at Arosa, Switzerland. They show unusual seasonal behavour in that the variance as well as the mean varies seasonally, and not in a way that can be removed by a transformation.

Table 12.1
Monthly Mean Thickness in Dobson Units of the Ozone Column at Arosa, Switzerland

Year	Jan	Feb	March	April	May	June	July	Aug	Sept	Oct	Nov	Dec
1926	-	-	-	-	-	-	312	300	281	267	295	325
1927	360	365	372	375	382	350	336	314	305	280	285	316
1928	324	325	347	359	373	331	308	295	288	268	279	313
1929	349	401	355	418	368	345	-	-	-	-	-	-
1930	-	-	-	-	-	-	291	299	-	-	-	-
1931	-	-	-	-	-	-	-	311	335	283	286	301
1932	318	347	370	394	360	347	334	299	292	287	293	281
1933	357	364	399	382	390	374	335	319	309	312	311	337
1934	334	321	392	358	365	355	328	321	282	287	291	297
1935	332	390	367	383	375	319	331	311	288	275	299	313
1936	329	393	398	384	373	352	328	315	303	310	298	307
1937	347	352	395	382	365	349	324	323	301	283	280	355
1938	337	370	325	392	384	336	325	325	296	280	285	299
1939	320	341	385	347	382	339	331	313	286	304	284	309
1940	387	400	418	430	403	388	346	323	310	292	302	338
1941	362	395	417	409	417	361	348	336	306	299	309	309
1942	400	422	373	408	376	347	325	309	284	272	298	313
1943	338	341	385	363	348	352	336	303	291	292	303	321
1944	300	365	385	360	349	351	319	306	290	293	298	320
1945	377	359	360	373	376	351	329	327	297	288	295	313
1946	336	352	380	361	355	344	318	307	276	291	297	316
1947	383	397	393	369	361	347	334	324	307	296	278	312
1948	341	371	348	374	353	345	345	311	299	281	286	321
1949	332	365	378	357	371	354	335	321	284	272	296	292
1950	352	365	365	382	374	354	322	316	292	288	287	340
1951	338	402	417	397	383	364	332	321	297	298	278	311
1952	378	384	411	386	385	359	341	320	317	297	302	332
1953	335	375	373	383	382	359	326	317	293	280	266	-
1954	-	-	373	415	389	362	348	329	305	285	291	284
1955	315	375	399	374	361	351	339	332	300	292	278	317
1956	341	402	381	395	365	360	327	308	287	284	286	312
1957	340	342	353	375	380	349	330	321	305	279	294	322
1958	361	351	411	417	369	369	349	330	306	312	305	326
1959	369	367	364	390	389	373	342	330	313	294	297	328
1960	349	397	405	400	382	353	339	315	306	299	284	332
1961	368	333	338	365	379	349	343	322	296	288	301	304
1962	352	362	428	400	363	349	338	302	296	271	298	309
1963	371	408	377	381	378	352	327	310	282	273	278	293
1964	301	335	347	378	363	330	326	323	299	288	274	310
1965	332	390	383	385	387	345	337	318	305	275	278	305
1966	354	338	401	380	372	-	-	326	307	282	321	319
1967	361	362	354	374	349	358	325	317	296	273	267	304
1968	342	383	376	379	349	351	337	335	308	275	270	331
1969	327	419	361	393	351	364	333	333	294	281	307	345
1970	331	417	414	419	389	360	331	323	291	278	271	309
1971	344	349	411	364	358	358	336	309	317	278	292	306

13. Downwind Effects from the Arizona Cloud Seeding Experiments

Contributor H. B. Osborn Southwest Watershed Research Center

Randomized silver-iodide seeding of summer convective clouds was carried out over the Santa Catalina Mountains in southern Arizona in 2 programs: the first from 1957 through 1960, and the second in 1961, 1962 and 1964. Experimental days were taken in pairs; the decision to seed or not to seed was made after a day was considered as seedable. The second day of the pair was the opposite of the first. Each 2-day set was considered independently. If the second and third days were not considered seedable, the set was omitted. Seedability depended primarily on the available moisture in the morning. Seeding was from aircraft upwind from the Santa Catalinas.

The results of the first 4-year program indicated a statistically nonsignificant decrease in seeded rainfall of about 30%. Primarily because of the negative results, the experimental design was changed for the second program. Changes involved stricter limits on what was considered a seedable day, more moisture needed to be present, adding more rain-gages to improve the estimates of rainfall amounts in the Santa Catalinas, assuming the original network might not be representative of the true rainfall, and seeding at varying altitudes depending on the height of the cloud base, seeding was just below the cloud base, rather than at a fixed altitude. Seeding for both programs began at 12:30 am and continued for 2 to 4 hours. Days were omitted if seeding for at least 2 hours was not possible, and this did happen quite a few times.

The results of the second program were similar to that of the first; seeded rainfall was about 30% less than unseeded rainfall.

In 1970, Dr. J. Neyman, Director, Statistical Laboratory, Department of Statistics, University of California, learned that the ARS-USDA operated a dense network of recording rain-gages on the Walnut Gulch Experimental Watershed about 70 miles south-southeast of the Santa Catalinas, and that many of these gages were in operation from 1957 through 1964. He contacted the ARS Southwest Watershed Research Center. I was the principal scientist from the ARS involved in the co-operative effort.

Without giving us any of the seeding information, other than the years when seeding was carried out, Dr. Neyman asked us to digitize hourly summer rainfall for all rain-gages on Walnut Gulch that were in continuous operation from 1957 through 1964. There were 26 such gages. When we had completed this step, we duplicated the cards and sent the original set to Dr. Neyman. Simultaneously, he sent us the seeding information so that we could carry out our own independent analysis of the data. He also described the method that he would use in his analysis in advance of receiving the rainfall data. I think this is an important point. Analyses that are designed after experiments are completed are always more susceptible to bias than those that are designed in advance of the experiment.

The experimental days were divided into 2 groups, depending on whether Walnut Gulch was upwind or downwind from the Santa Catalinas based on the 5:00 pm USWB radiosonde at Tucson. The alternative would have been to use the 5:00 pm radiosonde. These were the only upper-level wind records available. Rainfall amounts were determined for 24-hour periods from noon of the experimental day to noon of the next day. Almost all rainfall on Walnut Gulch occurs between noon and midnight, with the largest incidence of rainfall in the evening hours. For all experimental days, there was 40% less rainfall on Walnut Gulch on seeded days than on non-seeded days, which was statistically significant at the 2.5% level. For experimental days when Walnut Gulch was downwind from the Santa Catalinas, there was 70% less rainfall on seeded than on non-seeded days, which was statistically significant at the 1% level. A further breakdown indicated that differences were much greater on the experimental second days than on the experimental first days, although decreases were indicated on both days.

The results were unexpected to me. At that time, the general belief was that there could be no effect from convective cloud seeding for any appreciable distance from the target. At least the people to whom I had been talking felt that way.

We used the optimal $C(\alpha)$ test developed by Neyman and Scott (1965,1967) for these analyses. This test is particularly well suited to highly skewed distributions such as thunderstorm rainfall. However, other statistical tests could be used, and possibly compared.

We used 24 hours from noon to noon as our time unit for rainfall amounts. Since hourly amounts are available, it might be interesting to use shorter time units. For example, assuming the effect of seeding cannot be felt instantly 200 km away, that silver-iodide has a relatively short life in sunlight, and that most summer rainfall is in the late afternoon or early evening, one might feel more comfortable looking at rainfall amounts on experimental days between 1700 or 1800 hours and midnight. Also, if one ranks rainfall amounts for experimental days starting with the largest amount, other interesting tests and conclusions may result. The data are given in Table 13.1. Data are available for each hour for a period covering the experiment. Rainfall was recorded for each rain-gage in each area. The amount of detailed information is quite large and only a small initial portion is presented in the table.

Table 13.2 presents the hourly rainfall for two locations for the period of the experiment.

For further details on this problem, see Neyman and Osborn (1971) and Neyman, Osborn, Scott and Wells (1972).

References

Neyman, J. and Osborn, H.B. (1971). Evidence of widespread effects of cloud seeding at the two Arizona experiments. *Proc. Nat. Acad. Sci.* **68**, 649-652.

Neyman, J., Osborn, H.B., Scott, E.L. and Wells, M.A. (1972). Re-evaluation of the Arizona cloud seeding experiment. *Proc. Nat. Acad. Sci.* **69**, 1348-1352.

Neyman, J. and Scott, E. L. (1965). Asymptotically optimal tests of composite hypothesis for randomized experiments with non-controlled predictor variates. *Amer. Statist. Ass.* **60**, 699-721.

Neyman, J. and Scott, E. L. (1967). Note on techniques of evaluation of single rain stimulation experiments. *Proc. Fifth Berkeley Symp.* **5**, 371-384.

Table 13.1
Wind Direction and Experimental Treatment for Walnut Gulch and the Santa Catalina Mountains

	Wind direction at seeding level *					Wind direction at seeding level *			
Date	5 am	5 pm	Day of pair	Seed or not **	Date	5 am	5 pm	Day of pair	Seed or not **
57 07 08	174	177	1	1	58 08 04	135	200	1	1
57 07 09	160	114	2	0	58 08 05	58	118	2	0
57 07 10	175	63	1	1	58 08 06	58	6	1	1
57 07 11	177	142	2	0	58 08 07	159	157	2	0
57 07 12	288	170	1	0	58 08 12	76	83	1	1
57 07 13	195	33	2	1	58 08 13	120	75	2	0
57 07 16	140	142	1	1	58 08 14	98	95	1	1
57 07 17	124	151	2	0	58 08 15	129	122	2	0
57 07 18	219	225	1	0	58 08 16	97	125	1	1
57 07 19	197	217	2	1	58 08 18	125	95	2	0
57 07 27	184	188	1	0	58 08 19	105	160	1	1
57 07 29	183	171	2	1	58 08 20	45	220	2	0
57 07 30	224	125	1	1	58 08 21	192	127	1	1
57 07 31	33	46	2	0	58 08 22	8	315	2	0
57 08 01	49	53	1	1	58 08 23	177	20	1	0
57 08 02	175	245	2	0	58 08 25	327	59	2	1
57 08 09	74	211	1	1	58 08 27	43	350	1	0
57 08 10	103	122	2	0	58 08 28	122	40	2	1
57 08 12	166	234	1	1	58 08 29	280	337	1	1
57 08 13	257	218	2	0	58 08 30	22	66	2	0
57 08 14	302	0	1	0	59 07 07	195	273	1	1
57 08 15	61	197	2	1	59 07 08	212	126	2	0
57 08 16	72	35	1	1	59 07 09	100	58	1	0
57 08 17	93	60	2	0	59 07 10	86	68	2	1
57 08 19	172	147	1	0	59 07 13	80	124	1	0
57 08 20	176	199	2	1	59 07 14	85	108	2	1
57 08 21	214	205	1	0	59 07 15	32	102	1	0
57 08 22	140	219	2	1	59 07 16	35	60	2	1
57 08 23	257	176	1	0	59 07 17	68	29	1	0
57 08 24	243	245	2	1	59 07 18	44	46	2	1
57 08 26	226	213	1	1	59 07 20	45	93	1	1
57 08 27	265	218	2	0	59 07 21	30	55	2	0
58 07 16	212	200	1	0	59 07 23	83	82	1	1
58 07 17	209	284	2	1	59 07 24	97	85	2	0
58 07 18	110	100	1	1	59 07 28	20	70	1	1
58 07 19	283	349	2	0	59 07 29	82	159	2	0
58 07 24	193	170	1	1	59 07 30	175	127	1	1
58 07 25	180	245	2	0	59 07 31	68	114	2	0
58 07 30	238	229	1	0	59 08 03	174	167	1	0
58 07 31	190	117	2	1	59 08 04	148	260	2	1
58 08 01	94	140	1	0	59 08 05	128	181	1	0
58 08 02	8	99	2	1	59 08 06	182	161	2	1

Table 13.1 *(cont.)*

Date	Wind direction at seeding level * 5 am	Wind direction at seeding level * 5 pm	Day of pair	Seed or not **	Date	Wind direction at seeding level * 5 am	Wind direction at seeding level * 5 pm	Day of pair	Seed or not **
59 08 07	148	194	1	1	60 08 30	126	116	1	0
59 08 08	125	140	2	0	60 08 31	138	185	2	1
59 08 10	296	125	1	0	60 09 01	163	186	1	0
59 08 11	184	200	2	1	60 09 02	140	151	2	1
59 08 12	226	259	1	1	60 09 06	200	236	1	1
59 08 13	159	271	2	0	60 09 07	271	285	2	0
59 08 17	137	168	1	0	60 09 08	307	49	1	1
59 08 18	197	205	2	1	60 09 09	58	68	2	0
59 08 19	230	200	1	0	61 07 17	356	96	1	1
59 08 20	220	215	2	1	61 07 18	51	342	2	0
59 08 21	203	210	1	0	61 07 21	298	270	1	0
59 08 22	75	156	2	1	61 07 22	96	56	2	1
59 08 24	140	155	1	1	61 07 24	96	114	1	1
59 08 25	33	355	2	0	61 07 26	79	103	2	0
59 08 26	14	28	1	0	61 07 27	72	78	1	0
59 08 27	85	233	2	1	61 07 29	124	190	2	1
59 08 28	16	358	1	0	61 07 31	146	240	1	1
59 08 29	32	62	2	1	61 08 01	203	205	2	0
60 07 06	210	102	1	0	61 08 02	100	108	1	1
60 07 07	170	148	2	1	61 08 03	132	86	2	0
60 07 08	53	169	1	1	61 08 08	328	313	1	1
60 07 09	183	161	2	0	61 08 09	30	115	2	0
60 07 23	345	14	1	0	61 08 10	152	163	1	1
60 07 25	77	57	2	1	61 08 11	146	237	2	0
60 07 26	90	98	1	0	61 08 12	98	333	1	1
60 07 27	115	80	2	1	61 08 14	42	37	2	0
60 07 30	175	162	1	0	61 08 15	199	269	1	1
60 08 01	196	173	2	1	61 08 16	260	190	2	0
60 08 02	234	125	1	1	61 08 17	165	93	1	0
60 08 03	82	70	2	0	61 08 18	153	98	2	1
60 08 06	93	98	1	0	61 08 19	99	129	1	0
60 08 08	80	81	2	1	61 08 21	98	90	2	1
60 08 09	90	97	1	0	61 08 22	46	210	1	0
60 08 10	100	81	2	1	61 08 23	229	252	2	1
60 08 11	49	45	1	0	61 08 24	258	312	1	1
60 08 12	34	32	2	1	61 08 25	256	298	2	0
60 08 16	324	349	1	0	61 08 28	173	77	1	1
60 08 17	16	16	2	1	61 08 29	217	310	2	0
60 08 20	91	85	1	0	61 09 08	222	230	1	0
60 08 22	225	232	2	1	61 09 09	214	181	2	1
60 08 23	240	238	1	0	61 09 11	213	67	1	0
60 08 25	255	233	2	1	61 09 12	56	110	2	1
60 08 26	8	68	1	0	62 07 19	61	54	1	0
60 08 27	204	240	2	1	62 07 20	81	32	2	1

Table 13.1 *(cont.)*

Date	Wind direction at seeding level * 5 am	Wind direction at seeding level * 5 pm	Day of pair	Seed or not **	Date	Wind direction at seeding level * 5 am	Wind direction at seeding level * 5 pm	Day of pair	Seed or not **
62 07 21	155	200	1	0	64 07 17	215	47	2	1
62 07 23	310	326	2	1	64 07 18	35	60	1	0
62 07 24	7	300	1	1	64 07 20	113	350	2	1
62 07 25	8	71	2	0	64 07 21	83	296	1	1
62 07 26	33	51	1	0	64 07 22	36	270	2	0
62 07 27	65	51	2	1	64 07 24	28	91	1	1
62 07 31	4	15	1	0	64 07 25	136	156	2	0
62 08 01	21	130	2	1	64 07 29	100	284	1	0
62 08 20	155	162	1	1	64 07 30	140	142	2	1
62 08 21	163	233	2	0	64 08 03	169	178	1	0
62 09 04	40	101	1	0	64 08 04	140	122	2	1
62 09 05	106	66	2	1	64 08 05	27	141	1	0
64 07 08	31	200	1	1	64 08 07	162	295	2	1
64 07 09	227	238	2	0	64 08 08	35	329	1	0
64 07 11	328	310	1	0	64 08 10	120	340	2	1
64 07 13	305	340	2	1	64 08 13	265	330	1	0
64 07 14	135	267	1	0	64 08 14	352	322	2	1
64 07 15	255	295	2	1	64 08 17	54	103	1	1
64 07 16	343	200	1	0	64 08 19	238	276	2	0

* Directions; 0° true North

** Seed or not; 0 denotes absence of seeding, 1 denotes seeding

Table 13.2
Rainfall for Santa Catalina Mountains and Walnut Gulch

Site Code *	Date: Year	Month	Day	Indicator †	Rainfall for period ending at hour: 1	2	3	4	5	6	7	8	9	10	11	12
Santa Catalina Mountains (AZLB)																
AZLB 1	57	7	4	1	0	0	0	0	0	0	0	0	0	0	0	0
AZLB 1	57	7	4	2	0	0	0	0	0	0	0	5	0	0	0	0
AZLB 1	57	7	5	1	0	0	0	0	0	0	0	0	0	0	0	0
AZLB 1	57	7	5	2	0	0	0	0	0	0	0	0	10	29	0	0
AZLB 1	57	7	6	1	0	0	0	0	0	0	0	0	0	0	0	0
AZLB 1	57	7	6	2	0	0	0	0	0	0	0	0	5	8	0	0
AZLB 1	57	7	7	1	0	0	0	0	0	0	0	0	0	0	0	0
AZLB 1	57	7	7	2	0	2	0	0	0	0	0	0	0	0	0	0
AZLB 1	57	7	9	1	0	0	0	0	0	0	0	0	0	0	0	0
AZLB 1	57	7	9	2	0	0	0	0	6	2	0	0	0	0	0	0
AZLB 1	57	7	10	1	0	0	0	0	0	0	0	0	0	0	0	0
AZLB 1	57	7	10	2	0	0	4	0	0	0	0	0	0	0	0	0
AZLB 1	57	7	15	1	0	0	0	0	0	0	0	0	0	0	0	0
AZLB 1	57	7	15	2	0	0	0	0	0	0	0	0	6	2	0	0
AZLB 1	57	7	17	1	0	0	0	0	0	0	0	0	0	0	0	0
AZLB 1	57	7	17	2	0	15	0	0	0	0	0	0	0	0	0	0
AZLB 1	57	7	18	1	0	0	0	0	0	0	0	0	0	0	0	40
AZLB 1	57	7	18	2	0	0	0	0	0	0	0	0	0	0	0	0
AZLB 1	57	7	26	1	0	0	0	0	0	0	0	0	0	0	35	0
AZLB 1	57	7	26	2	0	0	0	0	0	0	0	0	0	0	0	0
Walnut Gulch (AZWG)																
AZWG 1	57	7	8	1	0	0	0	0	0	0	0	0	0	0	0	0
AZWG 1	57	7	8	2	0	0	0	0	0	0	0	0	0	2	8	16
AZWG 1	57	7	9	1	0	0	0	0	0	0	0	0	0	0	0	0
AZWG 1	57	7	9	2	0	0	0	15	0	0	0	0	0	0	0	0
AZWG 1	57	7	10	1	0	0	0	0	0	0	0	0	0	0	0	0
AZWG 1	57	7	10	2	0	0	0	22	0	0	0	0	0	0	0	0
AZWG 1	57	7	16	1	0	0	0	0	0	0	0	0	0	0	0	0
AZWG 1	57	7	16	2	0	21	0	0	0	0	0	0	0	0	0	0
AZWG 1	57	7	17	1	0	0	0	0	0	0	0	0	0	0	0	0
AZWG 1	57	7	17	2	0	0	0	0	1	3	0	0	0	0	0	0
AZWG 1	57	7	18	1	0	0	0	0	0	0	0	0	0	0	0	0
AZWG 1	57	7	18	2	0	40	10	0	0	0	0	0	0	0	0	0
AZWG 1	57	7	19	1	0	0	0	0	0	0	0	0	0	0	0	103
AZWG 1	57	7	19	2	0	0	0	0	0	0	0	0	0	0	0	0
AZWG 1	57	7	24	1	0	0	0	0	0	0	0	0	0	0	0	0
AZWG 1	57	7	24	2	0	0	0	0	0	0	0	0	2	2	2	1
AZWG 1	57	7	25	1	0	0	0	0	0	0	0	0	0	0	0	13
AZWG 1	57	7	25	2	2	0	0	0	0	0	0	0	0	0	0	0
AZWG 1	57	7	26	1	0	0	0	0	0	0	0	0	0	0	0	0
AZWG 1	57	7	26	2	3	0	0	0	0	0	0	0	0	0	0	0

* AZLB: Santa Catalina Mountains
AZWG: Walnut Gulch

† Indicator: 1, denotes am.; 2, denotes pm.

14. Half-Hourly Precipitation and Streamflow, River Hirnant, Wales, U.K., November and December, 1972

Source Welsh Water Authority, Dee and Clwyd Division, Shire Hall, Mold, Clwyd, U.K.

Contributor G. Weiss Tel-Aviv University

The Hirnant subcatchment of the River Dee has an area of 33.9 km^2 and is situated west of Bala Lake in North Wales. Impervious rocks, providing very little storage for rainfall, combine with steep slopes to give a fast streamflow response to rainfall. The river is gauged at Plas Rhiwaedog at a natural river section, to provide discrete time streamflow data at half-hourly intervals. The precipitation data consists of estimates of a real rainfall derived from six recording rain-gauges situated in and around the catchment. The present set of data, Table 14.1, is part of several years' data on the Hirnant and other subcatchments of the Dee collected and available at the Welsh Water Authority and at the Institute of Hydrology, Wallingford, U.K. The hydrolic control is a natural rock outcrop and is subject to minor changes from year to year. A real-time flow forecasting system has been operational on the River Dee since 1975 as part of extensive water supply and flood control schemes for the catchment.

Over the past twenty years, considerable research has been carried out in hydrology on the development of mathematical models of the rainfall run-off process, with the two-fold objective of obtaining a better understanding of the complexity of catchment response and of helping in the short-term management of water resources. In the latter context, rainfall run-off models can provide short-term forecasts of streamflow necessary for flood warning, flood control and river regulation; such forecasts, if they are to exceed the catchment *lag*, delay between rainfall and catchment response to it, will also depend on the availability of rain forecasts by meteorological methods.

A considerable number of rainfall run-off models have been developed to date. They can be roughly divided into three types:

(i) *Distributed physics based models:* equations of mass energy and momentum are used to describe the movement of water over the land surface and through the unsaturated and saturated zones. Such models are very much at the development stage at present (Abbot, O'Connell and Preissman, 1979).

(ii) *Lumped conceptual models:* simplified representations of the component processes of the rainfall run-off processes are used. These involve several inter-linked storage elements with simple budgeting rules to ensure mass balance. A forerunner of this type of model is the Stanford Watershed Model developed by Crawford and Linsley (1963).

(iii) *Input Output or Black-box models:* attempt to identify the relations between rainfall input and run-off output, without a physical interpretation. An example is the unit hydrograph model which postulates a linear relationship between the input and the output (Dooge, 1972).

These types of models are listed here by decreasing order of their reliance on physics and by increasing order of reliance on statistics. It must, however, be stressed that even the distributed physics based models will always involve some stochastic elements ensuing on the one hand from incompleteness of the model and on the other hand from the discrete temporal and spatial sampling of the processes and from measurement errors. At the other extreme, even the black-box models should retain some physical plausibility. Thus, although the unit hydrograph model is essentially equivalent to Box and Jenkins (1970) transfer function models, it restricts the possible transfer functions to be nonnegative and unimodal.

Several models were fitted to the present set of data. For a description of the models and a comparative study, see O'Connell (1980). The models include two nonlinear conceptual models, one fitted directly to the data and one fitted recursively using Kalman filter techniques, and two black-box models: a linear Kalman filter formulation and a self-tuning predictor formulation.

An inherent difficulty exists in modelling these data beyond the complexity of the relation between rainfall and run-off. Visual inspection of the precipitation data reveals that it consists of dry periods alternating with wet periods, and the wet periods themselves consist of few high peaks alternating with periods of lower rainfall; it is thus closer in nature to a marked point process than to a continuous-autoregressive-moving-average type stochastic process. The streamflow is a continuous process but it reflects the rainfall in gradual decrease corresponding to dry periods and in sharp rises corresponding to rainstorms. In all the models fitted to the data (O'Connell, 1980), the prediction errors also reflect the rainfall process, with much larger errors at periods of rain. Fitting of models has so far been based mainly on least squares techniques, which may not be suitable to the point process structure of the rain process. Better techniques to fit models and to deal with the nonhomogeneity of prediction errors await further research.

References

Abbot, M.B., O'Connell, P.E. and Preissman, A. (1979). The European Hydrologic System: An Advanced Hydrologic Mathematical Modelling System. Proc. XVIII IAHR Congress, Cagliari.

Box, G.E.P. and Jenkins, G.M. (1970). *Time Series Analysis Forecasting and Control.* Holden-Day: San Francisco.

Crawford, N.H. and Linsley, R.K. (1963). A conceptual model of the hydrologic cycle. IASH Symposium Surface Waters, Berkeley.

Dooge, J.C.I. (1972). Mathematical models of hydrologic series. In *Modelling of Water Resource Systems* Vol. 1 edited by A.K. Biswas. Montreal: Harvest House.

O'Connell, P.E. (Ed.) (1980). *Real Time Hydrological Forecasting and Control.* Wallingford, U.K.: Institute of Hydrology.

Table 14.1
Half-hourly Precipitation
and Streamflow, River Hirnant *

Rain	Flow	Rain	Flow	Rain	Rain	Rain	Flow
November 1 1972							
0.000	0.218	0.000	0.218	0.000	0.215	0.000	0.215
0.000	0.215	0.000	0.215	0.000	0.215	0.000	0.215
0.000	0.215	0.000	0.215	0.000	0.212	0.000	0.212
0.000	0.212	0.000	0.212	0.000	0.212	0.000	0.210
0.000	0.210	0.000	0.210	0.000	0.207	0.000	0.207
0.000	0.207	0.000	0.207	0.000	0.207	0.000	0.205
0.000	0.205	0.000	0.205	0.000	0.202	0.000	0.202
0.000	0.202	0.000	0.202	0.000	0.202	0.000	0.200
0.000	0.200	0.000	0.200	0.000	0.200	0.000	0.200
0.000	0.200	0.000	0.200	0.000	0.197	0.000	0.197
0.000	0.197	0.000	0.197	0.000	0.195	0.000	0.195
0.000	0.195	0.000	0.195	0.000	0.195	0.000	0.218
November 2 1972							
0.000	0.192	0.000	0.192	0.000	0.192	0.000	0.192
0.000	0.192	0.000	0.190	0.000	0.190	0.000	0.190
0.000	0.190	0.000	0.187	0.000	0.187	0.000	0.187
0.000	0.187	0.000	0.187	0.000	0.187	0.000	0.187
0.000	0.187	0.000	0.187	0.000	0.187	0.000	0.185
0.000	0.185	0.000	0.185	0.000	0.185	0.000	0.185
0.000	0.185	0.000	0.185	0.000	0.185	0.000	0.185
0.000	0.185	0.000	0.182	0.000	0.182	0.000	0.182
0.000	0.182	0.000	0.182	0.000	0.182	0.000	0.182
0.000	0.180	0.000	0.180	0.000	0.178	0.000	0.178
0.000	0.180	0.030	0.180	0.000	0.178	0.030	0.178
0.030	0.175	0.000	0.175	0.030	0.175	0.070	0.175
November 3 1972							
0.130	0.175	0.070	0.175	0.070	0.175	0.030	0.175
0.100	0.178	0.000	0.178	0.030	0.178	0.070	0.178
0.240	0.178	0.100	0.180	0.000	0.180	0.000	0.180
0.000	0.180	0.000	0.180	0.000	0.180	0.000	0.180
0.000	0.180	0.000	0.180	0.000	0.180	0.100	0.180
0.000	0.180	0.100	0.180	0.100	0.180	0.050	0.180
0.000	0.180	0.000	0.180	0.000	0.180	0.000	0.180
0.000	0.180	0.000	0.178	0.000	0.178	0.050	0.178
0.000	0.178	0.000	0.178	0.000	0.178	0.050	0.178
0.000	0.175	0.050	0.175	0.150	0.175	0.100	0.175
0.100	0.175	0.200	0.175	0.100	0.175	0.000	0.178
0.050	0.178	0.100	0.178	0.000	0.178	0.000	0.180
November 4 1972							
0.100	0.180	0.100	0.182	0.050	0.182	0.000	0.182
0.000	0.182	0.000	0.182	0.000	0.182	0.000	0.182
0.150	0.182	0.050	0.182	0.100	0.182	0.150	0.182
0.100	0.182	0.050	0.182	0.050	0.182	0.000	0.182
0.000	0.182	0.000	0.182	0.000	0.185	0.000	0.185
0.000	0.185	0.000	0.185	0.000	0.185	0.000	0.185
0.000	0.185	0.000	0.185	0.000	0.185	0.000	0.182
0.050	0.182	0.050	0.182	0.000	0.182	0.000	0.182
0.000	0.180	0.000	0.180	0.000	0.180	0.050	0.180
0.050	0.180	0.050	0.180	0.000	0.180	0.000	0.180
0.050	0.180	0.000	0.180	0.050	0.180	0.000	0.180
0.050	0.180	0.000	0.180	0.000	0.180	0.000	0.180

Table 14.1 *(cont.)*

Rain	Flow	Rain	Flow	Rain	Flow	Rain	Flow
			November 5 1972				
0.000	0.180	0.000	0.178	0.000	0.178	0.000	0.178
0.050	0.178	0.100	0.178	0.000	0.178	0.100	0.178
0.000	0.178	0.000	0.178	0.000	0.178	0.000	0.178
0.000	0.178	0.150	0.178	0.050	0.178	0.000	0.178
0.000	0.178	0.000	0.175	0.050	0.175	0.000	0.175
0.000	0.175	0.000	0.175	0.000	0.175	0.000	0.175
0.000	0.175	0.000	0.175	0.000	0.173	0.000	0.173
0.050	0.173	0.000	0.173	0.000	0.173	0.000	0.171
0.000	0.171	0.000	0.171	0.000	0.171	0.000	0.171
0.000	0.169	0.000	0.169	0.000	0.169	0.000	0.169
0.000	0.169	0.000	0.169	0.000	0.169	0.000	0.169
0.000	0.169	0.000	0.166	0.000	0.166	0.000	0.166
			November 6 1972				
0.000	0.166	0.000	0.166	0.000	0.166	0.000	0.166
0.000	0.166	0.000	0.166	0.000	0.166	0.000	0.166
0.000	0.166	0.000	0.166	0.000	0.166	0.000	0.166
0.000	0.166	0.000	0.166	0.000	0.166	0.000	0.166
0.000	0.166	0.000	0.166	0.040	0.166	0.000	0.166
0.040	0.166	0.000	0.166	0.040	0.166	0.000	0.166
0.000	0.166	0.000	0.166	0.000	0.166	0.000	0.166
0.000	0.166	0.000	0.166	0.000	0.166	0.000	0.166
0.000	0.166	0.000	0.166	0.000	0.166	0.000	0.166
0.000	0.166	0.000	0.166	0.000	0.164	0.160	0.164
0.040	0.164	0.000	0.164	0.000	0.166	0.360	0.166
0.200	0.166	0.000	0.166	0.200	0.173	1.360	0.182
			November 7 1972				
0.720	0.187	0.240	0.192	0.320	0.197	0.920	0.215
0.880	0.231	0.240	0.243	0.480	0.255	1.200	0.286
1.120	0.299	0.800	0.326	0.680	0.374	0.840	0.411
0.680	0.462	0.200	0.514	0.240	0.564	0.280	0.600
0.280	0.662	0.040	0.739	0.000	0.784	0.040	0.803
0.000	0.810	0.000	0.797	0.000	0.777	0.000	0.758
0.000	0.733	0.000	0.709	0.000	0.685	0.000	0.650
0.000	0.617	0.000	0.590	0.000	0.574	0.000	0.553
0.000	0.543	0.000	0.518	0.000	0.499	0.000	0.485
0.000	0.476	0.000	0.462	0.000	0.453	0.000	0.440
0.000	0.436	0.000	0.432	0.000	0.427	0.000	0.419
0.000	0.398	0.000	0.390	0.000	0.386	0.000	0.382
			November 8 1972				
0.000	0.374	0.000	0.371	0.080	0.367	0.040	0.363
0.000	0.359	0.000	0.355	0.040	0.352	0.040	0.348
0.000	0.341	0.000	0.326	0.000	0.326	0.000	0.326
0.000	0.326	0.000	0.323	0.000	0.323	0.000	0.319
0.000	0.319	0.000	0.316	0.000	0.316	0.000	0.312
0.000	0.312	0.000	0.309	0.000	0.309	0.000	0.305
0.000	0.305	0.000	0.302	0.000	0.302	0.000	0.299
0.000	0.299	0.000	0.295	0.000	0.295	0.000	0.295
0.000	0.292	0.000	0.292	0.000	0.289	0.080	0.289
0.000	0.289	0.000	0.286	0.040	0.286	0.040	0.286
0.000	0.286	0.000	0.282	0.080	0.282	0.000	0.282
0.000	0.282	0.040	0.282	0.000	0.279	0.040	0.279

Table 14.1 *(cont.)*

Rain	Flow	Rain	Flow	Rain	Flow	Rain	Flow
			November 9 1972				
0.120	0.279	0.400	0.282	0.560	0.286	0.800	0.289
0.760	0.295	1.240	0.305	1.280	0.323	1.200	0.359
0.960	0.406	0.560	0.476	0.760	0.553	0.480	0.667
0.200	0.987	0.360	1.322	0.200	1.427	0.480	1.506
1.080	1.547	1.120	1.610	0.880	1.674	1.240	1.786
1.120	2.036	1.480	2.405	1.600	2.966	1.600	3.511
1.360	4.161	1.160	4.793	0.960	5.289	1.360	5.656
2.200	6.018	0.840	6.719	1.160	7.236	0.600	7.351
0.640	7.122	0.480	6.610	0.200	6.247	0.080	5.795
0.000	5.346	0.080	4.991	0.000	4.582	0.000	4.194
0.000	3.937	0.000	3.689	0.000	3.496	0.080	3.380
0.040	3.210	0.000	3.087	0.000	2.993	0.000	2.888
			November 10 1972				
0.000	2.803	0.000	2.696	0.000	2.577	0.000	2.476
0.000	2.391	0.000	2.336	0.000	2.269	0.040	2.202
0.000	2.163	0.080	2.111	0.000	2.061	0.040	2.036
0.000	2.011	0.000	1.986	0.200	1.914	0.040	1.902
0.160	1.890	0.120	1.855	0.200	1.820	0.160	1.809
0.120	1.786	0.200	1.763	0.080	1.741	0.040	1.718
0.040	1.696	0.040	1.664	0.080	1.642	0.040	1.621
0.000	1.599	0.000	1.578	0.040	1.568	0.040	1.537
0.000	1.517	0.000	1.496	0.000	1.486	0.040	1.466
0.040	1.466	0.200	1.456	0.320	1.447	0.160	1.427
0.320	1.437	0.160	1.456	0.080	1.486	0.120	1.496
0.160	1.506	0.040	1.506	0.040	1.506	0.200	1.517
			November 11 1972				
0.160	1.517	0.240	1.517	0.520	1.517	0.440	1.527
0.360	1.547	0.320	1.558	0.080	1.589	0.080	1.621
0.160	1.653	0.120	1.674	0.040	1.674	0.040	1.664
0.160	1.642	0.120	1.621	0.200	1.599	0.080	1.578
0.120	1.558	0.120	1.547	0.120	1.537	0.080	1.527
0.040	1.517	0.000	1.506	0.120	1.496	0.000	1.486
0.120	1.476	0.080	1.466	0.000	1.456	0.160	1.447
0.000	1.417	0.120	1.417	0.000	1.398	0.040	1.388
0.000	1.388	0.160	1.379	0.320	1.369	0.160	1.360
0.400	1.360	0.880	1.388	0.560	1.437	0.760	1.578
0.520	1.786	0.920	2.036	0.600	2.336	0.720	2.726
0.880	3.020	0.520	3.238	0.200	3.309	0.080	3.309
			November 12 1972				
0.040	3.266	0.040	3.183	0.160	3.073	0.120	2.980
0.160	2.875	0.280	2.788	0.080	2.711	0.440	2.681
0.320	2.636	0.480	2.636	0.440	2.636	0.960	2.666
1.480	2.819	3.080	3.073	2.080	3.766	2.240	5.480
2.680	8.093	2.760	10.738	2.160	12.656	1.280	13.791
1.480	14.242	1.280	14.033	1.360	13.757	1.680	13.895
1.280	14.137	1.200	14.666	0.800	14.773	0.480	14.701
0.200	13.860	0.000	12.623	0.000	11.796	0.040	10.974
0.000	10.104	0.000	9.685	0.240	9.064	0.640	8.673
0.080	8.317	0.000	7.898	0.120	7.633	0.160	7.259
0.000	7.076	0.000	6.896	0.000	6.610	0.000	6.395
0.000	6.184	0.040	5.998	0.000	5.775	0.000	5.617

Table 14.1 *(cont.)*

Rain	Flow	Rain	Flow	Rain	Flow	Rain	Flow
			November 13 1972				
0.000	5.499	0.000	5.289	0.000	5.157	0.040	5.064
0.080	5.009	0.040	4.828	0.000	4.722	0.000	4.634
0.000	4.547	0.000	4.411	0.000	4.327	0.000	4.227
0.120	4.161	0.080	4.064	0.040	4.016	0.240	3.968
0.200	3.968	0.200	3.937	0.000	3.890	0.000	3.843
0.120	3.781	0.000	3.705	0.200	3.629	0.000	3.584
0.000	3.540	0.000	3.481	0.000	3.438	0.000	3.409
0.000	3.380	0.000	3.280	0.000	3.238	0.000	3.197
0.000	3.155	0.000	3.128	0.000	3.087	0.000	3.073
0.000	3.046	0.000	3.020	0.000	2.966	0.000	2.940
0.000	2.901	0.000	2.875	0.000	2.850	0.000	2.819
0.000	2.772	0.000	2.726	0.000	2.711	0.000	2.666
			November 14 1972				
0.000	2.606	0.000	2.577	0.040	2.533	0.000	2.490
0.040	2.462	0.050	2.433	0.000	2.405	0.050	2.378
0.150	2.364	0.100	2.350	0.000	2.350	0.050	2.336
0.000	2.323	0.000	2.282	0.050	2.255	0.000	2.229
0.000	2.202	0.000	2.176	0.000	2.150	0.000	2.137
0.000	2.111	0.000	2.099	0.000	2.073	0.000	2.048
0.000	2.036	0.000	2.023	0.000	2.011	0.000	1.999
0.050	1.986	0.000	1.926	0.000	1.914	0.000	1.890
0.000	1.878	0.000	1.867	0.000	1.843	0.000	1.832
0.000	1.820	0.000	1.797	0.000	1.786	0.000	1.774
0.000	1.763	0.000	1.752	0.000	1.730	0.000	1.730
0.000	1.718	0.000	1.696	0.000	1.685	0.000	1.674
			November 15 1972				
0.000	1.664	0.000	1.642	0.000	1.631	0.000	1.621
0.000	1.610	0.050	1.599	0.050	1.589	0.000	1.589
0.000	1.578	0.050	1.568	0.000	1.568	0.050	1.558
0.000	1.547	0.000	1.537	0.000	1.527	0.000	1.517
0.000	1.506	0.000	1.496	0.000	1.486	0.000	1.486
0.000	1.476	0.000	1.466	0.000	1.466	0.000	1.456
0.000	1.447	0.000	1.427	0.000	1.417	0.000	1.408
0.000	1.398	0.000	1.388	0.000	1.388	0.000	1.379
0.000	1.369	0.050	1.369	0.000	1.360	0.000	1.350
0.000	1.341	0.000	1.341	0.000	1.332	0.000	1.322
0.000	1.322	0.000	1.313	0.000	1.304	0.000	1.304
0.000	1.295	0.000	1.286	0.000	1.286	0.000	1.277
			November 16 1972				
0.000	1.268	0.000	1.268	0.000	1.259	0.050	1.259
0.000	1.250	0.000	1.250	0.050	1.241	0.050	1.232
0.050	1.232	0.000	1.223	0.050	1.223	0.000	1.223
0.000	1.215	0.000	1.215	0.050	1.206	0.000	1.189
0.000	1.171	0.000	1.163	0.000	1.155	0.000	1.146
0.000	1.146	0.000	1.138	0.000	1.129	0.000	1.129
0.000	1.121	0.000	1.113	0.000	1.105	0.000	1.105
0.000	1.097	0.000	1.088	0.000	1.088	0.000	1.080
0.000	1.072	0.000	1.064	0.000	1.056	0.000	1.048
0.000	1.048	0.000	1.041	0.000	1.033	0.000	1.033
0.000	1.033	0.000	1.033	0.000	1.025	0.000	1.025
0.000	1.025	0.000	1.017	0.000	1.017	0.000	1.017

Table 14.1 *(cont.)*

Rain	Flow	Rain	Flow	Rain	Flow	Rain	Flow
			November 17 1972				
0.000	1.010	0.000	1.010	0.000	0.871	0.000	1.002
0.000	0.994	0.000	0.994	0.050	0.987	0.000	0.964
0.000	0.957	0.050	0.957	0.080	0.957	0.000	0.949
0.000	0.949	0.000	0.949	0.000	0.942	0.000	0.942
0.000	0.935	0.000	0.935	0.000	0.927	0.000	0.920
0.000	0.920	0.000	0.913	0.000	0.913	0.000	0.906
0.000	0.899	0.000	0.899	0.000	0.892	0.000	0.885
0.000	0.885	0.000	0.878	0.000	0.871	0.000	0.871
0.000	0.864	0.000	0.864	0.000	0.857	0.000	0.857
0.000	0.857	0.000	0.850	0.000	0.850	0.000	0.843
0.000	0.843	0.000	0.843	0.000	0.843	0.000	0.836
0.000	0.836	0.000	0.836	0.000	0.830	0.000	0.830
			November 18 1972				
0.000	0.823	0.000	0.823	0.000	0.823	0.000	0.816
0.000	0.816	0.000	0.810	0.040	0.810	0.000	0.810
0.000	0.810	0.040	0.810	0.000	0.810	0.000	0.810
0.000	0.8[illegible]	0.000	0.803	0.000	0.803	0.000	0.803
0.000	0.803	0.000	0.797	0.000	0.797	0.000	0.797
0.000	0.797	0.000	0.790	0.000	0.790	0.040	0.790
0.040	0.790	0.200	0.797	0.320	0.803	0.400	0.823
0.600	0.836	0.600	0.864	0.720	0.906	0.600	0.949
0.600	0.987	0.360	1.056	0.520	1.121	0.480	1.189
0.360	1.268	0.440	1.322	0.120	1.369	0.280	1.398
0.040	1.398	0.080	1.398	0.040	1.388	0.120	1.369
0.080	1.341	0.080	1.322	0.080	1.304	0.080	1.286
			November 19 1972				
0.080	1.277	0.120	1.268	0.120	1.259	0.120	1.241
0.080	1.232	0.080	1.223	0.040	1.215	0.080	1.215
0.120	1.215	0.080	1.215	0.160	1.215	0.120	1.223
0.040	1.241	0.120	1.259	0.160	1.277	0.360	1.304
1.480	1.379	2.120	1.517	2.040	1.797	0.600	2.163
0.520	2.666	0.040	3.073	0.040	3.128	0.040	3.060
0.000	2.953	0.080	2.834	0.480	2.696	0.360	2.504
0.040	2.378	0.000	2.323	0.000	2.229	0.000	2.150
0.040	2.086	0.000	2.023	0.000	1.938	0.000	1.890
0.000	1.855	0.000	1.820	0.000	1.786	0.200	1.763
1.160	1.741	0.480	1.752	0.240	1.774	0.080	1.797
0.000	1.855	0.040	1.878	0.120	1.890	0.240	1.890
			November 20 1972				
0.040	1.878	0.000	1.855	0.000	1.843	0.040	1.820
0.040	1.809	0.640	1.797	1.520	1.832	0.680	1.902
0.120	1.999	0.280	2.215	0.680	2.364	0.800	2.462
0.120	2.577	0.000	2.651	0.040	2.696	0.000	2.681
0.000	2.592	0.000	2.533	0.880	2.462	0.120	2.433
0.840	2.419	0.760	2.433	0.480	2.533	0.520	2.666
0.360	2.862	0.320	3.020	0.560	3.155	0.120	3.252
0.280	3.280	0.240	3.280	0.040	3.266	0.160	3.238
0.320	3.210	0.240	3.169	0.360	3.141	0.160	3.114
0.200	3.100	0.240	3.100	0.040	3.087	0.000	3.073
0.040	3.060	0.000	3.033	0.000	2.980	0.000	2.927
0.000	2.888	0.000	2.862	0.000	2.834	0.000	2.788

Table 14.1 *(cont.)*

Rain	Flow	Rain	Flow	Rain	Flow	Rain	Flow
			November 21 1972				
0.000	2.741	0.000	2.711	0.000	2.666	0.000	2.621
0.000	2.592	0.000	2.548	0.000	2.519	0.000	2.490
0.000	2.462	0.000	2.448	0.000	2.419	0.000	2.391
0.000	2.378	0.360	2.364	0.520	2.364	0.200	2.364
0.000	2.364	0.080	2.378	0.040	2.364	0.000	2.364
0.040	2.350	0.200	2.336	0.240	2.336	0.240	2.336
0.480	2.336	0.680	2.350	0.440	2.391	1.240	2.490
1.040	2.636	0.720	2.803	0.240	2.914	0.120	3.006
0.030	3.046	0.200	3.060	0.570	3.046	0.730	3.033
0.730	3.033	0.530	3.073	0.170	3.128	0.570	3.183
0.530	3.309	0.330	3.423	0.070	3.584	0.100	3.720
0.030	3.750	0.130	3.674	0.000	3.599	0.000	3.511
			November 22 1972				
0.170	3.438	0.530	3.394	0.030	3.351	0.000	3.337
0.000	3.366	0.000	3.394	0.000	3.366	0.170	3.266
0.200	3.224	0.130	3.183	0.000	3.141	0.000	3.128
0.000	3.087	0.000	3.060	0.000	3.033	0.000	2.980
0.000	2.940	0.000	2.901	0.000	2.875	0.000	2.862
0.000	2.834	0.000	2.819	0.000	2.788	0.000	2.772
0.000	2.726	0.000	2.711	0.030	2.696	0.000	2.681
0.030	2.651	0.000	2.621	0.000	2.606	0.000	2.577
0.000	2.548	0.000	2.519	0.000	2.504	0.000	2.490
0.000	2.462	0.000	2.433	0.000	2.419	0.000	2.391
0.000	2.378	0.000	2.364	0.000	2.350	0.000	2.336
0.000	2.336	0.000	2.323	0.000	2.295	0.030	2.269
			November 23 1972				
0.000	2.255	0.000	2.229	0.000	2.215	0.000	2.202
0.030	2.189	0.000	2.176	0.000	2.163	0.100	2.150
0.070	2.137	0.030	2.124	0.000	2.111	0.030	2.099
0.030	2.086	0.000	2.061	0.030	2.048	0.000	2.048
0.000	2.036	0.000	2.023	0.000	2.023	0.000	2.011
0.030	1.999	0.000	1.950	0.000	1.938	0.000	1.914
0.000	1.902	0.070	1.890	0.030	1.878	0.000	1.867
0.000	1.855	0.000	1.843	0.000	1.832	0.000	1.820
0.000	1.809	0.000	1.797	0.000	1.786	0.000	1.774
0.000	1.763	0.000	1.752	0.000	1.741	0.000	1.730
0.000	1.730	0.000	1.718	0.000	1.707	0.000	1.696
0.000	1.674	0.000	1.664	0.000	1.653	0.000	1.642
			November 24 1972				
0.000	1.631	0.000	1.621	0.000	1.610	0.000	1.599
0.000	1.589	0.000	1.568	0.000	1.568	0.000	1.558
0.000	1.558	0.000	1.547	0.000	1.537	0.000	1.527
0.000	1.527	0.000	1.517	0.000	1.506	0.000	1.506
0.000	1.496	0.000	1.486	0.000	1.476	0.000	1.476
0.000	1.466	0.000	1.456	0.000	1.456	0.000	1.447
0.000	1.417	0.000	1.417	0.000	1.408	0.000	1.398
0.000	1.388	0.000	1.379	0.000	1.369	0.000	1.360
0.000	1.350	0.000	1.341	0.000	1.332	0.000	1.332
0.000	1.322	0.000	1.313	0.000	1.304	0.000	1.304
0.000	1.295	0.000	1.286	0.000	1.286	0.000	1.277
0.000	1.277	0.000	1.268	0.000	1.268	0.000	1.259

Table 14.1 *(cont.)*

Rain	Flow	Rain	Flow	Rain	Flow	Rain	Flow
			November 25 1972				
0.000	1.259	0.000	1.250	0.000	1.250	0.000	1.241
0.000	1.241	0.000	1.232	0.000	1.232	0.000	1.223
0.000	1.223	0.030	1.223	0.000	1.223	0.000	1.215
0.000	1.215	0.000	1.215	0.000	1.206	0.030	1.206
0.000	1.197	0.030	1.171	0.200	1.171	0.130	1.171
0.100	1.171	0.130	1.171	0.100	1.171	0.170	1.171
0.000	1.171	0.000	1.180	0.000	1.180	0.000	1.180
0.070	1.180	0.300	1.171	0.200	1.163	0.070	1.163
0.000	1.163	0.000	1.155	0.000	1.155	0.000	1.155
0.000	1.146	0.000	1.138	0.000	1.138	0.000	1.129
0.000	1.121	0.000	1.113	0.000	1.105	0.000	1.097
0.000	1.088	0.000	1.088	0.000	1.080	0.000	1.072
			November 26 1972				
0.000	1.064	0.000	1.048	0.000	1.048	0.000	1.041
0.000	1.033	0.000	1.033	0.000	1.033	0.000	1.025
0.030	1.025	0.030	1.025	0.000	1.025	0.000	1.017
0.000	1.017	0.000	1.017	0.000	1.017	0.000	1.010
0.000	1.010	0.000	1.010	0.000	1.002	0.000	1.002
0.000	0.994	0.000	0.987	0.000	0.972	0.000	0.964
0.000	0.957	0.000	0.957	0.000	0.957	0.070	0.949
0.000	0.949	0.000	0.949	0.000	0.942	0.000	0.942
0.000	0.942	0.000	0.935	0.000	0.935	0.000	0.927
0.000	0.927	0.000	0.920	0.000	0.920	0.000	0.913
0.000	0.913	0.000	0.913	0.000	0.906	0.000	0.906
0.000	0.899	0.000	0.892	0.000	0.892	0.000	0.885
			November 27 1972				
0.000	0.885	0.000	0.878	0.000	0.878	0.000	0.871
0.000	0.871	0.000	0.871	0.000	0.864	0.000	0.864
0.000	0.864	0.000	0.857	0.000	0.857	0.000	0.857
0.000	0.857	0.000	0.850	0.000	0.850	0.030	0.850
0.030	0.850	0.000	0.843	0.000	0.843	0.030	0.843
0.000	0.843	0.030	0.843	0.030	0.843	0.000	0.843
0.130	0.843	0.070	0.843	0.100	0.843	0.170	0.843
0.200	0.843	0.200	0.843	0.430	0.850	0.770	0.857
0.900	0.878	0.800	0.913	0.500	0.949	0.630	0.987
0.330	1.048	0.070	1.097	0.000	1.129	0.000	1.163
0.000	1.180	0.000	1.189	0.000	1.180	0.030	1.146
0.000	1.121	0.000	1.105	0.000	1.088	0.000	1.056
			November 28 1972				
0.000	1.041	0.000	1.025	0.030	1.017	0.030	1.010
0.100	1.010	0.170	1.002	0.130	1.002	0.100	0.994
0.030	0.987	0.030	0.964	0.000	0.950	0.000	0.949
0.000	0.949	0.000	0.942	0.000	0.935	0.000	0.927
0.000	0.920	0.000	0.913	0.000	0.906	0.000	0.892
0.000	0.885	0.000	0.885	0.000	0.878	0.000	0.871
0.000	0.871	0.000	0.864	0.000	0.857	0.000	0.857
0.000	0.850	0.000	0.850	0.000	0.843	0.000	0.843
0.000	0.843	0.000	0.843	0.000	0.836	0.000	0.836
0.000	0.836	0.000	0.830	0.000	0.830	0.000	0.823
0.100	0.823	0.070	0.823	0.300	0.823	0.370	0.823
0.330	0.830	0.370	0.836	0.600	0.857	0.330	0.871

Table 14.1 *(cont.)*

Rain	Flow	Rain	Flow	Rain	Flow	Rain	Flow
			November 29 1972				
0.030	0.899	0.130	0.920	0.100	0.942	0.070	0.949
0.200	0.957	0.400	0.964	0.230	0.964	0.200	0.964
0.030	0.964	0.170	0.972	0.200	0.972	0.200	1.010
0.200	1.025	0.030	1.056	0.200	1.072	0.170	1.080
0.070	1.088	0.100	1.097	0.100	1.097	0.170	1.097
0.170	1.097	0.170	1.105	0.030	1.113	0.570	1.121
0.300	1.138	0.230	1.163	0.200	1.180	0.800	1.215
0.470	1.277	0.730	1.360	0.600	1.447	0.600	1.537
0.730	1.653	1.070	1.797	0.530	1.962	0.200	2.150
0.200	2.282	0.230	2.364	0.000	2.391	0.000	2.364
0.000	2.323	0.000	2.255	0.000	2.163	0.000	2.086
0.000	2.036	0.000	1.926	0.000	1.878	0.000	1.820
			November 30 1972				
0.000	1.774	0.230	1.752	0.000	1.707	0.000	1.674
0.000	1.653	0.030	1.631	0.070	1.599	0.470	1.589
0.000	1.578	0.100	1.568	0.000	1.547	0.000	1.537
0.170	1.527	0.070	1.517	0.030	1.506	0.230	1.496
0.270	1.496	0.170	1.496	0.670	1.496	0.500	1.517
0.070	1.547	0.330	1.568	0.000	1.589	0.000	1.599
0.030	1.599	0.030	1.599	0.100	1.599	0.030	1.599
0.000	1.599	0.070	1.589	0.100	1.578	0.000	1.568
0.230	1.568	0.230	1.558	0.000	1.558	0.000	1.547
0.000	1.547	0.230	1.537	0.570	1.558	1.000	1.589
1.070	1.674	1.030	1.797	1.300	1.986	1.100	2.378
1.330	3.006	2.000	3.781	1.200	4.739	1.300	5.896
			December 1 1972				
0.630	6.523	1.200	6.829	1.870	7.515	2.630	9.011
2.600	11.274	1.560	14.068	1.200	15.722	1.200	15.946
1.120	15.279	0.400	14.737	0.600	13.791	1.280	12.591
0.160	12.142	0.000	11.859	0.000	10.885	0.000	10.047
0.040	9.548	0.160	8.802	0.000	8.418	0.080	7.946
0.000	7.705	0.000	7.305	0.000	7.144	0.000	6.874
0.160	6.697	0.680	6.588	0.520	6.588	1.040	6.763
0.640	6.896	0.320	6.941	0.080	6.829	0.040	6.653
0.000	6.373	0.000	6.184	0.440	6.080	0.280	6.059
0.040	5.896	0.000	5.735	0.000	5.617	0.000	5.441
0.000	5.308	0.000	5.176	0.000	5.101	0.000	5.046
0.240	4.918	0.320	4.900	0.000	4.828	0.040	4.757
			December 2 1972				
0.040	4.687	0.000	4.617	0.000	4.530	0.000	4.428
0.000	4.343	0.000	4.277	0.000	4.227	0.000	4.178
0.040	4.096	0.120	4.048	0.080	4.016	0.040	3.984
0.400	3.937	0.520	3.937	0.200	3.953	0.120	3.968
0.120	3.968	0.040	3.953	0.000	3.905	0.200	3.858
0.440	3.827	0.880	3.874	0.760	4.000	0.440	4.129
0.200	4.161	0.000	4.113	0.000	4.032	0.000	3.905
0.080	3.843	0.560	3.827	0.920	3.890	1.240	4.113
0.480	4.377	0.360	4.530	0.240	4.599	0.760	4.651
0.440	4.775	0.440	4.882	0.120	4.936	0.480	4.918
0.680	4.973	0.560	5.083	0.360	5.289	0.040	5.441
0.200	5.461	0.240	5.365	0.280	5.365	0.120	5.346

Table 14.1 *(cont.)*

Rain	Flow	Rain	Flow	Rain	Flow	Rain	Flow
December 3 1972							
0.000	5.251	0.000	5.120	0.000	5.064	0.000	4.864
0.000	4.739	0.120	4.651	0.080	4.582	0.120	4.496
0.280	4.479	0.080	4.462	0.000	4.445	0.000	4.411
0.000	4.377	0.000	4.360	0.000	4.327	0.080	4.277
0.120	4.260	0.200	4.227	0.040	4.211	0.040	4.129
0.000	4.064	0.000	4.032	0.000	3.968	0.080	3.921
0.200	3.890	0.040	3.874	0.000	3.858	0.000	3.827
0.000	3.796	0.000	3.766	0.040	3.689	0.000	3.659
0.120	3.629	0.080	3.614	0.040	3.584	0.160	3.570
0.000	3.555	0.040	3.525	0.200	3.511	0.440	3.496
0.600	3.511	0.400	3.570	0.480	3.659	0.400	3.750
0.440	3.874	1.040	4.096	1.960	4.582	2.600	5.916
December 4 1972							
2.800	8.443	2.040	11.672	0.600	13.283	0.200	13.485
0.320	12.429	0.560	11.580	0.560	10.562	0.240	10.504
0.480	10.019	0.360	9.851	0.240	9.575	0.320	9.038
0.560	8.932	0.200	8.570	0.200	8.342	0.480	8.093
0.440	8.118	0.400	8.217	0.160	8.242	0.200	8.192
0.160	8.118	0.080	7.971	0.320	7.849	0.240	7.729
0.160	7.468	0.240	7.305	0.160	7.190	0.080	6.941
0.240	6.807	0.040	6.653	0.320	6.588	0.160	6.437
0.280	6.373	0.320	6.331	0.360	6.352	0.520	6.331
0.280	6.331	0.040	6.289	0.160	6.184	0.040	6.080
0.040	5.957	0.160	5.856	0.720	5.795	1.720	5.977
1.840	6.523	1.640	7.468	1.560	8.828	1.960	10.533
December 5 1972							
1.440	12.142	0.640	12.983	0.280	13.017	0.520	12.852
1.840	12.885	1.760	13.553	1.920	15.243	1.480	18.013
1.160	19.848	1.360	23.095	2.240	22.486	0.720	23.001
0.080	24.588	0.000	23.142	0.200	20.325	0.760	18.958
0.440	17.254	0.120	15.833	0.120	15.316	0.400	14.701
0.240	14.347	0.000	13.757	0.000	13.116	0.200	12.688
0.440	12.269	0.000	12.110	0.000	11.953	0.120	11.580
0.400	11.274	0.280	11.064	0.720	10.856	0.120	10.621
0.280	10.331	0.000	9.935	0.000	9.630	0.280	9.331
1.240	9.439	0.200	9.575	0.040	9.277	0.440	9.011
0.440	8.932	0.360	8.880	0.480	8.932	0.320	8.932
0.200	8.750	1.120	8.724	3.200	9.575	3.560	12.301
December 6 1972							
2.360	15.499	0.280	16.824	0.000	16.059	0.200	14.737
0.200	13.485	0.080	12.526	0.000	11.672	0.080	11.154
0.320	10.708	0.120	10.360	0.240	10.019	0.600	9.823
0.040	9.657	0.120	9.575	0.440	9.412	0.120	9.331
0.000	9.143	0.040	8.932	0.040	8.647	1.000	8.519
0.960	8.596	0.200	8.673	0.840	8.698	0.600	8.802
0.120	8.906	0.080	8.854	0.120	8.776	0.200	8.647
0.000	8.418	0.040	8.192	0.000	7.922	0.000	7.729
0.000	7.421	0.600	7.259	0.800	7.236	0.280	7.236
0.440	7.190	0.360	7.167	0.280	7.144	0.320	7.144
1.040	7.190	0.680	7.610	0.400	7.898	0.280	8.044
0.560	8.118	0.560	8.192	0.200	8.242	0.080	8.242

Table 14.1 *(cont.)*

Rain	Flow	Rain	Flow	Rain	Flow	Rain	Flow
			December 7 1972				
0.080	8.168	0.240	7.995	0.280	7.873	0.320	7.801
0.280	7.729	0.560	7.681	0.080	7.539	0.080	7.351
0.040	7.167	0.040	6.963	0.080	6.829	0.360	6.719
1.000	6.697	1.160	6.852	0.480	7.076	0.240	7.167
0.040	7.144	0.120	7.008	0.000	6.829	0.040	6.653
0.200	6.588	0.160	6.502	0.560	6.416	0.920	6.502
0.160	6.610	0.000	6.545	0.000	6.395	0.040	6.247
0.000	6.121	0.000	5.977	0.000	5.835	0.000	5.735
0.000	5.617	0.000	5.558	0.000	5.422	0.000	5.327
0.000	5.251	0.000	5.157	0.000	5.120	0.000	5.064
0.000	4.918	0.000	4.864	0.000	4.793	0.000	4.739
0.000	4.687	0.240	4.651	0.200	4.634	0.080	4.617
			December 8 1972				
0.680	4.599	0.480	4.582	0.160	4.599	0.000	4.582
0.000	4.479	0.000	4.394	0.000	4.327	0.000	4.260
0.000	4.227	0.000	4.194	0.000	4.113	0.000	4.080
0.000	4.032	0.000	3.984	0.000	3.937	0.000	3.890
0.000	3.858	0.000	3.827	0.000	3.796	0.000	3.766
0.000	3.689	0.000	3.644	0.000	3.614	0.000	3.570
0.000	3.540	0.000	3.496	0.000	3.467	0.000	3.452
0.000	3.438	0.000	3.409	0.000	3.394	0.000	3.309
0.000	3.280	0.000	3.252	0.000	3.224	0.000	3.183
0.000	3.155	0.000	3.128	0.000	3.100	0.000	3.087
0.000	3.073	0.000	3.060	0.000	3.006	0.000	2.980
0.000	2.953	0.000	2.927	0.000	2.901	0.000	2.888
			December 9 1972				
0.000	2.862	0.000	2.850	0.000	2.819	0.000	2.788
0.000	2.757	0.050	2.741	0.100	2.726	0.200	2.726
0.100	2.711	0.300	2.696	0.150	2.696	0.100	2.696
0.300	2.696	0.500	2.696	0.700	2.726	0.450	2.819
0.400	2.888	0.150	2.940	0.050	2.966	0.000	2.980
0.000	2.980	0.000	2.940	0.000	2.901	0.000	2.875
0.000	2.834	0.000	2.788	0.000	2.741	0.000	2.711
0.000	2.636	0.000	2.606	0.000	2.548	0.150	2.519
0.100	2.504	0.950	2.519	0.550	2.592	0.300	2.651
0.000	2.651	0.000	2.636	0.000	2.592	0.000	2.548
0.000	2.504	0.000	2.462	0.050	2.419	0.250	2.405
0.650	2.391	0.150	2.405	0.000	2.405	0.000	2.405
			December 10 1972				
0.000	2.391	0.000	2.378	0.000	2.364	0.000	2.350
0.000	2.295	0.000	2.282	0.000	2.255	0.000	2.229
0.000	2.215	0.000	2.189	0.000	2.176	0.000	2.163
0.000	2.150	0.100	2.137	0.000	2.124	0.250	2.111
1.600	2.137	1.050	2.189	0.900	2.323	0.050	2.490
0.500	2.681	1.850	2.850	2.000	3.020	0.550	3.294
0.300	3.570	0.250	3.705	0.050	3.705	0.000	3.599
0.000	3.481	0.000	3.409	0.000	3.266	0.000	3.155
0.000	3.087	0.000	2.993	0.000	2.914	0.000	2.862
0.000	2.803	0.000	2.726	0.000	2.666	0.000	2.606
0.000	2.548	0.000	2.504	0.000	2.462	0.000	2.419
0.000	2.391	0.000	2.378	0.000	2.364	0.000	2.350

Table 14.1 *(cont.)*

Rain	Flow	Rain	Flow	Rain	Flow	Rain	Flow
			December	11 1972			
0.000	2.309	0.000	2.295	0.000	2.282	0.000	2.269
0.050	2.255	0.100	2.242	0.150	2.242	0.050	2.242
0.050	2.242	0.150	2.242	0.300	2.255	0.000	2.282
0.150	2.309	0.000	2.323	0.000	2.350	0.000	2.364
0.000	2.364	0.000	2.364	0.000	2.364	0.850	2.364
1.100	2.405	1.200	2.548	1.700	2.834	1.650	3.169
2.000	3.766	1.600	4.846	1.750	6.373	1.650	7.995
2.000	9.795	2.400	11.859	2.950	14.242	2.000	16.059
1.450	17.372	0.950	17.651	0.500	16.669	0.100	15.170
0.000	13.689	0.000	12.623	0.000	11.518	0.050	10.708
0.000	10.019	0.000	9.412	0.000	8.959	0.000	8.596
0.000	8.317	0.000	7.995	0.000	7.777	0.000	7.492
			December	12 1972			
0.000	7.213	0.000	7.076	0.000	6.807	0.000	6.675
0.000	6.566	0.000	6.373	0.000	6.247	0.000	6.101
0.000	5.957	0.000	5.835	0.000	5.715	0.000	5.617
0.000	5.538	0.000	5.384	0.000	5.270	0.000	5.176
0.000	5.120	0.000	5.064	0.000	4.918	0.000	4.828
0.000	4.757	0.000	4.704	0.000	4.651	0.000	4.599
0.000	4.496	0.050	4.428	1.300	4.428	0.350	4.547
0.450	4.634	0.300	4.722	0.450	4.775	0.550	4.828
1.000	4.955	1.050	5.213	0.650	5.636	0.750	5.957
1.000	6.142	1.250	6.416	1.100	6.874	0.800	7.445
0.650	8.020	0.900	8.367	0.700	8.519	0.050	8.519
0.000	8.317	0.000	7.898	0.000	7.398	0.000	6.941
			December	13 1972			
0.000	6.588	0.000	6.247	0.000	5.977	0.000	5.715
0.000	5.558	0.000	5.346	0.000	5.213	0.000	5.120
0.000	5.046	0.000	4.882	0.000	4.793	0.000	4.687
0.000	4.634	0.000	4.599	0.000	4.513	0.000	4.428
0.000	4.377	0.000	4.327	0.000	4.293	0.000	4.243
0.000	4.194	0.000	4.178	0.000	4.145	0.000	4.064
0.250	4.048	0.400	4.032	0.300	4.048	0.400	4.080
0.550	4.113	0.300	4.194	0.250	4.293	0.400	4.360
0.050	4.394	0.150	4.394	0.150	4.411	0.200	4.360
0.100	4.343	0.100	4.310	0.100	4.293	0.000	4.260
0.000	4.227	0.000	4.178	0.000	4.113	0.000	4.048
0.000	3.984	0.050	3.921	0.000	3.874	0.000	3.827
			December	14 1972			
0.000	3.781	0.000	3.750	0.000	3.659	0.000	3.614
0.000	3.570	0.000	3.525	0.000	3.496	0.000	3.452
0.000	3.438	0.000	3.423	0.000	3.394	0.000	3.351
0.000	3.309	0.000	3.280	0.000	3.252	0.000	3.224
0.000	3.210	0.000	3.183	0.000	3.169	0.050	3.141
0.000	3.128	0.000	3.114	0.000	3.087	0.000	3.073
0.000	3.060	0.000	3.046	0.000	3.006	0.000	2.980
0.000	2.953	0.000	2.927	0.000	2.901	0.000	2.888
0.000	2.862	0.000	2.850	0.000	2.834	0.000	2.819
0.000	2.788	0.000	2.757	0.000	2.741	0.000	2.711
0.000	2.696	0.000	2.681	0.000	2.636	0.000	2.621
0.000	2.592	0.000	2.562	0.000	2.548	0.000	2.519

Table 14.1 *(cont.)*

Rain	Flow	Rain	Flow	Rain	Flow	Rain	Flow
			December 15 1972				
0.000	2.490	0.000	2.476	0.000	2.448	0.000	2.419
0.000	2.391	0.000	2.378	0.000	2.364	0.000	2.350
0.000	2.336	0.000	2.309	0.000	2.295	0.000	2.282
0.000	2.269	0.000	2.255	0.000	2.229	0.000	2.215
0.000	2.189	0.000	2.176	0.000	2.163	0.000	2.150
0.000	2.137	0.000	2.124	0.000	2.111	0.000	2.099
0.000	2.086	0.000	2.061	0.000	2.048	0.000	2.048
0.000	2.036	0.000	2.023	0.000	2.023	0.000	2.011
0.000	1.999	0.000	1.974	0.000	1.950	0.000	1.938
0.000	1.914	0.000	1.902	0.000	1.890	0.040	1.878
0.040	1.867	0.000	1.867	0.000	1.855	0.000	1.843
0.000	1.832	0.000	1.820	0.000	1.809	0.000	1.797
			December 16 1972				
0.000	1.786	0.000	1.774	0.000	1.763	0.000	1.752
0.000	1.741	0.000	1.741	0.000	1.730	0.000	1.730
0.000	1.718	0.000	1.707	0.000	1.685	0.000	1.674
0.000	1.664	0.000	1.653	0.080	1.642	0.000	1.642
0.120	1.631	0.080	1.631	0.400	1.621	0.240	1.621
0.160	1.621	0.000	1.621	0.000	1.631	0.040	1.631
0.080	1.631	0.160	1.621	0.200	1.610	0.040	1.610
0.000	1.599	0.000	1.599	0.000	1.589	0.000	1.589
0.000	1.578	0.000	1.558	0.000	1.547	0.040	1.537
0.000	1.527	0.000	1.517	0.000	1.517	0.000	1.506
0.000	1.496	0.000	1.486	0.000	1.476	0.000	1.476
0.000	1.466	0.000	1.466	0.000	1.456	0.000	1.447
			December 17 1972				
0.040	1.437	0.000	1.417	0.000	1.408	0.000	1.408
0.000	1.398	0.000	1.398	0.000	1.388	0.000	1.379
0.000	1.369	0.000	1.360	0.000	1.350	0.000	1.350
0.000	1.341	0.000	1.332	0.000	1.332	0.000	1.322
0.000	1.313	0.000	1.304	0.000	1.304	0.000	1.295
0.000	1.286	0.000	1.286	0.040	1.277	0.000	1.277
0.000	1.268	0.000	1.268	0.000	1.259	0.000	1.259
0.000	1.250	0.000	1.250	0.000	1.241	0.000	1.232
0.000	1.232	0.000	1.223	0.000	1.223	0.000	1.215
0.000	1.215	0.000	1.206	0.000	1.189	0.000	1.180
0.000	1.171	0.000	1.163	0.000	1.155	0.000	1.155
0.000	1.146	0.000	1.138	0.000	1.138	0.000	1.129
			December 18 1972				
0.000	1.129	0.000	1.121	0.000	1.121	0.000	1.113
0.000	1.105	0.000	1.105	0.000	1.097	0.000	1.097
0.000	1.088	0.000	1.088	0.000	1.080	0.000	1.072
0.000	1.072	0.000	1.064	0.000	1.056	0.000	1.048
0.000	1.041	0.000	1.041	0.000	1.041	0.000	1.033
0.000	1.033	0.000	1.033	0.000	1.025	0.000	1.025
0.000	1.025	0.000	1.017	0.000	1.017	0.000	1.017
0.000	1.010	0.000	1.010	0.000	1.002	0.000	1.002
0.000	1.002	0.000	0.994	0.000	0.979	0.000	0.972
0.000	0.972	0.000	0.964	0.000	0.964	0.000	0.964
0.000	0.957	0.000	0.957	0.000	0.949	0.000	0.949
0.000	0.950	0.000	0.940	0.000	0.940	0.000	0.940

Table 14.1 *(cont.)*

Rain	Flow	Rain	Flow	Rain	Flow	Rain	Flow
			December	19 1972			
0.000	0.935	0.000	0.927	0.000	0.927	0.000	0.927
0.000	0.920	0.000	0.920	0.000	0.913	0.000	0.913
0.000	0.906	0.000	0.906	0.000	0.906	0.000	0.899
0.000	0.892	0.000	0.892	0.000	0.885	0.000	0.885
0.000	0.885	0.000	0.878	0.000	0.878	0.000	0.878
0.000	0.871	0.000	0.871	0.000	0.864	0.000	0.864
0.000	0.864	0.000	0.857	0.000	0.857	0.000	0.857
0.000	0.857	0.000	0.857	0.000	0.850	0.000	0.850
0.000	0.850	0.000	0.850	0.000	0.843	0.000	0.843
0.000	0.843	0.000	0.843	0.000	0.843	0.000	0.836
0.000	0.836	0.000	0.836	0.000	0.836	0.000	0.830
0.000	0.830	0.000	0.830	0.000	0.830	0.000	0.823
			December	20 1972			
0.000	0.823	0.000	0.823	0.000	0.823	0.000	0.816
0.000	0.816	0.000	0.816	0.000	0.816	0.040	0.810
0.040	0.810	0.000	0.810	0.000	0.810	0.000	0.810
0.000	0.810	0.000	0.803	0.000	0.803	0.000	0.803
0.000	0.803	0.000	0.803	0.000	0.797	0.000	0.797
0.000	0.797	0.000	0.790	0.000	0.790	0.000	0.790
0.000	0.790	0.000	0.784	0.000	0.784	0.000	0.784
0.000	0.777	0.000	0.777	0.000	0.777	0.000	0.777
0.000	0.771	0.000	0.771	0.000	0.771	0.000	0.771
0.000	0.764	0.000	0.764	0.000	0.764	0.000	0.764
0.000	0.758	0.000	0.758	0.000	0.758	0.000	0.758
0.000	0.758	0.000	0.752	0.000	0.752	0.000	0.752
			December	21 1972			
0.000	0.752	0.000	0.745	0.000	0.745	0.000	0.745
0.000	0.745	0.080	0.739	0.000	0.739	0.040	0.739
0.000	0.739	0.000	0.739	0.000	0.739	0.000	0.739
0.000	0.739	0.000	0.733	0.000	0.733	0.000	0.733
0.000	0.733	0.000	0.733	0.000	0.733	0.000	0.727
0.000	0.727	0.000	0.727	0.000	0.721	0.000	0.721
0.000	0.721	0.000	0.721	0.000	0.721	0.040	0.715
0.000	0.715	0.000	0.715	0.000	0.715	0.000	0.709
0.000	0.709	0.000	0.709	0.000	0.709	0.000	0.703
0.000	0.703	0.000	0.703	0.000	0.703	0.000	0.703
0.000	0.697	0.000	0.697	0.000	0.697	0.000	0.697
0.000	0.691	0.000	0.691	0.000	0.691	0.000	0.691
			December	22 1972			
0.000	0.691	0.000	0.691	0.000	0.685	0.000	0.685
0.000	0.685	0.000	0.685	0.000	0.685	0.000	0.679
0.000	0.679	0.000	0.673	0.000	0.673	0.000	0.673
0.000	0.667	0.000	0.662	0.000	0.662	0.000	0.656
0.000	0.656	0.000	0.656	0.000	0.656	0.000	0.650
0.000	0.650	0.000	0.650	0.000	0.644	0.000	0.644
0.000	0.644	0.000	0.639	0.000	0.639	0.000	0.639
0.000	0.639	0.000	0.633	0.000	0.633	0.000	0.633
0.000	0.633	0.000	0.628	0.000	0.628	0.000	0.628
0.000	0.628	0.000	0.622	0.000	0.622	0.000	0.622
0.000	0.622	0.000	0.617	0.000	0.617	0.000	0.617
0.000	0.617	0.000	0.611	0.000	0.611	0.000	0.611

Table 14.1 *(cont.)*

Rain	Flow	Rain	Flow	Rain	Flow	Rain	Flow
			December 23 1972				
0.000	0.611	0.000	0.611	0.000	0.606	0.000	0.606
0.000	0.606	0.000	0.606	0.000	0.600	0.000	0.600
0.000	0.600	0.000	0.600	0.000	0.595	0.000	0.595
0.000	0.595	0.000	0.595	0.000	0.595	0.320	0.595
0.120	0.595	0.160	0.595	0.320	0.595	0.240	0.600
0.640	0.606	0.360	0.611	0.120	0.617	0.000	0.628
0.000	0.644	0.000	0.656	0.000	0.662	0.000	0.667
0.000	0.667	0.000	0.667	0.000	0.662	0.000	0.650
0.000	0.644	0.000	0.639	0.000	0.633	0.000	0.622
0.000	0.617	0.000	0.611	0.000	0.611	0.000	0.606
0.000	0.600	0.000	0.600	0.000	0.595	0.000	0.595
0.000	0.590	0.000	0.590	0.000	0.584	0.000	0.584
			December 24 1972				
0.000	0.584	0.000	0.579	0.000	0.579	0.000	0.579
0.000	0.574	0.000	0.574	0.000	0.574	0.000	0.574
0.000	0.574	0.040	0.569	0.000	0.569	0.040	0.569
0.040	0.569	0.120	0.569	0.040	0.569	0.000	0.569
0.040	0.569	0.080	0.569	0.040	0.569	0.000	0.569
0.000	0.569	0.000	0.569	0.000	0.569	0.000	0.569
0.000	0.569	0.000	0.569	0.000	0.569	0.000	0.569
0.000	0.569	0.000	0.569	0.000	0.569	0.000	0.569
0.000	0.564	0.000	0.564	0.000	0.564	0.000	0.564
0.000	0.558	0.000	0.558	0.040	0.558	0.120	0.558
0.040	0.558	0.040	0.558	0.000	0.558	0.000	0.558
0.200	0.558	0.200	0.558	0.160	0.558	0.080	0.558
			December 25 1972				
0.080	0.558	0.040	0.564	0.000	0.564	0.000	0.564
0.000	0.564	0.000	0.564	0.000	0.564	0.000	0.564
0.000	0.564	0.000	0.564	0.000	0.564	0.000	0.564
0.000	0.564	0.000	0.558	0.000	0.558	0.000	0.558
0.000	0.558	0.000	0.558	0.000	0.553	0.000	0.553
0.000	0.553	0.000	0.553	0.000	0.553	0.000	0.553
0.000	0.548	0.000	0.548	0.000	0.548	0.000	0.548
0.000	0.543	0.000	0.543	0.000	0.538	0.000	0.538
0.000	0.538	0.000	0.533	0.000	0.533	0.000	0.533
0.000	0.528	0.000	0.528	0.000	0.528	0.200	0.528
0.240	0.528	0.000	0.528	0.000	0.528	0.000	0.528
0.000	0.528	0.000	0.528	0.000	0.528	0.000	0.528
			December 26 1972				
0.040	0.528	0.200	0.528	0.040	0.528	0.080	0.528
0.000	0.528	0.000	0.528	0.000	0.528	0.000	0.523
0.000	0.523	0.000	0.523	0.000	0.523	0.000	0.518
0.000	0.518	0.000	0.518	0.000	0.514	0.000	0.514
0.000	0.514	0.000	0.514	0.000	0.509	0.000	0.509
0.000	0.509	0.000	0.509	0.080	0.509	0.080	0.509
0.040	0.509	0.040	0.504	0.000	0.504	0.000	0.504
0.240	0.504	0.240	0.504	0.040	0.504	0.000	0.504
0.240	0.504	0.600	0.509	0.480	0.518	0.480	0.528
0.360	0.538	0.160	0.558	0.200	0.579	0.000	0.600
0.000	0.617	0.000	0.628	0.000	0.639	0.000	0.650
0.000	0.650	0.000	0.650	0.000	0.639	0.000	0.633

Table 14.1 *(cont.)*

Rain	Flow	Rain	Flow	Rain	Flow	Rain	Flow
			December 27 1972				
0.000	0.622	0.000	0.617	0.000	0.611	0.000	0.600
0.000	0.595	0.000	0.590	0.000	0.584	0.040	0.584
0.000	0.579	0.000	0.574	0.000	0.569	0.000	0.569
0.000	0.569	0.000	0.564	0.080	0.564	0.760	0.564
1.120	0.574	0.240	0.584	0.040	0.600	0.000	0.622
0.000	0.639	0.080	0.650	0.040	0.650	0.000	0.656
0.000	0.656	0.040	0.656	0.040	0.656	0.160	0.650
0.160	0.644	0.000	0.639	0.080	0.639	0.000	0.639
0.040	0.633	0.000	0.633	0.000	0.628	0.040	0.628
0.080	0.628	0.040	0.622	0.040	0.622	0.000	0.617
0.040	0.617	0.640	0.617	0.920	0.622	0.240	0.628
0.080	0.639	0.080	0.656	0.080	0.673	0.040	0.691
			December 28 1972				
0.160	0.703	0.040	0.715	0.080	0.721	0.000	0.727
0.000	0.727	0.000	0.727	0.000	0.727	0.040	0.721
0.040	0.721	0.040	0.715	0.000	0.709	0.000	0.703
0.000	0.697	0.000	0.697	0.000	0.691	0.000	0.691
0.000	0.679	0.000	0.673	0.000	0.673	0.000	0.662
0.000	0.656	0.000	0.650	0.000	0.644	0.000	0.639
0.040	0.639	0.000	0.633	0.000	0.628	0.000	0.622
0.000	0.622	0.000	0.617	0.000	0.611	0.000	0.611
0.000	0.606	0.000	0.606	0.000	0.600	0.000	0.600
0.000	0.595	0.000	0.595	0.000	0.590	0.000	0.590
0.000	0.590	0.080	0.584	0.040	0.584	0.000	0.584
0.000	0.579	0.000	0.579	0.000	0.574	0.000	0.574
			December 29 1972				
0.000	0.574	0.000	0.574	0.000	0.569	0.000	0.569
0.000	0.569	0.000	0.564	0.000	0.564	0.000	0.564
0.000	0.558	0.000	0.558	0.000	0.558	0.000	0.558
0.000	0.558	0.000	0.558	0.000	0.558	0.000	0.558
0.000	0.553	0.000	0.553	0.000	0.553	0.000	0.553
0.000	0.553	0.000	0.553	0.000	0.553	0.000	0.553
0.000	0.548	0.000	0.548	0.000	0.548	0.000	0.548
0.000	0.548	0.000	0.548	0.000	0.543	0.000	0.543
0.000	0.543	0.000	0.543	0.000	0.538	0.000	0.538
0.000	0.538	0.000	0.538	0.000	0.538	0.000	0.538
0.000	0.533	0.000	0.533	0.030	0.533	0.000	0.533
0.000	0.533	0.000	0.533	0.000	0.533	0.030	0.533
			December 30 1972				
0.000	0.528	0.000	0.528	0.000	0.528	0.000	0.528
0.000	0.528	0.000	0.528	0.000	0.528	0.000	0.528
0.000	0.528	0.000	0.523	0.000	0.523	0.000	0.523
0.000	0.523	0.000	0.523	0.000	0.523	0.030	0.523
0.000	0.523	0.000	0.523	0.000	0.518	0.000	0.518
0.000	0.518	0.000	0.518	0.000	0.518	0.000	0.518
0.000	0.518	0.000	0.518	0.000	0.518	0.030	0.514
0.000	0.514	0.000	0.514	0.030	0.514	0.000	0.514
0.000	0.514	0.000	0.514	0.000	0.514	0.000	0.514
0.000	0.514	0.000	0.514	0.000	0.509	0.000	0.509
0.000	0.509	0.030	0.509	0.030	0.509	0.000	0.509
0.000	0.509	0.000	0.509	0.000	0.509	0.000	0.509

Table 14.1 *(cont.)*

Rain	Flow	Rain	Flow	Rain	Flow	Rain	Flow
			December 31 1972				
0.000	0.509	0.030	0.509	0.000	0.509	0.000	0.509
0.000	0.509	0.000	0.504	0.000	0.504	0.000	0.504
0.000	0.504	0.000	0.504	0.000	0.504	0.000	0.504
0.000	0.504	0.000	0.499	0.000	0.499	0.000	0.499
0.000	0.499	0.000	0.499	0.000	0.499	0.000	0.499
0.030	0.499	0.070	0.499	0.000	0.499	0.030	0.499
0.000	0.499	0.030	0.499	0.000	0.499	0.000	0.499
0.000	0.499	0.000	0.499	0.000	0.495	0.000	0.495
0.000	0.495	0.000	0.495	0.000	0.495	0.000	0.495
0.000	0.495	0.000	0.495	0.000	0.495	0.000	0.495
0.000	0.495	0.000	0.490	0.000	0.490	0.000	0.490
0.000	0.490	0.000	0.490	0.000	0.490	0.000	0.490

* Data are recorded by rows.

15. The Rainfall at Adelaide, South Australia

Source Cornish, E.A. (1954). On the secular variation of rainfall at Adelaide. *Australian J. Phys.* 7, 334-346.

Contributor R.G. Jarrett CSIRO Melbourne

The rainfall at Adelaide is about 1530 mm each year, with about 70-80% of this falling in the winter months between April and October. The data used by Dr. E.A. Cornish consisted of 61 six-day totals for each year, with December 31 of the previous year being included in non-leap years. With the help of the Bureau of Meteorology, the data have been extended to cover the period 1839 to 1977 inclusive. The data are given in Table 15.1.

Cornish analyzed the data by first fitting orthogonal polynomials up to fifth order to each year. Each set of coefficients, the constants, the linear terms,...,was then examined separately for significant changes over years. The main result was a 23 year cycle in the linear coefficients, which was related to a cycle of amplitude of about 30 days in the onset of the winter rains. This cycle agrees with the known sunspot cycle. Another result was the gradual decrease in magnitude of the quadratic coefficient, indicating a gradual lengthening of the winter rains. A subsequent analysis in 1968 confirmed these results, although the 23 year cycle appears to be less well-defined over the last 30-40 years.

There are obviously many other ways in which the data can be analyzed. One of the major difficulties in any analysis, however, is the variability in the rainfall. Even using six-day totals, about 25% of the observations are zero and there are also some outliers, particularly due to thunderstorms during the summer months.

Table 15.1
Rainfall Data for Adelaide, South Australia

Year	Rainfall readings in six-day totals, mm.															
1839	8	0	22	7	8	0	0	0	0	45	6	42	0	22	12	3
	0	12	12	14	8	0	11	0	0	91	0	64	92	108	20	46
	106	41	0	104	25	7	86	111	158	8	10	3	18	106	34	21
	14	0	122	31	44	164	84	7	12	0	0	16	0			
1840	33	0	0	0	0	0	8	68	107	18	0	28	22	3	3	0
	0	50	37	13	20	67	0	0	79	3	0	132	137	44	12	22
	66	10	58	58	73	146	32	18	92	122	171	79	0	1	0	66
	120	0	4	0	7	0	12	0	0	0	20	111	251			
1841	0	45	0	0	0	0	0	0	11	24	56	22	0	0	3	125
	22	178	7	25	0	23	75	3	70	11	0	0	217	4	0	30
	18	0	18	50	80	78	39	16	34	5	125	11	58	10	5	40
	27	0	34	16	4	2	0	64	48	0	11	36	15			
1842	3	0	0	7	27	0	0	18	30	23	0	78	11	12	0	101
	64	1	2	13	0	49	71	85	0	0	24	108	0	60	76	45
	0	114	6	49	37	58	35	66	61	3	85	9	3	122	8	23
	142	20	35	4	0	57	42	50	95	0	0	0	0			
1843	0	9	12	0	0	4	0	5	0	45	0	0	59	0	0	0
	0	24	44	38	1	8	33	47	4	208	14	0	30	99	154	42
	14	76	31	87	44	68	1	37	20	12	53	11	27	5	22	55
	4	48	35	10	0	4	6	0	159	0	0	0	0			
1844	11	19	14	0	8	0	2	4	12	0	0	25	27	0	15	13
	6	74	55	20	9	5	85	31	31	31	0	0	57	57	103	92
	65	101	5	0	65	61	39	45	1	15	10	81	69	84	45	1
	1	18	15	28	0	31	32	0	0	75	8	0	0			
1845	0	4	9	0	0	3	12	0	0	17	0	3	0	0	23	6
	4	0	45	0	27	35	23	8	128	107	8	187	12	97	31	35
	18	81	45	23	50	92	35	106	35	0	103	42	9	58	0	81
	9	0	10	57	11	7	50	13	2	8	103	0	10			
1846	3	12	0	0	0	160	4	0	40	28	0	0	63	0	3	2
	9	24	8	209	100	9	0	145	28	76	101	3	17	46	55	131
	0	30	142	150	77	6	4	11	141	10	44	116	14	0	14	11
	1	143	55	0	10	211	74	0	1	0	5	0	145			
1847	162	0	0	0	0	3	0	0	0	0	83	5	0	0	73	0
	0	182	50	48	155	0	40	72	0	125	46	124	365	70	195	52
	39	155	98	81	70	0	21	0	5	0	35	15	3	189	0	0
	16	37	5	32	0	70	0	17	0	19	150	2	0			
1848	0	0	0	0	0	0	0	0	0	0	2	0	95	0	0	0
	48	27	0	0	20	110	93	0	0	0	13	8	47	20	139	14
	38	35	40	92	8	63	47	71	45	116	6	100	59	31	12	74
	18	117	59	33	25	101	82	1	22	6	2	0	37			
1849	0	0	0	0	0	0	28	0	1	0	7	10	32	12	0	0
	0	25	141	63	3	6	0	152	15	190	57	0	252	218	172	35
	53	124	0	66	134	14	22	46	99	19	78	55	48	48	9	27
	9	0	29	74	0	50	58	23	12	23	0	0	6			
1850	0	23	0	0	76	301	0	0	5	8	0	0	0	0	25	0
	0	81	0	9	0	44	78	10	48	22	155	133	17	50	10	43
	8	0	3	62	20	9	13	68	0	119	38	4	37	17	0	72
	0	0	100	65	0	0	0	15	0	68	0	0	92			

Table 15.1 *(cont.)*

Year	Rainfall readings in six-day totals, mm.															
1851	104	6	3	0	0	0	6	9	0	0	42	22	31	0	0	0
	23	14	70	45	192	191	55	34	153	32	27	169	119	19	68	71
	172	143	50	45	113	0	252	42	112	56	106	14	77	0	34	12
	0	70	30	3	59	253	7	7	4	11	7	0	3			
1852	25	25	25	50	25	0	0	40	0	0	0	0	0	0	3	0
	0	3	80	117	40	93	34	106	0	41	79	170	5	14	111	18
	190	53	54	60	165	149	65	117	117	10	43	19	27	112	50	7
	47	3	50	0	0	83	11	98	0	0	15	78	18			
1853	0	37	4	0	0	0	4	60	0	3	5	15	0	0	13	29
	24	138	5	207	459	14	103	10	121	15	24	64	127	17	45	41
	28	104	55	31	43	8	15	54	116	98	101	90	21	75	119	47
	0	0	0	26	0	0	0	16	0	0	0	82	0			
1854	0	0	0	0	9	1	0	0	0	3	3	13	10	6	73	0
	30	64	108	64	25	24	12	107	0	0	0	39	0	74	43	20
	29	57	97	15	30	49	48	89	38	11	50	4	4	144	10	2
	11	0	7	28	1	0	21	10	23	0	9	0	20			
1855	0	0	0	0	0	4	0	0	0	128	8	10	126	143	25	0
	3	0	68	39	0	36	9	72	187	147	51	43	0	20	20	123
	63	28	99	25	85	12	17	45	4	67	164	18	76	16	9	68
	0	92	10	0	43	8	0	0	0	47	57	0	0			
1856	0	81	0	0	7	0	250	0	0	0	0	0	23	5	15	123
	102	68	72	14	4	111	0	11	76	138	232	157	32	27	154	30
	33	7	21	38	37	26	83	10	21	41	45	37	16	15	146	6
	49	46	0	4	7	0	0	44	4	26	0	0	0			
1857	1	0	43	0	0	6	0	78	0	0	14	0	113	85	171	0
	0	0	22	81	55	0	0	25	13	16	55	91	149	75	113	13
	87	0	20	34	0	77	123	79	72	2	6	49	7	0	54	93
	7	9	48	0	25	85	22	0	0	52	0	0	51			
1858	22	39	25	0	0	0	0	21	242	4	0	2	5	0	16	8
	14	7	4	99	0	183	110	42	16	78	86	6	0	0	0	22
	105	54	120	22	29	60	25	6	31	87	54	2	46	9	0	0
	63	0	5	0	2	2	16	195	0	0	82	86	3			
1859	0	8	6	20	0	0	0	52	64	0	0	0	0	0	0	8
	45	20	0	0	16	3	289	88	39	44	51	99	0	24	23	14
	29	25	4	0	93	8	33	24	24	33	11	6	11	57	0	50
	2	27	24	44	0	0	19	0	43	8	0	0	0			
1860	0	6	3	0	9	0	0	0	0	0	0	52	100	0	55	318
	6	51	53	3	42	2	92	68	46	25	110	73	33	131	34	25
	43	39	0	0	0	0	73	3	0	19	9	34	69	52	111	18
	0	13	0	44	12	0	18	13	38	14	9	4	0			
1861	23	1	0	2	0	0	5	14	27	0	8	0	36	88	0	41
	2	96	0	2	67	250	44	38	64	21	18	71	40	27	149	51
	121	39	37	17	4	6	54	16	34	34	108	4	28	66	44	13
	18	33	80	4	6	0	11	132	5	37	175	49	0			
1862	0	0	10	11	0	0	70	0	0	0	13	25	10	2	0	3
	0	119	0	10	52	9	8	127	277	83	0	8	38	80	53	70
	53	180	92	81	91	193	1	32	35	12	83	7	44	0	0	0
	40	64	4	0	74	7	0	0	10	0	0	5	0			

Table 15.1 *(cont.)*

Year	Rainfall readings in six-day totals, mm.															
1863	0	0	0	28	56	0	0	0	38	0	31	6	3	0	0	20
	0	0	16	23	63	12	148	232	51	12	73	90	91	2	51	36
	52	56	106	124	81	14	85	73	64	21	77	24	8	28	24	37
	36	43	182	42	0	9	11	0	0	86	12	0	1			
1864	0	150	53	0	0	0	13	4	0	0	0	0	0	0	0	8
	34	26	30	22	45	44	172	7	0	64	75	9	129	27	0	176
	8	49	63	86	29	160	96	22	75	5	32	47	18	33	43	19
	65	12	1	10	0	0	0	3	3	5	2	9	0			
1865	0	0	0	2	0	1	5	20	0	0	44	0	50	0	2	0
	0	10	6	51	41	87	53	0	86	41	0	0	0	77	16	251
	44	142	69	39	0	33	33	43	14	96	29	0	22	58	9	13
	8	0	0	1	10	0	0	18	3	0	12	12	0			
1866	0	2	0	0	99	0	0	17	37	0	45	14	1	0	8	0
	0	22	0	3	16	91	182	36	189	22	34	72	0	100	20	92
	10	70	132	43	63	88	0	0	1	13	18	76	47	20	76	57
	20	109	0	4	15	0	37	3	0	10	0	0	0			
1867	0	0	16	0	9	0	0	33	9	103	3	63	0	0	100	0
	0	1	0	21	0	0	15	188	0	0	82	28	7	28	90	41
	51	49	54	4	56	39	0	5	20	67	100	37	90	75	136	94
	5	75	0	0	0	77	0	6	0	2	6	4	21			
1868	0	90	53	0	0	0	0	0	1	0	2	0	0	0	116	5
	178	17	0	0	3	0	0	70	23	0	51	140	224	79	65	70
	0	11	50	0	22	9	42	49	66	59	11	31	157	22	57	18
	16	29	33	4	36	4	26	0	10	36	12	0	0			
1869	9	0	2	7	3	14	27	2	8	0	0	29	8	74	81	12
	4	31	34	19	29	10	39	10	51	57	98	1	88	31	2	8
	11	29	26	91	0	14	30	31	19	15	0	36	0	53	81	48
	6	11	7	0	3	7	158	2	0	0	2	14	0			
1870	2	0	97	0	229	0	0	0	0	0	0	0	0	0	0	1
	4	30	11	5	13	10	50	16	21	18	60	130	25	197	13	11
	80	18	53	102	94	118	91	36	10	134	37	51	36	14	95	17
	36	145	77	0	80	0	35	10	0	0	15	0	59			
1871	32	179	38	0	25	0	106	0	0	0	11	51	8	0	4	0
	30	0	23	22	71	83	94	29	19	0	50	20	92	89	74	33
	13	52	154	3	113	10	32	6	59	60	4	82	18	23	21	77
	0	15	38	15	74	21	84	0	54	126	0	18	0			
1872	0	0	0	17	81	45	0	69	0	0	0	7	18	6	53	8
	20	0	10	18	87	58	100	103	9	44	47	11	175	33	92	27
	44	207	71	71	58	20	24	32	29	7	30	3	34	72	105	42
	16	0	11	0	135	24	42	17	0	14	8	9	6			
1873	0	0	0	8	52	0	3	19	83	0	0	0	57	11	0	0
	28	207	78	15	33	50	7	79	133	29	54	2	50	44	188	22
	63	0	18	34	63	93	62	8	28	99	64	13	58	0	22	29
	44	2	49	12	25	1	26	0	0	8	21	12	0			
1874	0	36	0	0	0	0	2	2	0	0	0	0	37	41	16	26
	0	50	28	0	58	188	51	134	0	2	59	25	77	16	38	45
	43	1	39	103	29	7	49	142	10	25	86	34	38	24	0	0
	14	79	0	0	14	15	0	0	0	7	8	25	0			

Table 15.1 *(cont.)*

Year	Rainfall readings in six-day totals, mm.															
1875	22	9	0	14	0	86	0	16	138	0	5	1	3	2	5	3
	89	0	28	115	347	305	99	2	3	75	158	117	55	27	0	28
	34	62	7	39	70	69	2	28	184	8	18	32	15	0	71	35
	91	1	49	16	21	0	23	10	59	35	30	19	140			
1876	0	0	4	10	0	5	46	0	0	0	0	10	14	0	21	54
	116	6	1	3	19	35	8	0	1	78	16	39	40	6	4	42
	129	2	60	6	102	0	10	41	8	0	7	79	17	9	7	38
	27	66	3	2	0	57	47	0	0	0	0	1	48			
1877	0	12	8	3	0	0	0	214	24	0	0	0	6	366	77	4
	2	18	96	45	39	25	96	179	113	0	34	76	5	57	33	24
	0	26	20	10	15	5	30	76	73	6	28	163	146	16	8	4
	161	4	60	33	0	1	2	7	13	1	10	18	1			
1878	0	0	0	0	0	5	10	36	11	13	367	0	44	0	48	6
	0	20	211	50	57	11	9	20	21	73	203	115	45	14	5	55
	30	132	33	11	7	28	28	31	5	30	54	7	80	7	10	78
	0	24	15	125	0	0	8	0	4	0	0	14	0			
1879	10	0	0	0	0	2	3	0	13	0	32	0	122	0	0	1
	0	1	0	117	14	20	9	80	238	7	43	27	0	57	93	182
	29	0	9	36	30	92	43	5	27	67	51	0	73	3	59	41
	30	2	18	6	116	30	12	11	28	20	0	50	111			
1880	76	0	0	0	0	0	0	0	0	64	4	52	0	191	0	18
	2	81	154	82	27	29	1	32	69	34	62	52	69	51	48	52
	23	41	97	40	115	73	10	38	17	81	64	29	35	8	78	0
	57	67	1	1	19	55	6	0	30	0	12	0	0			
1881	0	257	0	4	10	4	0	2	5	4	0	0	4	11	10	4
	1	40	0	80	3	98	8	17	45	128	82	91	78	16	89	23
	87	54	4	28	4	52	46	14	24	72	0	44	41	16	7	95
	12	1	1	4	4	30	18	2	0	3	15	8	5			
1882	0	9	0	6	9	0	0	0	0	0	7	0	17	0	26	0
	3	14	89	65	49	20	26	47	42	75	1	30	31	58	45	70
	73	0	34	104	26	72	10	121	38	29	19	4	4	50	47	0
	18	55	7	0	28	1	37	24	8	0	0	0	30			
1883	14	2	7	2	0	0	18	0	6	17	0	11	0	38	38	217
	33	0	3	1	130	53	107	107	216	96	7	73	130	7	51	172
	21	0	176	0	34	117	118	1	96	27	57	14	27	0	76	2
	34	29	39	2	59	4	116	1	11	53	3	23	0			
1884	3	0	11	84	73	0	0	1	11	0	0	6	0	0	169	3
	0	16	16	99	0	52	99	72	15	16	88	123	46	186	13	6
	0	33	0	0	39	24	11	35	15	7	43	27	160	12	24	65
	10	32	0	3	0	20	11	6	16	20	56	0	1			
1885	5	0	19	0	0	0	3	16	7	64	3	3	1	1	26	54
	9	28	0	6	24	5	44	49	91	24	24	181	28	72	107	81
	14	5	37	7	37	124	16	10	40	0	40	18	85	29	0	53
	3	21	26	0	3	0	0	0	0	0	1	45	0			
1886	36	11	8	8	2	22	18	0	0	7	0	0	0	0	0	9
	0	1	127	5	5	36	13	48	7	6	16	13	4	3	47	84
	12	62	64	25	41	90	111	13	35	25	13	28	1	46	17	3
	34	115	1	30	4	0	55	19	9	28	25	0	0			

Table 15.1 *(cont.)*

Year	Rainfall readings in six-day totals, mm.														
1887	0	0	69	0	0	3	41	6	0	0	17	2	1	1	11	32
	143	0	0	33	105	152	82	31	14	176	62	64	213	85	76	28
	2	37	35	134	22	17	9	38	70	89	16	7	93	122	13	8
	23	95	12	49	0	0	10	69	0	122	0	30	0			
1888	0	1	11	21	4	0	0	0	0	7	2	6	10	0	4	0
	8	0	1	0	80	26	54	2	33	55	42	20	184	1	14	50
	171	114	32	78	0	54	95	7	76	2	0	4	29	38	28	0
	0	0	0	12	53	0	0	0	0	1	13	1	13			
1889	273	0	0	16	10	23	0	0	0	0	49	1	7	4	0	251
	41	285	1	0	12	18	40	245	5	130	116	197	82	38	21	52
	7	0	31	22	159	44	53	0	104	11	54	11	41	48	99	5
	116	123	1	65	8	136	1	0	31	0	0	0	0			
1890	0	1	1	13	48	189	4	0	0	0	55	0	1	0	2	0
	70	0	0	30	0	0	52	91	21	1	151	157	81	6	82	195
	143	49	35	97	112	80	4	98	41	14	66	7	23	65	21	46
	51	52	84	33	59	38	89	0	1	0	0	12	7			
1891	48	0	0	1	6	0	0	0	8	0	0	22	0	0	34	25
	18	35	8	0	0	4	15	0	0	0	16	63	8	34	62	148
	34	25	35	7	0	17	20	112	12	13	30	17	5	146	5	19
	25	0	77	9	1	53	2	0	0	150	32	0	0			
1892	0	9	0	153	0	0	0	0	21	1	0	0	0	15	61	4
	35	0	59	62	32	8	33	30	106	38	39	74	99	17	0	153
	49	15	2	44	85	38	37	51	50	78	0	54	77	53	15	34
	124	83	44	0	1	22	22	10	2	91	14	0	9			
1893	0	2	0	1	0	0	0	0	0	0	0	0	0	0	54	53
	57	49	39	7	62	46	9	10	192	44	4	203	104	54	24	9
	88	70	24	66	59	13	0	28	121	86	56	1	170	15	21	70
	25	5	0	19	110	19	0	0	0	0	10	23	30			
1894	0	8	60	0	0	0	0	0	0	1	0	0	26	107	25	0
	0	2	215	45	16	0	61	20	30	205	15	5	14	23	18	15
	207	83	0	32	33	32	53	77	93	47	21	2	31	1	49	111
	56	79	0	21	2	0	0	0	0	6	2	120	9			
1895	122	0	0	0	0	0	0	0	2	0	2	6	171	0	0	55
	2	47	233	75	11	0	35	7	15	54	61	0	71	113	25	86
	131	142	49	83	21	14	64	83	6	49	13	68	6	13	5	23
	0	0	0	1	0	0	0	94	8	23	33	5	1			
1896	0	0	104	49	0	0	10	0	22	0	0	0	39	0	5	62
	191	5	29	12	46	22	40	38	0	114	0	19	95	85	2	93
	13	16	5	28	0	27	2	5	61	9	14	20	2	0	5	6
	22	0	1	0	0	4	0	48	0	1	0	0	146			
1897	1	64	1	0	0	8	0	81	36	54	18	25	8	1	5	0
	4	29	2	57	12	1	23	95	78	0	11	45	70	33	0	111
	18	34	5	45	32	69	35	57	129	55	0	15	9	87	5	12
	14	11	20	7	1	4	1	1	3	0	0	0	0			
1898	9	0	0	0	0	0	17	0	0	40	1	0	2	0	0	159
	3	121	48	10	23	0	317	38	0	69	99	143	18	22	9	15
	76	68	51	86	28	8	1	27	137	11	6	21	14	0	56	3
	12	19	116	0	4	1	81	33	38	0	0	15	0			

Table 15.1 *(cont.)*

Year	Rainfall readings in six-day totals, mm.															
1899	6	7	17	57	10	3	0	0	3	148	0	0	4	69	54	84
	3	0	17	103	71	0	69	96	0	25	161	30	29	45	32	1
	11	1	0	97	54	0	5	0	7	33	56	52	12	46	33	24
	26	28	3	125	1	34	33	0	1	7	0	24	27			
1900	0	3	0	23	32	9	0	0	0	53	36	1	127	65	1	153
	19	132	43	29	0	18	131	22	70	13	0	110	76	156	6	50
	33	1	44	70	62	163	46	49	53	8	53	34	20	1	12	2
	0	36	14	0	17	40	0	0	30	1	0	0	8			
1901	15	55	3	0	33	0	2	2	0	0	34	7	0	16	14	141
	0	4	49	0	0	12	0	1	92	130	144	118	112	0	12	28
	40	15	107	5	54	0	5	55	25	21	0	67	28	11	0	71
	8	48	32	0	52	12	3	21	0	0	0	56	41			
1902	47	7	1	5	11	7	0	3	0	25	0	0	0	15	85	0
	13	0	0	24	7	81	6	0	12	42	139	75	133	0	0	30
	12	20	52	55	21	1	0	58	23	4	79	49	12	26	11	82
	7	20	32	9	39	7	0	0	49	26	179	2	0			
1903	28	0	0	59	0	1	0	99	0	0	62	143	8	6	0	98
	21	20	6	132	8	56	0	98	8	59	178	78	52	20	74	50
	41	46	95	45	3	35	42	3	164	50	17	108	48	46	0	35
	29	3	62	0	11	10	46	127	0	76	2	0	39			
1904	148	0	0	84	22	0	2	1	6	18	0	0	26	14	0	0
	0	0	194	7	1	98	58	57	0	86	18	47	131	81	161	7
	111	8	101	0	16	37	21	66	58	21	20	18	11	0	46	27
	118	20	0	51	0	10	3	1	0	0	0	0	0			
1905	93	1	2	0	55	1	0	1	16	8	2	0	0	12	0	76
	115	0	171	4	0	58	0	22	238	72	89	118	32	88	40	142
	43	71	41	20	10	24	8	34	65	17	100	6	21	8	11	13
	81	84	95	0	15	0	0	3	1	1	0	0	0			
1906	0	0	0	0	0	0	0	0	7	23	0	4	0	163	51	16
	0	2	18	53	9	74	78	0	49	157	78	79	88	103	25	2
	53	130	52	76	107	4	71	158	73	45	57	21	134	51	75	34
	0	8	23	2	11	198	28	7	5	2	146	1	0			
1907	2	0	0	0	7	0	15	3	0	0	60	0	17	4	0	99
	19	10	74	23	0	13	0	212	15	130	52	43	9	0	47	98
	4	57	71	38	60	58	22	6	3	18	11	3	70	30	65	21
	8	0	53	92	0	7	49	5	3	0	3	0	69			
1908	0	0	0	0	0	35	41	3	0	1	119	86	38	0	25	10
	0	49	9	0	69	92	173	17	35	1	157	106	228	41	26	63
	8	26	5	8	7	5	56	127	98	50	36	58	46	35	26	2
	325	7	0	10	0	15	8	2	0	1	0	71	0			
1909	61	0	0	13	0	0	0	0	0	22	48	10	0	7	1	19
	2	4	87	214	47	63	182	0	0	119	0	84	73	58	2	39
	56	59	154	84	120	29	223	131	44	53	39	11	39	171	0	31
	29	10	21	0	162	71	25	25	2	0	1	24	0			
1910	0	0	0	2	0	5	1	0	0	0	372	38	0	0	0	0
	2	0	0	4	7	26	119	37	245	7	29	5	107	162	21	53
	95	70	157	22	39	2	43	56	22	59	125	81	13	26	45	26
	48	15	18	74	3	35	11	52	1	6	23	53	0			

Table 15.1 *(cont.)*

Year	Rainfall readings in six-day totals, mm.															
1911	0	0	17	0	0	48	41	8	0	33	1	1	83	2	1	7
	5	6	2	11	2	77	22	56	20	59	111	39	53	2	0	24
	71	45	41	19	0	54	8	10	100	0	36	33	70	141	5	16
	4	4	36	0	0	0	29	6	125	0	2	13	0			
1912	21	0	0	0	0	0	8	0	23	7	0	46	0	18	3	82
	74	11	9	0	0	34	0	1	5	79	3	70	7	266	81	53
	55	10	43	64	46	3	0	56	94	69	42	79	21	17	45	48
	3	0	0	106	1	82	12	23	114	23	0	0	0			
1913	18	0	0	0	0	13	0	235	9	0	24	25	13	56	1	0
	0	1	1	75	12	0	28	19	48	13	0	37	8	0	5	3
	28	20	9	47	42	22	91	1	17	39	30	107	91	2	0	53
	6	150	35	28	57	0	51	0	0	0	0	247	0			
1914	0	0	2	0	105	9	14	0	0	20	0	89	0	0	19	45
	6	35	85	5	15	5	75	0	34	7	12	47	1	3	3	44
	45	0	48	3	1	2	7	0	22	32	10	0	18	0	0	0
	0	7	9	3	9	22	171	0	0	14	23	22	0			
1915	0	0	0	49	0	0	2	0	1	10	7	7	0	0	0	0
	212	9	20	0	0	1	222	39	28	0	45	0	137	158	21	102
	6	125	0	92	130	17	0	24	36	55	131	82	29	34	14	10
	14	27	2	0	3	27	4	0	0	5	0	0	1			
1916	46	14	3	0	7	6	23	0	0	0	0	0	0	0	48	20
	5	29	19	77	0	0	52	4	0	194	57	28	150	390	116	118
	1	128	66	72	42	142	53	49	42	6	5	0	157	1	61	44
	10	70	8	66	69	95	53	2	11	138	2	0	17			
1917	6	2	8	26	0	0	0	185	46	10	0	4	103	1	141	25
	30	10	0	3	76	132	230	70	11	39	60	117	29	16	82	32
	48	212	31	50	5	26	150	8	75	89	156	80	37	9	81	16
	55	33	17	1	5	13	96	1	0	31	38	0	33			
1918	23	0	35	0	2	0	1	0	17	0	42	0	6	0	2	44
	0	0	12	32	0	83	99	95	51	8	33	32	117	89	70	57
	108	0	27	0	140	0	99	1	69	8	7	0	6	0	196	27
	9	20	20	0	0	14	0	2	3	33	0	25	0			
1919	0	31	0	0	2	22	16	198	52	0	3	2	0	0	5	16
	5	0	0	6	4	181	11	0	22	14	0	44	43	62	34	29
	61	1	5	76	96	113	0	56	11	118	124	10	22	31	48	13
	2	3	2	0	7	5	0	1	0	0	2	35	77			
1920	0	0	20	0	0	1	2	0	0	3	68	70	6	0	0	6
	1	32	0	1	22	84	143	4	0	357	57	98	101	77	46	93
	78	42	35	12	87	30	55	121	36	67	63	19	1	5	21	117
	0	144	4	22	74	13	0	121	16	0	17	2	176			
1921	159	0	0	0	0	5	0	0	10	168	0	0	0	0	37	1
	7	36	1	0	1	5	0	261	155	38	116	53	5	0	38	0
	46	18	54	195	6	39	10	25	17	51	28	169	21	43	38	93
	44	1	0	108	0	8	87	30	34	2	0	1	0			
1922	135	85	2	0	0	0	0	0	5	0	0	0	0	13	0	87
	2	2	3	57	99	115	15	0	104	109	11	30	9	59	90	219
	63	106	9	30	36	124	57	8	0	1	2	50	97	35	23	28
	0	78	17	8	0	0	0	0	3	102	52	106	34			

Table 15.1 *(cont.)*

Year	Rainfall readings in six-day totals, mm.															
1923	34	8	15	1	13	0	0	0	0	6	3	0	0	0	0	3
	0	0	0	0	61	195	13	66	132	77	191	120	85	71	142	67
	104	111	15	179	26	38	0	24	55	3	50	285	202	56	16	93
	34	2	64	2	9	34	0	1	35	94	142	2	0			
1924	11	25	0	20	15	2	5	117	137	3	108	0	0	0	99	0
	74	23	34	13	11	3	12	163	54	67	163	83	23	34	1	1
	12	16	26	23	19	105	3	70	7	43	137	79	55	27	21	7
	114	52	9	94	60	4	27	2	1	0	3	0	27			
1925	0	0	8	32	0	0	566	29	0	14	28	0	0	11	0	86
	10	0	11	1	0	55	76	96	75	105	45	0	1	88	48	16
	44	56	38	37	0	5	78	53	2	20	90	121	14	78	1	38
	52	2	2	0	0	39	0	0	0	5	0	15	0			
1926	0	2	0	0	0	0	33	0	1	70	0	0	1	0	0	26
	9	42	42	87	184	25	0	143	74	37	21	3	81	30	0	7
	29	46	89	134	39	49	13	102	119	23	13	55	122	19	160	72
	0	0	0	65	2	12	0	3	43	6	29	58	0			
1927	26	0	1	0	0	63	0	0	0	28	3	2	28	66	0	0
	0	0	2	14	29	3	54	49	84	5	42	11	96	2	24	111
	28	103	0	72	177	23	69	43	14	16	0	3	17	77	0	2
	4	8	109	17	0	0	21	73	72	0	0	0	1			
1928	0	3	0	90	8	146	0	105	0	0	0	4	0	85	16	4
	0	3	0	90	52	23	51	0	58	20	73	117	64	78	0	32
	109	23	108	7	1	3	5	47	30	51	25	0	31	99	119	19
	18	24	29	49	0	0	0	0	12	0	0	9	3			
1929	0	2	2	0	29	12	0	5	0	0	22	3	0	2	3	0
	4	0	20	18	27	60	9	39	5	122	67	0	141	10	91	63
	16	55	9	13	22	74	8	28	53	15	5	42	63	46	68	3
	4	4	19	6	40	32	18	20	9	0	0	1	322			
1930	2	0	0	0	0	0	4	0	39	0	0	3	0	0	2	14
	11	0	55	17	0	92	22	0	0	2	12	10	0	87	175	49
	51	26	91	101	101	108	36	19	51	29	59	77	53	4	69	4
	137	42	28	11	11	65	4	1	35	0	3	53	0			
1931	58	0	9	0	3	3	0	16	7	4	14	88	0	30	0	9
	0	38	1	42	91	21	69	50	51	20	245	87	76	125	57	46
	97	37	132	15	69	31	60	79	127	21	127	5	3	19	2	4
	35	0	0	20	18	49	0	0	0	0	0	22	0			
1932	0	15	0	0	0	78	15	19	28	0	0	51	18	0	71	202
	132	92	4	42	0	17	0	100	0	195	6	235	89	45	55	25
	39	31	70	44	78	129	0	35	53	35	16	56	34	54	28	101
	15	73	25	4	0	0	22	4	6	0	4	14	0			
1933	0	33	11	174	3	4	0	0	1	17	84	0	1	6	42	0
	169	26	0	3	123	17	125	5	244	54	5	13	36	51	11	7
	64	11	53	90	36	33	43	62	146	46	25	162	11	7	29	5
	0	22	1	0	0	9	0	14	0	41	7	4	26			
1934	42	4	0	0	0	0	1	13	0	0	0	0	23	22	21	25
	50	31	45	0	0	1	0	0	8	48	11	23	0	22	24	49
	1	28	2	89	45	180	31	30	40	114	82	109	28	70	33	69
	44	38	0	357	16	1	7	39	2	54	4	0	48			

Table 15.1 *(cont.)*

Year	Rainfall readings in six-day totals, mm.															
1935	70	8	0	0	0	0	0	1	1	0	0	0	97	140	9	5
	56	29	32	44	15	103	69	1	53	54	68	47	62	54	59	26
	90	28	37	152	116	24	0	42	53	1	48	181	0	7	11	42
	151	26	29	5	64	5	14	11	5	14	0	49	37			
1936	40	99	0	4	2	0	0	3	77	2	0	5	0	2	0	1
	89	0	31	0	17	50	36	0	0	189	17	28	29	44	45	15
	62	15	82	72	27	69	14	54	8	0	24	22	20	18	25	14
	99	4	84	10	0	41	0	3	51	29	252	10	0			
1937	11	3	0	106	122	0	0	38	41	0	2	0	0	0	103	0
	1	0	0	65	4	47	145	69	63	99	3	125	14	0	30	5
	21	70	67	42	2	132	130	33	74	0	51	141	2	25	14	13
	6	21	0	2	12	3	0	212	5	113	4	10	4			
1938	21	0	22	0	18	61	8	12	96	68	8	3	0	0	2	3
	0	393	10	175	18	7	33	0	32	81	31	59	7	46	38	0
	11	69	63	39	70	25	2	138	23	35	10	18	2	0	10	0
	0	15	45	13	9	0	0	28	0	13	0	6	30			
1939	25	0	105	0	0	3	0	63	7	115	0	112	0	1	2	32
	4	32	56	91	0	3	153	35	0	203	10	2	39	151	28	38
	21	62	11	40	69	16	66	89	81	20	52	5	3	1	49	34
	5	0	96	194	0	48	0	29	17	5	0	3	3			
1940	96	0	0	4	20	0	0	7	11	20	0	0	0	32	0	29
	56	60	0	121	98	9	23	9	5	20	22	6	30	3	18	116
	74	24	34	62	34	31	1	45	37	0	56	16	26	32	11	2
	6	30	2	25	95	0	55	0	3	2	83	10	5			
1941	0	0	8	112	211	71	0	0	1	5	125	0	28	15	15	51
	0	26	21	3	0	12	10	54	2	37	64	0	56	2	69	38
	190	75	49	2	28	5	101	9	22	180	90	4	65	118	27	0
	7	61	30	6	4	84	0	6	29	0	27	0	1			
1942	0	0	0	97	64	0	15	0	1	5	2	0	1	0	28	33
	0	0	6	147	230	37	82	14	88	48	68	140	81	135	87	39
	0	88	33	0	68	102	8	33	166	39	150	59	53	2	54	0
	21	0	39	68	0	3	43	15	35	0	4	10	3			
1943	0	33	32	19	0	0	22	233	63	1	5	1	0	0	2	9
	72	53	109	14	24	17	17	0	0	67	23	31	83	43	29	16
	25	23	108	67	53	3	8	49	6	0	17	45	81	42	44	2
	52	11	11	32	0	0	0	8	0	4	65	0	0			
1944	13	5	0	0	0	0	61	8	0	9	8	55	0	0	0	36
	106	12	11	26	204	121	27	37	24	0	18	34	14	0	65	96
	3	38	2	33	4	10	13	1	4	3	26	15	15	77	40	0
	17	40	26	85	32	1	25	0	14	64	23	112	0			
1945	0	53	0	0	0	71	0	16	48	1	0	0	0	6	17	0
	0	0	0	0	19	107	45	7	0	19	0	2	121	29	21	9
	12	18	4	61	37	59	49	104	19	30	0	127	58	9	11	2
	4	135	84	6	22	204	4	6	7	0	102	0	20			
1946	6	0	118	29	4	14	3	8	274	2	7	39	127	0	4	1
	4	8	88	38	5	0	168	0	36	73	106	29	11	2	36	13
	84	79	98	36	0	88	30	32	3	4	66	20	39	4	1	51
	9	47	0	7	40	28	1	49	45	3	5	110	27			

Table 15.1 *(cont.)*

Year	Rainfall readings in six-day totals, mm.															
1947	6	6	0	2	7	2	66	6	3	13	2	44	9	0	168	15
	68	0	51	37	0	18	1	20	73	12	57	4	24	93	57	74
	81	27	60	26	8	121	51	96	116	10	34	22	4	110	37	15
	108	26	3	82	6	2	2	35	4	19	31	115	0			
1948	13	1	0	0	0	1	0	0	5	16	26	1	0	0	0	0
	239	129	94	4	125	36	72	1	62	22	83	0	63	26	10	22
	2	33	59	31	0	18	79	113	51	24	0	0	9	11	1	205
	12	0	214	79	34	5	0	0	0	0	62	30	17			
1949	4	0	15	0	0	5	5	4	110	131	0	0	0	0	5	1
	10	0	11	0	79	4	74	45	3	43	11	1	59	8	2	1
	32	95	25	80	41	13	50	37	37	14	11	1	19	3	167	244
	20	15	96	53	30	16	71	2	0	1	5	0	14			
1950	0	4	0	0	0	86	0	0	1	1	1	0	0	0	23	0
	12	0	0	0	1	1	50	0	146	307	16	0	0	95	38	36
	0	18	12	28	70	4	63	1	72	0	4	75	21	22	77	41
	39	83	5	62	2	11	0	22	0	3	4	42	7			
1951	39	42	0	0	0	0	0	0	35	0	13	1	0	0	0	45
	37	72	61	22	3	102	190	37	114	1	95	107	15	61	18	177
	149	112	19	52	109	25	32	75	14	3	0	1	7	16	0	126
	24	201	0	25	20	2	1	6	0	105	51	39	44			
1952	12	0	0	32	93	0	17	7	0	0	0	4	3	0	0	0
	14	101	31	117	2	57	98	139	75	127	17	50	34	37	31	8
	65	11	0	14	41	26	16	65	1	6	10	51	8	42	1	22
	59	3	67	18	35	96	21	161	40	10	0	0	5			
1953	38	0	0	0	59	0	29	0	0	0	0	0	3	15	0	49
	0	0	39	37	0	0	18	90	2	3	69	146	101	64	20	146
	102	35	26	4	65	85	60	8	51	69	10	56	1	4	80	0
	40	23	4	93	10	20	6	22	133	36	0	11	19			
1954	0	0	0	0	70	0	0	0	0	0	0	0	13	34	0	0
	33	245	0	174	22	7	3	36	2	78	74	37	8	1	78	56
	31	24	43	6	61	35	2	26	46	15	6	29	1	0	54	0
	23	79	7	18	19	4	24	0	24	126	0	0	0			
1955	3	0	0	0	0	16	176	58	0	0	4	0	0	8	0	5
	0	119	13	18	47	38	118	27	147	177	4	165	133	72	3	18
	89	2	49	31	77	63	34	106	68	11	0	12	55	124	22	16
	12	3	75	35	65	38	28	16	44	1	1	0	12			
1956	22	0	13	115	0	0	19	0	0	0	107	2	0	15	0	19
	118	156	85	81	0	5	149	98	116	76	105	52	71	187	122	93
	42	46	49	36	9	69	30	64	59	48	76	25	56	15	6	22
	76	5	42	20	8	24	9	0	10	19	22	0	13			
1957	0	0	0	0	0	0	2	0	0	0	0	19	57	0	0	1
	37	5	30	10	114	0	13	56	0	1	0	0	237	5	35	118
	37	162	0	86	71	10	25	10	13	65	13	63	6	24	20	19
	3	133	6	43	8	71	6	5	2	1	0	21	12			
1958	0	0	0	2	20	7	0	0	17	0	0	28	0	2	73	0
	3	60	0	0	33	14	133	28	158	0	13	1	2	4	4	26
	120	100	42	31	84	86	29	30	8	62	134	71	21	11	66	87
	52	37	0	9	10	4	3	7	0	14	6	5	0			

Table 15.1 *(cont.)*

Year	Rainfall readings in six-day totals, mm.															
1959	2	0	0	0	27	0	6	76	0	1	4	2	0	0	14	87
	0	19	0	2	3	0	0	12	0	3	18	7	2	6	2	0
	12	22	108	75	45	69	0	1	0	0	16	77	1	42	30	0
	1	12	37	5	2	0	13	71	34	24	5	60	77			
1960	0	0	38	0	0	103	1	84	0	0	43	25	0	16	0	14
	6	21	60	118	67	58	190	0	97	76	35	111	1	2	20	33
	24	58	3	26	15	102	3	4	15	162	99	44	55	3	29	3
	44	0	9	5	347	16	11	0	0	0	4	0	7			
1961	15	7	0	0	4	0	28	0	1	2	0	0	0	21	7	83
	137	174	51	35	53	18	41	0	16	33	29	12	49	133	57	80
	26	1	31	21	15	10	67	14	59	13	22	34	0	2	0	20
	8	5	16	6	5	11	0	3	1	6	15	0	0			
1962	0	5	42	8	1	0	0	4	84	0	4	0	13	3	72	0
	0	1	0	8	62	13	64	127	106	90	47	13	25	19	42	0
	0	29	73	103	26	9	42	47	19	14	8	7	1	37	75	44
	94	36	1	24	4	0	25	0	19	6	0	2	198			
1963	13	4	273	1	23	1	7	2	0	0	0	0	0	0	3	0
	0	0	0	181	6	0	237	39	218	125	131	80	71	1	59	78
	95	52	87	61	32	49	16	141	6	33	68	19	5	18	0	0
	0	10	175	11	7	0	0	6	5	1	0	1	4			
1964	0	27	7	0	0	7	22	0	0	4	3	0	0	6	0	0
	88	3	129	12	95	5	6	0	92	30	0	5	71	18	79	12
	180	76	60	36	35	44	4	37	29	69	24	45	17	131	93	81
	2	10	13	123	0	265	16	23	1	2	2	53	0			
1965	0	0	0	6	3	0	0	0	0	0	8	23	6	0	0	24
	0	38	43	0	0	71	106	51	23	65	1	0	90	1	39	0
	106	21	34	19	18	76	80	23	4	59	4	41	1	11	2	13
	0	0	0	11	42	0	44	5	94	13	1	11	5			
1966	0	0	9	2	3	0	0	54	0	0	0	0	52	51	0	0
	2	5	13	6	138	46	0	29	6	98	0	99	76	24	53	22
	54	59	167	9	11	57	27	41	19	2	32	84	60	68	30	19
	6	1	70	10	36	0	0	53	200	4	28	1	13			
1967	0	0	62	0	2	0	0	11	0	132	7	0	1	3	1	25
	0	13	0	2	0	24	14	29	0	57	1	2	0	0	51	4
	14	78	9	65	29	88	59	31	9	29	12	40	0	16	31	12
	0	0	0	0	0	3	2	0	0	19	17	0	7			
1968	7	0	18	111	0	0	0	0	154	21	3	2	42	36	101	0
	0	2	73	77	172	5	4	101	161	34	94	46	76	0	59	23
	11	53	36	41	82	74	62	53	14	5	6	3	36	88	146	2
	15	56	0	135	6	22	32	3	45	0	0	124	0			
1969	0	0	16	0	15	0	334	1	60	0	0	6	33	15	7	1
	5	66	55	0	7	10	54	86	108	108	83	0	25	49	53	47
	70	101	12	67	38	15	0	27	61	77	98	7	32	0	0	0
	0	0	4	73	0	4	0	81	39	5	0	13	0			
1970	84	6	23	0	0	0	0	0	1	0	5	14	0	7	8	9
	24	0	122	71	70	31	65	50	30	53	43	42	40	52	58	66
	30	82	7	40	20	44	26	119	41	17	11	51	74	75	7	8
	8	4	4	11	9	128	2	0	99	6	1	0	3			

Table 15.1 *(cont.)*

Year	Rainfall readings in six-day totals, mm.															
1971	15	0	4	0	9	2	1	0	0	0	19	17	62	3	46	0
	0	158	135	310	153	43	8	32	114	79	53	23	43	54	2	1
	25	51	35	67	25	195	3	52	63	86	35	11	105	43	10	19
	30	25	33	134	5	64	26	0	5	66	0	18	9			
1972	13	152	0	4	0	0	0	135	51	0	0	0	5	0	0	0
	52	0	62	105	0	0	43	14	0	4	0	0	0	41	90	107
	94	46	10	33	32	94	30	0	50	106	39	5	3	0	15	41
	30	0	6	20	46	0	0	0	4	0	4	0	170			
1973	0	0	17	0	2	83	198	0	4	1	0	6	114	13	9	4
	0	37	169	24	48	44	65	15	75	69	87	121	20	22	0	0
	237	47	89	12	125	22	23	24	66	43	51	13	68	2	0	52
	248	3	21	2	72	0	0	35	37	0	48	14	61			
1974	0	0	27	0	76	339	0	0	35	0	0	0	1	5	22	17
	120	4	25	48	24	109	94	41	8	14	17	0	42	23	75	179
	49	72	38	72	17	57	22	29	50	18	31	74	69	62	138	7
	154	120	24	1	14	0	0	0	0	11	8	9	28			
1975	6	29	6	0	68	0	0	2	23	5	18	0	50	163	64	72
	26	0	0	3	8	142	108	3	90	43	2	5	0	21	2	57
	117	51	3	157	43	20	14	40	0	23	18	73	25	64	114	38
	31	72	54	38	9	1	17	0	13	6	2	0	1			
1976	0	50	0	5	1	0	7	0	41	204	2	0	0	2	1	6
	0	66	0	29	0	0	20	36	38	65	41	34	2	28	69	16
	3	0	0	47	12	42	8	79	1	2	35	14	12	39	129	8
	21	16	0	66	2	6	8	17	19	6	31	58	0			
1977	6	0	195	14	6	0	0	13	28	0	0	0	21	9	92	8
	72	8	13	0	0	63	2	35	24	80	6	36	51	31	30	1
	2	2	75	38	15	17	9	6	43	35	119	0	0	61	15	53
	49	6	31	0	46	0	1	47	36	0	4	9	8			

16. Particle Size Distribution of Soil Profiles

Source Nielsen, D.R., Biggar, J.W. and Erh, K.T. (1973). Spatial variability of field-measured soil-water properties. *Hilgardia* **42**, 215-259.

Contributor D.R. Nielsen University of California, Davis

The experiment to determine particle size distribution of soil profiles was conducted at the West Side Field Station of the University of California, located in Fresno County, forty miles southwest of Fresno. Fresno County is in the southernmost quarter of the central valley of California, which is an elongated trough paralleling the eastern and western boundaries of the state. The valley is 500 miles long in a north-south direction and averages about 40 miles in width. The valley is surrounded by mountains except for the outlet into San Francisco Bay through which the valley rivers drain.

The West Side Field Station is on an alluvial fan of Panoche soil series. Panoche soils have uniform profiles but a wide range of textures. They are light brownish, grey, calcareous, friable, and permeable throughout. The source of this soil is principally the softly consolidated calcareous and gypsiferous sandstone and shale on the eastern slope of the Coast Range. They are generally free of alkali or only slightly affected.

In order to determine the soil profile in the Panoche soil, twenty 6.5 - meter-square plots were randomly established over a 150 hectare site at the West Side Field Station; see Figure 16.1. Table 16.1 gives the values of the per cent sand, silt, and clay for twelve depths of each plot.

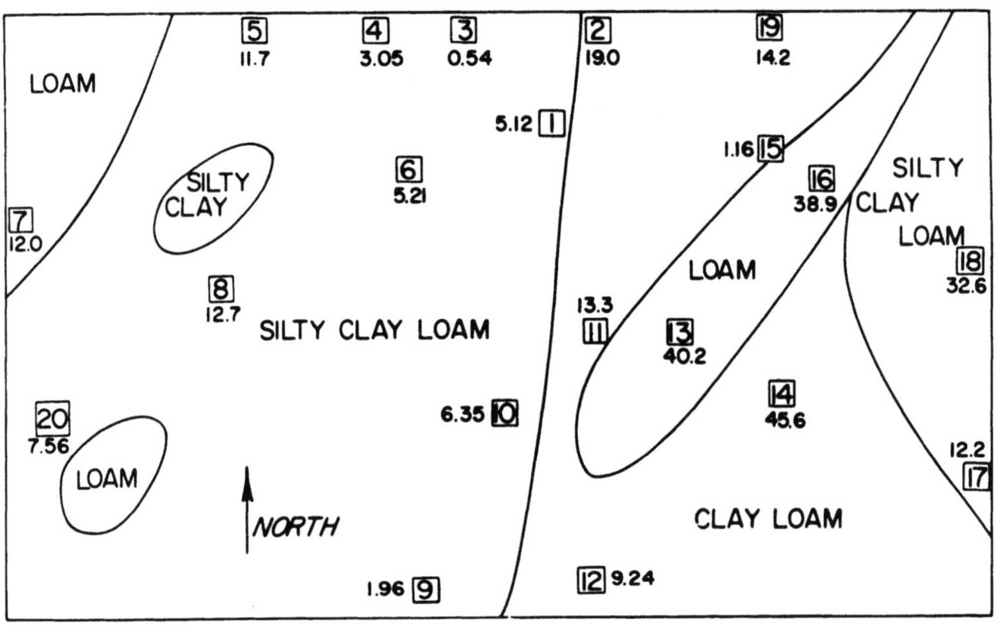

Figure 16.1
Field Site Showing Locations of the Twenty Plots

The number by each plot indicates the measured value of the steady-state infiltration (cm. / day). Textures at the soil surface of the Panoche soil are also indicated.

Table 16.1
Values of Per Cent of Sand, Silt and Clay at Twelve Depths for Twenty Plots *

	Plot 1			Plot 2			Plot 3			Plot 4		
Depth	Sand	Silt	Clay	Sand	Silt	Clay	Sand	Silt	Clay	Sand	Silt	Clay
1	27.3	25.3	47.4	40.29	20.35	39.36	12.7	30.3	57.0	7.86	27.89	64.24
2	31.2	25.4	43.4	43.67	19.38	36.95	11.6	31.6	56.8	7.06	25.56	67.38
3	25.0	26.6	48.4	43.88	22.47	33.65	9.9	33.3	56.8	8.28	32.25	68.48
4	30.1	27.4	42.6	50.12	18.14	31.74	19.3	35.1	45.6	2.96	29.68	67.36
5	33.6	25.1	41.4	53.07	15.56	31.37	18.4	33.1	48.5	12.36	33.68	53.97
6	34.9	24.2	40.7	41.99	19.79	38.22	25.7	25.4	49.0	8.04	26.56	64.44
7	25.2	28.4	46.4	31.42	23.58	45.00	15.0	30.0	55.0	12.18	28.66	59.16
8	23.6	31.4	44.8	27.40	25.60	47.00	13.2	31.5	55.3	1.01	39.27	59.72
9	33.4	30.3	36.5	18.08	31.46	50.46	16.1	28.9	55.0	3.50	29.20	67.30
10	36.4	28.3	35.7	26.90	27.04	46.06	13.8	31.7	54.5	0.46	24.42	75.12
11	36.4	29.8	33.7	27.92	30.72	41.36	6.9	45.1	44.3	6.06	27.86	66.56
12	20.7	36.7	42.6	44.96	25.28	29.76	13.1	37.6	49.3	22.05	30.82	47.12

	Plot 5			Plot 6			Plot 7			Plot 8		
Depth	Sand	Silt	Clay	Sand	Silt	Clay	Sand	Silt	Clay	Sand	Silt	Clay
1	16.1	24.2	59.7	10.4	27.8	61.8	19.0	33.5	47.5	15.5	34.4	50.2
2	15.6	27.2	57.2	10.9	31.0	58.1	27.0	31.1	42.0	19.9	28.3	51.8
3	14.2	26.8	59.0	12.0	33.9	54.1	30.0	29.6	40.4	14.2	33.5	51.8
4	6.8	23.9	69.3	8.7	32.2	59.1	24.3	34.5	41.2	14.8	34.9	50.3
5	9.8	47.9	42.3	23.9	21.2	54.6	33.2	29.6	37.2	20.8	31.9	47.3
6	14.3	30.4	55.3	18.3	27.6	54.1	27.5	37.6	34.9	11.9	38.8	49.2
7	12.3	34.4	53.3	16.6	28.0	55.4	24.2	38.6	37.2	8.6	39.2	52.2
8	12.7	32.8	54.5	13.6	29.4	57.0	18.0	41.8	40.2	4.7	39.0	56.3
9	14.2	33.6	52.2	2.8	28.6	68.6	16.2	39.6	44.2	4.9	40.7	54.4
10	11.9	32.6	55.5	9.3	24.8	65.9	3.8	34.1	61.9	2.6	36.6	60.8
11	5.6	36.2	58.2	22.5	27.3	50.2	1.0	39.1	59.9	3.9	42.8	52.3
12	5.6	33.4	61.0	8.2	34.4	57.4	0	30.1	69.9	4.4	40.8	54.8

	Plot 9			Plot 10			Plot 11			Plot 12		
Depth	Sand	Silt	Clay	Sand	Silt	Clay	Sand	Silt	Clay	Sand	Silt	Clay
1	21.4	27.8	50.8	19.4	25.1	55.5	39.4	25.5	35.6	32.3	32.7	35.0
2	22.2	30.8	47.0	19.5	27.5	53.0	39.8	23.6	36.6	34.5	21.2	44.3
3	20.5	31.1	48.4	20.5	33.0	46.5	35.9	25.5	38.6	36.4	20.3	43.3
4	28.3	29.4	42.3	24.0	27.5	48.5	35.1	25.6	38.8	33.9	21.5	44.6
5	25.6	32.6	41.8	15.3	32.7	52.0	40.9	24.8	34.3	30.7	24.6	45.2
6	18.2	37.0	44.8	15.4	35.7	48.9	42.6	23.6	33.8	20.6	28.6	50.8
7	20.2	30.3	49.3	9.1	39.4	51.5	41.4	25.8	32.8	13.4	35.8	50.8
8	19.8	36.5	43.8	12.1	34.9	53.0	36.3	26.9	36.8	19.8	30.4	48.3
9	18.7	35.5	45.8	13.0	36.8	50.0	34.6	28.6	36.6	24.6	30.1	45.3
10	30.2	40.0	29.8	16.8	29.4	53.7	27.1	35.1	36.8	34.1	28.1	37.8
11	29.5	36.0	34.5	13.5	31.5	55.0	33.1	34.6	32.3	33.6	29.1	37.3
12	18.9	36.1	45.0	3.2	44.4	52.4	38.4	32.5	29.1	26.7	37.7	35.6

Table 16.1 *(cont.)*

	Plot 13			Plot 14			Plot 15			Plot 16		
Depth	Sand	Silt	Clay	Sand	Silt	Clay	Sand	Silt	Clay	Sand	Silt	Clay
1	35.74	25.00	39.25	35.2	19.0	45.8	37.8	21.3	40.9	30.4	28.7	40.9
2	44.20	18.50	37.30	37.7	21.0	41.3	41.7	19.4	38.9	30.2	25.0	44.8
3	42.30	19.56	38.14	36.7	21.0	42.3	41.8	19.3	38.9	22.8	29.9	47.3
4	38.11	22.06	39.83	35.1	22.1	42.8	32.0	24.6	43.4	23.4	32.5	44.1
5	39.60	22.40	38.01	35.7	22.5	41.8	33.5	23.9	42.6	32.6	30.1	37.3
6	42.48	20.10	37.42	32.5	27.0	40.5	44.2	19.1	36.7	30.2	32.0	37.8
7	46.26	20.41	33.33	30.4	28.1	41.5	43.5	21.4	35.1	35.2	28.6	36.2
8	48.70	19.37	31.94	28.8	30.1	41.1	31.9	28.9	39.2	33.3	32.6	34.1
9	49.96	17.43	32.61	20.2	35.3	44.5	40.6	27.2	32.2	29.9	28.8	41.3
10	54.96	15.26	29.78	23.1	35.9	41.0	48.0	22.4	29.6	21.9	41.0	37.1
11	58.44	12.07	29.50	19.2	43.3	37.5	46.8	24.5	28.7	13.1	47.3	39.6
12	60.65	12.95	26.40	20.5	42.5	37.0	52.0	21.2	26.8	11.1	45.1	43.8

	Plot 17			Plot 18			Plot 19			Plot 20		
Depth	Sand	Silt	Clay	Sand	Silt	Clay	Sand	Silt	Clay	Sand	Silt	Clay
1	40.3	16.1	43.6	27.0	28.2	44.8	32.8	18.0	49.2	26.2	26.1	47.7
2	40.0	16.0	43.9	35.9	26.0	38.1	29.3	18.6	52.1	21.8	27.6	50.6
3	35.7	17.2	47.0	28.2	25.2	46.6	36.2	18.4	45.3	27.1	23.6	49.2
4	32.2	19.8	48.0	30.9	28.5	40.6	39.4	18.8	41.8	38.1	22.7	39.1
5	28.8	23.0	48.2	35.9	33.3	30.8	24.8	26.3	48.9	32.5	30.9	36.6
6	34.9	20.8	44.2	37.9	30.3	31.8	23.2	26.3	50.5	29.5	34.9	35.6
7	34.6	19.9	45.6	36.9	34.0	29.1	18.6	28.0	53.4	20.4	40.5	39.1
8	41.8	17.2	41.1	33.1	38.6	28.3	21.6	30.8	47.6	2.6	39.2	58.2
9	41.8	18.7	39.5	39.6	31.7	28.7	0.0	35.1	64.9	0.8	51.3	47.9
10	44.0	18.6	37.3	33.4	39.0	27.6	27.2	31.6	41.2	3.1	48.1	48.8
11	33.6	29.6	36.8	18.9	53.8	27.3	37.3	24.0	38.6	8.4	46.2	45.4
12	5.4	44.0	50.6	8.9	57.8	32.8	33.2	26.8	40.0	13.2	34.8	52.0

* Depth number refers to the intervals as follows
1, 0 to 15.2 cm.;
2, 15.2 to 30.5 cm.;
3, 30.5 to 45.7 cm,;
etc.

17. Identifying Groundwater Populations

Source Nichols, C.E., Kane V.E., Browning, M.T. and Cagle, G.W. (1976). Northwest Texas Pilot Geochemical Survey, Union Carbide, Nuclear Division Technical Report (K/UR-1).

Contributor V.E. Kane Union Carbide Corporation

The U.S. Department of Energy is directing a National Uranium Resource Evaluation Program in which an estimate of the U.S. uranium reserves will be obtained. A fundamental problem is identifying the populations in a group of samples so that background samples may be separated from samples anomalous in uranium and other elements. These samples represent a typical group of 127 groundwater samples, each having 12 measurements. The data are given in Table 17.1. An initial classification of the populations is given by the Groundwater Producing Horizon Code.

Which populations are comparable, and what samples are misclassified? Applicable statistical procedures may include: discriminant analysis, multivariate data transformations, test of normality and cluster analysis. Some data problems include censored measurements and a high correlation between several variables. Additionally, regression methods may be appropriate for determining what group of variables is related to uranium variability. Geographic analysis methods of trend surface analysis, contour plotting and Kriging may also be relevant.

Table 17.1
Northwest Texas Groundwater Survey from the U.S. Department of Energy Uranium Resource Evaluation Project *

Sample number	Latitude	Longitude	P_GC	U	AS	B	BA	MO	SE	V	SO4	T_AK	BC	CT	PH
2201	33.21	101.445	TPO	7.99	17.6	300	150	-4	0.4	100	35	278	157	640	7.60
2203	33.213	101.494	TPO	13.74	10.4	660	99	7	0.3	66	90	299	175	980	7.70
2205	33.148	100.537	PGWC	4.85	13.5	883	13	5	0.2	26	2575	42	22	6200	7.40
2210	33.164	100.608	PGWC	3.1	4.0	625	750	-4	-0.2	25	-30	175	101	540	7.65
2211	33.208	101.574	TPO	78	19.9	3125	100	150	0.2	500	1100	324	174	2700	7.10
2213	33.164	101.69	TPO	9.74	16.0	528	40	7	1.5	66	200	280	175	980	8.20
2215	33.168	101.747	TPO	6.9	12.0	600	200	40	0.4	100	435	300	166	1700	7.40
2217	33.17	101.506	TPO	21.73	12.2	5000	50	150	0.4	500	795	340	188	2200	7.50
2219	33.171	101.571	TPO	26.79	11.4	2000	40	80	0.2	150	1080	278	148	2600	7.20
2221	33.169	101.616	TPO	56.2	12.7	800	100	10	0.2	100	450	300	166	1700	7.35
2224	33.165	101.867	TPO	25.3	3.0	2000	50	25	-0.2	50	1950	200	109	4500	7.20
2226	33.166	102.051	TPO	4.42	10.3	300	200	-4	0.4	10	40	316	182	670	7.60
2228	33.169	101.986	TPO	29.75	21.4	564	49	7	0.2	111	975	280	170	2000	7.40
2230	33.253	101.501	TPO	22.32	19.4	1155	66	40	0.2	396	490	304	171	1300	7.40
2232	33.305	101.502	TPO	9.48	9.0	300	50	5	0.5	100	190	240	138	1700	7.30
2235	33.31	101.553	TPO	13.46	6.5	990	132	26	1.9	132	95	241	143	1400	7.80
2237	33.314	101.627	TPO	29.56	10.1	1500	200	60	0.2	200	350	300	151	1800	7.20
2239	33.312	101.688	TPO	13.39	8.7	660	40	13	0.2	99	270	308	167	1600	7.30
2241	33.308	101.757	TPO	20.96	9.7	2000	60	20	-0.2	150	440	292	162	1400	7.40
2243	33.314	101.809	TPO	26.67	6.4	990	99	7	-0.2	66	1220	292	153	2700	7.20
2245	33.314	101.926	TPO	52.47	9.7	2000	50	75	0.4	100	250	454	256	1500	7.60
2248	33.311	101.869	TPO	6.49	63.0	1500	150	200	2.2	600	100	332	198	1000	7.75
2250	33.312	101.978	TPO	15.78	15.5	1500	75	20	7.9	200	115	400	217	1100	7.30
2253	33.271	101.805	TPO	21.19	10.7	2000	100	200	0.7	200	435	312	178	1700	7.30
2255	33.167	101.45	TPO	13.16	18.2	559	33	9	0.5	72	115	300	179	1100	7.70
2256	33.152	101.798	TPO	12.33	7.5	744	372	12	5.1	124	325	199	108	1600	7.20
2258	33.068	101.859	TPO	5.73	6.1	500	100	5	0.4	75	60	310	176	800	7.50
2260	33.123	101.877	TPO	11.07	6.7	1000	150	10	-0.2	200	315	262	143	1500	7.30
2262	33.171	101.932	TPO	16.08	6.4	660	13	7	-0.2	66	720	282	162	2600	7.50
2264	33.168	101.921	TPO	-0.2	0.8	4980	-4	33	0.2	-4	1340	314	172	5600	7.70
2267	33.282	101.912	TPO	-0.2	73.5	2000	150	60	0.2	500	85	320	190	1100	8.00
2270	33.263	102.001	TPO	8.62	12.0	1250	75	50	2.8	250	155	345	190	1200	7.50
2272	33.123	101.915	TPO	11.43	28.0	1155	53	66	1.1	264	780	446	237	2000	7.00
2273	33.118	102.002	TPO	17.96	12.6	1750	60	40	0.6	200	1040	266	139	2400	7.10
2275	33.215	101.98	TPO	15.52	9.4	1500	50	100	0.8	200	120	420	219	1200	7.20
2276	33.228	101.805	TPO	21.49	6.2	837	36	7	0.8	45	1220	292	142	3200	7.05
2278	33.281	101.854	TPO	9.46	15.3	1155	66	66	6.4	132	440	328	190	1800	7.50
2281	33.217	100.588	PGWC	2.05	7.3	15000	-4	-4	-0.2	-4	3525	36	21	6700	7.60
2282	33.211	101.738	TPO	10.36	10.7	1750	20	20	6.0	200	1500	260	146	2200	7.30
2283	33.226	101.677	TPO	5.33	15.9	750	100	10	0.6	200	310	337	193	1400	7.20
2285	33.271	101.663	TPO	11.16	5.8	575	40	-4	1.7	34	780	220	122	1800	7.30
2287	33.272	101.622	TPO	0.87	1.0	7500	50	250	0.6	-4	525	350	215	2200	8.10
2289	33.258	101.559	TPO	7.8	8.6	660	132	13	1.9	132	60	270	142	1100	7.30
2321	33.443	101.339	TRD	6.69	1.3	281	112	-4	12.1	5	210	195	106	1800	7.20
2323	33.407	101.288	TRD	21.86	13.7	3000	200	80	0.5	200	60	730	438	1500	8.10
2325	33.416	101.261	TRD	20.28	2.8	1000	60	20	-0.2	40	780	300	168	3500	7.40
2330	33.443	101.203	TRD	13.68	2.4	642	23	43	0.3	-4	140	412	230	1200	7.60
2332	33.411	101.145	TRD	2.91	1.4	2170	74	-4	1.7	124	630	276	126	2900	7.00
2333	33.426	101.005	POQ	125.2	2.9	600	40	200	0.4	20	880	404	199	2300	7.00
2345	33.436	101.355	TRD	58.27	13.1	7500	200	300	-0.2	500	860	480	262	5600	7.30

Table 17.1 *(cont.)*

Sample number	Lati-tude	Longi-tude	P_GC	U	AS	B	BA	MO	SE	V	SO4	T_AK	BC	CT	PH
2353	33.291	101.44	TRD	83.44	15.3	2617	7	53	-0.2	88	2000	394	209	5500	7.30
2359	33.774	101.256	TPO	8.9	9.2	204	102	-4	-0.2	136	45	298	159	610	7.10
2361	33.797	101.244	TPO	11.98	11.7	340	68	7	2.8	102	35	326	175	720	7.40
2365	33.68	101.172	TRD	6.72	4.5	340	136	10	-0.2	68	35	276	152	610	7.30
2379	33.625	100.815	TRD	9.86	-0.5	164	58	-4	-0.2	5	170	232	129	1100	7.20
2383	33.641	100.765	POQ	15.12	1.0	996	-4	16	-0.2	32	1250	166	105	2500	7.80
2384	33.682	100.763	POQ	70.39	-0.5	830	-4	-4	-0.2	32	2250	300	181	5200	7.70
2385	33.656	100.764	POQ	21.27	-0.5	4000	-4	50	-0.2	50	2125	204	113	4300	7.20
2386	33.475	100.972	POQ	58.2	2.0	119	166	-4	-0.2	166	30	310	190	620	8.10
2390	33.347	100.935	POQ	25.04	-0.5	756	13	25	-0.2	12	2100	228	121	3600	7.30
2393	33.357	100.903	POQ	5.54	1.2	169	8	-4	-0.2	18	1800	234	119	2200	7.10
2395	33.314	100.926	POQ	14.04	0.8	2000	-4	50	-0.2	150	1325	124	81	2800	8.00
2401	33.675	100.718	POQ	6.02	1.0	1250	750	25	0.3	50	380	354	196	1200	7.20
2406	33.619	100.706	POQ	11.98	-0.5	399	10	7	0.2	11	2300	140	82	3200	8.50
2409	33.621	100.903	POQ	15.3	0.5	756	180	12	0.3	25	145	402	211	1100	7.10
2411	33.598	100.976	TRD	6.98	-0.5	145	74	-4	0.3	9	90	286	159	840	7.40
2413	33.477	101.195	TRD	13.63	1.0	2490	560	32	-0.2	16	165	410	240	1200	8.20
2415	33.324	101.112	TRD	16.35	0.9	830	160	16	-0.2	-4	1040	325	202	2200	8.20
2417	33.256	101.039	POQ	19.85	-0.5	5000	-4	100	0.3	-4	2325	76	47	9000	8.10
2421	33.256	101.075	POQ	19.41	1.5	498	16	16	1.0	32	470	286	165	6400	7.70
2425	33.315	101.022	POQ	6.63	0.7	996	-4	17	-0.2	100	1775	92	54	3300	8.20
2432	33.578	101.047	TRD	2.32	0.6	2000	40	40	-0.2	-4	125	324	189	1000	8.70
2433	33.612	101.1	TRD	1.94	0.5	1660	-4	33	-0.2	-4	260	268	155	4800	7.90
2434	33.636	101.1	TRD	6.0	2.4	300	150	20	0.2	100	45	230	130	520	7.90
2437	33.637	101.115	TPO	1.52	4.4	68	510	-4	-0.2	20	40	260	143	420	7.80
2444	33.6	101.1	TRD	4.12	2.2	136	170	7	0.3	102	55	242	136	490	7.50
2447	33.59	100.829	POQ	33.98	0.7	1000	10	20	-0.2	40	840	260	141	2700	7.30
2449	33.571	100.878	POQ	17.64	1.3	300	20	10	10.3	10	495	352	168	2300	7.00
2450	33.564	101.009	POQ	18.63	2.9	300	100	20	0.7	30	60	264	157	780	7.60
2451	33.596	100.995	TRD	8.02	-0.5	204	68	7	0.5	14	65	260	154	700	7.90
2453	33.604	101.033	TRD	7.91	-0.5	1000	200	60	0.8	10	50	278	147	580	8.80
2459	33.532	100.939	POQ	56.86	1.4	1000	60	60	0.2	60	465	400	220	2200	7.70
2470	33.393	100.786	POQ	53.74	0.7	250	-4	-4	-0.2	25	2425	406	196	3900	7.00
2473	33.267	100.753	PGWC	8.25	3.6	500	10	-4	0.2	60	1950	142	73	2500	7.10
2479	33.519	100.774	POQ	33.46	-0.5	562	6	-4	-0.2	21	1845	278	138	3100	7.20
2480	33.475	100.715	POQ	38.2	0.9	1860	-4	12	-0.2	50	3210	240	133	5600	7.20
2485	33.4	100.706	PGWC	2.82	2.5	520	39	-4	0.2	10	795	104	62	1400	7.50
2489	33.268	100.675	PGWC	4.16	1.3	496	-4	-4	-0.2	25	2055	90	53	2700	7.90
2497	33.544	100.584	PGWC	18.66	1.8	566	7	-4	0.2	11	2235	98	56	2800	8.10
3406	33.302	100.542	PGWC	12.72	-0.5	719	13	12	-0.2	29	1605	170	101	2700	7.50
3407	33.239	100.319	PGEB	3.82	-0.5	1328	-4	-4	-0.2	17	990	186	113	3900	7.70
3408	33.233	100.566	PGWC	8.75	-0.5	1240	-4	-4	-0.2	-4	1900	40	24	3100	7.70
3410	33.226	100.532	PGWC	2.29	-0.5	1660	-4	-4	-0.2	-4	1850	92	46	2800	7.10
3412	33.241	100.508	PGWC	7.22	1.3	2500	-4	-4	-0.2	50	2200	80	47	2800	7.50
3416	33.259	100.443	PGWC	9.76	-0.5	5000	-4	-4	-0.2	-4	2350	46	22	13000	7.20
3421	33.291	100.431	PGWC	7.72	3.5	1241	9	11	-0.2	24	2350	74	43	5300	7.20
3436	33.358	100.475	PGWC	27.38	5.8	7500	-4	100	0.4	200	2100	65	35	5000	8.40
3437	33.322	100.435	PGWC	5.14	3.3	664	17	17	-0.2	166	1900	214	111	2800	7.00
3453	33.319	100.326	PGEB	7.92	1.7	1000	40	-4	-0.2	60	500	226	125	2200	7.40
3455	33.358	100.317	PGEB	11.12	2.9	1500	20	10	-0.2	40	580	234	144	1700	7.90
3459	33.428	100.36	PGEB	24.72	1.4	2500	-4	25	-0.2	50	2600	146	76	4100	7.30
3462	33.355	100.342	PGEB	9.44	-0.5	1000	10	10	-0.2	20	900	226	121	2000	7.20

Table 17.1 *(cont.)*

Sample number	Lati- tude	Longi- tude	P_GC	U	AS	B	BA	MO	SE	V	SO4	T_AK	BC	CT	PH
3465	33.424	100.401	PGWC	47.78	4.6	7500	-4	-4	-0.2	25	3600	252	139	4500	7.40
3480	33.416	100.436	PGEB	35.3	5.4	2346	15	19	7.4	50	3900	82	48	5200	8.50
3482	33.404	100.424	PGEB	18.06	3.0	1000	-4	10	-0.2	80	3000	82	48	2900	7.80
3493	33.486	100.459	PGWC	35.92	3.1	1860	-4	12	-0.2	50	1900	136	76	2900	7.70
3495	33.282	100.318	PGEB	3.2	-0.5	500	40	-4	-0.2	5	225	302	174	1400	7.30
3497	33.274	100.264	PGEB	1.22	-0.5	830	133	-4	1.7	33	275	280	159	1200	7.20
3498	33.311	100.253	PGEB	7.15	-0.5	600	20	-4	-0.2	10	1150	340	169	3000	7.00
3500	33.34	100.268	PGEB	9.99	1.3	578	13	8	-0.2	15	2550	141	81	3000	7.50
3504	33.347	100.283	PGEB	6.08	-0.5	1500	20	-4	-0.2	10	1050	260	134	3000	7.10
3506	33.319	100.353	PGEB	10.5	2.6	100	-4	-4	-0.2	-4	2500	200	109	3000	7.60
3508	33.283	100.367	PGWC	21.36	1.4	500	60	-4	-0.2	10	620	248	147	1500	7.80
3511	33.218	100.268	PGEB	20.46	-0.5	1100	6	7	-0.2	9	2900	284	148	5000	7.00
3515	33.171	100.365	PGWC	4.97	2.3	6250	50	100	-0.2	50	2700	44	24	5000	8.20
3518	33.213	100.389	PGWC	0.74	1.0	248	99	-4	-0.2	25	310	212	103	1500	7.10
3520	33.21	100.482	PGWC	5.96	5.4	400	75	5	-0.2	50	460	160	94	1000	7.40
3521	33.178	100.444	PGWC	5.04	0.8	10	-4	-4	-0.2	-4	2000	186	96	2000	7.20
3522	33.141	100.392	PGWC	14.98	0.5	620	-4	-4	-0.2	25	3000	260	117	3800	6.70
3526	33.538	100.442	PGWC	6.09	1.6	830	-4	-4	-0.2	66	1850	210	110	2600	7.10
3528	33.536	100.425	PGWC	11.05	2.1	664	100	-4	-0.2	66	2500	156	87	2500	7.30
3529	33.49	100.437	PGWC	5.31	1.8	4980	-4	66	-0.2	17	2850	46	29	3400	7.70
3543	33.561	100.416	PGWC	13.36	-0.5	164	7	27	-0.2	59	2050	176	104	3200	7.50
3545	33.495	100.479	PGWC	22.27	1.4	1500	-4	-4	-0.2	50	2650	96	52	2700	7.40
3551	33.529	100.384	PGWC	28.31	3.4	996	-4	17	-0.2	33	3600	120	68	6500	7.30
3557	33.51	100.412	PGEB	21.95	-0.5	2170	-4	25	-0.2	25	2750	48	29	3000	7.80
3561	33.531	100.351	PGEB	20.87	0.5	4000	-4	-4	-0.2	-4	4000	352	182	14000	7.10

* Variable	Code (Units)
Producing Horizon	P_GC[(a)]
Uranium	U (ppb)[(b)]
Arsenic	AS (ppb)
Boron	B (ppb)
Barium	BA (ppb)
Molybdenum	MO (ppb)
Selenium	SE (ppb)
Vanadium	V (ppb)
Sulfate	S04 (ppm)
Total Alkalinity	T_AK (ppm)
Bicarbonate	BC (ppm)
Conductivity	CT (μmhos/cm)
ph	PH

(a)

Groundwater Producing Horizon Code	Geological Unit Name
TPO	Ogallala Formation
TRD	Dockum Formation
POQ	Quartermaster Group
PGWC	Whitehorse and Cloud Chief Group
PGEB	El Reno Group and Blaine Formation

(b)
Negative concentrations denote censored values (e.g., -4 means < 4).

18. Soil Data from the Province of Murcia, Spain

Source Wright, R.L. and Wilson, S.R. (1979). On the analysis of soil variability, with an example from Spain. *Geoderma* **22**, 297-313.

Contributor S.R. Wilson The Australian National University

The objective of this study was to compare and to classify soil mapping units on the basis of their lateral variabilities for a single property, or for several properties. The area investigated consisted of a sequence of eight contiguous sites extending over the crest and flank of a low rise in a valley plain underlain by marl near Albudeite in the province of Murcia, Spain; see Figure 18.1. This relatively simple situation was considered appropriate for this study's objective. The geomorphological *sites* were the primary mapping units adopted and were small areas of ground surface of uniform shape internally and delimited by relative discontinuities externally. Following the delimitation of the sites, as described in Wright and Wilson (1979), soil samples were obtained in each site at 11 random points within a 10m × 10m area centred on the mid-point of the site. All samples were taken from the same depth, chosen after due consideration of various factors affecting the area and detailed by Wright and Wilson (1979). The soil properties considered appropriate for the study's objective were the silt content and the clay content, expressed as percentages of the total silt, clay and sand content. The data are given in Table 18.1.

One approach may be to first differentiate the mapping units into sets with different variances for the selected property, or variances and covariances for more than one property, checking the assumption of normality, then using a fixed effects analysis of variance model to distinguish, within any set having the same variance, those units having different mean values, or using a multivariate analysis of variance analysis to distinguish within any set having not distinguishably different variance-covariance matrices, those units having different centroid values. The investigated mapping units form three distinctly different sets of groups according to whether the clay content, silt content, or the co-variability of silt and clay, is adopted as the classificatory criterion. The results of the analysis of variance and multivariate analysis of variance give three further sets of contrasted groups when the spatial distribution of the mapping units is taken into account.

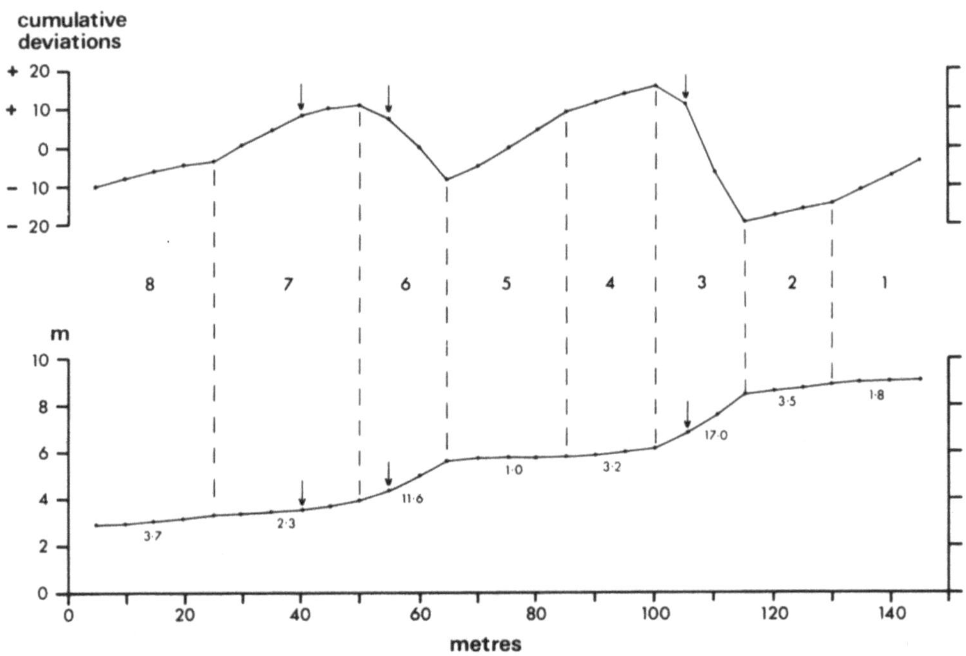

Figure 18.1
Gradient Profile for the Eight Contiguous Sites

The top graph shows cumulative deviations from the mean gradient of the whole profile. Arcs in the graph represent trends of relatively constant gradient, or gradational change, and therefore correspond to the geomorphological sites, as defined. Inflections between arcs reflect relative discontinuities in the rate of change of gradient, and thus identify boundaries between sites. The arrows indicate slope irregularities within sites. Slope profile, bottom graph, for the eight sites, 1-8 with mean gradient values in percent, vertical exaggeration × 2.5. Dots indicate survey stations.

Table 18.1
Silt and Clay Content
for Eight Contiguous Sites*

Site no.	ST †											
1	1	46.2	36.0	47.3	40.8	30.9	34.9	39.8	48.1	35.6	48.8	45.2
	2	30.3	27.6	40.9	32.2	33.7	26.6	26.1	34.2	25.4	35.4	48.7
2	1	40.0	48.9	48.7	44.5	30.3	40.1	46.4	42.3	34.0	41.9	34.1
	2	35.9	32.8	36.5	37.7	34.3	35.1	36.2	37.7	28.4	28.4	39.8
3	1	41.9	40.7	44.0	40.7	32.3	37.0	44.3	41.8	41.4	41.5	29.7
	2	34.0	36.6	40.0	30.1	38.6	30.0	28.7	34.5	34.6	34.7	32.8
4	1	41.1	40.4	39.9	41.1	31.9	43.0	42.0	40.3	42.2	50.7	33.4
	2	48.3	49.6	40.4	43.0	49.0	49.1	39.6	44.5	42.2	38.2	59.6
5	1	48.6	50.2	51.2	47.0	42.8	46.6	46.7	48.3	47.1	48.8	38.3
	2	44.3	45.1	44.4	44.7	52.1	42.0	41.8	46.0	46.3	37.1	54.9
6	1	43.7	41.0	44.4	44.6	35.7	50.3	44.5	42.5	48.6	48.5	35.8
	2	37.0	31.3	34.1	29.7	39.1	36.1	32.4	33.4	38.3	36.7	39.9
7	1	47.0	46.4	46.3	47.1	36.8	54.6	43.0	43.7	43.7	45.1	36.1
	2	38.3	35.4	42.6	38.3	45.4	34.0	44.6	38.6	41.9	44.0	54.3
8	1	48.0	47.9	49.9	48.2	40.6	49.5	46.4	47.7	48.9	47.0	37.1
	2	40.1	38.6	38.1	39.8	46.0	39.4	41.6	41.4	39.2	39.5	52.2

* Near Albudeite (latitude 38°2′N, longitude 1°25′W)
† ST; 1 = Silt, 2 = Clay

19. Lamoka Lake Site Determinations

Source Long, A. and Rippeteau, B. (1974). Testing contemporaneity and averaging radiocarbon dates. *Amer. Antiquity* **39**, 205-215. [Permission obtained from the publisher.]

Contributor S.R. Wilson The Australian National University

A radiocarbon age (RC) determination is usually presented by the radiocarbon dating laboratory in the form A$\pm$E, where A is the estimate of the radiocarbon age, bp, and E is the standard deviation due to counting error. The counting error is taken to be normally distributed. If one is making inferences in real time from different samples then it is necessary to take into account the conversion of conventional radiocarbon dates (5568 half-life) denoted bp, to calendar or tree-ring dates, denoted BP. However, if one has two or more radiocarbon ages bp from the same sample then to check their agreement with each other it is not necessary to consider this conversion. Such considerations are often ignored in the evaluation of a series of dates. For example, Long and Rippeteau (1974) considered the 8 dates presented in Table 19.1. This gives us seven dates, one for each of the seven different samples. Thus one can check whether these estimates may be dating the same event in real time, after taking into account the calibration error due to the need to convert from years bp to BP . Ward and Wilson (1978) have re-analyzed these dates, under the assumption that the calibration error, in the years bp, is normally distributed and independent of the counting error. They found no statistical evidence to doubt the consistency of C-288 and M-26, and therefore combined these two dates. After analyzing the seven samples, and making the assumption concerning the calibration error, they found some borderline evidence that C-367 may be aberrant.

References

Clark, R.M. (1975). A calibration curve for radiocarbon dates. *Antiquity* **49**, 251-266.

Ward, G.K. and Wilson, S.R. (1978). Procedures for comparing and combining radiocarbon age determinations: a critique. *Archaeometry* **20**, 19-37.

Table 19.1
Lamoka Lake Site Determinations

Sample number	Value of RC determination bp	Counting error variance as reported	Calibration error variance *
C-288	2419	200^2	
M-26	2485	200^2	Not applicable
combined C-288,M-26	2452	141^2	50^2
C-367	3433	250^2	60^2
M-195	2575	200^2	50^2
M-911	2521	150^2	50^2
M-912	2451	125^2	50^2
Y-1279	2550	80^2	50^2
Y-1280	2540	80^2	50^2

* These values are based on values recommended by Clark (1975).

20. Variations in the Earth's Rotation Rate, 1820-1970

Source Luo Shi-fang, Liang Shi-guang, Ye Shu-hua, Yan Shao-zhong and Li Yuan-xi (1977). Analysis of periodicity in the irregular rotation of the earth. *Chinese Astronomy* **1**, 221-227. Also in *Acta Astron. Sinica* **15** (1974), 79-85.

Contributor G. Tunnicliffe Wilson University of Lancaster

Variations in the rate of rotation of the earth have been measured for many years with increasing precision as new instruments became available. The quality of the primary sources of the observations used to compile these data varies considerably over the 151 year period from 1820-1970. Luo, Liang, Ye, Yan and Li (1977) take their data from Brouwer (1952) and in order to produce a plot of homogeneous appearance, different sets of smoothing have been applied over the periods 1820-1922 and 1923-1956. The data are reproduced in Table 20.1.

The resulting series appears to have a slight trend with marked variations about this trend which are of unknown cause. Luo *et al* were bold enough to claim detection of 12 different cyclical periods in the data, and the speculation on the possible coincidence with the periods of the planets, the moon and the sunspots was reported by Gribbin (1977).

It is questionable whether the data contain true cycles or have a continuous spectrum, and the challenge lies in finding the best model for a series which may have a mixed spectrum.

Another analysis claiming to find cycles in the earth's rotation is presented by Currie (1981).

References

Brouwer, D. (1952). A study of the changes in the rate of rotation of the earth. *Astron. J.* **57**, 125-146.

Currie, R.G. (1981). Solar cycle signal in earth rotation: Nonstationary behavior. *Science* **24**, 386-389.

Gribbin, J. (1977). New Chinese results tie up sun cycles and earth weather. *New Scientist* **76**, 703.

Table 20.1
Changes in the Earth's Rotation, Day Length, 1821-1970*

Year	Change	Year	Change	Year	Change	Year	Change
1821	-217	1861	-83	1901	384	1941	141
1822	-177	1862	-104	1902	415	1942	150
1823	-166	1863	-93	1903	421	1943	157
1824	-136	1864	-88	1904	402	1944	143
1825	-110	1865	-75	1905	392	1945	138
1826	-95	1866	-80	1906	387	1946	137
1827	-64	1867	-101	1907	391	1947	151
1828	-37	1868	-156	1908	396	1948	151
1829	-14	1869	-226	1909	400	1949	136
1830	-25	1870	-293	1910	391	1950	111
1831	-51	1871	-333	1911	361	1951	105
1832	-62	1872	-347	1912	328	1952	105
1833	-73	1873	-329	1913	296	1953	110
1834	-88	1874	-279	1914	282	1954	104
1835	-113	1875	-205	1915	269	1955	92
1836	-120	1876	-131	1916	256	1956	96
1837	-83	1877	-86	1917	225	1957	115
1838	-33	1878	-59	1918	202	1958	144
1839	-19	1879	-48	1919	193	1959	126
1840	21	1880	-35	1920	205	1960	131
1841	17	1881	-30	1921	201	1961	112
1842	44	1882	-12	1922	178	1962	119
1843	44	1883	11	1923	139	1963	139
1844	78	1884	57	1924	130	1964	183
1845	88	1885	92	1925	101	1965	206
1846	122	1886	86	1926	67	1966	231
1847	126	1887	53	1927	22	1967	244
1848	114	1888	26	1928	2	1968	239
1849	85	1889	6	1929	12	1969	263
1850	64	1890	-12	1930	26	1970	273
1851	55	1891	-35	1931	21		
1852	51	1892	-31	1932	10		
1853	40	1893	0	1933	-11		
1854	30	1894	36	1934	-12		
1855	14	1895	54	1935	-15		
1856	1	1896	65	1936	6		
1857	1	1897	104	1937	22		
1858	-4	1898	166	1938	51		
1859	-13	1899	248	1939	78		
1860	-56	1900	318	1940	111		

*Units are in 0.00001 of a second.

21. Association of Pairs of Radio Sources with Peculiar Galaxies

Source Arp, H. (1967). Peculiar galaxies and radio sources. *Astrophys. J.* **148**, 321-365.

Contributor V. Stagg University of Toronto

Arp (1967) argued that there exists a definite association between radio sources and peculiar galaxies of his earlier paper (Arp, 1966). He advocated that for two radio sources lying relatively close to one another, and often possessing similar radio brightness or flux strength, there exists a peculiar galaxy, approximately equidistant from each and unusually close to the line joining the pair, forming an angle of approximately 180°. This is true for particular peculiar galaxies, but can be supported as a general theory only by randomly selecting a sample area of the sky and studying the objects therein. The area was centred on the celestial equator between right ascension ("longitude") 10h and 16h, and between declination ("latitude") -16° and +16°, thus avoiding any curvature effect of the lines of right ascension. All peculiar galaxies and radio sources within the boundaries were enumerated and the distances, r_1, from each galaxy to the nearest source, and r_2, to the second nearest source, are recorded in degrees in Table 21.1. The radio brightnesses, s_1 and s_2, of the closest and second closest radio sources, respectively, are also given. Object numbers are denoted by N, ranging from 1 to 33 for peculiar galaxies, and from 34 to 56 for radio sources.

The hypothesis of no association between the radio sources and peculiar galaxies can be tested by the fit of the distribution of the observed angles, θ, formed by each radio source pair and the corresponding peculiar galaxy, to the uniform distribution on (0°,180°). These angles are given in Table 21.2. It seemed appropriate to use a chi-squared test, based on observed and expected frequencies, to test this hypothesis, specifically, H_0: $p_1 = 0.75$, $p_2 = 0.25$, where p_1 is the probability that the angles θ will fall between 0° and 135°, and p_2 is the probability they will fall between 135° and 180°.

The test may be repeated for triples of a peculiar galaxy and its two nearest radio sources for which the distance ratio, r_1/r_2, is rather close to 1, i.e. $0.580 \leqslant r_1/r_2 \leqslant 1$.

The theory of association also involves the concept of similar radio brightness of the two radio sources in a triple. The theory predicts that the ratio $\min(s_1,s_2)/\max(s_1,s_2)$ is approximately 1. The population frequency distribution of the ratio $\min s_i / \max s_i$ for all possible pairs of radio sources in the sample was calculated and the frequencies and probabilities, p_i (i=1,...,4), associated with the four intervals (0,.25], (.25,.50], (.50,.75], (.75,1], respectively, were obtained. Then, another chi-squared test was conducted to study the agreement of the distribution of the actual sample ratios, $\min(s_1,s_2)/\max(s_1,s_2)$, with the population distribution.

Finally, the distribution of the distance ratio, $v = r_1/r_2$, was looked at. Arp's belief was that v should be close to 1 for the majority of sample pairs. Two annuli, centred at a peculiar galaxy, were considered. The inner annulus had radius r_1, and r_2 was the radius to the outer annulus. The joint density function of r_1 and r_2 was calculated based on Poisson probabilities, having first found the probability of there being exactly two radio sources within the region encompassed by the annulus; one at a distance r_1 from the peculiar galaxy, and the other at r_2. Note that the area within the inner annulus is $2\pi r_1 \Delta r_1$, and within the second it is $2\pi r_2 \Delta r_2$. The joint density function in terms of r_1 and v was then found. From this, the marginal density function of $v = r_1/r_2$ was calculated. A plot of the distribution function of v, $F(v)$, against the angle θ further illuminated the situation.

References

Arp, H. (1966). Atlas of peculiar galaxies. *Astrophys. J. Suppl.* **14**, 1-20.

Bennett, A.S. (1962). The revised 3C catalogue of radio sources. *Mem. Roy. Astron. Soc.* **68**, 163-172.

Table 21.1
Distances, Flux Strengths and Angles for Each Sample Triple of a Peculiar Galaxy and its Two Nearest Radio Sources

Peculiar Galaxy	Nearest Radio Source			Second Nearest Radio Source			Angle
N	N	r_1	s_1 *	N	r_2	s_2 *	θ **
1	38	2.44174	9.0	37	6.16574	16.5	155°
2	42	4.49679	11.5	43	9.76465	12.5	120°
3	36	8.93193	20.0	35	9.58292	13.0	73°
4	45	3.87620	44.0	43	4.14237	12.5	27°
5	37	0.46529	16.5	36	4.97004	20.0	39°
6	39	3.56775	44.0	55	4.49999	970.0	125°
7	46	5.88302	10.5	45	6.28243	44.0	92°
8	37	6.62825	16.5	38	7.30228	9.0	106°
9	48	4.86427	28.0	49	6.48760	43.0	3°
10	42	2.41571	11.5	44	10.3503	9.5	175°
11	37	7.02915	16.5	38	11.0947	9.0	50°
12	47	7.61620	12.5	51	13.6702	40.0	70°
13	47	6.35220	12.5	46	10.1908	10.5	43°
14	44	3.78154	9.5	46	7.57606	10.5	6°
15	45	4.39004	44.0	43	5.66480	12.5	16°
16	40	6.95931	67.0	41	8.19752	9.5	61°
17	45	2.64569	44.0	43	4.55482	12.5	2°
18	42	4.49679	11.5	43	9.76465	12.5	120°
19	38	4.82249	9.0	36	5.95545	20.0	107°
20	36	14.2331	20.0	35	14.4627	13.0	63°
21	55	0.70711	970.0	54	0.93263	18.0	49°
22	55	0.26926	970.0	54	1.76161	18.0	4°
23	55	1.34272	970.0	54	2.67972	18.0	20°
24	55	1.86681	970.0	54	3.18760	18.0	17°
25	55	3.54158	970.0	56	4.63789	9.5	76°
26	41	8.71678	9.5	42	15.3585	11.5	67°
27	56	6.16198	9.5	55	11.8746	970.0	21°
28	56	10.7765	9.5	43	11.5823	12.5	162°
29	52	1.27283	9.5	34	10.7225	24.0	46°
30	53	7.21000	13.5	38	7.31921	9.0	86°
31	53	7.21000	13.5	38	7.32918	9.0	86°
32	53	3.19690	13.5	38	10.5987	9.0	69°
33	50	8.21913	11.0	49	9.38855	43.0	5°

* The unit of flux strength is the jansky (1 Jy = 10^{-26} watts/m^2/Hz).

** The angle θ was computed from the formula:

$$\cos\theta = \{(x_2-x_1)(x_3-x_2)+(y_2-y_1)(y_3-y_2)\}/(r_1 r_2),$$

where (x_1,y_1), (x_2,y_2), (x_3,y_3) are the celestial co-ordinates, converted into degrees of the nearest radio source, the peculiar galaxy and the second nearest radio source, respectively, for each sample triple.

22. Motions of Stars in a Star Cluster, M92

Source Cudworth, K.M. (1976). Membership and internal motions in the globular cluster M92. *Astronomical J.* **81**, 975-982.

Contributor K.M. Cudworth University of Chicago

Observational tests of theories of stellar evolution are typically based on star clusters, where many stars of essentially equal ages and at approximately the same distance from the earth can be studied and compared. In such studies, however, one must remove the contamination of foreground and background stars which are unrelated to the cluster. This is done well *via* stellar motions, since cluster members should move together whereas the non-members should show a wide variety of motions. The stars of the globular cluster, M92, are among the oldest stars in our Galaxy. The globular cluster M92, also referred to as NGC 6341, located at 17 h 15 m 6 + 43 12' (1950) is one of the most metal-poor objects in our galaxy. As such it has been the subject of numerous investigations.

Cudworth (1976) measured angular motions of stars in the neighbourhood of this cluster and found that about 134 of the stars he measured were probably cluster members. Several stars discussed in earlier studies as peculiar members were shown to be non-members. The dispersion of the motions among the probable members was used to estimate the mass of the cluster.

A method of deriving the probability that a star is a cluster member was developed by Vasilevskis, Klemola and Preston (1958). Other clustering techniques may be used.

Table 22.1 lists the star numbers in the system of Sandage and Walker (1966), Barnard (1931), Nassau (1938), Zinn, Newell and Gibson (1972), or the variable star number (Hogg, 1973). The stars are listed in numerical order rather than increasing right ascension to facilitate the use of the table. The papers referred to give identification charts for all the stars except stars C1-C41 which are given in Cudworth (1976). Table 22.1 also gives the x and y coordinates of each star in seconds of arc relative to the cluster centre. This coordinate system agrees closely with that in the variable-star list; see Hogg (1973). The next four columns give the centennial relative proper motions and their standard errors. *Proper motions* are the angular motions of stars across the sky due to effects other than the motion of the earth. These primarily reflect the stars' own velocities. The adjective *centennial* indicates that the units are arc seconds per century, rather than per year; the adjective *relative* indicates that these proper motions are accurate relative to one another but that the zero point of this system of motions may not coincide with zero velocity in any absolute sense. The correction from relative proper motions to so-called *absolute* proper motions is still a controversial problem. Fortunately, such a correction is not needed for cluster membership determinations. The final columns list the V, yellow, magnitudes, B-V colours, membership probabilities P_c, and notes leading to additional names for several stars. Most of the probabilities listed as 99 were actually greater than or equal to 99.5 but were rounded downward in order to avoid implying absolute certainty of cluster membership.

The intensity of starlight is normally measured in magnitudes, on a logarithmic scale where one magnitude represents a factor of 2.512... = $100^{0.2}$ in intensity. The zero point was set by historical designation of the 20 brightest stars in the sky as first magnitude. Positive numbers indicate fainter stars. Magnitudes are now best determined *via* careful photoelectric measurements through well-defined filters. The V, yellow, and B, blue, filters are commonly used. Thus B-V represents the difference in the magnitudes measured through these filters, or the logarithm of the ratio of intensity. It, therefore, measures stellar colour; very blue stars having negative B-V and red stars with B-V>1.0.

References

Barnard, E.E. (1931). Micrometric observations of star clusters. *Publ. Yerkes Obs.* **6**.

Hogg, H.S. (1973). A third catalogue of variable stars in globular clusters comprising 2119 entries. *Publ. David Dunlap Obs.* **3**, No. 6.

Nassau, J. (1938). A study of the globular cluster M92. *Astrophys. J.* **87**, 361-366.

Sandage, A.R. and Walker, M.F. (1966). Three-colour photometry of the bright stars in the globular cluster M92. *Astrophys. J.* **143**, 313-326.

Vasilevskis, S., Klemola, A. and Preston, G. (1958). Relative proper motions of stars in the region of the open cluster NGC 6633. *Astron. J.* **63**, 387-395.

Zinn, R., Newell, E.B. and Gibbson, J.B. (1972). A Search for UV-bright stars in 27 globular clusters. *Astron. Astrophys.* **18**, 390-402.

Table 22.1
Proper Motions, Photometry and Membership Probabilities in M92

ID		X (arc sec)	Y (arc sec)	μ_x ("/cent.)	ϵ ("/cent.)	μ_y ("/cent.)	ϵ ("/cent.)	V	B-V	P_c	Notes
I-1		464	174	1.260	0.026	0.942	0.031	15.00	0.63	0	
I-2		437	161	0.251	0.034	-0.107	0.015	15.11	0.82	0	
I-10		318	76	-0.065	0.050	-0.120	0.026	15.47	0.02	24	
I-13		252	105	0.788	0.028	0.354	0.025	14.65	0.53	0	
I-14		235	108	-0.036	0.028	-0.013	0.023	14.74	0.73	99	
I-21		224	36	0.038	0.024	-0.011	0.037	15.23	0.10	99	
I-22		245	32	0.050	0.060	-0.074	0.039	15.41	0.72	92	
I-38		154	67	0.057	0.040	0.017	0.028	15.15	0.17	99	
I-40		168	36	0.008	0.019	0.029	0.023	14.78	0.73	99	
I-67	B117	152	13	-0.049	0.019	0.004	0.026	13.32	0.92	99	
I-68		157	2	-0.010	0.030	0.021	0.027	14.61	0.75	99	
II-2		405	370	0.219	0.039	-0.219	0.031	14.66	0.58	0	
II-5		344	351	-2.488	0.024	0.594	0.035	12.08	0.83	0	
II-6		474	328	0.052	0.048	-0.035	0.040	15.14	0.75	99	
II-12		245	240	0.019	0.032	-0.038	0.030	14.68	0.80	99	
II-18		335	229	-0.551	0.029	-4.949	0.028	14.04	0.92	0	
II-23		153	250	-0.044	0.039	0.012	0.031	15.24	0.13	99	
II-24		163	229	0.004	0.021	-0.018	0.024	14.54	0.78	99	
II-25		212	216	0.667	0.019	-0.769	0.054	15.25	0.66	0	
II-28		162	206	-0.006	0.043	0.013	0.021	15.10	0.45	99	
II-39		158	169	0.001	0.015	-0.044	0.027	14.38	0.82	99	
II-53	B106	63	95	0.014	0.030	-0.059	0.021	12.35	1.18	99	
II-66		112	179	-0.040	0.030	-0.061	0.031	15.02	0.09	99	
II-70	B112	102	152	-0.023	0.025	-0.004	0.018	13.10	0.99	99	*
II-77		144	135	0.034	0.012	-0.054	0.022	14.21	0.74	99	
II-89	B111	93	143	0.004	0.013	0.021	0.025	14.03	0.78	99	
II-120		127	76	-0.041	0.020	0.006	0.020	14.59	0.74	99	
II-121	B114	120	78	-0.006	0.020	-0.035	0.033	13.79	0.71	99	
III-4		139	389	0.046	0.025	-0.007	0.013	14.18	0.73	99	
III-11		132	281	0.023	0.036	0.026	0.026	15.23	0.69	99	
III-12		162	288	0.759	0.029	-0.655	0.033	13.29	0.77	0	
III-13		160	279	0.113	0.025	-0.001	0.092	12.03	1.33	62	
III-65	B108	76	153	-0.016	0.023	0.014	0.030	12.49	1.19	99	*
III-81	B81	16	166	0.035	0.015	0.028	0.014	14.37	0.65	99	
III-82	B70	9	153	-0.065	0.020	0.021	0.024	13.33	0.97	99	*
III-88		43	136	-0.101	0.025	-0.048	0.017	14.75	0.72	86	
III-96		44	118	-0.004	0.027	-0.020	0.025	15.08	0.69	99	
III-98		31	125	0.018	0.015	-0.022	0.011	14.50	0.73	99	
III-109		0	142	-0.018	0.038	0.027	0.021	14.93	0.72	99	
IV-2	B6	-88	421	-0.024	0.024	-0.024	0.026	13.58	0.91	99	

Table 22.1 *(cont.)*

ID		X	Y	μ_x	ϵ	μ_y	ϵ	V	B-V	P_c	Notes
		(arc sec)		("/cent.)		("/cent.)					
IV-10	B5	-101	361	-0.023	0.015	0.033	0.013	13.42	0.95	99	
IV-13		-33	346	0.005	0.062	0.080	0.057	15.38	0.69	97	
IV-17		-70	292	-0.083	0.033	-0.041	0.042	15.51	0.03	97	
IV-27		-5	251	-0.011	0.035	-0.061	0.048	15.20	0.18	99	
IV-40		-74	217	0.024	0.021	0.017	0.019	13.94	0.80	99	
IV-78		-76	140	0.037	0.024	0.043	0.025	15.15	0.18	99	
IV-79	B12	-70	134	-0.029	0.014	0.036	0.010	13.47	0.92	99	
IV-87		-12	149	-0.089	0.024	0.068	0.028	15.10	0.70	51	
IV-94	B36	-15	125	-0.038	0.028	-0.043	0.020	13.06	0.96	99	
IV-99		-37	124	-0.116	0.024	0.014	0.041	15.12	0.15	61	
IV-114		-81	320	0.001	0.021	-0.039	0.030	13.87	0.87	99	
V-2		-336	353	0.697	0.017	1.004	0.014	13.50	1.05	0	
V-7	V14	-315	246	0.661	0.027	1.270	0.042	14.63		0	
V-44		-207	127	0.112	0.038	0.028	0.051	15.30	0.70	63	
V-45		-190	152	0.130	0.033	-0.018	0.026	12.83	1.04	13	
V-55		-176	112	-0.010	0.036	0.007	0.040	15.30	0.16	99	
V-69		-137	136	-0.011	0.026	0.050	0.018	14.60	0.76	99	
V-78		-117	141	0.003	0.021	0.005	0.022	14.48	0.60	99	
V-117		-93	90	-0.017	0.029	0.073	0.020	15.30	0.05	97	
VI-2		-365	181	0.140	0.037	0.374	0.048	15.45	0.68	0	
VI-3		-323	175	0.042	0.053	0.045	0.045	15.33	0.09	99	
VI-6		-438	155	1.438	0.030	-0.365	0.019	13.16	0.84	0	
VI-7	ZNG4	-419	98	0.069	0.019	-0.104	0.028	13.58	0.69	15	
VI-10		-478	8	-0.046	0.058	0.057	0.029	15.24	0.13	98	
VI-18		-324	34	-0.031	0.019	-0.004	0.018	13.77	0.86	99	
VI-36		-258	55	0.039	0.027	-0.128	0.033	15.47	0.71	6	
VI-55		-207	86	-0.006	0.019	-0.013	0.035	15.25	-0.04	99	
VI-61		-182	71	0.045	0.035	0.044	0.019	15.31	0.14	99	
VI-67		-179	54	0.017	0.045	-0.042	0.046	15.14	0.23	99	
VI-74		-207	6	0.623	0.026	-0.333	0.021	14.25	0.56	0	
VII-1		-508	-49	0.173	0.040	1.339	0.026	14.58	0.89	0	
VII-5		-459	-180	0.097	0.060	-0.812	0.026	15.08	0.66	0	
VII-10		-363	-31	-0.030	0.023	0.010	0.023	13.70	0.84	99	
VII-12		-366	-167	0.369	0.021	-0.990	0.033	14.70	0.77	0	
VII-15		-341	-71	-0.124	0.034	-0.032	0.046	15.04	0.74	48	
VII-18		-329	-112	-0.037	0.038	-0.024	0.014	12.19	1.30	99	
VII-36		-248	-31	-0.051	0.025	0.008	0.023	15.38	0.09	99	
VII-39		-257	-37	0.065	0.037	0.071	0.026	14.48	0.74	92	
VII-66		-206	-105	-0.050	0.033	-0.036	0.036	15.09	0.72	99	
VII-67	N15	-215	-113	-0.004	0.024	0.046	0.036	14.70	0.72	99	

Table 22.1 *(cont.)*

ID		X	Y	μ_x	ϵ	μ_y	ϵ	V	B-V	P_c	Notes
		(arc sec)		("/cent.)		("/cent.)					
VII-68		-170	-7	0.020	0.021	0.026	0.024	14.26	0.73	99	
VII-79		-179	-86	0.005	0.020	0.023	0.014	14.21	0.73	99	
VII-80		-171	-92	0.017	0.015	0.011	0.012	13.92	0.71	99	
VII-102		-133	-8	0.064	0.036	0.031	0.037	15.33	0.06	99	
VII-122	B4	-110	-47	0.040	0.029	0.005	0.034	12.29	1.18	99	
VII-123		-119	-47	0.062	0.028	-0.007	0.032	14.32	0.76	99	
VIII-10		-224	-282	0.522	0.029	2.121	0.035	15.08	1.37	0	
VIII-12		-267	-232	-0.052	0.022	-0.004	0.015	14.76	0.71	99	
VIII-15		-230	-269	0.151	0.027	1.047	0.027	14.28	0.67	0	
VIII-24		-188	-136	0.015	0.014	-0.046	0.017	14.13	0.74	99	
VIII-30		-171	-153	-0.017	0.049	0.019	0.035	15.29	0.04	99	
VIII-43		-127	-202	-0.024	0.019	0.003	0.021	14.62	0.75	99	
VIII-44		-133	-187	0.024	0.020	0.035	0.016	14.06	0.83	99	
IX-2		-139	-371	-0.035	0.031	0.040	0.036	14.69	0.77	99	
IX-5	ZNG3	-104	-413	-0.334	0.016	0.148	0.029	13.14	0.61	0	
IX-6		-90	-425	-0.040	0.035	0.002	0.031	14.65	0.82	99	
IX-10		-43	-368	-0.028	0.023	0.057	0.016	14.64	0.74	99	
IX-12		-163	-280	-0.040	0.019	0.034	0.024	14.54	0.59	99	
IX-13	NH	-127	-243	0.021	0.015	-0.004	0.015	14.03	0.85	99	
IX-30		-17	-306	0.415	0.021	0.004	0.019	14.07	0.80	0	
IX-31		-32	-263	0.387	0.034	0.018	0.037	15.09	1.11	0	
IX-49	B8	-79	-179	0.011	0.019	0.025	0.016	13.89	0.84	99	
IX-77		-20	-146	-0.036	0.024	0.033	0.024	14.19	0.79	99	
IX-81	B21	-39	-139	-5.114	0.074	-6.217	0.029	12.92	0.83	0	*
IX-89	B38	-12	-129	-0.036	0.020	0.029	0.025	14.18	0.74	99	
X-3		48	-352	-0.027	0.025	0.017	0.022	14.63	0.73	99	
X-4		66	-383	0.502	0.023	-0.737	0.028	12.86	1.16	0	
X-9		4	-287	-0.072	0.032	-0.066	0.032	15.07	0.61	95	
X-10		24	-303	0.578	0.014	-0.582	0.014	14.36	0.61	0	
X-22		3	-224	-0.020	0.045	-0.026	0.028	15.13	0.20	99	
X-28		87	-194	-0.013	0.016	0.000	0.009	14.62	0.72	99	
X-49	B103	57	-151	0.026	0.022	0.007	0.030	12.16	1.19	99	*
X-65		42	-130	-0.023	0.024	-0.004	0.019	14.42	0.72	99	
X-66		40	-140	0.000	0.044	0.012	0.030	15.05	0.16	99	
XI-2		178	-311	-0.432	0.021	-0.078	0.021	14.26	0.73	0	
XI-3		253	-306	0.077	0.025	-0.362	0.013	13.55	0.89	0	
XI-4		433	-254	0.975	0.023	-2.349	0.018	14.28	0.72	0	
XI-8		287	-213	-0.122	0.030	-0.797	0.015	14.07	0.59	0	
XI-13		177	-213	0.014	0.027	0.015	0.022	14.82	0.54	99	
XI-14		160	-204	0.052	0.016	0.005	0.020	13.81	0.87	99	

Table 22.1 *(cont.)*

ID		X	Y	μ_x	ϵ	μ_y	ϵ	V	B-V	P_c	Notes
		(arc sec)		("/cent.)		("/cent.)					
XI-19	B115	122	-203	0.016	0.021	-0.061	0.023	12.87	1.05	99	
XI-27		154	-118	-0.013	0.024	-0.023	0.021	14.49	0.66	99	
XI-29		148	-121	-0.023	0.013	0.004	0.020	14.40	0.62	99	
XI-38		114	-150	0.008	0.025	0.042	0.020	14.59	0.76	99	
XI-50		90	-118	0.907	0.036	-2.129	0.016	14.44	0.88	0	
XI-70	B113	114	-83	-0.004	0.010	-0.002	0.016	14.11	0.75	99	
XI-80	B109	78	-108	-0.054	0.030	0.037	0.016	13.04	1.00	99	
XII-1		407	-138	-0.036	0.055	-0.032	0.040	15.11	0.18	99	
XII-2		433	-66	0.433	0.020	1.687	0.029	12.89	0.67	0	
XII-5		323	-90	0.039	0.019	-0.075	0.025	14.88	0.76	95	
XII-7		291	-120	0.047	0.028	-0.043	0.025	14.63	0.75	99	
XII-8		264	-154	-0.017	0.024	-0.016	0.016	12.76	1.06	99	
XII-9		246	-125	0.004	0.026	0.015	0.042	15.12	0.15	99	
XII-10		242	-114	0.092	0.036	0.028	0.040	15.22	0.15	92	
XII-18	B119	163	-78	0.023	0.024	-0.020	0.022	14.36	0.72	99	
XII-26		234	-14	0.062	0.040	-0.005	0.037	15.27	0.10	99	
XII-31	B120	167	-27	-0.025	0.024	-0.024	0.024	13.95	0.85	99	
XII-34	B118	155	-33	0.010	0.024	-0.019	0.025	13.45	0.89	99	
XII-45	B116	144	-48	0.010	0.022	-0.029	0.011	14.08	0.76	99	
X1		155	589	0.911	0.034	0.751	0.033	14.94	0.70	0	
X2		242	551	1.169	0.031	-1.450	0.046	12.27	1.07	0	
X3		362	539	-0.719	0.024	1.299	0.016	13.32	0.92	0	
X4		191	655	0.445	0.020	-1.069	0.024	13.54	0.96	0	
X5		-236	490	-0.131	0.029	0.477	0.028	14.22	0.58	0	
X6		349	658	0.516	0.036	-0.654	0.027	14.96	0.67	0	
X7		460	700	-0.030	0.028	0.079	0.060	15.06	0.74	94	
X8		646	701	0.888	0.038	-1.094	0.045	11.96	0.98	0	
X9		628	813	-0.020	0.056	-0.202	0.037	12.48	0.51	0	
X10		693	814	0.677	0.022	-0.968	0.027	14.05	0.73	0	
X11		579	-45	0.306	0.025	-0.774	0.028	14.43	0.55	0	
X12		634	-78	0.676	0.016	-0.738	0.022	14.22	0.59	0	
X13		640	-190	-0.116	0.072	-0.553	0.048	10.99	0.77	0	
X14		752	-174	1.318	0.055	0.158	0.053	10.94	0.40	0	
X15		728	-88	1.660	0.020	-0.347	0.013	13.46	0.50	0	
X16		709	-138	1.808	0.042	0.640	0.018	14.40	0.73	0	
X17		651	-324	0.605	0.036	0.033	0.040	14.30	0.51	0	
X18	B23	-32	443	0.616	0.024	-0.595	0.027	13.35	0.69	0	
X19		-107	506	0.868	0.043	-0.862	0.031	14.24	0.57	0	
X21		-166	462	0.669	0.028	-0.410	0.025	13.64	0.65	0	
B2	ZNG1	-119	-153	1.057	0.022	-0.209	0.024	12.05		0	

Table 22.1 *(cont.)*

ID		X	Y	μ_x	ϵ	μ_y	ϵ	V	B-V	P_c	Notes
		(arc sec)		("/cent.)		("/cent.)					
B3		-119	5	0.055	0.025	0.056	0.028	14.16		98	
B7		-80	80	0.041	0.027	0.060	0.026	14.09		99	
B13		-58	-108	-0.090	0.020	-0.025	0.032	12.90		96	
B15		-55	81	0.030	0.014	-0.011	0.030	13.95		99	
B18		-44	-117	0.045	0.016	0.022	0.022	13.76		99	
B40		-9	95	0.006	0.012	0.053	0.021	13.89		99	
B51		-2	84	0.017	0.020	0.037	0.019	13.83		99	
B52	V15	0	79	-0.105	0.049	-0.017	0.014	13.12		88	*
B71		9	126	0.025	0.018	-0.028	0.012	13.66		99	
B110		88	93	0.000	0.028	-0.022	0.022	14.09		99	
V1	N7	128	41	-0.001	0.032	0.017	0.031	15.05		99	
V3	N1	54	253	0.037	0.050	-0.031	0.048	15.09		99	
V4	N8	-76	58	0.063	0.022	0.006	0.028	14.90		99	
V8	N4	209	208	0.020	0.024	-0.036	0.035	15.29		99	
V13	N14	153	-60	0.019	0.024	0.022	0.031	15.19		99	
N16		-90	13	0.054	0.013	0.010	0.037	14.71		99	
NI		-54	181	-0.011	0.043	0.017	0.043	15.31		99	
NJ		49	-166	0.046	0.036	0.007	0.043	15.18		99	
C1		-695	-90	0.851	0.021	-1.098	0.048	15.02		0	
C2		-693	-301	0.870	0.030	-1.833	0.023	12.91		0	
C3		-671	-433	-0.415	0.038	-0.206	0.038	15.34		0	
C4		-597	143	-0.143	0.087	0.031	0.041	15.45		3	
C5		-594	-169	1.024	0.060	-0.088	0.028	14.98		0	
C6		-594	7	0.503	0.022	-3.386	0.033	12.02		0	
C7		-549	307	-0.874	0.030	0.351	0.027	13.99		0	
C8		-451	-405	-0.483	0.023	0.376	0.021	13.70		0	
C9		-385	-338	0.216	0.027	1.616	0.057	11.90		0	
C10		-138	-48	-0.068	0.026	0.000	0.016	14.09		99	
C11		-124	-118	-0.001	0.030	-0.030	0.027	14.76		99	
C12		-88	-61	0.058	0.034	0.024	0.025	14.65		99	
C13		-88	-97	0.070	0.023	-0.024	0.022	14.45		98	
C14		-87	-85	0.034	0.028	-0.028	0.019	14.41		99	
C15		-83	1	0.017	0.024	-0.011	0.023	14.26		99	
C16		-82	10	0.013	0.025	-0.026	0.019	14.32		99	
C17		-82	49	-0.045	0.025	0.034	0.029	14.96		99	
C18		-65	-94	0.036	0.015	-0.006	0.028	14.70		99	
C19		-49	-95	0.000	0.017	0.001	0.025	14.65		99	
C20		-36	-69	0.055	0.018	0.013	0.019	14.04		99	
C21		-36	-86	0.004	0.029	0.004	0.027	14.70		99	
C22		-30	-81	0.036	0.029	0.006	0.018	14.22		99	

Table 22.1 *(cont.)*

ID	X (arc sec)	Y	μ_x ("/cent.)	ϵ	μ_y ("/cent.)	ϵ	V	B-V	P_c	Notes
C23	2	87	0.003	0.011	-0.001	0.022	14.38		99	
C24	23	618	-0.137	0.031	0.499	0.061	15.41		0	
C25	26	-76	-0.009	0.022	-0.002	0.019	13.79		99	
C26	38	106	-0.014	0.024	-0.047	0.021	14.67		99	
C27	48	49	-0.022	0.030	-0.022	0.023	13.84		99	
C28	53	60	-0.026	0.017	-0.029	0.019	14.21		99	
C29	54	77	0.008	0.023	0.054	0.034	14.92		99	
C30	57	-493	-0.007	0.034	-0.015	0.044	15.07		99	
C31	63	69	-0.063	0.021	0.031	0.025	13.39		99	
C32	73	-24	-0.012	0.026	-0.041	0.023	14.59		99	
C33	236	-80	-0.044	0.048	-0.005	0.038	14.98		99	
C34	312	-424	0.013	0.055	0.015	0.039	15.18		99	
C35	603	244	0.430	0.026	0.235	0.024	14.56		0	
C36	624	131	-0.319	0.056	-2.443	0.028	15.35		0	
C37	629	117	0.043	0.028	-0.028	0.033	15.03		99	
C38	653	21	0.203	0.083	-1.053	0.058	15.41		0	
C39	683	233	-1.642	0.026	0.934	0.031	12.87		0	
C40	747	-255	0.093	0.072	-0.456	0.036	15.36		0	
C41	763	-217	-0.032	0.064	0.146	0.021	15.35		0	

* Notes

II-70 = Ne
III-65 = Nc
III-82 = Nf
IX-81 = Nd = ZNG2
X-49 = Nb
B2 = Na
B52 = N12

23. Motions and Distances of Planetary Nebulae

Source Cudworth, K. M. (1974). New proper motions, statistical parallaxes and kinematics of planetary nebulae. *Astronomical J.* **79**, 1384-1395.

Contributor K. M. Cudworth University of Chicago

Planetary nebulae are symmetrical clouds of gas surrounding very hot stars. They probably represent a late stage in the evolution of many stars like the sun. The distances of these objects from us, and hence their intrinsic luminosities, are rather poorly estimated. Cudworth measured the angular motions of these objects across the sky and adopted other data from Perek and Kohoutek (1967). Application of the technique known in astronomy as *statistical parallax,* essentially the method of least squares, yielded distances for the nebulae. Table 23.1 presents the angular motions, μ_α and μ_δ, together with their mean errors and the sources for the proper motions of each nebula. Table 23.1 also lists whether the nebula is optically thick, TK, or thin, TN, the extinction parameter *C*, a measure of distance *d* and the class obtained from Greig (1971,1972).

Cudworth's main conclusions were that the distances of the nebulae were much larger than previously believed. The suspicion of a relationship to stars like the sun was strengthened.

Cudworth also attempted an iterative maximum likelihood statistical parallax. This solution did not converge, possibly because of the very non-uniform distribution of the nebulae in the sky.

References

Anderson, C. M. (1934). The proper motions of thirty-three planetary nebulae. *Lick Obs. Bull.* **17**, 21-32.

Greig, W. E. (1971). The morphological classification of symmetrical nebulae. *Astron. Astrophys.* **10**, 161-174.

Greig, W. E. (1972). Spatial and kinematic parameters of binelsalous, centric and annular nebulae. *Astron. Astrophys.* **18**, 70-78.

Perek, L. and Kohoutek, L., (1967). *Catalogue of Galactic Planetary Nebulae.* Czechoslovakian Acad. Science.

van Maanen, A. (1933). Investigations on proper motions. *Astrophys. J.* **77**, 186-194.

Table 23.1
Combined Absolute Proper Motions and Other Statistical Parallax Input Data

Nebula		μ_α*	Mean err.	μ_δ*	Mean err.	μ Source**	Thick or thin	C	d (pc)	Class
NGC	40	-52	28	25	29	1,2,3	TN	0.77	1783	B
NGC	246	-204	10	-112	42	1,2	TN	0.00	557	C
NGC	650-1	102	49	-94	45	2	TN	0.20	1055	B
IC	1747	58	41	-21	42	1	TK	0.99	3170	B
IC	351	-39	28	-92	37	1	TN	0.27	7680	C
NGC	1501	-5	10	-53	10	2,3	TN	0.94	1776	C
NGC	1514	-29	17	-24	36	1,2,3	TN	0.52	1084	C
NGC	1535	89	44	-15	14	1,2	TN	0.16	3126	C
J	320	-2	40	16	47	1,2	TN	0.19	9252	C
IC	418	4	31	88	52	1,2	TK	0.26	760	B
NGC	2022	48	52	-72	23	2,3	TN	0.60	3558	C
A	14	-12	36	-18	41	1				
NGC	2371-2	-4	63	-109	55	1,2,3	TN	0.09	2536	B
NGC	2392	39	67	-34	45	1,2,3	TN	0.02	1982	C
NGC	2438	-69	40	-11	45	1,2	TN	0.38	1712	C
A	24	-65	20	44	39	1	TN	0.17	525	B
NGC	2610	96	107	93	89	2	TN	0.46	2992	C
A	33	-173	39	63	24	1	TN	0.02	915	C
NGC	3242	-3	51	-31	39	1,2	TN	0.02	1697	C
NGC	3587	-3	27	-14	40	1	TN	0.00	793	C
NGC	4361	-42	49	-46	40	1	TN	0.01	1455	C
IC	3568	14	36	-12	42	1	TN	0.25	3756	C
A	36	189	37	-29	36	1				
NGC	6058	-56	30	-91	33	1,2,3	TN	0.00	5023	B
IC	4593	-78	21	-8	34	1	TN	0.13	4301	C
NGC	6210	-43	25	-82	75	1,3	TK	0.11	1663	C
NGC	6309	71	37	133	41	1	TN	0.79	4549	C
A	43	18	21	-3	28	1	TN	0.16	2460	B
NGC	6543	-21	40	-30	15	1,2,3	TK	0.00	1112	C
NGC	6572	-58	75	-8	69	3	TK	0.32	904	C
NGC	6629	-17	12	71	10	1,2	TN	0.83	3391	C
A	45	59	16	-2	31	1				
NGC	6644	-159	54	-65	61	1	TK	0.40	3945	C
IC	4732	34	10	44	42	1,2	TK	0.65	4909	C
NGC	6720	10	10	14	23	1,2,3	TN	0.10	1264	B
NGC	6751	-7	12	-40	10	1,2	TN	0.98	2983	B
M1-	67	-39	10	-91	25	1	TN	0.21	891	C
IC	4846	-18	38	-76	27	1	TK	0.45	4909	C
NGC	6781	-76	11	-37	54	1,2	TN	0.33	1482	B
NGC	6790	-57	16	39	13	1,2	TK	0.13	3597	C
NGC	6804	-95	10	-94	10	1,2,3	TN	0.95	2398	C
NGC	6807	-51	85	122	60	1	TK	0.75	4797	C
BD+30	3639	-47	12	-205	40	3	TK	0.44	1074	B

Table 23.1 *(cont.)*

Nebula		μ_α*	Mean err.	μ_δ*	Mean err.	μ Source**	Thick or thin	C	d (pc)	Class
NGC	6818	100	56	-58	39	1	TK	0.04	1954	B
NGC	6826	-96	10	-92	29	1,2	TN	0.00	2265	C
NGC	6833	112	23	10	33	1	TK	0.23	5508	C
NGC	6853	129	20	11	14	1,2,3	TN	0.00	383	B
NGC	6891	-10	10	-71	31	1	TN	0.21	4861	C
NGC	6894	17	58	-66	12	2,3	TN	0.93	2157	C
IC	4997	145	37	44	72	1	TK	0.00	3097	C
NGC	6905	-53	16	-67	10	1,2,3	TN	0.35	2291	B
NGC	7008	42	49	-92	39	1,2,3	TN	0.61	1272	C
NGC	7009	13	62	-5	39	1,2	TN	0.12	1887	C
NGC	7026	-2	10	-2	31	1,2,3	TK	0.79	2000	B
NGC	7027	-26	22	-134	45	3	TK	1.19	514	C
IC	5117	179	30	-13	43	1	TK	1.27	2880	C
A	78	14	28	-39	1					
NGC	7139	-57	104	-75	88	2	TN	0.83	1815	B
IC	5217	1	13	10	19	3	TN	0.32	4742	C
NGC	7293	402	28	-10	19	1,2	TN	0.07	212	B
NGC	7662	5	36	-11	39	2,3	TK	0.10	1535	C
A	82	-96	42	-83	36	1				

* Units of 0.0001/yr.
**Sources are from 1. Cudworth (1974), 2. Anderson (1934) and 3. van Maanen (1933).

24. Quality Control Data in Clinical Chemistry

Source National Quality Control Scheme, Queen Elizabeth Hospital Birmingham

Contributor M.J.R. Healy London School of Hygiene and Tropical Medicine

Every two weeks, a specimen from a large homogeneous pool of serum is sent out to a large number of laboratories who perform up to 15 separate analyses. The data consist of ten sets of results from some 400 - 500 laboratories; not all laboratories perform all analyses, and not every laboratory provided ten reports. Different laboratories used different methods for a given analysis. These methods have been grouped together into method-groups, there being 4 to 12 method-groups for a particular analysis. A laboratory did not necessarily stick to a particular method-group for a given analysis over all ten occasions. The data consist of ten sets of 500 records giving the laboratory number, the fifteen results and their method-groups, not all different, for each laboratory. Tables 24.1 and 24.2 present the data for the first two analyses by 100 laboratories for each of the ten occasions. *

There are consistent differences among method-groups in most analyses, and the interest of the data lies largely in determining these in the presence of a substantial fraction, for example 2%, of apparently aberrant values. The identification of such aberrant values for further study is also of interest.

* The complete set of data is available.

Table 24.1
Results of Measurements by Different Methods of Analysis on Sodium Content of a Large Homogeneous Pool of Serum Analyzed by 100 Laboratories *

Laboratory	Measurements									
1	140 5	143 5	141 5	137 5	132 5	157 5	143 5	149 5	118 5	145 5
2	138 5	144 5	144 5	139 5	133 5	159 5	139 5	141 5	124 5	145 5
3	139 8	147 8	145 8	141 8	135 8	157 8	145 6	143 6	123 6	146 6
4	139 5	147 5	142 5	137 5	132 5	158 5	144 5	139 5	121 5	144 5
5	141 4	146 4	142 4	138 4	133 4	158 4	143 4	0 4	122 4	0 4
6	143 8	146 8	143 8	139 8	134 8	159 8	145 8	142 8	124 8	146 8
7	146 5	143 5	143 5	138 5	132 5	157 5	143 5	141 5	125 5	145 5
8	141 9	148 9	143 9	139 9	133 9	151 9	145 9	138 9	124 9	141 9
9	0 0	0 0	0 0	0 0	0 0	0 0	0 0	0 0	0 0	0 0
10	0 5	0 5	0 5	0 5	0 5	0 5	0 5	0 5	0 5	0 5
11	140 6	145 6	142 6	140 6	132 6	155 6	145 6	141 6	126 6	146 6
12	0 0	0 0	0 0	0 0	0 0	0 0	0 0	0 0	0 0	0 0
13	140 5	143 5	142 5	139 5	135 5	158 5	0 5	141 5	126 5	146 5
14	142 5	0 5	141 5	134 5	0 5	157 5	141 5	137 5	125 5	145 5
15	140 5	146 5	142 5	139 5	137 5	154 5	143 5	133 5	125 5	145 5
16	141 5	145 5	145 5	139 5	134 5	158 5	143 5	141 5	126 5	145 5
17	140 9	144 9	141 9	139 9	132 9	155 9	141 9	141 9	0 9	147 9
18	140 5	138 5	144 5	138 5	131 5	159 5	141 5	142 5	120 5	144 5
19	140 5	145 5	144 5	141 5	136 5	155 5	141 5	141 5	126 5	146 5
20	142 6	0 6	140 6	136 6	0 6	159 6	143 6	143 6	123 6	149 6
21	140 5	144 5	143 5	137 5	133 5	157 5	143 5	143 5	125 5	144 5
22	0 5	0 5	0 5	0 5	0 5	0 5	0 5	0 5	0 5	0 5
23	140 5	145 5	141 5	138 5	135 5	158 5	140 5	141 5	0 5	141 5
24	0 4	147 4	143 4	141 4	0 4	161 4	0 4	143 4	0 4	0 4
25	141 5	143 5	144 5	140 5	132 5	158 5	141 5	142 5	126 5	147 5
26	140 5	145 5	140 5	137 5	132 5	157 5	142 5	143 5	122 5	144 5
27	141 6	146 6	142 6	138 6	133 6	159 6	141 6	142 6	125 6	145 6
28	0 6	144 6	0 6	0 6	130 6	155 6	0 6	0 6	127 6	143 6
29	137 7	142 7	145 7	138 7	130 7	156 7	143 7	143 7	124 7	143 7
30	139 5	143 5	142 5	136 5	132 5	157 5	142 5	140 5	128 5	139 5
31	140 5	145 5	141 5	137 5	132 5	0 5	142 5	0 5	125 5	146 5
32	140 8	144 8	143 8	139 8	0 8	157 8	145 8	141 8	125 8	145 8
33	141 6	151 6	143 6	140 6	130 6	0 6	0 6	0 6	0 6	145 6
34	141 8	144 8	142 8	137 8	132 8	158 8	142 8	0 8	124 8	143 8
35	138 5	146 5	142 5	0 5	134 5	152 5	140 5	141 5	122 5	145 5
36	143 5	149 5	144 5	139 5	135 5	157 5	147 5	144 5	126 5	148 5
37	0 5	146 5	142 5	139 5	134 5	160 5	144 5	143 5	127 5	143 5
38	142 5	144 5	142 5	138 5	135 5	156 5	142 5	142 5	125 5	146 5
39	0 9	146 9	143 9	0 9	133 9	154 9	140 9	139 9	125 9	145 9
40	140 8	144 8	140 8	139 8	131 8	158 8	0 8	141 8	125 8	145 8
41	140 6	145 6	145 6	139 6	133 6	155 6	144 6	140 6	125 6	146 6
42	140 5	146 5	143 5	136 5	132 5	157 5	141 5	140 5	123 5	143 5
43	140 3	147 3	144 3	138 3	132 3	160 3	147 6	138 6	124 6	146 6
44	140 6	151 6	144 6	138 6	131 6	156 6	144 6	142 6	130 6	141 6
45	0 6	145 6	142 6	138 6	132 6	159 6	146 6	143 6	126 6	150 6
46	139 5	146 5	143 5	139 5	133 5	158 5	0 5	0 5	124 5	146 5
47	140 5	146 5	142 5	0 5	132 5	156 5	143 5	141 5	0 5	144 5
48	139 5	145 5	143 5	133 5	135 5	0 5	143 5	142 5	0 5	145 5
49	141 6	146 6	143 6	139 6	133 6	0 6	143 6	141 6	124 6	147 6

Table 24.1 *(cont.)*

Laboratory	Measurements									
50	0 5	144 5	139 5	0 5	132 5	156 5	145 5	141 5	0 5	0 5
51	140 9	145 9	0 9	135 9	131 9	160 9	143 9	138 9	120 9	142 9
52	142 7	146 7	143 7	140 7	134 7	157 7	145 7	143 6	127 6	146 6
53	141 6	148 6	141 6	139 6	133 6	160 6	144 6	143 6	124 6	148 6
54	141 5	0 5	142 5	138 5	132 5	156 5	145 5	142 5	119 5	0 5
55	144 5	147 5	142 5	140 5	135 5	156 5	144 5	142 5	126 5	144 5
56	143 8	144 8	144 8	139 8	132 8	0 8	142 8	0 8	126 8	145 8
57	142 6	145 6	140 6	138 6	132 6	156 6	145 6	143 6	126 6	144 6
58	144 6	145 6	141 6	0 6	0 6	157 6	145 6	139 6	0 6	0 6
59	143 6	146 6	147 6	140 6	0 6	159 6	147 6	142 5	123 5	148 5
60	140 5	146 5	0 5	137 5	131 5	157 5	146 5	141 5	124 5	147 5
61	138 5	144 5	140 5	137 5	134 5	155 5	145 5	140 5	127 5	142 5
62	140 8	145 8	143 8	138 8	134 8	158 8	0 8	141 8	124 8	145 8
63	142 4	144 4	139 4	140 4	135 4	156 4	141 4	145 4	122 4	145 4
64	136 5	144 5	142 5	137 5	0 5	159 5	144 5	0 5	0 5	145 5
65	137 5	147 5	142 5	138 5	0 5	145 5	144 5	141 5	0 5	0 5
66	140 6	146 6	143 6	136 6	135 6	161 6	145 6	140 6	0 6	145 6
67	139 5	144 5	142 5	137 5	131 5	153 5	143 5	139 5	125 5	143 5
68	144 3	144 3	143 3	141 3	131 3	157 3	143 3	142 3	126 3	146 3
69	142 5	145 5	144 5	138 5	134 5	154 5	145 5	144 5	124 5	0 5
70	0 5	145 5	144 5	0 5	0 5	0 5	144 5	0 5	127 5	147 5
71	138 8	146 8	141 8	143 8	136 8	0 8	144 8	141 8	0 8	145 8
72	0 9	145 9	0 9	137 9	137 9	0 9	141 9	0 9	0 9	0 9
73	141 5	141 5	141 5	140 5	134 5	155 5	143 5	139 5	0 5	144 5
74	139 9	143 9	141 9	139 9	132 9	158 9	143 9	140 9	122 9	143 9
75	0 3	0 3	0 3	0 3	0 3	0 3	0 3	0 3	0 3	0 3
76	139 8	0 8	143 8	139 8	132 8	158 8	140 8	140 8	125 8	145 8
77	138 5	147 5	143 5	137 5	132 5	0 5	140 5	141 5	126 5	146 5
78	141 5	145 5	144 5	140 5	132 5	0 5	143 5	141 5	125 5	145 5
79	139 5	144 5	142 5	137 5	132 5	155 5	140 5	141 5	0 5	145 5
80	0 6	148 6	0 6	140 6	134 6	155 6	142 6	140 6	0 6	0 6
81	139 5	142 5	142 5	135 5	133 5	159 5	139 5	138 5	122 5	143 5
82	0 3	0 3	0 3	0 3	0 3	0 3	0 3	0 3	0 3	0 3
83	142 6	145 6	146 6	140 6	138 6	159 6	141 6	142 6	127 6	143 6
84	0 0	0 0	0 0	0 0	0 0	0 0	0 0	0 0	0 0	0 0
85	142 8	0 8	0 8	140 8	0 8	153 8	143 8	0 8	128 8	143 8
86	0 3	0 3	0 3	0 3	0 3	0 3	0 3	0 3	0 3	0 3
87	140 6	145 6	144 6	138 6	133 6	157 6	146 6	143 6	124 6	146 6
88	139 6	150 6	142 6	138 6	134 6	156 6	146 6	143 6	125 6	144 6
89	143 6	143 6	142 6	136 6	134 6	155 6	142 6	142 6	123 6	0 6
90	141 5	144 5	141 5	139 5	143 5	158 5	141 5	142 5	125 5	146 5
91	141 9	143 9	142 9	139 9	132 9	151 9	141 9	142 9	127 9	142 9
92	131 5	144 5	141 5	136 5	133 5	156 5	138 5	146 5	124 5	148 5
93	142 6	143 6	142 6	139 6	133 6	159 6	0 6	142 6	0 6	148 6
94	140 4	148 4	144 4	0 4	136 4	159 4	141 4	142 4	126 4	147 4
95	138 5	145 5	142 5	138 5	132 5	156 5	142 5	140 5	124 5	143 5
96	143 5	147 5	145 5	141 5	132 5	158 5	141 5	141 5	124 5	144 5
97	141 5	143 5	144 5	139 5	134 5	155 5	140 5	141 5	124 5	145 5
98	141 9	145 9	143 9	0 9	134 9	0 9	143 9	141 9	124 9	146 9
99	142 5	146 5	0 5	139 5	134 5	153 5	0 5	141 5	131 5	145 5
100	0 5	145 5	144 5	137 5	133 5	159 5	141 5	143 5	125 5	147 5

* The integer beside each observation denotes the method.
0 denotes missing value.

Table 24.2
Results of Measurements by Different Methods of Analysis on Potassium Content of a Large Homogeneous Pool of Serum Analyzed by 100 Laboratories *

Laboratory	Measurements									
1	5.4 5	4.5 5	4.8 5	4.0 5	3.5 5	5.6 5	5.2 5	4.8 5	7.5 5	6.2 5
2	5.4 5	4.5 5	4.9 5	4.1 5	3.6 5	5.7 5	5.1 5	4.6 5	7.7 5	6.1 5
3	5.1 8	4.4 8	4.8 8	4.2 8	3.6 8	5.5 8	5.2 6	4.5 6	7.4 6	6.2 6
4	5.3 5	4.5 5	4.9 5	4.2 5	3.5 5	5.7 5	5.2 5	4.6 5	7.7 5	6.1 5
5	5.3 4	4.5 4	4.8 4	4.1 4	3.5 4	5.6 4	5.2 4	0.0 4	0.0 4	0.0 4
6	5.3 8	4.5 8	4.8 8	4.0 8	3.5 8	5.5 8	5.2 8	4.5 8	7.5 8	6.1 8
7	5.4 5	4.4 5	4.8 5	4.1 5	3.5 5	5.6 5	5.1 5	4.6 5	7.5 5	6.1 5
8	5.3 9	4.6 9	4.8 9	4.1 9	3.5 9	5.4 9	5.1 9	4.5 9	7.4 9	5.9 9
9	0.0 0	0.0 0	0.0 0	0.0 0	0.0 0	0.0 0	0.0 0	0.0 0	0.0 0	0.0 0
10	0.0 5	0.0 5	0.0 5	0.0 5	0.0 5	0.0 5	0.0 5	0.0 5	0.0 5	0.0 5
11	5.2 6	4.5 6	4.8 6	4.0 6	3.4 6	5.6 6	5.1 6	4.6 6	7.6 6	6.1 6
12	0.0 0	0.0 0	0.0 0	0.0 0	0.0 0	0.0 0	0.0 0	0.0 0	0.0 0	0.0 0
13	5.3 5	4.5 5	4.9 5	4.2 5	3.6 5	5.7 5	0.0 5	4.5 5	7.8 5	6.2 5
14	5.4 5	0.0 5	5.0 5	3.9 5	0.0 5	5.5 5	5.2 5	4.5 5	7.7 5	6.1 5
15	5.2 5	4.4 5	4.8 5	4.0 5	3.4 5	5.6 5	5.0 5	4.3 5	7.5 5	5.9 5
16	5.4 5	4.6 5	5.0 5	4.1 5	3.6 5	5.7 5	5.1 5	4.6 5	7.6 5	6.0 5
17	5.4 9	4.4 9	4.6 9	4.0 9	3.6 9	5.5 9	5.0 9	4.6 9	0.0 9	6.0 9
18	5.3 5	4.3 5	4.8 5	4.0 5	3.4 5	5.5 5	5.1 5	4.6 5	7.1 5	6.1 5
19	5.1 5	4.4 5	5.0 5	4.2 5	3.6 5	5.7 5	5.1 5	4.6 5	7.7 5	6.2 5
20	5.4 6	0.0 6	4.8 6	4.1 6	0.0 6	5.5 6	5.1 6	4.7 6	7.6 6	6.2 6
21	5.3 5	4.5 5	4.8 5	4.0 5	3.5 5	5.6 5	5.0 5	4.7 5	7.5 5	6.0 5
22	0.0 5	0.0 5	0.0 5	0.0 5	0.0 5	0.0 5	0.0 5	0.0 5	0.0 5	0.0 5
23	5.2 5	4.5 5	4.7 5	4.0 5	3.6 5	5.5 5	5.2 5	4.5 5	0.0 5	6.0 5
24	0.0 4	4.4 4	4.8 4	3.9 4	0.0 4	5.7 4	0.0 4	4.6 4	0.0 4	0.0 4
25	5.3 5	4.3 5	5.0 5	4.1 5	3.7 5	5.8 5	5.1 5	4.6 5	7.6 5	6.1 5
26	5.3 5	4.6 5	4.8 5	4.0 5	3.5 5	5.6 5	5.1 5	4.7 5	7.7 5	6.1 5
27	5.3 6	4.5 6	4.8 6	4.1 6	3.5 6	5.7 6	5.0 6	4.7 6	7.6 6	6.1 6
28	0.0 6	4.6 6	0.0 6	0.0 6	3.5 6	5.6 6	0.0 6	0.0 6	7.8 6	6.0 6
29	5.3 7	4.5 7	4.9 7	4.0 7	3.5 7	5.5 7	5.2 7	4.7 7	7.5 7	6.0 7
30	5.1 5	4.4 5	4.7 5	3.9 5	3.4 5	5.5 5	5.0 5	4.5 5	7.4 5	5.9 5
31	5.2 5	4.5 5	4.8 5	4.2 5	3.7 5	0.0 5	5.1 5	0.0 5	7.6 5	5.9 5
32	5.2 8	4.4 8	4.7 8	4.0 8	0.0 8	5.6 8	5.2 8	4.5 8	7.5 8	6.0 8
33	5.2 6	4.6 6	4.8 6	4.1 6	3.5 6	0.0 6	0.0 6	0.0 6	0.0 6	6.1 6
34	5.4 8	4.4 8	4.9 8	4.0 8	3.5 8	5.7 8	5.2 8	0.0 8	7.6 8	6.1 8
35	5.3 5	4.5 5	4.8 5	3.9 5	3.5 5	5.6 5	5.0 5	4.6 5	7.9 5	6.0 5
36	5.4 5	4.7 5	4.9 5	4.2 5	3.6 5	5.7 5	5.3 5	4.7 5	7.6 5	6.3 5
37	0.0 5	4.5 5	4.8 5	4.1 5	3.5 5	5.6 5	5.2 5	4.7 5	7.8 5	5.9 5
38	5.2 5	4.4 5	4.8 5	4.0 5	3.5 5	5.6 5	5.2 5	4.5 5	7.4 5	6.1 5
39	0.0 9	4.6 9	4.8 9	0.0 9	3.6 9	5.7 9	5.2 9	4.5 9	7.4 9	6.1 9
40	5.2 8	4.5 8	4.7 8	3.9 8	3.5 8	5.5 8	0.0 8	4.5 8	7.5 8	6.1 8
41	5.4 6	4.6 6	4.9 6	4.1 6	3.5 6	5.5 6	5.1 6	4.6 6	7.5 6	5.9 6
42	5.3 5	4.6 5	4.8 5	4.0 5	3.6 5	5.8 5	5.1 5	4.7 5	7.4 5	6.1 5
43	5.3 3	4.5 3	4.8 3	4.0 3	3.6 3	5.7 3	5.4 6	4.5 6	7.6 6	6.1 6
44	5.6 6	4.8 6	5.0 6	4.2 6	3.8 6	5.7 6	5.1 6	4.6 6	8.0 6	6.0 6
45	0.0 6	4.5 6	4.8 6	4.0 6	3.5 6	5.7 6	5.3 6	4.5 6	7.6 6	6.1 6
46	5.4 5	4.6 5	5.0 5	4.1 5	3.5 5	5.5 5	0.0 5	0.0 5	7.9 5	6.1 5
47	5.4 5	4.4 5	4.8 5	0.0 5	3.4 5	5.4 5	5.1 5	4.5 5	0.0 5	6.0 5
48	5.2 5	4.4 5	4.9 5	4.0 5	3.6 5	0.0 5	5.1 5	4.5 5	0.0 5	6.1 5
49	5.3 6	4.5 6	4.8 6	4.0 6	3.4 6	0.0 6	5.1 6	4.6 6	7.7 6	6.1 6

Table 24.2 *(cont.)*

Laboratory	Measurements									
50	0.0 5	4.6 5	4.9 5	0.0 5	3.6 5	5.7 5	5.2 5	4.6 5	0.0 5	0.0 5
51	5.2 9	4.6 9	0.0 9	4.0 9	3.5 9	5.6 9	5.2 9	4.5 9	7.3 9	5.9 9
52	5.4 7	4.6 7	4.9 7	4.1 7	3.5 7	5.7 7	5.2 7	4.6 6	7.7 6	6.1 6
53	5.3 6	4.5 6	4.6 6	4.0 6	3.5 6	5.7 6	5.1 6	4.6 6	7.4 6	6.1 6
54	5.4 5	0.0 5	5.0 5	4.0 5	3.6 5	5.6 5	5.2 5	4.5 5	7.7 5	0.0 5
55	5.4 5	4.7 5	5.0 5	4.3 5	3.5 5	5.4 5	5.2 5	4.7 5	7.6 5	6.0 5
56	5.4 8	4.4 8	4.9 8	4.0 8	3.5 8	0.0 8	4.9 8	0.0 8	7.5 8	5.9 8
57	5.3 6	4.6 6	4.8 6	4.1 6	3.5 6	5.5 6	5.2 6	4.5 6	7.6 6	6.1 6
58	5.3 6	4.3 6	4.7 6	0.0 6	0.0 6	5.6 6	5.2 6	4.6 6	0.0 6	0.0 6
59	5.4 6	4.6 6	5.0 6	4.2 6	0.0 6	5.7 6	5.4 6	4.6 5	7.7 5	6.3 5
60	5.3 5	4.6 5	0.0 5	4.1 5	3.4 5	5.5 5	5.1 5	4.5 5	7.7 5	6.1 5
61	5.4 5	4.6 5	4.9 5	4.2 5	3.6 5	5.6 5	5.3 5	4.6 5	7.7 5	6.2 5
62	5.4 8	4.7 8	5.0 8	4.0 8	3.4 8	5.5 8	0.0 8	4.3 8	7.8 8	6.0 8
63	5.7 4	4.6 4	4.7 4	4.1 4	3.5 4	4.7 4	5.2 4	4.7 4	7.7 4	6.2 4
64	5.5 5	4.6 5	4.8 5	4.1 5	0.0 5	5.8 5	5.2 5	0.0 5	0.0 5	6.3 5
65	5.0 5	4.7 5	4.8 5	4.0 5	0.0 5	5.6 5	5.1 5	4.5 5	0.0 5	0.0 5
66	5.2 6	4.4 6	4.8 6	4.0 6	3.6 6	5.6 6	5.2 6	4.5 6	0.0 6	5.9 6
67	5.5 5	4.5 5	4.9 5	4.1 5	3.5 5	5.7 5	5.2 5	4.6 5	7.8 5	6.2 5
68	5.2 3	4.5 3	5.0 3	4.1 3	3.5 3	5.5 3	5.1 3	4.5 3	7.4 3	6.1 3
69	5.4 5	4.6 5	4.8 5	4.0 5	3.5 5	5.6 5	5.3 5	4.6 5	7.6 5	0.0 5
70	0.0 5	4.5 5	4.8 5	0.0 5	0.0 5	0.0 5	5.2 5	0.0 5	7.6 5	6.2 5
71	5.1 8	4.4 8	4.7 8	4.0 8	3.7 8	0.0 8	5.1 8	4.6 8	0.0 8	6.1 8
72	0.0 9	4.7 9	0.0 9	4.1 9	3.6 9	0.0 9	4.9 9	0.0 9	0.0 9	0.0 9
73	4.0 5	4.6 5	4.9 5	4.0 5	3.6 5	5.6 5	5.1 5	4.6 5	0.0 5	6.1 5
74	5.1 9	4.3 9	4.8 9	4.1 9	3.5 9	5.6 9	5.1 9	4.5 9	7.5 9	6.0 9
75	0.0 3	0.0 3	0.0 3	0.0 3	0.0 3	0.0 3	0.0 3	0.0 3	0.0 3	0.0 3
76	4.9 8	0.0 8	4.7 8	3.9 8	3.4 8	5.4 8	4.9 8	4.6 8	7.4 8	5.9 8
77	5.1 5	4.5 5	4.9 5	4.1 5	3.5 5	0.0 5	5.1 5	4.7 5	7.8 5	6.2 5
78	5.4 5	4.5 5	4.9 5	4.2 5	3.6 5	0.0 5	5.0 5	4.5 5	7.4 5	6.0 5
79	5.2 5	4.5 5	4.8 5	4.1 5	3.6 5	5.5 5	5.1 5	4.6 5	0.0 5	6.2 5
80	0.0 6	4.7 6	0.0 6	4.2 6	3.6 6	5.6 6	5.2 6	4.5 6	0.0 6	0.0 6
81	5.4 5	4.5 5	5.1 5	4.1 5	3.6 5	5.8 5	5.0 5	4.5 5	7.6 5	6.1 5
82	0.0 3	0.0 3	0.0 3	0.0 3	0.0 3	0.0 3	0.0 3	0.0 3	0.0 3	0.0 3
83	5.2 6	4.3 6	4.6 6	3.8 6	3.3 6	5.5 6	4.8 6	4.4 6	7.2 6	5.8 6
84	0.0 0	0.0 0	0.0 0	0.0 0	0.0 0	0.0 0	0.0 0	0.0 0	0.0 0	0.0 0
85	5.4 8	0.0 8	0.0 8	4.6 8	0.0 8	5.6 8	5.3 8	0.0 8	7.3 8	6.2 8
86	0.0 3	0.0 3	0.0 3	0.0 3	0.0 3	0.0 3	0.0 3	0.0 3	0.0 3	0.0 3
87	5.3 6	4.5 6	4.8 6	4.0 6	3.5 6	5.5 6	5.2 6	4.6 6	7.5 6	6.0 6
88	5.2 6	4.5 6	4.7 6	4.0 6	3.5 6	5.5 6	5.2 6	4.5 6	7.5 6	5.9 6
89	5.3 6	4.5 6	4.9 6	4.0 6	3.6 6	5.5 6	5.1 6	4.5 6	7.6 6	0.0 6
90	5.4 5	4.6 5	4.9 5	4.1 5	3.6 5	5.7 5	5.0 5	4.6 5	7.7 5	6.1 5
91	5.4 9	4.5 9	4.9 9	4.2 9	3.6 9	5.6 9	5.2 9	4.7 9	7.6 9	6.0 9
92	5.6 5	4.5 5	4.9 5	4.1 5	3.5 5	5.6 5	5.0 5	4.5 5	7.5 5	6.1 5
93	5.4 6	4.5 6	4.8 6	4.1 6	3.5 6	5.6 6	0.0 6	4.6 6	0.0 6	6.2 6
94	5.2 4	4.5 4	4.7 4	0.0 4	3.6 4	5.7 4	5.0 4	4.6 4	7.6 4	5.8 4
95	5.2 5	4.5 5	4.7 5	4.0 5	3.5 5	5.6 5	5.0 5	4.4 5	7.4 5	6.0 5
96	5.4 5	4.5 5	3.8 5	4.1 5	5.2 5	5.9 5	4.8 5	4.4 5	7.7 5	6.0 5
97	5.2 5	4.6 5	4.8 5	4.1 5	3.5 5	5.6 5	5.0 5	4.5 5	7.5 5	6.3 5
98	5.3 9	4.4 9	4.8 9	0.0 9	3.6 9	0.0 9	5.0 9	4.5 9	7.5 9	6.0 9
99	5.3 5	4.4 5	0.0 5	4.0 5	3.4 5	5.9 5	0.0 5	4.7 5	7.9 5	6.2 5
100	0.0 5	4.5 5	4.9 5	4.1 5	3.4 5	5.6 5	5.0 5	4.6 5	7.8 5	6.3 5

* The integer beside each observation denotes the method.
0 denotes missing value.

25. Calcium Assay

Source and Contributor M.J.R. Healy London School of Hygiene and Tropical Medicine

The data relate to a chemical assay of calcium discussed in Brown, Healy and Kearns (1981). A set of standard solutions is prepared and these and the unknowns are read on a spectrophotometer in arbitrary units. A straight-line response curve is fitted to the standards and the values of the unknowns are read off from this. The preparation of the standard and unknown solutions involves a fair amount of laboratory manipulation, and the actual concentrations of the standards may differ slightly from their target values, the very precise instrumentation being capable of detecting this. The target values are 2.0, 2.0, 2.5, 3.0, 3.0 mmol. per litre; the 'duplicates' are made up independently. The sequence of reading the standards and unknowns is repeated four times. Two specimens of each unknown are included in each assay, and the four sequences of readings are done twice, first with the flame conditions in the instrument optimized and then with a slightly weaker flame.

The data relate to assays on the above pattern of a set of 6 unknowns performed by 12 laboratories A, B, C, D, E, F, G, J, K, L, M and N. An important aspect of the analysis is the complex error structure. In particular, there appears to be among-laboratory variation which is too big to be accounted for by any of the known factors.

The data are given in Table 25.1. They consist of two replicates for each laboratory and four readings of each solution. The standards are identified as 2.0A, 2.0B, 2.5, 3.0A, 3.0B; the unknowns are identified as as U1, U2, W1, W2, Y1, Y2. Target values, obtained by an independent method, are available for these unknowns as follows: U = 2.63, W = 2.269, Y = 2.800. Unknown W was a liquid serum and there was some question of the homogeneity of the calcium concentration from vial to vial.

A target for analysis might be the drafting of a protocol for the statistical results which would give a guarantee of a stated margin of error or a claim that no such guarantee can be given.

References

Brown, S.S, Healy, M.J.R. and Kearns, M. (1981). Report on the interlaboratory trial of the reference method for the determination of total calcium in serum. *J. Clin. Chem. Clin. Biochem.* **19**, 395-426.

Table 25.1
Chemical Assay of Calcium Measurements from Twelve Laboratories for Six Unknowns and Three Standards

Replicate	Solution	Measurements			
		Laboratory A, Replicate 1			
A1	W1	1206	1202	1202	1201
A1	2.0A	1068	1071	1067	1066
A1	W2	1194	1193	1189	1185
A1	2.0B	1072	1068	1064	1067
A1	U1	1387	1387	1384	1380
A1	2.5	1333	1321	1326	1317
A1	U2	1394	1390	1383	1376
A1	3.0A	1579	1576	1578	1572
A1	Y1	1478	1480	1473	1466
A1	3.0B	1579	1571	1579	1567
A1	Y2	1483	1477	1482	1472
		Laboratory A, Replicate 2			
A2	W1	1124	1133	1139	1133
A2	2.0A	1002	1001	1002	1003
A2	W2	1116	1125	1125	1117
A2	2.0B	1008	1006	1009	1005
A2	U1	1304	1306	1305	1302
A2	2.5	1248	1248	1251	1246
A2	U2	1301	1305	1308	1305
A2	3.0A	1487	1479	1483	1486
A2	Y1	1382	1388	1390	1385
A2	3.0B	1478	1485	1484	1484
A2	Y2	1386	1391	1385	1385
		Laboratory B, Replicate 1			
B1	W1	1017	1017	1012	1020
B1	2.0A	910	916	915	921
B1	W2	1012	1018	1015	1023
B1	2.0B	913	923	914	921
B1	U1	1188	1199	1197	1202
B1	2.5	1129	1148	1136	1147
B1	U2	1186	1196	1193	1199
B1	3.0A	1359	1378	1370	1373
B1	Y1	1263	1280	1280	1279
B1	3.0B	1349	1361	1359	1363
B1	Y2	1259	1269	1259	1265
		Laboratory B, Replicate 2			
B2	W1	965	961	960	962
B2	2.0A	869	869	875	870
B2	W2	970	971	968	964
B2	2.0B	870	874	872	877
B2	U1	1129	1133	1126	1134
B2	2.5	1076	1084	1091	1082
B2	U2	1131	1128	1124	1123
B2	3.0A	1297	1294	1292	1300
B2	Y1	1207	1204	1206	1210
B2	3.0B	1290	1286	1281	1291
B2	Y2	1206	1203	1191	1202

Table 25.1 *(cont.)*

Replicate	Solution	Measurements			
		Laboratory C, Replicate 1			
C1	W1	1090	1098	1090	1100
C1	2.0A	969	975	969	972
C1	W2	1088	1092	1087	1085
C1	2.0B	969	960	960	966
C1	U1	1270	1261	1261	1269
C1	2.5	1196	1196	1209	1200
C1	U2	1261	1268	1270	1273
C1	3.0A	1451	1440	1439	1449
C1	Y1	1352	1349	1353	1343
C1	3.0B	1439	1433	1433	1445
C1	Y2	1349	1353	1349	1355
		Laboratory C, Replicate 2			
C2	W1	875	877	875	883
C2	2.0A	763	764	768	770
C2	W2	864	870	867	867
C2	2.0B	769	769	768	763
C2	U1	1006	1015	1010	1013
C2	2.5	957	957	962	958
C2	U2	1007	1007	1011	1010
C2	3.0A	1152	1154	1154	1158
C2	Y1	1083	1079	1079	1085
C2	3.0B	1158	1152	1156	1158
C2	Y2	1078	1079	1073	1085
		Laboratory D, Replicate 1			
D1	2.0A	1122	1117	1119	1120
D1	W2	1256	1254	1256	1263
D1	W1	1260	1251	1252	1264
D1	2.0B	1122	1110	1111	1116
D1	U2	1453	1447	1451	1455
D1	2.5	1386	1381	1381	1387
D1	U1	1450	1446	1448	1457
D1	3.0A	1656	1663	1659	1665
D1	Y2	1543	1548	1543	1545
D1	3.0B	1658	1658	1661	1660
D1	Y1	1545	1546	1548	1544
		Laboratory D, Replicate 2			
D2	W2	1128	1136	1134	1131
D2	2.0A	1003	1011	1014	1015
D2	W1	1131	1137	1140	1132
D2	2.0B	1008	1009	1011	1008
D2	U2	1322	1324	1325	1321
D2	2.5	1254	1259	1256	1252
D2	U1	1324	1322	1321	1319
D2	3.0A	1505	1507	1501	1500
D2	Y2	1403	1405	1406	1391
D2	3.0B	1500	1503	1497	1496
D2	Y1	1397	1409	1404	1391

Table 25.1 *(cont.)*

Replicate	Solution	Measurements			
		Laboratory E, Replicate 1			
E1	W1	1548	1552	1569	1560
E1	2.0A	1357	1366	1376	1364
E1	W2	1557	1559	1556	1567
E1	2.0B	1366	1362	1367	1368
E1	U1	1783	1800	1784	1797
E1	2.5	1704	1701	1716	1711
E1	U2	1802	1797	1800	1818
E1	3.0A	2026	2021	2031	2038
E1	Y1	1913	1931	1937	1941
E1	3.0B	2050	2054	2048	2068
E1	Y2	1918	1925	1935	1948
		Laboratory E, Replicate 2			
E2	W1	1548	1550	1548	1578
E2	2.0A	1366	1352	1368	1360
E2	W2	1563	1558	1569	1573
E2	2.0B	1359	1355	1368	1383
E2	U1	1782	1795	1804	1812
E2	2.5	1711	1724	1716	1712
E2	U2	1791	1813	1807	1826
E2	3.0A	2028	2030	2037	2049
E2	Y1	1914	1925	1936	1947
E2	3.0B	2026	2051	2066	2063
E2	Y2	1932	1937	1931	1953
		Laboratory F, Replicate 1			
F1	W2	1511	1499	1485	1495
F1	2.0A	1362	1352	1351	1359
F1	W1	1531	1521	1521	1535
F1	2.0B	1354	1346	1348	1354
F1	U1	1767	1752	1749	1765
F1	2.5	1684	1678	1682	1685
F1	U2	1785	1767	1768	1766
F1	3.0A	2024	2004	2000	2020
F1	Y1	1860	1862	1858	1866
F1	3.0B	1997	1992	1975	1993
F1	Y2	1882	1871	1876	1873
		Laboratory F, Replicate 2			
F2	W2	1338	1335	1331	1328
F2	2.0A	1195	1196	1203	1199
F2	W1	1362	1357	1362	1355
F2	2.0B	1215	1203	1208	1206
F2	U1	1575	1570	1561	1571
F2	2.5	1489	1483	1481	1487
F2	U2	1560	1550	1544	1556
F2	3.0A	1783	1770	1762	1768
F2	Y1	1668	1658	1653	1654
F2	3.0B	1766	1750	1749	1748
F2	Y2	1664	1664	1657	1647

Table 25.1 *(cont.)*

Replicate	Solution	Measurements			
		Laboratory G, Replicate 1			
G1	W1	1186	1194	1201	1198
G1	2.0A	1082	1092	1091	1098
G1	W2	1190	1198	1207	1199
G1	2.0B	1086	1098	1096	1096
G1	U1	1381	1379	1382	1385
G1	2.5	1378	1375	1382	1376
G1	U2	1386	1390	1388	1385
G1	3.0A	1660	1665	1674	1672
G1	Y1	1539	1534	1547	1542
G1	3.0B	1654	1655	1665	1669
G1	Y2	1532	1527	1526	1530
		Laboratory G, Replicate 2			
G2	W1	1166	1169	1162	1165
G2	2.0A	1062	1072	1069	1067
G2	W2	1172	1186	1178	1170
G2	2.0B	1071	1076	1074	1071
G2	U1	1350	1347	1354	1350
G2	2.5	1348	1350	1348	1351
G2	U2	1352	1347	1352	1351
G2	3.0A	1602	1618	1616	1610
G2	Y1	1508	1510	1511	1508
G2	3.0B	1611	1616	1604	1609
G2	Y2	1490	1486	1496	1486
		Laboratory J, Replicate 1			
J1	W1	2327	2321	2325	2321
J1	2.0A	2047	2050	2042	2049
J1	W2	2341	2349	2347	2330
J1	2.0B	2043	2042	2041	2034
J1	U1	2664	2665	2656	2657
J1	2.5	2539	2547	2565	2550
J1	U2	2664	2651	2668	2648
J1	3.0A	3026	3029	3049	3026
J1	Y1	2831	2843	2834	2799
J1	3.0B	3037	3024	3038	3014
J1	Y2	2799	2806	2813	2786
		Laboratory J, Replicate 2			
J2	W1	2330	2320	2300	2317
J2	2.0A	2068	2071	2046	2049
J2	W2	2346	2338	2311	2336
J2	2.0B	2041	2063	2035	2044
J2	U1	2686	2656	2651	2665
J2	2.5	2578	2564	2553	2558
J2	U2	2660	2662	2656	2669
J2	3.0A	3061	3073	3031	3041
J2	Y1	2869	2816	2817	2843
J2	3.0B	3064	3046	3042	3047
J2	Y2	2824	2825	2807	2821

Table 25.1 *(cont.)*

Replicate	Solution	Measurements			
		Laboratory K, Replicate 1			
K1	W1	1186	1192	1190	1190
K1	2.0A	1067	1085	1078	1076
K1	W2	1221	1228	1227	1234
K1	2.0B	1076	1089	1075	1088
K1	U1	1435	1435	1434	1435
K1	2.5	1361	1358	1347	1363
K1	U2	1438	1428	1423	1426
K1	3.0A	1643	1645	1641	1640
K1	Y1	1520	1523	1527	1521
K1	3.0B	1623	1630	1641	1638
K1	Y2	1527	1530	1533	1530
		Laboratory K, Replicate 2			
K2	W1	1158	1146	1139	1132
K2	2.0A	1038	1037	1024	1033
K2	W2	1186	1178	1174	1181
K2	2.0B	1035	1037	1027	1036
K2	U1	1370	1380	1375	1374
K2	2.5	1277	1279	1271	1284
K2	U2	1354	1346	1369	1357
K2	3.0A	1543	1545	1557	1547
K2	Y1	1438	1438	1438	1432
K2	3.0B	1538	1533	1531	1540
K2	Y2	1451	1434	1438	1431
		Laboratory L, Replicate 1			
L1	W1	1017	1016	1017	1020
L1	2.0A	798	796	802	807
L1	W2	1014	1017	1020	1023
L1	2.0B	805	797	799	801
L1	U1	1127	1130	1128	1137
L1	2.5	1003	999	997	1001
L1	U2	1162	1158	1157	1153
L1	3.0A	1201	1199	1197	1206
L1	Y2	1172	1171	1160	1167
L1	3.0B	1208	1201	1199	1195
L1	Y1	1175	1173	1167	1169
		Laboratory L, Replicate 2			
L2	W1	1019	1020	1011	1022
L2	2.0A	804	807	799	796
L2	W2	1020	1025	1014	1015
L2	2.0B	798	797	795	801
L2	U1	1129	1120	1130	1131
L2	2.5	997	995	1001	995
L2	U2	1154	1162	1156	1159
L2	3.0A	1196	1195	1206	1201
L2	Y2	1169	1167	1164	1162
L2	3.0B	1202	1197	1197	1199
L2	Y1	1180	1168	1172	1170

Table 25.1 *(cont.)*

Replicate	Solution	Measurements			
		Laboratory M, Replicate 1			
M1	W1	1146	1161	1150	1152
M1	2.0A	1068	1048	1042	1052
M1	W2	1183	1164	1158	1165
M1	2.0B	1053	1052	1054	1052
M1	U1	1360	1370	1344	1340
M1	2.5	1292	1314	1301	1310
M1	U2	1363	1367	1348	1333
M1	3.0A	1573	1560	1552	1561
M1	Y1	1494	1473	1466	1443
M1	3.0B	1577	1574	1568	1549
M1	Y2	1448	1476	1449	1440
		Laboratory M, Replicate 2			
M2	2.0A	1047	1062	1055	1042
M2	W1	1158	1176	1190	1164
M2	W2	1171	1189	1157	1137
M2	2.0B	1057	1062	1045	1048
M2	U1	1352	1364	1353	1355
M2	2.5	1312	1308	1304	1297
M2	U2	1366	1360	1356	1347
M2	3.0A	1561	1559	1554	1539
M2	Y1	1450	1460	1447	1425
M2	3.0B	1560	1575	1548	1545
M2	Y2	1462	1462	1449	1440
		Laboratory N, Replicate 1			
N1	W1	1338	1346	1354	1356
N1	2.0A	1204	1208	1215	1214
N1	W2	1357	1369	1377	1365
N1	2.0B	1203	1204	1210	1206
N1	U1	1557	1561	1580	1568
N1	2.5	1496	1501	1501	1508
N1	U2	1560	1565	1577	1569
N1	3.0A	1789	1792	1811	1802
N1	Y1	1659	1660	1664	1673
N1	3.0B	1785	1782	1793	1801
N1	Y2	1652	1648	1663	1665
		Laboratory N, Replicate 2			
N2	W1	0678	0684	0691	0682
N2	2.0A	0610	0616	0620	0610
N2	W2	0688	0691	0700	0685
N2	2.0B	0606	0614	0615	0605
N2	U1	0787	0792	0801	0783
N2	2.5	0759	0761	0763	0748
N2	U2	0790	0793	0799	0782
N2	3.0A	0909	0917	0916	0899
N2	Y1	0841	0843	0845	0829
N2	3.0B	0908	0912	0902	0893
N2	Y2	0838	0847	0838	0830

26. Determination of Dunham Coefficients for the Ground State of D_2

Source Bredohl, H. and Herzberg, G. (1973). The Lyman and Werner bands of deuterium. *Canadian J. Physics* **51**, 867-887.

Contributors I. Dabrowski and G. Herzberg National Research Council of Canada

The spectrum of any atomic system is determined by its energy levels. For a diatomic molecule in a given electronic state these energy levels may be expressed by

$$G(v)+F_v(J)=\sum_{ik} Y_{ik}(v+1/2)^i\, J^k(J+1)^k, \tag{1}$$

where $G(v)$ is the vibrational energy, v = 0,1,2,... the vibrational quantum number, $F_v(J)$ the rotational energy in the vibrational level v, and J = 0,1,2,... is the rotational quantum number (i,k = 0,1,...). For the example of the D_2, deuterium, molecule the observed energies $G(v)+F_v(J)$ are listed in Table 26.1.

The quantities of greatest interest are Y_{10} and Y_{01} since they are very nearly equal to ω_e, the vibrational frequency for infinitesimal amplitude, and B_e the rotational constant in the equilibrium position which is directly related to the equilibrium inter-nuclear distance r_e [$B_e = h/(8\pi^2 c\mu r_e)$, μ = reduced mass, h = Planck's constant, c = speed of light].

The most direct way of evaluating the constants Y_{ik} is of course a least squares solution of (1) for the whole set of experimental energies in Table 26.1. The values of the Y_{ik} are found to vary slightly with the order of the polynomial and with the number of levels included in the fit.

Often there is a need for intermediate information, *viz.* the rotational constants B_v, D_v and coefficients of higher terms for a given vibrational level defined by

$$F_v(J)=B_v\, J(J+1)-D_v\, J^2(J+1)^2+\cdots \tag{2}$$

and the vibrational intervals

$$\Delta G(v+1/2)=G(v+1)-G(v)$$

with

(3) $$G(v)=\omega_e(v+1/2)-\omega_e x_e(v+1/2)^2+\cdots,$$

where in usually sufficient approximation

(4) $$B_v=B_e-\alpha_e(v+1/2)+\cdots=Y_{01}+Y_{11}(v+1/2)+\cdots,$$

(5) $$D_v=D_e+\beta_e(v+1/2)+\cdots=Y_{02}+Y_{12}(v+1/2)+\cdots.$$

In Table 26.2 the B_v and D_v obtained separately for each v, by least squares, from the data of Table 26.1 are listed. These can now be used as input data in the least squares solution of (4) and (5) to determine B_e,α_e,D_e,β_e. In the present case of D_2, the values 30.443_6, 1.078_6, 0.011345, -2.220×10^{-4} were obtained.

In Table 26.3, the ΔG values obtained from the observed energy values for $J = 0$ are given which can be used as input data in the least squares solution of (3) to determine the vibrational constants ω_e, $\omega_e x_e$, ... In the present case the values $\omega_e=3115.50, \omega_e x_e=61.82, \omega_e y_e=0.562$ were obtained.

Exactly the same problem exists for another isotope of hydrogen, HD. The data for this molecule are in Dabrowski and Herzberg (1976). For discussion of the application of robust regression to a related problem, see Beaton and Tukey (1974).

References

Beaton, A.E. and Tukey, J.W. (1974). The fitting of power series, meaning polynomials, illustrated on band-spectroscopic data. *Technometrics* **16**, 147-185.

Dabrowski, I. and Herzberg, G. (1976). The absorption and emission spectra of HD in the vacuum ultraviolet. *Canadian J. Physics* **54**, 525-567.

Table 26.1
Observed Energy Levels *

$J\backslash v$	0	1	2	3	4	5
0	0.00	2993.60	5868.46	8625.71	11267.80	13795.72
1	59.79	3051.28	5924.01	8679.24	11319.31	13845.24
2	179.01	3166.28	6034.87	8786.11	11422.22	13944.08
3	357.25	3338.21	6200.37	8945.73	11575.75	14091.66
4	593.64	3566.22	6420.27	9157.42	11779.47	14287.52
5	887.08	3849.22	6693.10	9420.29	12032.36	14530.05
6	1236.38	4186.02	7017.74	9733.13	12332.65	14818.70
7	1640.09		7392.64	10094.22	12680.61	15 152.64
8	2096.36		7815.90	10504.21	13075.19	15534.12
9	2603.45					

$J\backslash v$	6	7	8	9	10
0	16210.27	18511.38	20698.87	22771.48	24727.63
1	16257.74	18556.89	20742.36	22813.01	24766.99
2	16352.62	18647.73	20829.13	22895.66	24845.50
3	16494.18	18783.33	20958.68	23019.07	24962.70
4	16681.90	18963.14	21130.45	23182.73	25117.97
5	16914.73	19186.16	21343.41	23385.54	25310.49
6	17191.91	19451.20	21596.41	23626.47	25538.88
7		19756.83	21888.53	23904.55	25802.65
8		20102.38	22217.90		
9			22583.96		

$J\backslash v$	11	12	13	14	15
0	26564.54	28278.54	29864.94	31317.47	32628.33
1	26601.74	28313.52	29897.56	31347.67	32655.89
2	26675.98	28383.31	29962.67	31407.82	32710.71
3	26786.70	28487.43	30059.77	31497.48	32792.44
4	26933.40	28625.17	30188.24	31616.05	32900.30
5	27115.07	28795.86	30347.22	31762.67	33033.59
6	27330.96	28998.22	30535.55	31936.15	33191.13
7	27579.49	29231.37	30752.37	32135.44	33371.70
8			30995.98	32359.24	33573.58
9			31264.70		

$J\backslash v$	16	17	18	19	20	21
0	33787.61	34782.80	35598.11	36213.83	36605.18	36746.82
1	33812.24	34804.24	35615.97	36227.52	36613.81	36748.79
2	33861.33	34846.91	35651.46	36254.62	36630.62	
3	33934.39	34910.38	35704.05	36294.53	36654.89	
4	34030.66	34993.82	35772.89	36346.23	36685.32	
5	34149.45	35096.41	35857.00	36408.60	36719.85	
6	34289.37	35216.75	35954.99	36479.87		
7	34449.19	35353.39	36065.24			
8	34627.80					

* The values are given in $(\mathrm{cm})^{-1}$.

Table 26.2
Rotational Constants in the Ground State

v	B_v $\pm 0.0020(\mathrm{cm}^{-1})$	D_v $\pm 0.00030(\mathrm{cm}^{-1})$
0	29.9084	0.0113_7
1	28.8485	0.0108_5
2	27.8226†	0.01094†
3	26.7966	0.01057
4	25.7863	0.01034
5	24.7730	0.00995
6	23.7756	0.00980
7	22.7787	0.00968
8	21.7679	0.00950
9	20.7459	0.00935
10	19.7018	0.00929

† These values are much less accurate than the others because of the poor quality of most of the bands with $v'' = 2$.

Table 26.3
Vibrational Intervals in the Ground State

v	$\Delta G(v + 1/2)$
0	2993.57
1	2874.38
2	2757.18
3	2642.20
4	2528.08
5	2414.52
6	2301.21
7	2187.51
8	2072.61
9	1956.14
10	1836.87

27. Clock Intercomparisons

Source The National Research Council of Canada, Time and Frequency Section, Physics Division

Contributors H. Daams and D. Morris National Research Council of Canada

These data on clock intercomparisons give the time difference between secondary cesium clocks in 0.01 microseconds. Tables 27.1 to 27.3 present the data. Each line in the tables represents the data for one week; no observations were taken on week-ends. The last two columns indicate on what days the clock output jumped and by what amount. All successive data should be corrected for this. These jumps are caused by spurious pulses in divider chains and are, therefore, always integral amounts of 0.2 microseconds. Missing data during a weekday indicate that no measurements were taken because of failure in equipment.

The data form a time series giving phase or time values of two clocks relative to each other, each clock having a certain offset which changes slowly; The first difference, therefore, increases slowly; this increment is the frequency difference. After the slow variation has been removed, for example by fitting a second-degree or third-degree polynomial, the result may be described by a random component, the parameters of which depend on the type of clock used.

The clocks used are HP1, HP2 and HP4. The clocks are subject to a slow drift on which noise is superimposed. The clocks were compared in the following pairs:

HP4 *vs* HP1, June 29, 1970 - March 31, 1971;
HP2 *vs* HP4, June 29, 1970 - April 2, 1971;
HP2 *vs* HP1, August 17, 1970 - June 4, 1971.

The best procedure is first to take the difference between readings to find out if there are any improper data recorded. If this is the case, these data are replaced by interpolating between their neighbours.

Table 27.1
Time Differences between Clocks HP4 and HP1, June 29, 1970 to March 31, 1971*

Day of Week					Shifts	
1	2	3	4	5	Day	Amount
99986817	9986787		99986750	99986726		
99986641	9986627	99986594	99986619	99986566		
99986489	9986436	99986415	99986402	99986453		
99986307	9986255	99986239	99986213	99986203		
99986227	9986186	99986194	99986173	99986131		
	9986056	99986000	99985963	99985914		
99985851			99984869	99984887	4	-1000
99984931	9984884	99984860	99984844	99984898		
99984841	9984831	99984812	99984834	99984784		
99984703	9984677	99984639	99984608	99984563		
	9984461	99984441	99984415	99984353		
99984263	9984247			99984242		
99984094	9984056	99983986	99983995	99983928		
99983880	9983850	99983809	99983766	99983704		
99983636	9983577	99983544	99983518	99983464		
	9983378	99983345	99983318	99983298		
99983165	9983161	99983067	99983106	99983018		
99982927	9982877	99982876	99982866	99982825		
99982791	9982723	99982749	99982668	99982673		
99982559	9982490		99982512	99982469		
99982378	9982359	99982321	99982277	99982293		
99982199	9982135	99982114	99982065	99982036		
99981963	9981897	99981897	99981847	99981771		
99981683	9981653	99981609	99981586	99981562		
99981501	9981466	99981430	99981406	99981397		
99981351	9981342	99981321	99981273			
	9981135	99981138	99981131			
99981017	9981017	99981011	99980984	99980956		
99980922	9980879	99980874	99980826	99980848		
99980796	9980753	99980701	99980663	99980663		
99980590	9980582	99980544	99980541	99980499		
99980449	9980438	99980415	99980363	99980380		
99980320	9980314	99980283	99980256	99980230		
99980140	9980142	99980092	99980074	99980073		
99979955	9979918	99979900	99979852	99979855		
99979805	9979768	99979682	99979670	99979695		
99979620	9979571	99979564	99979535	99979541		
99979461	9979424	99979427	99979406	99979413		
99979436	9979458	99979439	99979414	99979432		
99979415	9979407					

*Units are in 0.01 microseconds.

Table 27.2
Time Differences between Clocks HP2 and HP4, June 29, 1970 to April 2, 1971*

Day of Week					Shifts	
1	2	3	4	5	Day	Amount
3502935	3502971		3503033	3503062		
3503150	3503178	3503203	3503230	3504262	5	+1000
3504341	3504373	3504391	3504419	3504442		
3504544	3504577	3504611	3504646	3504665		
3504703	3504716	3504731	3504767	3504809		
	3505957	3505992	3506033	3506077	2	+1000
3506191	3506231	3514255	3514269	3514278	3	+8000
3516303	3516319	3516332	3516342	3516352	1	+2000
3516391	3516401	3516418	3516428	3516431		
3516573	3516609	3516648	3516683	3516724		
	3516879	3516919	3516958	3516990		
3517101	3517130			3517233		
3517350	3517390	3517432	3517469	3518501	5	+1000
3520610	3520645	3520682	3520726	3520766	1	+2000
3520882	3520927	3521965	3522003	3522041	3	+10
	3522188	3522225	3522259	3522296		
3522421	3522463	3522510	3522547	3522587		
3522712	3522752	3522779	3522804	3522836		
3522934	3522964	3523001	3523033	3523068		
3523185	3523219		3523289	3523322		
3523429	3523460	3523500	3523540	3523579		
3523699	3523742	3523782	3523822	3523865		
3523985	3524025	3524070	3524113	3524159		
3524277	3524323	3524365	3524400	3524437		
3524546	3524581	3524601	3524632	3524661		
3524746	3524774	3524804	3524833			
	3524975	3525007	3525031			
3525140	3525173	3525201	3525226	3525244		
3525319	3525350	3525378	3525407	3525430		
3525512	3525549	3525580	3525611	3526649	5	+1000
3526740	3526772	3526802	3526828	3526853		
3526937	3526961	3526995	3527018	3527046		
3527128	3527154	3527193	3527230	3527265		
3527384	3527425	3527471	3527507	3527545		
3527666	3527709	3527751	3527791	3527831		
3527938	3527975	3528008	3528039	3528076		
3528184	3528222	3528255	3528290	3528323		
3528428	3528488	3528482	3528494	3528503		
3528531	3533558	3533559	3533567	3533578	2	+5000
3533623	3533633	3533644	3533659	3533672		

*Units are in 0.01 microseconds.

Table 27.3
Time Differences between Clocks HP2 and HP1, August 17, 1970 to June 4, 1971*

Day of Week					Shifts	
1	2	3	4	5	Day	Amount
8398780	8398778	8398768	8398763	8398761		
8398735	8398728	8398718	8398712	8398707		
8398684	8398676	8398670	8398662	8398656		
	8398633	8398624	8398615	8398608		
8398591	8398579			8398566		
8398532	8398521	8398507	8398496	8397491	5	-1000
8395482	8395476	8395468	8395457	8395449	1	-2000
8395430	8395423	8394417	8394410	8394398	3	-1000
	8394376	8394373	8394369	8394361		
8394351	8394340	8394329	8394321	8394311		
8394285	8394277	8394266	8394264	8394255		
8394237	8394233	8394229	8394222	8394216		
8394189	8394181		8394155	8394148		
8394122	8394114	8394104	8394097	8394085		
8394053	8394041	8394030	8394020	8394012		
8393983	8393975	8393962	8393947	8393937		
8393914	8393895	8393883	8393879	8393874		
8393846	8393834	8393829	8393818	8393811		
8393783	8393774	8393761	8393751			
	8393714	8393706	8393702			
8393670	8393659	8393653	8393642	8393638		
8393620	8393606	8393592	8393583	8393578		
8393542	8393531	8393525	8393517	8392503	5	-1000
8392476	8392464	8392458	8392453	8392444		
8392418	8392412	8392401	8392394	8392381		
8392353	8392339		8392309	8392300		
8392270	8392258	8392243	8392233	8392224		
8392193	8392176	8392165	8392156	8392145		
8392110	8392098	8392089	8392076	8392066		
8392015	8392002	8391994	8391980	8391969		
8391944	8391935	8391912	8391908	8391894		
8391872	8386858	8386842	8386836	8386824	2	-5000
8386784	8386779	8386788	8386755	8386740		
8386703	8386690	8386675	8386663			
	8386593	8386579	8386567	8386553		
8386513	8386497	8386480	8386471	8386456		
8386410	8386398	8386382	8386367	8386353		
8386313	8386299	8386286	8386273	8386260		
8386221	8386204	8386185	8386163	8386142		
8386109	8386099	8386086	8386072	8386058		
	8386002	8385994	8385980	8385965		
8385922	8385910	8385896	8385884	8385875		

*Units are in 0.01 microseconds.

28. An Evaluation of Cryogenic Flow Meters

Source Joiner, B.L. (1977). Evaluation of cryogenic flow meters: An example in non-standard experimental design and analysis. *Technometrics* **19**, 353-379

Contributor B.L. Joiner University of Wisconsin

Liquid nitrogen, liquid oxygen and other liquefied gases are sold widely. At the time of the present experiment there were no satisfactory meters available for measuring the flow of such fluids. Also there was no national reference standard available by which the meters that were used could be evaluated. The United States National Bureau of Standards had designed and built a cryogenic flow test facility but its stability and accuracy were not known. An experiment was designed to test the accuracy of the meters and the flow velocity. Thus an important difference of this experiment from many is that it was necessary to not only evaluate the object being measured but also the measuring device. The new facility was a complex apparatus with a variety of components each of which might be subject to some error. It was necessary to evaluate accuracy as well as precision.

The meters were subjected to two types of test, *rangeability* and *stability*. The purpose of the rangeability test was to subject the meter to a variety of conditions and to observe its response to the various conditions. The purpose of the stability test was to test the performance of the meter over a period of time during which a large amount of fluid was metered (approximately 240,000 gallons in 80 hours). The rangeability test was a designed experiment so that the effect of various factors could be separated, but the stability test was taken at whatever conditions were convenient for the operator. Each meter was subjected to a rangeability test both before and after the stability test.

The factors varied in the rangeability test were the temperature, pressure and flow rate of the fluid. The time-order in which the measurements were to be made was believed to be a potentially important variable, but was not one that could be economically randomized due to practical considerations associated with the time required to change temperature and pressure.

The data are given in Tables 28.1, 28.2 and 28.3. Some results are missing. These runs were made but in most cases their readings were not recorded primarily because either of the electronic counters failed to trigger a command or the control system failed to function properly. A few of the missing observations in the first rangeability test correspond to readings at a flow rate of 10 gallons per minute which turned out to be too slow, leading to excessive scatter in the readings.

Table 28.1
Rangeability and Stability Test Readings
From Cryogenic Flow Meters on Liquid Nitrogen
First Rangeability Test *

Day	Run No.	Temp.	Press. lbs./sq.in.	Dens.	Code	Wt. lbs.	Apparent wt.	Time sec.	Pulses	Ullage temp.
-12	1	85.54	75.1	6.42	9.39	577.82	581.70	104.13	82.0	140.53
-12	2	85.54	74.3	6.42	9.24	577.77	581.69	102.14	82.0	129.53
-12	3	85.63	85.2	6.42	10.96	577.03	581.52	101.82	82.0	122.17
-12	4	85.08	85.2	6.44	11.52	578.07	576.49	129.77	81.0	115.36
-12	5	84.5	89.1	6.46	12.65	577.93	578.68	128.98	81.0	105.54
-12	6	84.57	88.7	6.46	12.53	570.79	571.28	126.80	80.0	100.33
-12	15	89.08	113.8	6.28	11.04	579.41	582.56	133.16	84.0	98.85
-12	16	89.27	114.6	6.27	10.95	578.57	581.84	132.10	84.0	103.31
-12	17	89.34	112.4	6.27	10.62	577.75	581.51	128.47	84.0	98.68
-12	18	89.82	96.9	6.24	8.36	574.74	579.24	183.29	84.0	95.90
-12	19	90.38	98.5	6.22	8.00	572.43	577.00	181.44	84.0	98.59
-12	20	90.86	99.8	6.19	7.68	583.76	588.71	185.92	86.0	98.25
-2	4	80.25	75.7	6.63	14.79	406.84	403.20	124.25	55.0	150.43
-2	5	80.02	75.8	6.64	15.04	399.28	396.41	120.58	54.0	146.09
-2	6	80.13	77.4	6.64	15.22	398.94	396.15	122.33	54.0	144.41
-2	10	80.69	99.6	6.62	17.83	412.54	409.67	89.59	56.0	145.66
-2	11	80.67	97.7	6.62	17.62	404.28	402.40	89.49	55.0	134.47
-2	12	80.69	96.7	6.62	17.47	411.27	409.65	91.18	56.0	126.44
-2	13	80.74	113.7	6.62	19.38	404.95	402.38	72.12	55.0	139.21
-2	14	80.71	112.0	6.62	19.21	411.24	409.73	73.45	56.0	127.19
-2	15	80.64	110.5	6.62	19.12	411.19	409.89	74.44	56.0	116.33

Table 28.1 *(cont.)*

Day	Run No.	Temp.	Press. lbs./sq.in.	Dens.	Code	Wt. lbs.	Apparent wt.	Time sec.	Pulses	Ullage temp.
-1	8	86.97	100.4	6.36	11.65	532.95	534.49	94.87	76.0	89.00
-1	9	87.77	100.3	6.33	10.83	536.77	538.60	96.26	77.0	90.58
-1	10	87.91	99.5	6.32	10.60	536.33	538.09	96.99	77.0	93.17
-1	11	88.05	109.5	6.32	11.60	522.68	523.78	152.79	75.0	97.20
-1	12	88.44	111.9	6.30	11.47	527.12	529.39	156.36	76.0	95.88
-1	13	88.97	114.2	6.28	11.21	531.68	534.45	159.42	77.0	96.52
0	4	83.28	99.2	6.52	15.20	491.14	489.69	203.13	68.0	124.80
0	5	83.07	97.5	6.53	15.19	491.49	490.30	207.02	68.0	117.55
0	6	82.84	96.1	6.53	15.25	491.77	490.98	206.67	68.0	112.76
0	7	82.65	88.3	6.54	14.39	492.70	491.45	149.56	68.0	106.39
0	8	82.77	87.5	6.54	14.16	492.85	491.09	145.19	68.0	104.23
0	9	82.77	85.1	6.54	13.81	492.50	491.06	144.48	68.0	101.54
0	10	82.65	74.6	6.54	12.18	522.38	520.18	112.58	72.0	97.61
0	11	82.35	74.0	6.55	12.37	522.88	521.13	112.86	72.0	97.83
0	12	82.65	74.6	6.54	12.18	528.72	527.40	114.45	73.0	96.94
0	13	82.38	63.1	6.55	10.14	565.93	564.33	98.79	78.0	93.87
0	14	82.45	62.4	6.55	9.92	557.78	556.85	97.61	77.0	94.42
0	15	82.75	62.2	6.53	9.57	563.99	563.03	99.07	78.0	93.95

*

Day reading was taken using December 31, 1969 as reference point, i.e. 0.
Run numbers are within days.
Density of the liquid in pounds per gallon computed from temperature and pressure.
Code: this is the amount by which the temperature of the liquid is below the boiling point of the liquid at that pressure.
Apparent weight of the accumulated liquid as registered by the flow meter.
Time is time elapsed.
Pulses of the photoelectric cell counting revolutions of the flow-meter shaft.
Ullage temperature: the temperature of the gaseous helium above the liquid hydrogen in the weighing tank.

Table 28.2
Rangeability and Stability Test Readings From Cryogenic Flow Meters on Liquid Nitrogen Second Rangeability Test *

Day	Run No.	Temp.	Press. lbs./sq.in.	Dens.	Code	Wt. lbs.	Apparent wt.	Time sec.	Pulses	Ullage temp.
21	1	85.58	110.6	6.42	14.18	531.96	532.42	98.14	75.0	114.62
21	2	85.35	111.5	6.43	14.51	593.80	597.21	255.26	84.0	102.94
21	3	84.87	112.1	6.45	15.06	405.75	406.51	111.89	57.0	99.53
21	4	85.10	100.6	6.44	13.53	531.60	533.93	226.43	75.0	109.37
21	5	84.82	98.0	6.45	13.49	462.30	463.53	85.74	65.0	102.61
21	6	85.17	99.3	6.44	13.31	587.21	590.60	132.66	83.0	99.59
21	7	85.26	87.0	6.43	11.59	461.11	462.09	101.19	65.0	106.58
21	8	85.28	86.6	6.43	11.51	404.53	405.15	119.82	57.0	101.91
21	9	85.08	85.6	6.44	11.57	581.29	583.61	104.94	82.0	107.84
21	10	84.89	76.2	6.45	10.23	398.49	398.93	124.25	56.0	106.05
21	11	84.80	75.6	6.45	10.21	525.43	527.47	94.38	74.0	104.94
21	12	85.03	77.6	6.44	10.34	462.29	462.65	192.04	65.0	106.36
21	13	89.71	100.8	6.25	8.95	396.78	400.33	71.59	58.0	108.72
21	14	90.12	101.1	6.23	8.57	457.68	461.10	208.67	67.0	107.82
21	15	90.28	100.9	6.22	8.38	577.14	584.30	187.76	85.0	103.81
21	16	90.19	112.2	6.23	9.74	409.71	412.88	93.22	60.0	115.17
21	17	90.28	113.3	6.22	9.78	522.25	529.54	245.08	77.0	110.76
21	18	90.51	114.6	6.21	9.70	468.54	473.76	137.57	69.0	112.23
22	1	79.83	109.4	6.66	19.81	469.01	463.32	102.40	63.0	105.68
22	2	79.86	110.6	6.65	19.92	408.51	404.44	167.89	55.0	119.98
22	3	80.53	109.9	6.63	19.17	589.95	585.95	107.28	80.0	112.25
22	4	80.69	98.1	6.62	17.65	409.02	402.35	88.63	55.0	116.10
22	5	80.09	99.4	6.64	18.41	459.86	455.19	138.43	62.0	109.61
22	6	80.04	97.0	6.65	18.16	532.75	528.72	93.44	72.0	113.00
22	7	79.97	87.6	6.65	16.98	583.88	580.24	170.46	79.0	102.29
22	8	80.09	86.8	6.64	16.75	466.70	462.40	80.69	63.0	107.87

Table 28.2 *(cont.)*

Day	Run No.	Temp.	Press. lbs./sq.in.	Dens.	Code	Wt. lbs.	Apparent wt.	Time sec.	Pulses	Ullage temp.
22	9	80.23	89.1	6.64	16.93	531.87	528.05	226.67	72.0	102.85
22	10	80.46	76.2	6.63	14.68	466.97	461.28	197.12	63.0	100.11
22	11	80.34	75.3	6.63	14.63	589.27	586.14	131.77	80.0	99.64
22	12	80.27	75.1	6.63	14.66	400.55	395.81	69.76	54.0	94.07
22	13	80.11	60.1	6.64	11.74	525.43	520.74	150.59	71.0	101.21
22	14	80.11	61.3	6.64	12.01	408.21	403.40	70.39	55.0	103.94
22	15	80.27	63.8	6.63	12.39	589.63	586.23	260.55	80.0	96.63
22	16	87.63	111.2	6.34	12.20	520.59	518.28	158.46	74.0	110.51
22	17	87.47	111.1	6.35	12.35	465.26	462.74	82.27	66.0	112.37
22	18	87.54	112.8	6.34	12.47	589.34	588.71	131.81	84.0	108.99
22	19	87.43	100.2	6.34	11.17	520.64	518.82	117.10	74.0	110.83
22	20	87.31	100.1	6.35	11.27	583.41	582.37	106.28	83.0	106.60
22	21	87.91	100.5	6.32	10.72	402.03	398.34	181.19	57.0	106.66
22	22	87.77	85.7	6.33	8.90	533.91	531.37	242.65	76.0	104.27
22	23	87.75	87.9	6.33	9.24	472.54	468.55	104.32	67.0	108.97
22	24	87.84	88.8	6.32	9.28	410.32	405.37	123.83	58.0	108.41
23	1	82.52	63.4	6.54	10.05	536.23	534.94	164.92	74.0	140.54
23	2	82.56	70.3	6.54	11.43	470.57	469.82	104.51	65.0	131.35
23	3	82.84	68.2	6.53	10.73	583.30	584.44	291.66	81.0	121.98
23	4	82.65	76.3	6.54	12.49	462.79	462.40	83.20	64.0	125.82
23	5	82.79	77.5	6.53	12.57	585.27	584.74	232.50	81.0	115.62
23	6	82.70	74.4	6.54	12.09	526.24	527.25	117.22	73.0	116.48
23	7	82.70	86.0	6.54	14.01	463.49	462.38	135.30	64.0	108.07
23	8	83.00	89.8	6.53	14.25	525.66	526.47	215.23	73.0	118.86
23	9	82.63	86.6	6.54	14.16	412.26	411.99	73.59	57.0	113.90
23	10	82.49	97.7	6.55	15.79	585.28	586.12	244.34	81.0	106.57
23	11	82.52	98.6	6.55	15.88	412.49	412.40	91.27	57.0	103.67

* See footnote of Table 28.1.

Table 28.3
Rangeability and Stability Test Readings
From Cryogenic Flow Meters on Liquid Nitrogen
Stability Test *

Day	Run No.	Temp.	Press. lbs./sq.in.	Dens.	Code	Wt. lbs.	Apparent wt.	Time sec.	Pulses	Ullage temp.
6	1	78.88	82.3	6.69	17.27	438.19	436.06	78.93	59.0	120.55
6	2	79.02	83.0	6.68	17.24	482.08	480.03	85.79	65.0	106.15
6	3	79.00	83.2	6.68	17.29	481.02	480.09	84.92	65.0	103.07
6	4	79.35	86.7	6.67	17.47	480.92	479.16	84.73	65.0	84.84
6	5	78.51	85.2	6.70	18.08	483.38	481.47	84.79	65.0	88.64
6	6	77.89	83.7	6.73	18.48	477.74	475.75	83.22	64.0	89.25
6	7	78.00	80.0	6.72	17.79	433.28	430.83	78.38	58.0	85.03
6	8	78.07	82.0	6.72	18.03	433.08	430.68	74.64	58.0	87.98
6	9	78.10	82.4	6.72	18.07	432.96	430.63	75.26	58.0	87.87
6	10	78.95	89.7	6.69	18.29	415.72	413.79	72.50	56.0	82.82
6	11	79.07	90.5	6.68	18.28	422.49	420.90	73.51	57.0	86.80
6	12	79.21	90.8	6.68	18.18	415.00	413.19	72.35	56.0	87.54
7	1	87.31	97.0	6.35	10.89	517.44	519.17	92.59	74.0	95.46
7	2	87.77	98.6	6.33	10.63	515.46	517.59	93.71	74.0	98.38
7	3	87.89	98.0	6.32	10.44	522.00	524.17	95.24	75.0	97.39
7	4	87.75	106.3	6.33	11.55	515.48	517.80	159.48	74.0	99.41
7	5	87.36	109.1	6.35	12.25	517.36	519.20	157.83	74.0	96.54
7	6	87.13	108.5	6.36	12.42	518.07	519.99	158.85	74.0	99.55
7	7	79.76	104.0	6.66	19.27	516.82	514.95	100.14	70.0	99.87
7	8	79.42	105.7	6.67	19.82	517.37	516.02	101.44	70.0	101.82
7	9	79.39	106.5	6.67	19.93	517.39	516.10	101.60	70.0	103.06
7	10	79.51	107.0	6.67	19.87	406.67	405.24	79.08	55.0	102.46
7	11	79.53	107.6	6.67	19.91	413.88	412.55	80.41	56.0	100.39
7	12	79.49	107.5	6.67	19.95	413.79	412.66	80.25	56.0	96.75
7	13	79.51	107.3	6.67	19.90	509.76	508.39	99.88	69.0	90.74
7	14	79.46	108.2	6.67	20.05	502.47	501.17	98.50	68.0	96.52
7	15	79.46	107.9	6.67	20.01	509.73	508.53	99.21	69.0	92.53
7	16	79.51	108.0	6.67	19.98	509.57	508.40	99.91	69.0	91.38
7	17	79.76	108.3	6.66	19.76	494.00	492.93	95.99	67.0	87.91
7	18	81.13	109.5	6.60	18.52	497.52	496.27	88.92	68.0	83.64
7	19	79.56	104.8	6.67	19.57	494.82	493.49	88.60	67.0	85.15
7	20	78.61	102.2	6.70	20.22	497.76	496.19	88.24	67.0	85.85
8	1	79.69	94.1	6.66	18.13	495.45	492.98	91.09	67.0	115.55
8	2	79.81	94.3	6.65	18.04	494.39	492.64	90.76	67.0	109.69
8	3	79.90	94.5	6.65	17.98	487.24	485.03	89.79	66.0	102.59
9	1	81.22	87.8	6.60	15.75	490.68	488.46	87.29	67.0	143.33
9	2	81.50	88.2	6.59	15.53	496.47	494.92	88.58	68.0	131.99
9	3	81.73	88.6	6.58	15.36	488.34	486.97	87.49	67.0	122.64
9	4	82.03	90.4	6.57	15.31	495.23	493.35	212.45	68.0	110.55

Table 28.3 *(cont.)*

Day	Run No.	Temp.	Press. lbs./sq.in.	Dens.	Code	Wt. lbs.	Apparent wt.	Time sec.	Pulses	Ullage temp.
9	5	81.82	90.5	6.57	15.53	487.74	486.72	199.38	67.0	107.80
9	6	81.87	91.0	6.57	15.55	494.77	493.85	210.06	68.0	104.87
9	7	81.80	88.4	6.57	15.26	494.88	494.02	108.84	68.0	101.56
9	8	81.64	88.4	6.58	15.42	495.67	494.51	109.81	68.0	100.05
9	9	81.59	88.7	6.58	15.51	495.49	494.65	109.40	68.0	98.90
9	14	81.59	92.4	6.58	16.02	495.10	494.69	151.85	68.0	86.51
9	15	81.71	94.2	6.58	16.14	487.54	487.10	148.84	67.0	90.39
9	16	81.78	94.9	6.58	16.16	494.46	494.17	151.62	68.0	93.55
9	17	81.78	98.6	6.58	16.62	487.16	486.95	98.48	67.0	85.08
9	18	82.05	101.0	6.57	16.63	493.71	493.41	94.06	68.0	89.86
9	19	82.19	101.3	6.56	16.53	493.36	493.00	93.71	68.0	89.41
9	20	82.26	101.3	6.56	16.46	493.19	492.79	92.57	68.0	89.68
13	1	81.24	73.8	6.59	13.43	504.46	502.80	88.62	69.0	99.26
13	2	81.41	73.9	6.59	13.29	503.95	502.31	88.92	69.0	98.79
13	3	81.52	73.8	6.58	13.15	496.39	494.69	87.48	68.0	97.69
13	11	78.65	75.6	6.70	16.36	512.37	510.59	91.27	69.0	89.23
13	12	78.51	75.9	6.70	16.55	512.79	511.00	91.02	69.0	91.82
13	13	78.37	76.2	6.71	16.75	513.07	511.41	91.41	69.0	92.15
13	14	78.98	81.9	6.68	17.11	511.14	509.69	92.21	69.0	83.65
13	15	79.09	82.9	6.68	17.15	510.46	509.36	92.32	69.0	87.15
13	16	79.18	83.5	6.68	17.15	503.03	501.71	91.02	68.0	87.51
14	1	82.33	90.5	6.55	15.02	508.02	506.93	90.72	70.0	131.25
14	2	82.56	90.5	6.54	14.79	507.13	506.22	90.98	70.0	118.27
14	3	82.63	90.9	6.54	14.78	506.61	506.00	90.75	70.0	111.47
14	4	80.97	83.5	6.61	15.37	504.79	503.76	88.53	69.0	86.94
14	5	81.08	84.3	6.60	15.38	504.29	503.42	88.59	69.0	88.95
14	6	81.13	85.0	6.60	15.44	511.59	510.58	89.83	70.0	89.51
14	7	81.66	87.1	6.58	15.21	509.27	508.97	89.15	70.0	84.50
14	8	81.82	88.4	6.57	15.24	508.87	508.48	88.55	70.0	87.86
14	9	82.01	89.5	6.57	15.21	508.30	507.92	88.59	70.0	88.10
14	10	82.88	94.6	6.53	15.01	505.62	505.26	90.77	70.0	86.50
14	14	83.23	92.3	6.52	14.36	504.35	504.15	90.99	70.0	99.83
15	1	81.54	93.1	6.59	16.15	503.24	502.12	91.91	69.0	106.76
15	2	81.54	93.1	6.59	16.15	510.43	509.39	93.22	70.0	106.76
15	3	81.57	93.0	6.58	16.11	510.20	509.32	93.46	70.0	102.15
15	4	81.57	94.5	6.58	16.31	509.98	509.34	94.82	70.0	85.86
15	5	81.59	95.3	6.58	16.39	509.72	509.28	95.21	70.0	88.64
15	6	81.68	96.0	6.58	16.38	509.22	509.00	95.25	70.0	88.68
15	7	82.49	99.4	6.55	16.00	506.43	506.54	94.35	70.0	85.31
15	8	82.61	100.7	6.54	16.04	506.15	506.20	91.14	70.0	87.62
15	9	82.63	101.1	6.54	16.06	505.78	506.13	91.14	70.0	88.10
15	10	84.80	115.8	6.46	15.57	505.17	506.63	112.44	71.0	87.90
15	11	85.12	120.0	6.45	15.75	510.58	512.77	97.59	72.0	91.02

Table 28.3 *(cont.)*

Day	Run No.	Temp.	Press. lbs./sq.in.	Dens.	Code	Wt. lbs.	Apparent wt.	Time sec.	Pulses	Ullage temp.
15	12	85.21	105.4	6.44	13.98	510.93	512.26	95.60	72.0	98.78
15	13	85.84	109.8	6.41	13.84	508.06	510.27	95.42	72.0	88.88
15	14	86.00	111.3	6.41	13.84	507.27	509.76	95.09	72.0	92.06
16	1	79.97	92.7	6.65	17.67	506.60	506.85	90.89	69.0	111.00
16	2	80.37	92.0	6.63	17.18	513.55	512.99	92.92	70.0	106.57
16	3	80.37	91.7	6.63	17.14	506.72	505.66	91.92	69.0	102.66
16	4	80.62	88.6	6.62	16.46	505.37	504.86	91.01	69.0	85.34
16	5	80.67	89.3	6.62	16.51	512.44	512.04	92.19	70.0	88.50
16	6	80.78	89.6	6.62	16.44	512.38	511.69	92.42	70.0	88.91
16	7	81.61	92.7	6.58	16.02	502.00	501.90	91.16	69.0	84.70
16	8	81.78	94.4	6.58	16.09	508.71	508.70	92.08	70.0	88.11
16	9	81.91	95.5	6.57	16.09	515.36	515.54	93.57	71.0	89.06
16	10	83.09	99.2	6.52	15.37	511.25	511.88	92.91	71.0	87.05
16	11	83.23	99.8	6.52	15.30	510.75	511.44	92.64	71.0	91.03
16	12	83.32	99.8	6.52	15.21	510.82	511.15	92.99	71.0	90.60
19	1	77.47	68.9	6.74	16.25	507.51	506.54	88.86	68.0	147.56
19	2	77.61	68.9	6.74	16.11	507.02	506.14	89.03	68.0	143.52
19	3	77.65	69.3	6.73	16.15	506.57	506.01	89.00	68.0	142.36
19	4	77.98	73.6	6.72	16.66	512.11	512.55	90.82	69.0	88.38
19	5	78.17	74.7	6.72	16.69	504.78	504.59	88.38	68.0	90.19
19	6	78.19	75.3	6.71	16.78	512.31	511.95	89.54	69.0	90.22
19	7	78.98	77.3	6.68	16.35	509.81	509.64	89.73	69.0	82.64
19	8	77.93	75.7	6.72	17.10	506.08	505.27	88.44	68.0	84.31
19	9	77.17	74.6	6.75	17.67	508.44	507.47	88.62	68.0	85.35
19	10	77.00	79.6	6.76	18.72	507.97	507.98	88.76	68.0	82.01
19	11	77.07	80.7	6.76	18.83	508.65	507.80	88.94	68.0	85.01
19	12	77.10	81.4	6.76	18.91	515.15	515.20	90.09	69.0	85.59
20	1	77.91	65.0	6.72	14.99	506.73	505.23	89.46	68.0	111.92
20	2	77.98	65.7	6.72	15.07	506.52	505.04	90.10	68.0	107.46
20	3	78.05	66.2	6.72	15.11	506.26	504.84	89.44	68.0	103.85
20	4	76.91	69.4	6.76	16.90	502.79	500.68	87.41	67.0	85.88
20	5	76.80	69.7	6.77	17.08	510.06	508.48	89.07	68.0	86.64
20	6	76.84	77.9	6.77	18.59	510.67	508.43	88.90	68.0	87.54
20	7	77.21	78.3	6.75	18.29	509.11	507.37	88.72	68.0	82.52
20	8	77.21	79.4	6.75	18.47	509.01	507.38	88.99	68.0	86.01
20	9	77.31	80.0	6.75	18.47	501.61	499.66	87.54	67.0	86.30
20	10	78.19	84.2	6.72	18.25	505.85	504.62	88.60	68.0	85.42
20	11	78.30	84.5	6.71	18.18	505.59	504.29	89.41	68.0	86.35
20	12	78.33	84.6	6.71	18.17	505.46	511.64	89.38	69.0	86.37
20	12	78.33	84.6	6.71	18.17	505.46	511.64	89.38	69.0	86.37

* See footnote of Table 28.1.

29. Stress-Rupture Life of Kevlar 49/Epoxy Spherical Pressure Vessels

Source Barlow, R.E., Toland, R.H. and Freeman, T. (1984). A Bayesian analysis of stress-rupture life of Kevlar/epoxy spherical pressure vessels. In *Proceedings of the Canadian Conference in Applied Statistics,* 1981, edited by T.D. Dwivedi. New York: Marcel Dekker. [The work was performed at the Lawrence Livermore National Laboratory, University of California, and supported by the U.S. Department of Energy.]

Contributor R.E. Barlow University of California, Berkeley

A study of the lifetimes of Kevlar 49/epoxy spherical pressure vessels that are subjected to a constant sustained pressure until vessel failure, commonly known as static fatigue or stress-rupture, has been made. The NASA space shuttle uses Kevlar/epoxy spherical pressure vessels in a sustained pressure mode throughout the usage life of the vessel, and several commercial applications, such as fire-fighters' air-breathing apparatus, are also subject to this service condition. The study was done to generate baseline data on vessel life under pressure and to predict vessel life and design reliability. Two sets of data were used :

(i) Kevlar/epoxy strand-life data, Table 29.1; and

(ii) Kevlar/epoxy spherical pressure vessel-life data, Table 29.2.

It was observed that the hazard rates for both data sets are time-dependent at low stress levels; thus the exponential lifetime distribution is not valid.

The Kevlar 49/epoxy strands were based on 59,064 hours, about 7 years, of data. There is only about one year of life test experience on spherical vessels.

Table 29.1
Ordered Time to Failure
of Kevlar 49/Epoxy Strands Tested
at Various Stress Levels *

Rank	T_F	Rank	T_F	Rank	T_F	Rank	T_F
		90% Stress Level, of 101					
1	0.01	26	0.24	51	0.80	76	1.45
2	0.01	27	0.24	52	0.80	77	1.50
3	0.02	28	0.29	53	0.83	78	1.51
4	0.02	29	0.34	54	0.85	79	1.52
5	0.02	30	0.35	55	0.90	80	1.53
6	0.03	31	0.36	56	0.92	81	1.54
7	0.03	32	0.38	57	0.95	82	1.54
8	0.04	33	0.40	58	0.99	83	1.55
9	0.05	34	0.42	59	1.00	84	1.58
10	0.06	35	0.43	60	1.01	85	1.60
11	0.07	36	0.52	61	1.02	86	1.63
12	0.07	37	0.54	62	1.03	87	1.64
13	0.08	38	0.56	63	1.05	88	1.80
14	0.09	39	0.60	64	1.10	89	1.80
15	0.09	40	0.60	65	1.10	90	1.81
16	0.10	41	0.63	66	1.11	91	2.02
17	0.10	42	0.65	67	1.15	92	2.05
18	0.11	43	0.67	68	1.18	93	2.14
19	0.11	44	0.68	69	1.20	94	2.17
20	0.12	45	0.72	70	1.29	95	2.33
21	0.13	46	0.72	71	1.31	96	3.03
22	0.18	47	0.72	72	1.33	97	3.03
23	0.19	48	0.73	73	1.34	98	3.34
24	0.20	49	0.79	74	1.40	99	4.20
25	0.23	50	0.79	75	1.43	100	4.69
						101	7.89

Table 29.1 *(cont.)*

Rank	T_F	Rank	T_F	Rank	T_F	Rank	T_F
			80% Stress Level, of 100				
1	1.8	26	84.2	51	152.2	76	285.9
2	3.1	27	87.1	52	152.8	77	292.6
3	4.2	28	87.3	53	157.7	78	295.1
4	6.0	29	93.2	54	160.0	79	301.1
5	7.5	30	103.4	55	163.6	80	304.3
6	8.2	31	104.6	56	166.9	81	316.8
7	8.5	32	105.5	57	170.5	82	329.8
8	10.3	33	108.8	58	174.9	83	334.1
9	10.6	34	112.6	59	177.7	84	346.2
10	24.2	35	116.8	60	179.2	85	351.2
11	29.6	36	118.0	61	183.6	86	353.3
12	31.7	37	122.3	62	183.8	87	369.3
13	41.9	38	123.5	63	194.3	88	372.3
14	44.1	39	124.4	64	195.1	89	381.3
15	49.5	40	125.4	65	195.3	90	393.5
16	50.1	41	129.5	66	202.6	91	451.3
17	59.7	42	130.4	67	220.2	92	461.5
18	61.7	43	131.6	68	221.3	93	574.2
19	64.4	44	132.8	69	227.2	94	653.3
20	69.7	45	133.8	70	251.0	95	663.0
21	70.0	46	137.0	71	266.5	96	669.8
22	77.8	47	140.2	72	267.9	97	739.7
23	80.5	48	140.9	73	269.2	98	759.6
24	82.3	49	148.5	74	270.4	99	894.7
25	83.5	50	149.2	75	272.5	100	974.9

Table 29.1 *(cont.)*

Rank	T_F	Rank	T_F	Rank	T_F	Rank	T_F
			70% Stress Level, of 49				
1	1051	14	5817	27	9711	40	12044
2	1337	15	5905	28	9806	41	13520
3	1389	16	5956	29	10205	42	13670
4	1921	17	6068	30	10396	43	14110
5	1942	18	6121	31	10861	44	14496
6	2322	19	6473	32	11026	45	15395
7	3629	20	7501	33	11214	46	16179
8	4006	21	7886	34	11362	47	17092
9	4012	22	8108	35	11604	48	17568
10	4063	23	8546	36	11608	49	17568
11	4921	24	8666	37	11745		
12	5445	25	8831	38	11762		
13	5620	26	9106	39	11895		
			60% Stress Level, of 47				
1	13872	11	28512	21	36240	31	41610
2	18024	12	28676	22	36480	32	41620
3	18948	13	29784	23	36648	33	42384
4	21960	14	29784	24	38640	34	42720
5	22608	15	31188	25	38976	35	44616
6	25100	16	31644	26	39144	36	49080
7	25536	17	32760	27	39480	37	49416
8	27216	18	33480	28	39648	38	51168
9	27744	19	35880	29	39648		
10	27840	20	36024	30	41520		
			50% Stress Level, of 48				
1	30960						
2	32040						
3	57984						

* T_F denotes time to failure in hours.

Table 29.2

Ordered Time to Failure of Kevlar 49/Epoxy Vessels Tested at Various Stress Levels *

Rank	Vessel number	T_F	Rank	Vessel number	T_F
		86% Stress Level, of 39			
1	65	2.2	20	51	55.4
2	197	4.0	21	208	61.2
3	210	4.0	22	138	87.5
4	205	4.6	23	245	98.2
5	202	6.1	24	83	101.0
6	176	6.7	25	36	111.4
7	209	7.9	26	166	144.0
8	140	8.3	27	31	158.7
9	66	8.5	28	147	243.9
10	42	9.1	29	124	254.1
11	49	10.2	30	13	444.4
12	87	12.5	31	244	590.4
13	135	13.3	32	241	638.2
14	220	14.0	33	17	755.2
15	94	14.6	34	21	952.2
16	170	15.0	35	23	1108.2
17	90	18.7	36	122	1148.5
18	46	22.1	37	109	1569.3
19	217	45.9	38	105	1750.6
			39	120	1802.1
		80% Stress Level, of 24			
1	80	19.1	13	69	544.9
2	75	24.3	14	237	554.2
3	79	69.8	15	29	664.5
4	60	71.2	16	61	694.1
5	77	136.0	17	110	876.7
6	55	199.1	18	12	930.4
7	57	403.7	19	185	1254.9
8	44	432.2	20	118	1275.6
9	15	453.4	21	119	1536.8
10	74	514.1	22	26	1755.5
11	191	514.2	23	247	2046.2
12	183	541.6	24	108	6177.5

Table 29.2 *(cont.)*

Rank	Vessel number	T_F	Rank	Vessel number	T_F
		74% Stress Level, of 24			
1	174	225.2	10	35	3708.9
2	200	503.6	11	231	4908.9
3	93	1087.7	12	70	5556.0
4	39	1134.3	13	164	6271.1
5	50	1824.3	14	236	7332.0
6	45	1920.1	15	241	7918.7
7	56	2383.0	16	184	7996.0
8	243	2442.5	17	234	9240.3
9	15	3708.9	18	240	9973.0
		68% Stress Level, of 21			
1	204	4000			
2	218	5376			
3	163	7320			
4	94	8616			
5	142	9120			

* T_F denotes time to failure in hours.

30. A Chemical Reaction

Source Box, G.E.P. and Youle, P.V. (1955). The exploration and exploitation of response surfaces: An example of the link between the fitted surface and the basic mechanism of the system. *Biometrics* **11**, 287-323. [With permission of the Biometric Society.]

Contributor G. E. P. Box University of Wisconsin

Consider a chemical reaction of the type $A + B \rightarrow C$ followed by $A + C \rightarrow D$ in which two reactants A and B formed a mixture of C and D. The object was to obtain the maximum for C subject to the condition that the yield of D should not exceed 20% since more than this amount would cause difficulty in purification. The quantity of B used was kept constant throughout, the factors varied being temperature, T, in degrees Centigrade, the percent concentration of A, and the time in hours of the reaction.

The data are given in Table 30.1, namely the values for the uncoded and coded variables, the estimated fraction of the unchanged starting material, the estimated fraction converted to the desired product and the estimated fraction occurring as an unwanted by-product.

When a second-degree polynomial was fitted to the yields of the desired product for experiments 1-15, a possible planar ridge system was indicated. Therefore, experiments 16-19 were carried out on the estimated maximum plane. In spite of the great differences in the actual conditions employed these experiments gave yields close to the maximum value of about 60% in accordance with prediction. These observations were then included in the calculation of the second-degree model. Later consideration was given to fitting a model based on reaction kinetics.

Further discussion of the analysis of these data is given in Box (1954), Davies (1956) and Box (1960).

References

Box, G. E. P. (1954). The exploration and exploitation of response surfaces: some general considerations and examples. *Biometrics* **10**, 16-60.

Box, G. E. P. (1960). Fitting empirical data. *Ann. New York Acad. Sci,* **86**, 792-816.

Davies, O. L. (1956). *The Design and Analysis of Industrial Experiments,* 2nd. edition. Edinburgh: Oliver and Boyd.

Table 30.1
Temperature, Concentration, Reaction Time and Yield of a Chemical Reaction

	Levels of variables			Observed yields		
Expt.	Temp.	Conc.	Time	Unchanged	Converted	Unwanted
1	162	23	3	41.5	45.9	11.2
2	162	23	8	33.8	53.3	11.2
3	162	30	5	27.7	57.5	12.7
4	162	30	8	21.7	58.8	16.0
5	172	25	5	19.9	60.6	16.2
6	172	25	8	15.0	58.0	22.6
7	172	30	5	12.2	58.6	24.5
8	172	30	8	4.3	52.4	38.0
9	167	27.5	6.5	19.3	56.9	21.3
10	177	27.5	6.5	6.4	55.4	30.8
11	157	27.5	6.5	37.6	46.9	14.7
12	167	32.5	6.5	18.0	57.3	22.2
13	167	22.5	6.5	26.3	55.0	18.3
14	167	27.5	9.5	9.9	58.9	28.0
15	167	27.5	3.5	25.0	50.3	22.1
16	177	20	6.5	14.1	61.1	23.0
17	177	20	6.5	15.2	62.9	20.7
18	160	34	7.5	15.9	60.0	22.1
19	160	34	7.5	19.6	60.6	19.3

31. Product Preferences

Contributor M.B. Carroll General Food Corporation

1. The purpose of this study was to determine whether a cost reduced product was equally preferred to the current product. A Paired Comparison Central Location Preference test was run where 200 judges were asked their preference on the pair of samples, coded 27 and 45. One half of the group tasted sample 27 first and 45 second, the other half tasted the samples in the reverse order. Besides being asked their preference, each judge was asked to rate six qualities of these samples on a 1 to 6 scale. The data are given in Table 31.1.

2. These data are from a three sample Multiple Paired Comparison Central Location Preference test, where two experimentals, coded 43 and 96, were compared to a competitor's product, coded 67. Twenty-eight judges sampled each of the six possible pairings for a total of 168 observations and indicated their overall preference, and their preference based on two selected attributes of these samples. They also rated each sample in the pairing on eleven attributes, using a scale of 1 to 6. The data are given in Table 31.2.

Table 31.1
Qualities and Preferences of Two Products Determined by 200 Judges

Judge	Sample 1	Sample 2	Attribute scores * Sample 1	Attribute scores * Sample 2	Preference **
1	27	45	3 3 3 4 2 4	5 4 3 4 2 4	1
2	27	45	4 4 3 3 2 4	5 3 3 5 5 5	1
3	27	45	5 4 3 3 1 4	5 4 3 4 1 4	1
4	27	45	5 5 2 2 3 2	3 4 3 3 3 3	2
5	27	45	3 3 3 3 3 3	2 2 3 3 3 3	2
6	27	45	4 3 4 2 3 3	5 4 3 4 1 4	1
7	27	45	2 3 4 3 3 3	4 3 3 4 3 3	1
8	27	45	1 3 4 3 3 3	5 3 3 4 2 4	1
9	27	45	5 4 4 5 3 3	4 6 5 5 3 3	1
10	27	45	3 2 3 3 3 3	1 3 3 3 3 3	2
11	27	45	5 5 3 5 2 4	4 4 3 4 3 4	2
12	27	45	3 3 3 3 3 3	4 4 3 2 2 2	1
13	27	45	4 3 3 4 3 4	5 4 2 4 1 4	1
14	27	45	5 3 3 4 4 2	5 5 3 2 4 4	3
15	27	45	3 4 4 3 3 3	2 4 4 3 3 3	2
16	27	45	3 3 3 4 3 3	5 5 3 5 3 2	1
17	27	45	2 2 3 3 3 3	5 5 4 4 1 4	1
18	27	45	5 4 3 2 5 2	4 4 3 3 4 3	2
19	27	45	2 4 5 4 3 3	6 6 5 5 2 5	1
20	27	45	4 3 3 4 2 4	3 3 3 3 3 3	2
21	27	45	4 4 4 2 4 2	4 4 3 3 4 2	2
22	27	45	4 3 3 4 2 4	5 3 3 5 2 5	1
23	27	45	3 3 3 3 2 4	4 3 3 2 2 3	1
24	27	45	4 3 3 3 3 3	4 4 3 3 3 3	1
25	27	45	2 4 3 4 2 4	5 4 4 5 2 4	1
26	27	45	4 4 4 4 4 3	2 4 3 4 3 4	1
27	27	45	5 5 3 4 2 4	6 5 3 5 5 5	1
28	27	45	1 1 4 3 3 3	4 5 4 4 2 4	1
29	27	45	3 2 3 3 2 3	1 1 3 3 3 3	2
30	27	45	2 2 3 3 3 3	5 3 3 4 4 2	1
31	27	45	5 4 3 4 2 4	4 4 3 3 2 4	2
32	27	45	5 5 3 5 4 3	5 5 3 4 3 4	2
33	27	45	4 3 4 4 3 4	3 2 3 3 2 3	2
34	27	45	5 4 3 4 3 2	3 3 3 3 3 3	2
35	27	45	3 4 3 3 3 3	5 4 3 3 3 3	2
36	27	45	4 4 3 3 3 3	5 4 3 4 1 4	1
37	27	45	1 2 4 3 3 3	4 2 4 2 4 2	1
38	27	45	5 5 4 4 3 4	3 3 3 3 3 3	2
39	27	45	5 5 4 4 3 4	2 3 4 4 2 3	2
40	27	45	2 2 3 3 3 3	3 3 3 2 2 4	1
41	27	45	5 1 3 2 2 1	4 4 3 2 4 2	1
42	27	45	5 4 4 4 3 4	6 6 4 2 3 3	1
43	27	45	4 4 3 4 5 4	5 5 0 5 5 4	1
44	27	45	4 4 3 3 3 3	3 2 3 3 3 3	2
45	27	45	1 1 3 3 3 3	1 1 3 3 3 3	2
46	27	45	2 2 4 4 3 3	5 3 4 3 3 3	1
47	27	45	3 3 4 4 2 4	3 4 2 4 3 3	1
48	27	45	4 4 3 3 3 3	4 4 3 3 3 3	2
49	27	45	4 2 3 3 3 3	5 3 3 2 2 2	1
50	27	45	4 3 3 4 2 4	6 4 3 5 2 4	1

Table 31.1 *(cont.)*

			Attribute scores *		
Judge	Sample 1	Sample 2	Sample 1	Sample 2	Preference **
51	27	45	4 4 3 3 3 3	5 4 3 5 2 4	1
52	27	45	4 3 4 4 3 4	4 3 3 3 5 3	2
53	27	45	3 3 3 3 3 3	4 4 3 2 4 2	1
54	27	45	3 4 3 2 2 3	6 5 4 2 5 1	1
55	27	45	1 1 3 3 3 3	5 4 4 4 3 4	1
56	27	45	2 1 4 4 3 3	1 1 3 3 3 3	2
57	27	45	3 4 4 3 3 2	4 4 4 2 3 2	1
58	27	45	5 3 4 4 1 5	2 2 3 3 3 3	2
59	27	45	3 3 3 3 5 2	6 3 3 1 5 1	1
60	27	45	5 4 4 5 2 4	4 4 3 4 2 4	2
61	27	45	5 4 4 5 2 4	5 4 4 5 2 5	1
62	27	45	5 5 4 5 5 4	4 4 4 3 3 3	2
63	27	45	5 4 4 4 3 4	5 4 3 4 0 4	3
64	27	45	6 6 5 1 5 1	3 3 3 3 3 3	2
65	27	45	2 2 3 3 3 3	2 3 3 3 3 3	3
66	27	45	1 3 5 3 5 5	4 4 4 1 4 2	1
67	27	45	5 6 2 2 5 4	6 5 2 1 1 5	1
68	27	45	4 3 3 3 2 3	6 4 3 1 3 2	1
69	27	45	3 6 5 3 3 3	3 5 4 3 3 3	3
70	27	45	4 3 3 3 2 3	4 3 3 3 3 2	1
71	27	45	1 1 3 3 3 3	4 4 3 4 2 4	1
72	27	45	3 4 4 3 3 3	5 3 3 5 2 4	1
73	27	45	4 5 3 3 3 3	5 5 3 4 2 4	1
74	27	45	5 4 3 2 4 1	4 4 4 3 3 3	2
75	27	45	4 5 3 3 4 5	3 3 3 3 3 3	2
76	27	45	4 5 4 4 3 4	4 4 3 3 3 3	2
77	27	45	3 4 4 3 3 3	2 3 3 3 3 3	2
78	27	45	4 4 4 4 4 3	6 5 4 5 4 4	1
79	27	45	2 2 3 4 3 4	2 3 3 4 3 4	3
80	27	45	4 4 3 4 2 4	3 3 3 3 3 3	2
81	27	45	2 2 3 3 3 3	4 2 4 2 4 2	1
82	27	45	2 2 3 3 3 3	6 4 3 5 5 5	1
83	27	45	2 1 3 3 3 3	3 2 2 2 4 2	1
84	27	45	1 1 3 3 3 3	1 1 3 3 3 3	3
85	27	45	3 3 3 4 2 3	4 3 2 3 4 3	1
86	27	45	4 6 4 4 3 4	4 2 3 3 3 3	2
87	27	45	3 3 3 3 3 3	4 3 3 4 3 4	1
88	27	45	3 3 3 4 2 3	2 3 3 3 3 3	2
89	27	45	4 5 4 4 3 2	2 3 3 3 3 3	2
90	27	45	2 2 3 3 3 3	5 2 4 4 3 4	1
91	27	45	3 3 3 3 3 4	5 4 3 4 2 4	1
92	27	45	2 5 5 4 3 5	4 5 5 4 2 4	2
93	27	45	4 4 3 4 3 3	4 4 4 3 2 3	1
94	27	45	5 4 4 4 2 4	5 3 3 2 2 3	3
95	27	45	4 4 3 3 3 3	3 3 3 3 3 3	2
96	27	45	4 3 4 3 2 3	6 6 4 4 2 5	1
97	27	45	3 4 3 4 2 3	5 4 2 2 1 2	1
98	27	45	5 5 3 5 1 4	6 5 4 4 2 4	1
99	27	45	3 3 3 3 3 3	5 4 3 4 2 4	1
100	27	45	4 3 4 4 3 5	3 3 3 3 3 3	1
101	45	27	4 4 3 2 5 3	3 3 3 3 3 3	2
102	45	27	3 4 2 3 3 3	4 3 3 3 3 3	1

Table 31.1 *(cont.)*

Judge	Sample 1	Sample 2	Attribute scores * Sample 1	Attribute scores * Sample 2	Preference **
103	45	27	2 2 4 3 2 4	2 2 4 4 2 4	1
104	45	27	4 4 4 4 3 3	3 3 3 3 2 3	2
105	45	27	4 4 3 3 2 3	5 4 3 2 1 5	1
106	45	27	3 1 3 4 3 3	4 3 3 4 3 4	1
107	45	27	4 4 3 3 3 3	4 4 3 3 3 3	3
108	45	27	5 1 3 4 3 3	4 1 3 5 3 3	1
109	45	27	5 1 3 4 3 3	3 3 3 3 3 3	2
110	45	27	4 3 3 4 3 4	2 3 4 3 3 3	2
111	45	27	6 5 4 5 4 4	5 4 3 4 3 3	2
112	45	27	4 4 3 2 3 3	5 4 3 2 3 2	1
113	45	27	5 4 3 4 3 3	4 4 4 3 4 3	2
114	45	27	3 3 3 3 3 3	4 4 2 2 3 2	1
115	45	27	3 3 3 3 3 3	3 3 3 3 3 3	2
116	45	27	1 3 3 3 3 3	5 4 3 2 4 2	1
117	45	27	5 5 3 4 2 4	5 5 3 3 2 4	1
118	45	27	4 4 4 4 3 4	3 3 3 3 2 3	2
119	45	27	5 2 3 3 4 3	4 5 3 4 5 1	2
120	45	27	5 4 3 3 3 3	2 2 3 3 3 3	2
121	45	27	5 5 4 4 3 5	6 5 3 4 3 4	1
122	45	27	4 2 3 2 2 2	1 1 3 3 3 3	2
123	45	27	4 1 3 3 3 3	0 3 3 2 4 1	1
124	45	27	5 5 4 4 2 4	5 5 4 4 2 4	3
125	45	27	5 4 4 4 2 4	4 4 4 4 3 3	2
126	45	27	3 4 3 4 3 3	5 4 4 4 3 5	1
127	45	27	3 3 4 4 5 4	2 3 3 3 3 3	2
128	45	27	4 3 3 3 3 2	3 3 3 3 3 3	2
129	45	27	3 3 3 4 3 4	3 3 3 3 2 3	2
130	45	27	2 2 3 3 3 3	4 3 3 3 3 3	2
131	45	27	3 2 3 4 3 3	4 4 3 2 2 4	1
132	45	27	1 1 3 3 5 3	3 1 3 2 4 2	1
133	45	27	5 5 4 4 2 4	5 5 4 4 2 4	2
134	45	27	2 2 3 3 3 3	4 4 4 4 2 4	1
135	45	27	3 4 3 4 3 3	3 2 3 2 2 3	2
136	45	27	4 4 3 3 3 3	4 4 3 3 2 3	1
137	45	27	4 5 4 2 3 2	3 2 3 3 3 3	2
138	45	27	3 3 3 3 3 3	4 4 3 2 4 2	1
139	45	27	3 3 3 3 3 3	4 4 3 2 4 2	1
140	45	27	5 5 3 2 4 3	3 3 3 3 3 3	2
141	45	27	6 4 3 2 1 1	6 6 4 2 1 2	3
142	45	27	1 1 3 3 3 3	4 3 3 3 3 3	1
143	45	27	1 4 4 3 3 3	4 2 3 2 4 2	1
144	45	27	1 1 3 4 3 3	4 1 3 4 3 3	1
145	45	27	2 3 3 3 3 3	4 3 3 2 4 2	1
146	45	27	4 4 4 4 3 4	3 4 3 3 3 3	2
147	45	27	5 5 4 2 2 2	3 4 3 3 3 3	2
148	45	27	2 2 3 3 3 3	5 4 4 4 4 4	1
149	45	27	2 1 3 2 3 3	6 1 3 1 4 4	1
150	45	27	4 5 4 4 2 4	4 4 3 5 2 4	1
151	45	27	5 4 3 5 4 4	4 3 3 3 3 3	2
152	45	27	3 4 4 3 2 3	6 5 4 5 3 4	1
153	45	27	4 5 4 3 3 3	5 4 3 3 2 5	1

Table 31.1 *(cont.)*

			Attribute scores *		
Judge	Sample 1	Sample 2	Sample 1	Sample 2	Preference **
154	45	27	3 3 3 3 3 4	5 3 3 4 2 2	1
155	45	27	6 4 3 1 5 1	5 4 5 2 2 2	2
156	45	27	2 2 3 3 3 3	4 3 4 2 3 2	1
157	45	27	5 5 4 4 3 3	4 4 4 3 3 3	2
158	45	27	4 5 4 2 3 3	3 5 3 3 3 3	2
159	45	27	2 2 4 3 3 3	3 3 4 3 3 3	3
160	45	27	3 4 4 3 4 2	2 4 3 3 3 3	2
161	45	27	4 4 3 3 4 2	3 3 3 3 3 3	2
162	45	27	3 4 3 4 3 3	4 5 5 2 4 2	1
163	45	27	4 3 4 2 4 2	4 4 3 2 2 3	3
164	45	27	4 4 3 4 2 4	4 4 3 4 2 4	1
165	45	27	3 5 4 3 3 3	5 5 4 4 4 4	1
166	45	27	5 4 4 4 5 2	4 4 3 3 3 3	2
167	45	27	2 1 3 3 3 3	5 4 4 4 4 2	1
168	45	27	5 2 3 2 5 2	1 1 3 3 3 3	2
169	45	27	4 1 3 4 3 3	5 5 2 2 3 2	1
170	45	27	5 5 3 4 3 3	6 5 3 4 2 4	1
171	45	27	2 1 3 3 3 3	3 3 3 3 3 3	3
172	45	27	5 2 3 4 2 4	4 3 3 3 2 3	2
173	45	27	6 4 3 5 5 5	4 3 3 3 3 3	2
174	45	27	2 3 2 5 2 2	5 4 4 5 5 2	1
175	45	27	3 4 4 4 3 3	4 4 4 3 2 3	1
176	45	27	5 5 2 5 3 4	5 5 2 5 2 4	1
177	45	27	1 2 3 3 3 3	5 2 4 2 2 1	1
178	45	27	3 3 3 4 3 3	4 2 3 3 3 3	1
179	45	27	1 1 3 3 3 3	2 2 3 3 3 3	1
180	45	27	4 4 3 2 5 2	5 4 4 2 5 2	1
181	45	27	4 4 3 3 3 5	5 4 3 2 3 2	1
182	45	27	4 4 4 4 3 4	4 4 4 4 4 4	1
183	45	27	5 4 3 2 4 2	5 4 3 2 2 3	2
184	45	27	3 2 3 3 3 3	6 4 4 2 4 2	1
185	45	27	3 5 3 4 3 3	5 4 3 4 4 2	1
186	45	27	5 5 3 4 5 2	5 5 3 4 3 3	2
187	45	27	3 3 3 4 3 3	5 4 3 4 4 2	1
188	45	27	4 4 3 3 3 3	5 4 3 1 2 4	1
189	45	27	4 4 3 3 3 3	4 4 3 3 3 3	2
190	45	27	3 2 3 3 5 3	5 4 4 4 5 5	1
191	45	27	6 2 3 1 5 1	5 2 3 2 5 1	2
192	45	27	5 2 3 5 1 5	5 2 3 4 2 2	3
193	45	27	4 4 3 3 3 3	4 4 3 2 3 2	1
194	45	27	3 3 3 4 2 3	5 4 3 2 4 2	1
195	45	27	3 4 4 3 3 3	3 3 4 3 3 3	2
196	45	27	4 4 3 3 3 5	5 5 2 1 5 1	1
197	45	27	4 4 4 4 3 3	3 3 3 3 3 3	2
198	45	27	4 4 4 4 3 3	4 4 4 3 3 3	2
199	45	27	4 4 4 4 3 3	3 4 3 3 2 3	2
200	45	27	2 2 3 3 3 3	4 4 3 2 4 2	1

* Rated on 1-6 scale (1, excellent,...,6, poor)

** Preference : 1, prefers first sample tasted
2, prefers second sample tasted
3, no preference

Table 31.2
Qualities and Preferences of Two Experimental Products Compared to a Competitor's Product by 28 Judges

Judge	Sample 1	Sample 2	Attribute scores * Sample 1	Attribute scores * Sample 2	Overall performance	Preference based on two selected attributes **	
117	43	67	1 3 1 3 2 2 3 3 2 2 2	3 3 2 3 1 3 3 3 1 2 3	2	2	2
127	43	67	4 4 1 2 2 2 3 3 4 1 2	3 3 2 3 3 2 2 2 2 3 3	1	1	1
132	43	67	4 3 1 3 2 2 3 3 4 1 1	1 2 1 4 3 5 4 4 3 4 3	1	1	1
135	43	67	4 2 3 3 4 5 5 3 2 1 1	2 3 1 4 2 2 2 3 2 2 3	2	1	1
144	43	67	4 5 2 2 2 2 3 3 1 1 3	4 3 1 2 2 1 3 3 4 1 3	2	2	2
154	43	67	1 4 1 2 1 1 3 3 3 1 1	3 4 1 2 2 1 3 3 3 1 1	1	1	1
68	43	67	4 4 3 3 2 4 2 2 1 2 4	2 3 3 4 2 2 3 3 4 2 2	2	1	2
69	43	67	2 5 1 3 2 2 3 3 3 1 3	3 3 3 3 3 3 4 3 3 1 2	1	1	1
75	43	67	4 2 3 4 4 4 3 2 1 5 3	4 2 1 3 3 4 4 3 1 1 2	2	2	2
78	43	67	4 2 5 2 5 5 1 2 4 5 5	5 4 1 2 1 1 3 3 3 1 1	2	2	2
79	43	67	4 3 1 2 2 1 4 3 2 1 1	5 2 4 2 2 2 2 2 1 4 5	1	2	3
83	43	67	2 3 2 4 4 4 4 4 2 3 4	1 2 5 5 5 5 2 2 5 5 1	1	1	3
90	43	67	1 6 1 2 2 1 3 3 4 4 3	3 2 2 3 4 2 1 1 4 5 4	1	2	1
102	43	67	1 5 1 2 2 1 3 3 3 2 3	5 3 2 4 4 3 4 3 3 4 3	1	1	1
107	43	67	1 4 2 3 2 2 3 3 1 3 2	4 3 3 3 3 3 4 2 2 4 2	1	2	2
6	43	67	1 3 3 5 4 4 4 4 1 2 2	4 4 2 2 2 2 3 3 2 2 3	2	1	2
10	43	67	1 3 3 4 3 3 4 3 2 3 1	2 4 3 4 2 2 4 3 3 3 3	2	1	1
11	43	67	4 3 2 3 3 2 3 3 2 3 3	3 2 2 2 3 2 3 3 2 3 3	1	2	1
15	43	67	4 5 1 2 2 2 3 3 1 1 2	2 2 2 4 4 4 4 5 3 4 2	1	1	1
18	43	67	2 4 1 2 3 2 3 3 2 2 2	4 3 1 3 3 2 4 3 2 3 2	1	2	2
21	43	67	4 3 1 2 4 3 4 3 4 1 1	2 2 3 4 4 5 5 1 4 5 1	1	1	1
22	43	67	1 4 1 2 1 1 3 3 4 1 1	3 4 1 2 3 2 3 3 1 2 2	1	1	1
26	43	67	4 3 2 2 2 2 3 3 4 1 2	2 2 4 3 4 4 4 1 4 5 4	1	1	1
27	43	67	4 4 3 3 3 4 4 3 1 2 1	4 4 1 3 2 4 3 3 2 2 3	2	2	2
30	43	67	1 2 3 3 4 2 4 3 4 3 3	5 5 1 2 2 2 3 3 1 1 2	2	2	2
44	43	67	1 4 2 3 3 2 4 4 1 2 2	4 1 4 5 4 3 1 1 4 4 4	1	1	1
60	43	67	1 3 2 3 3 2 4 3 2 3 2	5 5 1 2 2 1 4 3 1 2 2	2	2	2
62	43	67	2 4 2 1 2 1 1 3 2 1 5	4 6 1 1 2 1 3 3 5 1 3	2	2	2
110	67	43	1 6 1 2 1 1 3 3 4 1 2	2 1 4 5 5 5 5 4 1 5 3	1	1	1
113	67	43	4 4 3 3 3 4 4 2 1 3 3	2 2 4 5 5 5 5 2 4 4 3	1	1	1
123	67	43	1 3 1 2 2 2 3 3 2 3 3	2 1 4 5 5 4 1 2 1 5 5	1	1	1
125	67	43	1 5 1 1 2 2 3 3 1 1 3	4 3 3 3 4 4 4 4 1 4 2	1	1	1
137	67	43	1 3 2 2 2 5 4 3 2 5 3	3 3 3 4 4 3 5 2 3 5 1	1	1	1
145	67	43	2 6 2 4 4 5 4 2 2 2 2	2 6 2 4 3 5 2 3 2 1 4	2	2	2
156	67	43	2 4 1 3 2 2 3 3 2 2 4	4 3 1 4 3 2 3 3 2 1 2	2	2	2
160	67	43	1 4 1 1 1 1 1 3 1 1 1	2 1 3 3 4 4 4 2 2 5 5	1	1	1
162	67	43	4 4 1 4 2 3 3 3 3 2 3	2 3 2 4 2 4 4 5 4 1 3	1	1	3
67	67	43	4 3 1 2 2 2 4 2 4 5 1	2 2 3 5 5 5 5 3 4 5 1	1	1	1
86	67	43	1 5 1 1 1 1 3 3 5 1 3	4 3 2 3 2 2 3 3 2 3 2	1	1	1
87	67	43	1 4 1 1 1 1 3 3 3 1 1	2 1 3 4 5 5 5 1 1 5 3	1	1	1

Table 31.2 *(cont.)*

Judge	Sample 1	Sample 2	Attribute scores * Sample 1	Attribute scores * Sample 2	Overall performance	Preference based on two selected attributes **	
94	67	43	1 2 3 4 4 5 5 3 3 4 1	3 2 2 3 3 2 4 2 3 3 2	2	2	2
95	67	43	4 3 5 5 5 4 2 2 1 3 3	1 1 5 5 5 3 5 5 1 1 2	1	1	3
100	67	43	1 2 3 5 5 4 2 1 2 3 5	1 3 2 4 3 4 2 2 2 3 4	2	2	2
105	67	43	4 4 1 2 2 1 3 5 5 2 1	3 3 1 2 3 2 4 5 2 2 1	1	1	1
3	67	43	4 2 3 5 3 5 1 2 1 5 4	6 6 1 2 2 2 3 3 2 2 2	2	2	2
13	67	43	1 1 4 5 4 5 4 1 2 5 4	2 1 3 4 4 5 4 4 2 4 3	2	1	1
25	67	43	4 3 2 3 2 2 3 2 1 2 3	2 3 1 3 2 2 4 3 1 2 3	2	1	1
36	67	43	1 5 1 2 1 1 3 3 5 2 2	2 3 3 4 3 4 4 2 2 3 4	1	1	1
40	67	43	2 3 2 1 3 2 2 3 3 3 3	2 3 4 4 4 4 4 4 1 5 1	1	1	1
42	67	43	4 5 1 5 1 1 3 3 1 1 1	2 3 1 3 2 2 4 4 1 1 2	1	2	2
46	67	43	4 4 2 3 3 4 2 2 3 5 5	5 5 1 2 2 2 3 3 2 2 1	2	2	2
47	67	43	1 4 1 2 1 1 3 3 1 1 2	4 4 1 1 1 2 1 2 2 1 1	1	1	2
52	67	43	4 2 3 2 3 2 4 2 4 3 3	3 2 3 4 3 2 4 2 3 3 3	1	1	1
54	67	43	4 3 3 4 3 2 4 3 2 2 2	3 3 2 4 3 2 3 3 2 2 3	2	1	1
58	67	43	1 3 3 3 4 5 4 3 1 2 3	2 2 3 5 5 5 5 4 5 5 1	1	1	1
65	67	43	4 4 2 1 2 2 3 3 4 2 3	3 3 2 3 3 2 3 3 2 3 3	1	1	1
115	43	96	4 3 1 2 2 1 1 3 5 1 5	3 3 1 3 3 5 4 5 1 1 1	2	1	1
126	43	96	1 4 2 4 3 2 3 3 4 2 3	4 2 3 2 4 4 4 3 2 3 1	1	2	2
130	43	96	4 4 1 2 2 1 3 3 4 2 2	2 2 4 5 5 4 4 1 1 5 1	1	1	1
139	43	96	4 3 2 4 4 3 2 2 1 4 5	4 4 2 2 2 2 3 3 4 2 2	2	2	2
148	43	96	4 4 1 2 1 1 3 3 5 2 2	5 2 1 1 3 2 2 5 2 5 3	1	1	1
159	43	96	4 2 3 3 3 3 2 2 2 4 4	3 4 1 2 2 2 3 3 1 1 3	2	2	2
161	43	96	4 3 1 2 1 3 3 3 4 2 2	2 1 3 5 4 4 5 4 4 5 1	1	1	1
168	43	96	4 3 2 3 2 2 3 3 2 2 2	4 4 2 3 3 2 3 3 4 2 2	3	1	2
74	43	96	1 3 2 3 4 4 5 3 1 3 2	4 3 1 5 4 5 5 3 1 4 1	1	2	3
80	43	96	4 3 1 4 3 2 3 2 1 3 3	3 2 1 3 4 5 2 2 1 4 4	1	1	1
82	43	96	4 2 1 1 3 2 1 1 2 1 3	6 6 1 1 1 1 3 3 3 1 1	2	1	2
85	43	96	4 3 3 4 5 4 4 4 2 2 2	3 3 2 4 4 2 4 2 2 4 3	2	2	2
93	43	96	1 6 1 2 1 1 3 3 3 1 1	6 6 1 1 2 1 3 3 4 2 2	2	2	2
96	43	96	4 3 2 3 2 2 3 3 4 2 3	3 3 2 5 4 5 5 3 1 4 3	1	1	3
98	43	96	1 3 1 2 3 2 3 3 2 2 2	5 5 1 1 2 1 3 3 5 1 1	2	2	2
103	43	96	1 3 3 4 4 4 4 2 1 3 4	3 2 3 4 5 5 5 4 4 1 1	1	3	3
17	43	96	1 1 5 3 2 2 1 1 1 5 5	2 3 2 2 3 2 3 3 2 2 2	2	1	1
20	43	96	1 5 3 2 1 2 3 3 2 1 3	2 2 1 1 1 1 2 2 4 4 5	1	1	2
28	43	96	4 6 3 3 3 3 3 3 2 2 3	5 5 1 2 2 1 3 3 3 2 3	2	2	2
31	43	96	2 4 3 4 3 4 3 3 1 3 1	5 5 1 1 2 1 3 3 4 1 1	2	2	2
35	43	96	2 1 2 4 4 5 4 2 4 5 4	3 4 2 4 3 4 4 3 3 2 3	2	2	2
39	43	96	4 3 2 4 4 2 5 3 2 1 1	2 4 1 2 1 1 3 3 5 1 3	2	2	2
43	43	96	4 2 2 4 3 4 2 2 4 4 5	4 4 2 2 3 2 3 1 1 1 1	2	1	2
45	43	96	2 4 1 2 2 1 3 3 2 1 2	4 2 2 3 4 3 4 4 2 5 1	1	1	1
50	43	96	1 2 3 3 3 3 2 2 2 5 5	3 4 1 3 2 1 3 3 1 2 2	2	1	1
56	43	96	2 4 2 3 3 2 4 3 1 1 2	3 2 3 3 2 2 3 2 3 2 3	2	2	2
59	43	96	1 4 1 2 2 1 3 3 4 2 3	4 3 3 4 4 4 4 3 2 4 2	1	1	1
64	43	96	1 3 2 3 2 1 3 2 1 2 2	4 3 2 4 4 4 4 5 2 2 3	1	1	1

Table 31.2 *(cont.)*

Judge	Sample 1	Sample 2	Attribute scores * Sample 1	Attribute scores * Sample 2	Overall performance	Preference based on two selected attributes **	
109	96	43	4 2 3 4 4 4 4 3 1 1 1	3 5 1 3 2 3 3 3 1 1 3	2	2	2
118	96	43	1 5 1 2 1 2 3 3 2 1 1	2 1 3 5 4 3 3 2 4 5 4	1	1	2
122	96	43	2 3 2 4 3 4 4 3 3 2 3	1 3 3 4 4 4 3 2 3 4 3	1	2	3
124	96	43	4 3 2 2 2 4 2 2 2 2 4	1 1 2 5 3 4 2 3 4 3 2	1	1	2
128	96	43	1 5 1 2 1 1 3 3 2 2 3	4 3 2 3 4 4 4 4 2 4 2	1	1	1
140	96	43	1 3 3 3 4 4 3 2 1 5 4	1 1 3 5 4 3 5 4 3 2 1	1	2	1
142	96	43	1 2 2 2 3 2 4 3 5 3 3	3 1 2 5 5 1 5 5 5 3 1	1	1	2
147	96	43	2 3 2 2 3 1 2 2 3 2 4	2 3 3 3 3 3 3 3 3 3 3	1	1	1
149	96	43	3 5 1 2 1 4 3 5 1 1 3	2 2 4 5 4 1 1 1 5 5 5	1	2	2
155	96	43	2 3 3 4 2 3 3 3 3 3 3	3 2 3 2 2 2 2 2 4 4 4	1	2	2
165	96	43	4 4 1 4 3 4 3 3 5 1 2	3 1 5 5 5 1 2 2 1 5 5	1	2	2
166	96	43	4 3 2 2 2 2 3 3 2 2 3	2 2 3 4 4 4 4 4 3 4 2	1	1	1
167	96	43	4 5 1 1 2 1 3 3 2 1 2	2 1 3 3 5 4 2 2 1 5 5	1	1	1
70	96	43	1 3 1 2 2 1 3 3 1 2 3	3 4 1 1 2 2 3 3 1 4 2	2	1	1
73	96	43	4 3 1 2 1 1 2 3 4 1 4	4 2 2 4 3 2 2 2 2 4 3	1	1	1
84	96	43	4 4 1 4 2 1 3 3 4 1 3	3 3 1 5 4 2 4 3 2 2 3	1	1	1
2	96	43	1 3 3 2 4 2 4 3 2 3 3	2 3 3 4 4 4 4 2 3 5 3	1	1	1
4	96	43	4 1 3 5 3 1 1 1 4 5 5	4 3 1 3 3 2 2 2 2 3 4	2	2	2
5	96	43	4 3 3 4 4 2 3 6 2 3 1	4 4 3 3 2 4 3 3 2 1 3	2	2	2
7	96	43	4 6 1 2 1 1 3 2 5 1 2	5 6 1 1 1 1 3 3 5 1 1	2	2	2
8	96	43	1 3 2 3 3 2 3 3 2 3 2	4 3 3 3 4 4 4 5 2 4 2	1	1	1
9	96	43	4 3 1 2 5 3 5 4 2 2 1	4 3 1 1 2 2 3 3 1 1 3	2	2	2
16	96	43	2 3 1 2 2 1 3 3 4 1 2	2 2 3 5 5 4 5 1 3 5 5	1	1	1
51	96	43	4 1 3 5 5 5 5 5 3 5 5	2 2 2 5 5 5 5 4 5 4 2	2	2	3
57	96	43	4 2 4 4 3 4 2 2 1 3 3	5 5 1 3 1 2 3 3 1 1 2	2	2	2
61	96	43	4 2 2 2 4 4 3 3 4 4 2	4 1 2 2 3 3 2 1 4 2 4	1	1	2
63	96	43	4 3 1 2 1 1 3 3 1 2 3	2 1 5 5 5 5 5 1 5 5 5	1	1	1
111	67	96	1 5 1 2 2 3 3 3 3 2 1	2 3 4 5 4 4 5 2 2 5 4	1	1	1
114	67	96	1 3 2 4 3 2 1 1 1 3 3	4 4 4 4 4 2 3 3 1 5 3	1	2	3
116	67	96	2 2 4 3 5 5 2 1 4 5 5	3 3 2 5 5 5 5 3 2 2 3	1	1	1
120	67	96	4 4 2 3 2 2 3 5 4 2 3	2 3 2 4 4 4 4 3 3 4 1	1	1	1
121	67	96	4 3 2 3 3 4 4 2 2 2 3	3 3 1 3 2 2 3 3 3 4 3	1	1	2
129	67	96	4 3 2 3 2 2 3 3 2 3 3	3 4 2 3 3 2 3 3 2 3 3	2	1	1
133	67	96	4 3 2 4 4 4 5 3 2 3 2	3 4 1 2 2 2 3 3 4 1 3	2	2	2
134	67	96	1 4 1 2 4 2 4 3 2 1 2	3 2 2 3 3 2 4 3 3 1 2	1	1	1
143	67	96	1 4 1 2 2 1 2 3 2 1 2	2 3 3 3 2 3 2 2 2 3 3	1	1	1
146	67	96	2 6 1 2 2 1 3 3 5 1 2	2 1 5 5 5 2 5 5 3 5 1	1	1	1
151	67	96	1 2 3 5 4 5 5 5 2 4 2	1 1 5 5 5 5 5 5 5 1 1	1	1	1
152	67	96	2 2 3 3 4 4 2 2 1 4 5	2 2 3 4 5 5 4 5 3 5 3	1	1	1
158	67	96	4 2 3 4 4 5 4 3 3 5 3	4 4 1 2 2 2 3 3 3 3 3	2	2	2
163	67	96	1 5 1 2 2 1 3 3 3 1 1	4 4 2 2 2 2 2 2 2 3 4	1	2	1
164	67	96	4 3 2 2 3 3 4 3 2 3 2	1 3 3 5 4 5 4 2 1 5 3	1	1	1
66	67	96	4 5 1 2 2 1 3 3 2 2 2	3 2 2 3 4 4 2 2 2 4 4	1	1	1
72	67	96	1 3 1 3 3 4 4 3 1 3 3	6 2 2 3 4 2 4 3 2 3 3	1	1	2

Table 31.2 *(cont.)*

Judge	Sample 1	Sample 2	Attribute scores * Sample 1	Attribute scores * Sample 2	Overall performance	Preference based on two selected attributes **
76	67	96	1 6 1 2 2 1 3 3 2 1 3	3 3 2 3 2 1 2 2 3 3 4	1	2 2
81	67	96	1 2 4 4 3 2 5 3 1 2 3	4 2 3 4 4 4 3 3 2 3 2	2	2 2
89	67	96	4 3 1 3 2 2 3 3 2 2 2	3 4 2 3 2 2 3 3 2 2 3	1	1 1
97	67	96	1 4 1 2 2 1 3 3 4 1 3	2 3 4 4 4 5 2 2 1 5 5	1	1 1
99	67	96	2 3 3 5 3 5 4 3 1 2 2	3 3 3 3 3 4 3 3 1 2 2	2	2 2
108	67	96	4 1 4 5 4 4 5 3 1 2 2	3 2 2 3 4 5 4 2 1 2 3	2	2 3
14	67	96	4 5 1 2 1 1 3 3 1 1 2	5 5 1 3 2 1 3 3 1 1 1	1	2 2
23	67	96	4 2 1 5 4 5 5 3 2 2 2	3 3 4 3 3 2 4 3 2 3 2	2	2 2
33	67	96	1 3 1 1 2 2 2 2 4 4 4	2 1 5 4 4 4 4 3 1 4 2	1	1 1
37	67	96	4 4 3 2 2 1 3 3 4 1 3	2 3 3 3 3 4 4 3 2 5 3	1	1 1
49	67	96	1 3 2 3 4 5 5 3 3 3 2	2 4 1 1 1 2 3 3 2 1 3	2	2 2
112	96	67	4 3 2 3 3 2 2 2 4 3 4	4 3 2 5 5 4 4 4 4 2 1	1	1 1
119	96	67	4 6 2 4 3 5 3 3 1 3 3	1 1 4 4 5 3 3 1 5 5 3	1	1 1
131	96	67	1 3 1 2 2 2 2 3 2 2 4	2 2 2 3 3 3 3 3 4 1 3	1	1 1
136	96	67	1 4 2 2 2 1 4 3 3 2 3	4 3 3 2 2 1 2 2 4 4 4	1	2 2
138	96	67	1 5 2 2 1 1 3 3 5 1 2	6 6 1 1 1 1 3 3 5 1 1	2	2 3
141	96	67	4 1 4 4 5 5 5 4 5 5 1	4 4 3 3 4 5 5 4 2 4 3	2	2 3
150	96	67	4 4 1 2 1 5 3 3 4 2 2	1 3 1 3 1 2 3 3 4 1 2	1	1 2
153	96	67	1 3 1 3 2 4 4 2 1 2 2	2 4 1 5 3 4 4 3 1 1 2	1	1 1
157	96	67	1 4 1 2 2 1 3 3 4 5 3	2 3 3 4 3 2 4 2 2 3 2	1	2 1
71	96	67	4 3 1 1 2 1 3 3 4 1 2	1 1 5 5 5 5 5 5 5 5 5	1	1 1
88	96	67	2 4 1 4 2 2 3 2 1 1 2	4 4 1 1 1 3 3 3 4 1 2	2	2 2
91	96	67	2 6 1 1 1 1 3 3 5 1 1	3 3 2 3 3 2 2 2 3 4 4	1	1 2
92	96	67	4 5 1 1 2 1 2 2 2 3 5	6 5 1 2 2 2 3 3 3 3 3	2	2 3
101	96	67	4 2 1 5 4 1 5 3 4 2 1	6 3 1 3 3 2 3 2 3 3 3	1	1 1
106	96	67	4 2 3 3 2 2 4 3 4 5 4	2 1 4 5 4 4 4 3 5 5 1	1	1 1
1	96	67	2 4 3 5 3 2 4 3 2 3 3	2 4 2 4 2 4 3 3 2 2 3	2	1 1
12	96	67	4 1 3 5 5 3 5 3 3 2 1	3 3 3 4 3 3 3 3 2 2 3	2	2 3
19	96	67	4 5 2 2 2 1 2 2 2 2 4	2 3 3 4 4 4 3 2 1 1 4	1	1 1
24	96	67	1 2 3 4 5 5 5 2 1 4 1	2 1 4 5 5 4 4 1 1 5 3	1	1 1
29	96	67	1 3 3 3 3 2 4 3 2 3 2	4 3 3 2 3 2 3 3 2 3 3	2	2 3
32	96	67	1 5 1 3 2 2 3 3 2 2 3	2 2 2 1 2 2 1 2 4 4 5	1	2 1
34	96	67	4 2 3 2 4 3 4 3 2 5 1	2 3 2 4 3 3 3 3 2 2 2	1	1 1
38	96	67	1 4 1 3 2 2 3 3 4 2 3	2 2 3 4 4 3 4 4 2 4 2	1	1 1
41	96	67	1 4 1 2 2 1 4 3 4 1 3	1 1 2 5 4 5 5 2 5 5 1	1	1 1
48	96	67	4 4 1 5 2 4 1 2 1 2 4	6 6 1 3 2 2 3 3 1 2 3	2	2 2
53	96	67	4 4 3 3 2 2 3 2 2 2 3	3 1 5 5 4 5 5 1 2 5 1	1	1 1
55	96	67	1 4 1 2 2 1 4 3 1 1 1	2 1 3 3 3 2 4 2 2 5 3	1	2 1

* Rated on 1-6 scale (1, excellent,...,6, poor)

** Preference : 1, prefers first sample tasted
2, prefers second sample tasted
3, no preference

32. Incidence of Malignant Melanoma After Peaks of Sunspot Activity

Source Houghton, A., Munster, E.W. and Viola, M.V. (1978). Increased incidence of malignant melanoma after peaks of sunspot activity. *The Lancet,* April 8, 759-760.

Contributor A.N. Houghton Memorial Sloan-Kettering Cancer Center

The aetiology of melanoma is complex and may include the influences of trauma, heredity and hormonal activity (Lee, 1975). In particular, exposure to solar radiation may be involved in the pathogenesis of melanoma. Melanoma is more common in fair-skinned individuals (Lancaster and Nelson, 1957) and most frequent in skin sites exposed to the sun (Davis, Herron and McLeod, 1966). In white populations melanoma is more common in areas closer to the equator where the intensity of solar radiation is higher (Elwood, Lee, Walter, Mo and Green, 1974). Data from various parts of the world suggest that the incidence of melanoma is increasing (Burbank, 1971; Lee and Carter, 1970; Houghton, Flannery and Viola, 1980).

The data in Table 32.1, giving age-adjusted melanoma incidence, are from the Connecticut Tumor Registry from 1936-1972. Connecticut has the longest record of state population-based cancer statistics in the United States of America. Table 32.1 also includes the sunspot relative number.

Houghton, Munster and Viola (1978) have shown that the age-adjusted incidence rate for malignant melanoma in the state of Connecticut has risen since 1935 and that superimposed on the rise are 3-5 year periods in which the rise in the rate of incidence is excessive. These periods have a cycle of 8-11 years and follow times of maximum sunspot activity. The relationship between solar cycles and melanoma supports the hypothesis that melanoma is related to sun exposure and provides evidence that solar radiation may trigger the development of clinically apparent melanoma.

References

Burbank, F. (1971). *In Patterns of Cancer Mortality in the United States* 1950-1967 No. 4 Cancer Inst. Monograph, p.33.

Davis, N.C., Herron, J.J. and McLeod, C.R. (1966). Malignant melanoma in Queensland. Analysis of 400 skin lesions. *Lancet* **ii**, 407-410.

Elwood, J.M., Lee, J.A.H., Walter, S.D., Mo, T. and Green, A. (1974). Relationship of melanoma and other skin cancer mortality to latitude and ultraviolet radiation in the United States and Canada. *Inst. J. Epidemiol.* **3**, 325-332.

Houghton, A.N., Flannery, J. and Viola, M.V. (1980). Malignant melanoma in Connecticut and Denmark. *Inst. J. Cancer* **25**, 95-104.

Lancaster, H.O. and Nelson, J. (1957). Sunlight as a cause of melanoma: A clinical survey. *Med. J. Aust.* **1**, 452-456.

Lee, J.A.H. (1975). Current evidence about the causes of malignant melanoma. In *Progress in Clinical Cancer,* pp. 151-161.

Lee, J.A.H. and Carter, A.P. (1970). Secular trends in mortality from malignant melanoma *J. Natl. Cancer Inst.* **45**, 91-97.

Table 32.1
Age-adjusted Melanoma Incidence from the Connecticut Tumor Registry from 1950-1972 *

Year	Male	Total	Sunspot relative number
1936	1.0	0.9	40
1937	0.8	0.8	115
1938	0.8	0.8	100
1939	1.4	1.3	80
1940	1.2	1.4	60
1941	1.0	1.2	40
1942	1.5	1.7	23
1943	1.9	1.8	10
1944	1.5	1.6	10
1945	1.5	1.5	25
1946	1.5	1.5	75
1947	1.6	2.0	145
1948	1.8	2.5	130
1949	2.8	2.7	130
1950	2.5	2.9	80
1951	2.5	2.5	65
1952	2.4	3.1	20
1953	2.1	2.4	10
1954	1.9	2.2	5
1955	2.4	2.9	10
1956	2.4	2.5	60
1957	2.6	2.6	190
1958	2.6	3.2	180
1959	4.4	3.8	175
1960	4.2	4.2	120
1961	3.8	3.9	50
1962	3.4	3.7	35
1963	3.6	3.3	20
1964	4.1	3.7	10
1965	3.7	3.9	15
1966	4.2	4.1	30
1967	4.1	3.8	60
1968	4.1	4.7	105
1969	4.0	4.4	105
1970	5.2	4.8	105
1971	5.3	4.8	80
1972	5.3	4.8	65

* Incidence is the number of cases per 100,000 population.

33. Supplemental Ascorbate, Vitamin C, in the Supportive Treatment of Cancer

Source Cameron, E. and Pauling, L. (1978). Supplemental ascorbate in the supportive treatment of cancer: reevaluation of prolongation of survival times in terminal human cancer. *Proc. Natl. Acad. Sci. U.S.A.* **75**, 4538-4542.

Contributor L. Pauling Linus Pauling Institute of Science and Medicine

A study was made of the survival times of 100 terminal cancer patients who were given supplemental ascorbate, Vitamin C, as part of their routine management and 1000 matched controls, similar patients who had received the same treatment except for the ascorbate. The object of the investigation was to determine whether supplemental ascorbate prolongs the survival times of patients with terminal human cancer.

The clinical trial involved a treated group of 100 patients with terminal cancer of various kinds and a control group of 1000 untreated and matched patients, 10 for each treated patient. The treated group began ascorbate treatment, usually 10g/day, by intravenous infusion for about 10 days and orally thereafter at the time in the progress of their disease when in the considered opinion of at least two independent clinicians the continuance of any conventional form of treatment would offer no further benefit.

The original group of matched controls was selected by members of the Medical Records Staff in the Vale of Leven District General Hospital, Loch Lomondside, Scotland. It was obtained by a random search of the case record index of similar patients by the same clinicians in Vale of Leven District General Hospital over the last 10 years. For each treated patient, 10 controls were found of the same sex, within five years of the same age, and who had suffered from cancer of the same primary organ and histological tumour type. These 1000 cancer patients comprise the control group. The controls received the same treatment as the ascorbate-treated patients except for the ascorbate. They did not receive a placebo. The presentation date of untreatability corresponds to the date when ascorbate supplementation was initiated in the treated group. Comparable survival times of the 10 matched controls could then be calculated. The presentation date of untreatability can be influenced by many factors in the individual patients, but it is contended that the use of 1000 controls managed by the same clinicians in the same hospital over the last 10 years provides a sound basis for this comparative study.

It is believed that the ascorbate-treated patients represent a random selection of all of the terminal patients in the hospital, even though no formal randomization process was used. In the random selection three patients were excluded because supplemental ascorbate had been deliberately discontinued by order of another physician and five were excluded because matched controls could not be found for them. Patients suspected or known to have voluntarily discontinued ascorbate treatment have been retained in the group, as have those who died from some cause other than cancer. No patient was excluded because of short survival time.

The first study was reported by Cameron and Pauling (1976). In order to check the results, a fresh selection of a set of 1000 matched controls was carried out. The data. from the 1978 paper, which overlap largely with the older set, are given in Table 33.1. Table 33.1 gives the survival times of the ascorbate-treated patients after the date of first hospital attendance for the cancer that became untreatable, their survival times after the date of untreatability, and the corresponding mean values for the matched controls. In Cameron and Pauling (1976) the survival times after the date of untreatability of each of the 1100 subjects of the first study are given. There appear to be some differences in survival times of patients with different primary cancers.

Table 33.2, which has not been published elsewhere, gives for 11 patients in Table 33.1 (numbers 28, 33, 34, 35, 36, 37, 38, 84, 90, 92,99) with advanced cancer of the colon and each of the corresponding controls, the time in days between the date of first hospital attendance and date of untreatability.

References

Cameron, E. and Pauling, L. (1976). Supplemental ascorbate in the supportive treatment of cancer: Prolongation of survival times in terminal human cancer. *Proc. Natl. Acad. Sci. U.S.A.* 73, 3685-3689.

Table 33.1
Comparison of Times of Survival of 100 Cancer Patients Who Received Ascorbate and 1000 Matched Patients with No Treatment, by Site of Primary Cancer *

Case	Sex	Age, yrs.	Survival, days† A	B	C	D
			Stomach			
1	F	61	124	264	124	38
2	M	69	42	62	12	18
3	F	62	25	149	19	36
4	F	66	45	18	45	12
5	M	63	412	180	257	64
6	M	79	51	142	23	20
7	M	76	1112	35	128	13
8	M	54	46	299	46	51
9	M	62	103	85	90	10
10	F	69	876+	69	876+	19
11	M	46	146	361	123	52
12	M	57	340	269	310	28
95	F	59	396	130	359	55
			Bronchus			
13	M	74	81	72	74	33
14	M	74	461	134	423	18
15	M	66	20	84	16	20
16	M	52	450	98	450	58
17	F	48	246	48	87	13
18	F	64	166	142	115	49
19	M	70	63	113	50	38
20	M	77	64	90	50	24
21	M	71	155	30	113	18
22	M	70	859+	56	857+	18
23	M	39	151	260	38	34
24	M	70	166	116	156	20
25	M	70	37	87	27	27
91	M	55	223	69	218	32
93	M	74	138	100	138	27
97	M	69	72	315	39	39
98	M	73	245	188	231	65
			Colon			
28	F	76	248	292	135	18
29	F	58	377	492	50	30
30	M	49	189	462	189	65
31	M	69	1843	235	1267	17
32	F	70	180	294	155	57
33	F	68	537	144	534	16
34	M	50	519	643	502	25
35	F	74	455	301	126	21
36	M	66	406	148	90	17
37	F	76	365	641	365	42
38	F	56	942	272	911	40
77	M	65	776+	198	743+	14
84	F	74	372	37	366	28
90	M	58	163	199	156	31
92	F	60	101	154	99	28
99	M	77	20	649	20	33
100	M	38	283	162	274	80
			Rectum			
39	F	56	185	422	62	38
40	F	75	479	82	226	10
41	F	57	875	551	437	62
42	M	56	115	140	85	13
43	M	68	362	106	122	36
44	M	54	241	645	198	80
45	M	59	2175	407	759	64
			Pancreas			
86	M	77	465	56	342	20
87	M	67	27	60	27	24
88	F	60	83	99	16	58

Table 33.1 *(cont.)*

Case	Sex	Age, yrs.	Survival, days† A	B	C	D
			Ovary			
46	F	49	1234	307	320	64
47	F	68	89	690	88	21
48	F	52	201	285	196	129
49	F	67	356	244	337	58
50	F	56	2970	371	154	66
96	F	51	456	368	91	70
			Breast			
51	F	56	1235	796	10	45
52	F	57	24	977	24	68
53	F	53	1581	1623	580	23
54	F	68	1166	555	747	12
55	F	68	40	1304	39	52
56	F	53	727	1165	87	28
57	F	75	3808	675	633	62
58	F	68	791	871	251	75
59	F	55	1804	916	389	104
60	F	43	3460+	1311	2270+	41
61	F	53	719	978	322	64
			Bladder			
62	M	93	4288	464	260	29
63	F	70	3658	694	305	22
64	F	77	51	221	37	21
65	F	72	278	490	109	16
66	M	44	548	433	37	11
67	M	64	1607+	484	1320+	14
68	M	63	1250+	152	419+	32
			Kidney			
71	F	71	205	332	8	91
72	F	63	538	377	96	47
73	F	51	203	147	190	35
74	M	53	296	500	64	34
75	M	57	870	299	260	19
76	M	73	331	585	326	37
78	M	69	1685	1056	46	15
79	M	74	2060+	647	2060+	44
			Gall-bladder			
69	F	71	31	91	21	17
70	M	67	256	169	245	68
			Esophagus			
26	M	69	199	103	60	25
27	F	80	838	90	44	11
			Reticulum cell sarcoma			
80	M	44	1664+	367	1659+	23
81	M	65	86	427	86	45
			Prostate			
82	M	48	1306	467	255	77
89	M	68	331	944	122	15
			Uterus			
83	F	62	1273	497	86	44
			Brain			
85	M	49	47	296	34	59
			Chronic lymphatic leukemia			
94	F	59	3281	237	889	30

* The + following the survival time of an ascorbate-treated patient indicates that the patient was alive on 15 May 1978.

† A, survival time of the ascorbate patient, measured from the date of first hospital attendance for the cancer that reached the terminal stage; B, the corresponding mean survival time of that patient's 10 matched controls; C and D, survival times for ascorbate patient and controls, respectively, measured from the dates of untreatability.

Table 33.2
The Time in Days between Date of First Hospital Attendance and Date of "Untreatability" for 11 Patients with Advanced Cancer of the Colon

Case no.	Sex	Age, yrs	Time to "untreatability", days											
			Ten matched controls										Mean	Ascorbate treated patient
28	F	76	0	0	834	134	7	0	555	872	2	339	274.3	113
33	F	68	212	125	162	11	239	262	182	69	0	19	128.1	3
34	M	50	1363	0	25	3755	1010	0	8	2	6	16	618.5	17
35	F	74	10	46	45	0	- *	249	30	1831	187	402	280.0	329
36	M	66	0	21	0	26	176	8	0	4	7	1075	131.7	316
37	F	76	279	1835	725	0	43	419	0	1838	583	265	598.7	0
38	F	56	196	475	294	110	869	101	217	13	5	43	232.3	31
84	F	74	14	45	0	3	1	0	0	9	12	4	8.8	6
90	M	58	3	1	0	194	22	281	690	29	0	472	169.2	7
92	F	60	10	464	0	580	17	6	0	3	156	23	125.9	2
99	M	77	721	51	2	2	808	30	13	4530	0	5	616.2	0

* Uncertainty about the date of untreatability.

34. Incidence of Byssinosis: A Cross-Sectional Occupational Health Study

Source Higgins, J.E. and Koch, G.G. (1977). Variable selection and generalized chi-square analysis of categorical data applied to a large cross-sectional occupational health survey. *Int. Statist. Rev.* **45**, 51-62.

Contributor F. Yates Rothamsted Experimental Station

In 1973 a large cotton textile company made a study to investigate the prevalence of byssinosis, a form of pneumoconiosis to which workers exposed to cotton dust are subject. It was desired to determine the extent to which byssinosis is explained by such variables as sex, race, length of employment, smoking habit and dustiness of work place.

The study is based on the incidence of byssinosis amongst 5419 workers classified by five factors:

Type of work place	1 (most dusty), 2 (less dusty), 3 (least dusty).
Employment, years	<10, 10-19, 20 - .
Smoking	Smoker or non-smoker in the last five years.
Sex	Male, female.
Race	White, other.

The data, therefore, constitute a $3 \times 3 \times 2 \times 2 \times 2 \times 2$ contingency table, but this is better thought of as a $3 \times 3 \times 2 \times 2 \times 2$ table of percentage incidences, together with a table of the numbers on which the percentages are based. Table 34.1 presents the data.

The overall incidences for each factor separately suggest that all the factors, including sex and race, were affecting the incidence. In the paper by Higgins and Koch (1977), the basic six-dimensional contingency table was analyzed. Yates (1981) considers the five-dimensional table of percentage incidence and uses the log log transformation in his analysis. It is of interest to compare the two analyses.

References

Yates, F. (1981). *Sampling Methods for Censuses and Surveys,* 4th edition. London: Griffin.

Table 34.1
Incidence of Byssinosis

Employment, years	Smoking	Sex	Race, White, W, Other, O	Workspace 1 Byssinosis		Workspace 2 Byssinosis		Workspace 3 Byssinosis	
				Yes	No	Yes	No	Yes	No
<10	Yes	M	W	3	37	0	74	2	258
<10	Yes	M	O	25	139	0	88	3	242
<10	Yes	F	W	0	5	1	93	3	180
<10	Yes	F	O	2	22	2	145	3	260
<10	No	M	W	0	16	0	35	0	134
<10	No	M	O	6	75	1	47	1	122
<10	No	F	W	0	4	1	54	2	169
<10	No	F	O	1	24	3	142	4	301
10 to 19	Yes	M	W	8	21	1	50	1	187
10 to 19	Yes	M	O	8	30	0	5	0	33
10 to 19	Yes	F	W	0	0	1	33	2	94
10 to 19	Yes	F	O	0	0	0	4	0	3
10 to 19	No	M	W	2	8	1	16	0	58
10 to 19	No	M	O	1	9	0	0	0	7
10 to 19	No	F	W	0	0	0	30	1	90
10 to 19	No	F	O	0	0	0	4	0	4
⩾ 20	Yes	M	W	31	77	1	141	12	495
⩾ 20	Yes	M	O	10	31	0	1	0	45
⩾ 20	Yes	F	W	0	1	3	91	3	176
⩾ 20	Yes	F	O	0	1	0	0	0	2
⩾ 20	No	M	W	5	47	0	39	3	182
⩾ 20	No	M	O	3	15	0	1	0	23
⩾ 20	No	F	W	0	2	3	187	2	340
⩾ 20	No	F	O	0	0	0	2	0	3

35. Inter and Intra Individual Variation of Blood Glucose Levels

Source O'Sullivan, J.B. and Mahan, C.M. (1966). Glucose tolerance test: Variability in pregnant and non-pregnant women. *American J. Clin. Nutr.* **19**, 345-351.

Contributor C. M. Mahan Tufts University

The most commonly used diagnostic aid for early diabetes mellitus is the oral glucose tolerance test. The test, however, is subject to considerable variation due to differences in individual rates of gastrointestinal absorption of the glucose challenge. This problem is accentuated in the pregnant woman; yet pregnancy is the ideal period during which to identify the potentially diabetic female. The additional variation during pregnancy has been considered by some, see Burt (1962), to be reason to exclude use of the test.

Data previously obtained from registrants for pre-natal care at Boston City Hospital between July 1955 and June 1960 were available for estimating the components of variance of blood sugars in the pregnant and non-pregnant states. Every n^{th} patient to register had been requested to return for an oral glucose test. If a selected patient had failed to return by the thirty-seventh week of pregnancy, a replacement was selected at random from among the new registrants of that time period. In total, 241 candidates were systematically selected from the 14,300 screenees. Of these, 142 candidates and 76 random replacements completed glucose tolerance tests, to give a pool of 218 women. The number of repeat pregnancies in subsequent years varied widely as did the number of tests performed at annual intervals in the non-pregnant state.

The data presented are derived from the sample of 218 women. Fifty-three of these had annual glucose tolerance tests in the non-pregnant state over the six year period following their index glucose tolerance test. Their six fasting blood sugars and one hour post 100 grams glucose blood sugar concentration during the non-pregnant state are presented in Table 35.1. Fifty-two of the 218 women had at least two repeat pregnancies during which the glucose tolerance test was performed in the third trimester. Their fasting and one hour blood glucose levels on the three occasions are presented in Table 35.2. Unfortunately the number of women who fell into both subsets was too small for the more desirable analytic design.

Of interest is an assessment of intra individual variation of the fasting blood sugars observed ante and post partum contrasted with the intra individual variation after the glucose load.

References

Burt, R.L. (1962). Glucose tolerance in pregnancy. *Diabetes* **11**, 227-229.

Table 35.1
Six Annual Fasting and One Hour Post Glucose Blood Sugars of 53 Non-pregnant Women

Fasting glucose blood test						One hour post glucose blood glucose concentration					
1	2	3	4	5	6	1	2	3	4	5	6
45	66	83	71	76	64	59	73	60	85	97	64
59	59	81	63	72	70	56	59	74	82	81	87
71	69	89	82	70	86	103	91	89	98	80	134
80	84	104	86	76	87	104	109	173	136	87	121
76	85	85	69	81	73	53	73	85	73	84	63
65	96	80	77	76	77	80	103	86	70	76	81
65	85	75	86	90	60	90	72	75	99	94	67
76	82	80	81	85	77	85	116	83	81	95	98
72	80	85	85	89	97	92	43	105	89	94	108
82	90	87	72	79	69	93	106	103	96	123	108
83	75	70	95	81	84	85	65	98	107	90	127
64	80	75	87	85	76	94	76	104	100	120	111
91	73	75	89	73	80	95	68	100	141	89	83
76	69	65	64	77	76	94	99	73	87	101	84
87	85	65	84	70	59	76	59	115	83	63	53
81	82	83	85	89	69	97	135	116	137	144	116
71	91	85	79	89	92	85	98	62	100	113	91
89	84	62	80	76	73	78	48	75	115	76	67
74	77	75	83	74	86	104	79	119	106	79	92
65	86	68	84	88	75	48	56	31	121	78	80
79	75	92	79	96	86	79	69	100	134	107	182
68	63	70	73	70	88	58	57	48	87	97	98
71	87	69	81	76	96	89	87	72	77	92	106
88	74	88	85	84	83	69	93	94	112	84	60
84	80	85	73	91	81	100	116	87	123	130	109
83	83	86	86	78	88	108	130	108	114	114	131
89	77	73	90	88	84	90	121	100	144	198	136
79	69	69	85	83	76	64	75	76	109	116	99
71	78	75	87	73	67	83	80	80	89	88	92
103	93	104	90	90	94	111	129	96	124	118	114
106	93	106	77	78	101	132	110	97	136	97	103
83	87	87	81	82	94	114	82	94	77	107	64
84	70	65	72	64	84	76	70	56	51	73	84
74	85	99	90	89	89	73	105	123	119	97	121
81	76	93	88	84	94	81	75	80	90	133	109
70	87	85	83	80	83	94	96	78	72	92	96
65	87	83	76	73	73	74	89	87	87	80	106
89	79	80	80	73	79	75	66	84	102	90	99
70	91	78	79	75	90	109	124	107	129	139	127
80	86	80	77	76	84	92	109	96	116	112	119
84	82	85	78	84	87	114	100	94	96	96	108
77	82	82	78	67	78	86	93	115	89	85	125
65	84	83	69	82	77	47	94	86	99	96	106
46	85	83	94	88	86	74	77	93	85	107	118
80	90	90	84	75	74	75	68	70	78	99	94
70	79	84	84	83	79	48	92	76	110	74	103
75	83	74	75	75	88	84	98	76	57	127	99
45	65	63	64	73	77	95	89	94	73	118	100
101	87	81	85	88	77	99	103	88	82	109	84
71	70	72	83	67	80	82	80	76	87	71	78
77	89	80	88	81	102	95	74	100	94	92	115
109	81	73	79	73	71	125	84	95	105	104	132
74	74	73	77	48	73	74	94	114	110	120	115

Table 35.2
Three Fasting and One Hour Post Glucose Blood Sugars of 52 Women Observed in 3 Pregnancies *

Fasting blood glucose test			One hour post glucose blood glucose concentration		
1	2	3	1	2	3
60	69	62	97	69	98
56	53	84	103	78	107
80	69	76	66	99	130
55	80	90	80	85	114
62	75	68	116	130	91
74	64	70	109	101	103
64	71	66	77	102	130
73	70	64	115	110	109
68	67	75	76	85	119
69	82	74	72	133	127
60	67	61	130	134	121
70	74	78	150	158	100
66	74	81	150	131	142
83	70	74	99	98	105
68	66	90	119	85	109
78	63	75	164	98	138
103	77	77	160	117	121
77	68	74	144	71	153
66	77	68	77	82	89
70	70	72	114	93	122
75	65	71	77	70	109
91	74	93	118	115	150
66	75	73	170	147	121
75	82	76	153	132	115
74	71	66	143	105	100
76	70	64	114	113	129
74	90	86	73	106	116
74	77	80	116	81	77
67	71	69	63	87	70
78	75	80	105	132	80
64	66	71	83	94	133
67	71	69	63	87	70
78	75	80	105	132	80
64	66	71	83	94	133
71	80	73	81	87	86
63	75	73	120	89	59
90	103	74	107	109	101
60	76	61	99	111	98
48	77	75	113	124	97
66	93	97	136	112	122
74	70	76	109	88	105
60	74	71	72	90	71
63	75	66	130	101	90
66	80	86	130	117	144
77	67	74	83	92	107
70	67	100	150	142	146
73	76	81	119	120	119
78	90	77	122	155	149
73	68	80	102	90	122
72	83	68	104	69	96
65	60	70	119	94	89
52	70	76	92	94	100

* Measurements are in mg/100 ml.

36. Chemical and Overt Diabetes

Source Reaven, G.M. and Miller, R.G. (1979). An attempt to define the nature of chemical diabetes using a multidimensional analysis. *Diabetologia* **16**, 17-24.

Contributors Rupert G. Miller and Jerry W. Halpern Stanford University

Reaven and Miller (1979) examined the relationship between chemical subclinical and overt nonketotic diabetes in 145 non-obese adult subjects.

The three primary variables used in the analysis and presented in Table 36.1 are glucose intolerance, insulin response to oral glucose and insulin resistance. The first two variables are the areas under the straight line connecting glucose and insulin levels, respectively, determined from blood samples drawn during a three hour glucose tolerance test following an oral administration of a glucose load. Insulin resistance is measured by the steady state plasma glucose (SSPG) determined after chemical suppression of endogenous insulin secretion. In addition, the relative weight and fasting plasma glucose were measured for each individual in the study conducted at the Stanford Clinical Research Center and are included in Table 36.1.

These multidimensional data were visually inspected with the aid of the Prim 9 program at the Stanford Linear Accelerator Computation Center. The three primary variables have a configuration resembling a boomerang with a fat middle and two wings, or a donkey's head with floppy ears. From the clinical point of view the middle points represent the normal subjects and the two wings, or ears, the chemical and overt diabetic subjects.

Reaven and Miller applied a variant of a clustering algorithm due to Friedman and Rubin (1967) to the three primary variables to form clusters of "normal", "chemical diabetic", and "overt diabetic" subjects. The classifications of the subjects generated by the computer clustering were compared with the classifications obtained by current medical criteria. The latter classification for each patient is given in the last column of Table 36.1. A number of different clustering routines were tried on this same data set by Symons (1981).

References

Friedman, H. P. and Rubin, J. (1967). On some invariant criteria for grouping data. *J. Amer. Statist. Assoc.* **62**, 1159-1178.

Symons, M. J. (1981). Clustering criteria and multivariate normal mixtures. *Biometrics* **37**, 35-43.

Table 36.1
Measures of Blood Glucose and Insulin Levels, Relative Weights and Clinical Classifications

Patient number	Relative weight	Fasting plasma glucose	Glucose area	Insulin area	SSPG	Clinical classification *
1	0.81	80	356	124	55	3
2	0.95	97	289	117	76	3
3	0.94	105	319	143	105	3
4	1.04	90	356	199	108	3
5	1.00	90	323	240	143	3
6	0.76	86	381	157	165	3
7	0.91	100	350	221	119	3
8	1.10	85	301	186	105	3
9	0.99	97	379	142	98	3
10	0.78	97	296	131	94	3
11	0.90	91	353	221	53	3
12	0.73	87	306	178	66	3
13	0.96	78	290	136	142	3
14	0.84	90	371	200	93	3
15	0.74	86	312	208	68	3
16	0.98	80	393	202	102	3
17	1.10	90	364	152	76	3
18	0.85	99	359	185	37	3
19	0.83	85	296	116	60	3
20	0.93	90	345	123	50	3
21	0.95	90	378	136	47	3
22	0.74	88	304	134	50	3
23	0.95	95	347	184	91	3
24	0.97	90	327	192	124	3
25	0.72	92	386	279	74	3
26	1.11	74	365	228	235	3
27	1.20	98	365	145	158	3
28	1.13	100	352	172	140	3
29	1.00	86	325	179	145	3
30	0.78	98	321	222	99	3
31	1.00	70	360	134	90	3
32	1.00	99	336	143	105	3
33	0.71	75	352	169	32	3
34	0.76	90	353	263	165	3
35	0.89	85	373	174	78	3
36	0.88	99	376	134	80	3
37	1.17	100	367	182	54	3
38	0.85	78	335	241	175	3
39	0.97	106	396	128	80	3
40	1.00	98	277	222	186	3
41	1.00	102	378	165	117	3
42	0.89	90	360	282	160	3
43	0.98	94	291	94	71	3
44	0.78	80	269	121	29	3
45	0.74	93	318	73	42	3
46	0.91	86	328	106	56	3

Table 36.1 *(cont.)*

Patient number	Relative weight	Fasting plasma glucose	Glucose area	Insulin area	SSPG	Clinical classification *
47	0.95	85	334	118	122	3
48	0.95	96	356	112	73	3
49	1.03	88	291	157	122	3
50	0.87	87	360	292	128	3
51	0.87	94	313	200	233	3
52	1.17	93	306	220	132	3
53	0.83	86	319	144	138	3
54	0.82	86	349	109	83	3
55	0.86	96	332	151	109	3
56	1.01	86	323	158	96	3
57	0.88	89	323	73	52	3
58	0.75	83	351	81	42	3
59	0.99	98	478	151	122	2
60	1.12	100	398	122	176	3
61	1.09	110	426	117	118	3
62	1.02	88	439	208	244	2
63	1.19	100	429	201	194	2
64	1.06	80	333	131	136	3
65	1.20	89	472	162	257	2
66	1.05	91	436	148	167	2
67	1.18	96	418	130	153	3
68	1.01	95	391	137	248	3
69	0.91	82	390	375	273	3
70	0.81	84	416	146	80	3
71	1.10	90	413	344	270	2
72	1.03	100	385	192	180	3
73	0.97	86	393	115	85	3
74	0.96	93	376	195	106	3
75	1.10	107	403	267	254	3
76	1.07	112	414	281	119	3
77	1.08	94	426	213	177	2
78	0.95	93	364	156	159	3
79	0.74	93	391	221	103	3
80	0.84	90	356	199	59	3
81	0.89	99	398	76	108	3
82	1.11	93	393	490	259	3
83	1.19	85	425	143	204	2
84	1.18	89	318	73	220	3
85	1.06	96	465	237	111	2
86	0.95	111	558	748	122	2
87	1.06	107	503	320	253	2
88	0.98	114	540	188	211	2
89	1.16	101	469	607	271	2
90	1.18	108	486	297	220	2
91	1.20	112	568	232	276	2
92	1.08	105	527	480	233	2
93	0.91	103	537	622	264	2
94	1.03	99	466	287	231	2
95	1.09	102	599	266	268	2
96	1.05	110	477	124	60	2

Table 36.1 *(cont.)*

Patient number	Relative weight	Fasting plasma glucose	Glucose area	Insulin area	SSPG	Clinical classification *
97	1.20	102	472	297	272	2
98	1.05	96	456	326	235	2
99	1.10	95	517	564	206	2
100	1.12	112	503	408	300	2
101	0.96	110	522	325	286	2
102	1.13	92	476	433	226	2
103	1.07	104	472	180	239	2
104	1.10	75	455	392	242	2
105	0.94	92	442	109	157	2
106	1.12	92	541	313	267	2
107	0.88	92	580	132	155	2
108	0.93	93	472	285	194	2
109	1.16	112	562	139	198	2
110	0.94	88	423	212	156	2
111	0.91	114	643	155	100	2
112	0.83	103	533	120	135	2
113	0.92	300	1468	28	455	1
114	0.86	303	1487	23	327	1
115	0.85	125	714	232	279	1
116	0.83	280	1470	54	382	1
117	0.85	216	1113	81	378	1
118	1.06	190	972	87	374	1
119	1.06	151	854	76	260	1
120	0.92	303	1364	42	346	1
121	1.20	173	832	102	319	1
122	1.04	203	967	138	351	1
123	1.16	195	920	160	357	1
124	1.08	140	613	131	248	1
125	0.95	151	857	145	324	1
126	0.86	275	1373	45	300	1
127	0.90	260	1133	118	300	1
128	0.97	149	849	159	310	1
129	1.16	233	1183	73	458	1
130	1.12	146	847	103	339	1
131	1.07	124	538	460	320	1
132	0.93	213	1001	42	297	1
133	0.85	330	1520	13	303	1
134	0.81	123	557	130	152	1
135	0.98	130	670	44	167	1
136	1.01	120	636	314	220	1
137	1.19	138	741	219	209	1
138	1.04	188	958	100	351	1
139	1.06	339	1354	10	450	1
140	1.03	265	1263	83	413	1
141	1.05	353	1428	41	480	1
142	0.91	180	923	77	150	1
143	0.90	213	1025	29	209	1
144	1.11	328	1246	124	442	1
145	0.74	346	1568	15	253	1

* Clinical classification: 1, overt diabetic; 2, chemical diabetic; 3, normal.

37. The Maternal Age Distribution of Patients with Down's Syndrome

Source Moran, P.A.P. (1974). Are there two maternal age groups in Down's Syndrome? *Brit. J. Psychiat.* **124**, 453-455.

Contributor P.A.P. Moran The Australian National University

Several authors have observed a bimodality in the age distribution of mothers of patients with Down's Syndrome suggesting that the distribution is the mixture of two different distributions with different etiologies.

Moran suggests that the attempted division of the maternal age distribution is not logically necessary. If the probability of a normal mother producing a child with Down's Syndrome increases smoothly but sufficiently fast with the age of the mother, bimodality may result without a mixture of distributions. This suggestion is strengthened by the fact that Penrose's conjectural groupings of the different known chromosomal aberrations which are known to cause Down's Syndrome results in two groups whose relative frequencies are inconsistent with the relative frequencies which would be needed to produce the observed bimodality (Penrose, 1954). The data in Table 37.1 were given originally by Collmann and Stoller (1962) and include all cases of Down's Syndrome in Victoria, Australia from 1942 to 1957. Similar results for other countries are given in Penrose and Smith (1966).

References

Collmann, R.D. and Stoller, A. (1962). A survey of mongoloid births in Victoria, Australia, 1942-1957. *Amer. J. Pub. Health* **57**, 813-829.

Penrose, L.S. (1954). Mongolian idiocy (Mongolism) and maternal age. *Ann. New York Acad. Sci.* **57**, 494-502.

Penrose, L.S. and Smith, G.F. (1966). *Down's Anomaly.* London: Churchill.

Table 37.1
Age of Mothers of Patients with Down's Syndrome for Births in Australia from 1942 to 1952

Age group, years	Total no. of births	No. of mothers of patients
20 or less	35,555	15
20-24	207,931	128
25-29	253,450	208
30-34	170,970	194
35-39	86,046	297
40-44	24,498	240
45 or over	1,707	37

38. Procedures for the Detection of Muscular Dystrophy Carriers

Contributor M. Percy Mount Sinai Hospital Toronto

Duchenne Muscular Dystrophy, DMD, is a genetically transmitted disease, passed from a mother to her children. Affected female offspring usually suffer no apparent symptoms and may unknowingly carry the disease. Male offspring with the disease die at a young age. Not all cases of the disease come from an affected mother. A fraction, perhaps one third, of the cases arise spontaneously, to be genetically-transmitted by an affected female. This is the most widely held view at present. The incidence of DMD is about 1 in 10,000 male births. The population risk that a woman is a DMD carrier is about 1 in 3,300.

The detection problem arises when, for example, a woman suspects that she may be a carrier. This suspicion may start with the detection of an affected male child in the family. Although carriers of DMD usually have no physical symptoms, they tend to exhibit elevated levels of certain serum enzymes or proteins, such as creatine kinase (CK), hemopexin (H), lactate dehydroginase (LD) and pyruvate kinase (PK). The levels of these enzymes may also depend on age and season.

A program to develop an effective method for screening female relatives of boys with DMD began under the direction of Dr. M. Thompson, the Hospital for Sick Children, Toronto. The purpose of the program was to develop a procedure for informing a woman of the likelihood that she is a carrier, based on measurements of serum markers and her family pedigree.

Levels of the enzymes were measured in known carriers and in a group of non-carriers using standard laboratory procedures. The determinations are made by rather delicate, indirect measurement. The data are presented in Table 38.1. In this table, the measurements from separate blood samples are identified by a hospital ID number. In some cases, more than one sample was obtained from an individual. This observation number is given as well as the date the sample was taken.

The first two serum markers (CK and H) may be measured rather inexpensively from frozen serum. The second two (PK and LD) require fresh serum. Do the second two increase the detection rate in an important way? Should age and season be taken into account? The water supply of the laboratory was changed in the course of the study. This may have affected the measurements of some of the markers.

Table 38.1
Measurements of Levels of Creatine Kinase, Hemopexin, Lactate Dehydroginase and Pyruvate Kinase Enzymes

Obs. No.	Hospital ID	Age	Date M	Date Y	CK	H	PK	LD
					Measurements *			
			Normals					
1	1007	22	6	79	52.0	83.5	10.9	176
1	786	32	8	78	20.0	77.0	11.0	200
1	778	36	7	78	28.0	86.5	13.2	171
1	1306	22	11	79	30.0	104.0	22.6	230
1	895	23	1	78	40.0	83.0	15.2	205
1	987	30	5	79	24.0	78.8	9.6	151
1	789	27	8	78	15.0	87.0	13.5	232
1	825	30	11	78	22.0	91.0	17.5	198
1	1296	25	10	79	42.0	65.5	13.3	216
1	906	26	2	79	130.0	80.3	17.1	211
2	933	26	3	79	48.0	85.2	22.7	160
3	1246	27	7	79	31.0	86.5	6.9	162
4	671	25	10	77	41.0	87.3	15.0	
1	818	26	10	78	47.0	53.0	14.6	131
2	921	27	3	79	36.0	56.0	18.2	105
3	1253	27	7	79	24.0	57.5	5.6	130
1	941	31	4	79	34.0	92.7	7.9	140
2	1287	31	9	79	38.0	96.0	12.6	158
1	1301	35	10	79	40.0	104.6	16.1	209
1	948	28	4	79	59.0	88.0	9.9	128
2	1259	28	8	79	75.0	81.0	10.1	177
3	1289	28	9	79	72.0	66.3	16.4	156
1	773	27	7	78	42.0	77.0	15.3	163
2	934	27	3	79	30.0	80.2	8.1	100
3	1022	28	6	79	24.0	87.0	3.5	132
1	667	31	11	77	29.0	94.0	11.8	
1	801	24	9	78	26.0	84.5	20.7	145
1	1262	23	8	79	65.0	75.0	19.9	187
1	938	27	3	79	34.0	86.3	11.8	120
1	903	25	2	79	37.0	73.3	13.0	254
1	916	34	3	79	73.0	57.4	7.4	107
2	1249	34	7	79	87.0	76.3	6.0	87
1	1218	25	7	79	35.0	71.0	8.8	186
1	779	20	7	78	31.0	61.5	9.9	172
1	990	20	5	79	62.0	81.0	10.2	181
1	1001	31	6	79	48.0	79.0	16.8	182
1	1250	31	7	79	40.0	82.5	6.4	151
1	1219	26	7	79	55.0	85.5	10.9	216
1	1248	26	7	79	32.0	73.8	8.6	147
1	1307	21	11	79	26.0	79.3	16.4	123
1	769	27	6	78	25.0	91.0	10.3	135

Table 38.1 *(cont.)*

Obs. No.	Hospital ID	Age	Date M	Date Y	Measurements * CK	H	PK	LD
2	1016	27	6	79	22.0	84.0	2.8	145
1	774	26	7	78	30.0	76.0	17.1	145
2	917	26	3	79	35.0	76.7	10.9	105
3	1255	26	7	79	34.0	78.0	8.0	140
1	813	31	10	78	27.0	90.0	15.6	167
2	920	31	3	79	22.0	71.5	11.8	98
3	1012	31	6	79	22.0	73.5	5.1	184
1	1285	25	9	79	72.0	80.5	12.0	225
1	669	22	12	77	22.0	85.5	15.0	
1	782	35	7	78	51.0	70.0	16.6	146
2	909	36	2	79	30.0	66.7	15.3	124
3	1247	36	7	79	23.0	66.3	4.4	142
1	943	33	4	79	67.0	98.0	9.3	225
1	804	27	9	78	50.0	69.0	15.1	160
2	924	27	3	79	92.0	68.0	16.5	115
1	1305	36	10	79	55.0	78.2	21.8	188
1	765	25	6	78	38.0	82.0	15.8	161
1	657	27	10	77	22.0	99.0	10.8	
1	918	33	3	79	27.0	100.0	10.3	169
1	673	26	12	77	28.0	93.5	7.0	
1	790	22	8	78	34.0	84.0	12.0	175
2	908	23	2	79	44.0	81.3	10.5	159
3	1244	25	7	79	32.0	86.5	6.7	149
1	1298	22	10	79	35.0	59.4	11.3	130
1	949	31	5	79	35.0	90.3	15.3	124
1	791	33	8	78	31.0	75.5	13.7	160
2	824	33	11	78	25.0	78.9	12.2	127
1	1003	27	6	79	52.0	77.0	17.9	198
1	819	30	10	78	34.0	75.0	15.4	171
1	1292	20	10	79	53.0	93.2	22.3	349
1	914	30	2	79	69.0	66.7	8.7	119
2	1220	30	7	79	25.0	70.5	5.3	123
1	989	27	5	79	24.0	89.5	16.1	176
1	1281	35	8	79	21.0	108.5	9.8	148
1	768	26	6	78	51.0	82.0	12.9	149
2	1011	27	6	79	37.0	77.3	3.9	141
1	771	24	7	78	24.0	82.0	14.2	123
2	910	25	2	79	30.0	77.0	16.2	124
3	1245	26	7	79	34.0	81.3	9.7	158
1	802	20	9	78	22.0	102.0	10.3	177
1	1295	22	10	79	32.0	79.2	5.8	190
1	675	38	12	77	45.0	108.0	13.7	
1	1294	34	10	79	24.0	70.4	10.6	181
1	781	22	7	78	20.0	72.0	11.9	110
1	762	22	6	78	34.0	91.0	14.5	144
1	1290	24	9	79	25.0	92.0	14.0	166

Table 38.1 *(cont.)*

Obs. No.	Hospital ID	Age	Date M	Date Y	Measurements * CK	H	PK	LD
1	831	38	12	78	26.0	109.0	8.9	163
2	896	39	1	79	28.0	102.3	17.1	146
3	919	39	3	79	21.0	92.4	10.3	197
4	927	39	3	79	23.0	111.5	10.0	133
5	940	39	4	79	26.0	92.6	12.3	196
6	947	39	4	79	25.0	98.7	10.0	174
7	1010	39	6	79	21.0	93.2	5.9	181
1	899	32	2	79	56.0	72.0	9.9	227
2	1303	34	10	79	48.0	83.0	13.7	228
1	763	22	6	78	51.0	91.0	12.7	149
1	926	39	3	79	18.0	95.0	11.3	66
1	1252	33	7	79	28.0	104.0	6.9	169
1	1310	33	11	79	41.0	105.5	15.1	252
1	1017	20	6	79	40.0	81.0	6.1	167
1	776	22	7	78	21.0	74.5	12.2	163
1	901	25	2	78	95.0	69.8	7.3	169
2	1009	25	6	79	59.0	72.5	10.7	314
1	913	36	2	79	40.0	72.7	7.0	131
2	1260	36	8	79	30.0	79.5	11.9	130
1	767	22	6	78	48.0	76.0	16.6	133
1	1019	39	6	79	39.0	88.5	7.6	168
1	936	30	3	79	30.0	82.7	18.1	124
1	1261	37	7	79	38.0	85.0	21.6	198
1	929	27	3	79	27.0	87.2	12.5	99
2	1254	28	7	79	32.0	76.3	5.6	159
1	911	29	2	79	74.0	80.4	8.9	207
2	1015	30	6	79	33.0	86.0	3.8	149
1	777	31	7	78	34.0	80.5	11.1	149
2	904	31	2	79	45.0	86.5	10.8	169
3	1021	32	6	79	52.0	79.0	10.7	187
1	789	32	7	78	28.0	82.5	17.4	144
2	829	32	12	78	35.0	97.0	14.5	137
1	764	25	6	78	37.0	93.0	15.3	167
2	907	26	2	79	44.0	81.3	15.3	166
3	1014	26	6	79	68.0	82.8	11.9	177
1	811	37	10	78	97.5	34.0	12.0	203
1	785	25	7	78	37.0	98.0	16.4	198
2	1217	25	7	79	34.0	92.0	12.1	217
1	766	20	6	78	30.0	80.0	12.9	129
1	770	25	6	78	37.0	98.0	11.7	177
2	682	24	10	77	26.0	94.2	11.7	
1	798	34	9	78	24.0	100.5	14.0	231
1	942	32	4	79	41.0	78.5	10.9	191
2	1024	32	6	79	43.0	87.5	6.0	136
1	810	32	10	78	30.0	90.5	15.3	136
2	1258	33	7	79	30.0	85.0	11.4	176
3	1300	33	10	79	43.0	88.5	20.3	175

Table 38.1 *(cont.)*

Obs. No.	Hospital ID	Age	Date M	Date Y	Measurements * CK	H	PK	LD
			Carriers					
1	1027	30	10	78	167.0	89.0	25.6	364
1	1013	41	10	78	104.0	81.0	26.8	245
1	1324	22	8	79	30.0	108.0	8.8	284
2	1332	22	8	79	44.0	104.0	17.4	172
1	966	20	10	78	65.0	87.0	23.8	198
1	979	42	9	78	440.0	107.0	20.2	239
1	1327	59	8	79	58.0	88.2	11.0	259
1	978	35	9	78	129.0	93.1	18.3	188
2	1290	36	6	79	104.0	87.5	16.7	256
3	1139	35	2	79	122.0	88.5	21.6	263
4	1487	36	11	79	144.0	24.4		329
1	1193	29	4	79	265.0	83.5	16.1	136
2	1513	30	12	79	510.0	60.2		272
1	1208	27	4	79	285.0	79.5	36.4	245
2	1395	27	9	79	25.0	91.0	49.1	209
1	1209	28	4	79	124.0	92.0	32.2	298
1	947	29	8	78	53.0	76.0	14.0	174
2	1153	30	2	79	46.0	71.0	16.9	197
3	1311	30	7	79	40.0	85.5	12.7	201
4	1325	30	8	79	41.0	90.0	9.7	342
5	1536	31	1	80	45.0	13.8		217
1	923	31	6	78	657.0	104.0	110.0	358
2	1156	32	2	79	465.0	86.5	63.7	412
3	1266	32	5	79	485.0	83.5	73.0	382
4	1496	32	12	79	610.0	111.7		593
1	1135	37	2	79	168.0	82.5	23.3	261
1	914	38	6	78	286.0	109.5	31.9	260
2	1124	39	1	79	388.0	91.0	41.6	204
3	1398	39	9	79	148.0	105.2	18.8	221
1	913	34	6	78	73.0	105.5	17.0	285
2	1223	35	4	79	36.0	92.8	22.0	308
3	1531	36	1	80	55.0	20.7		262
1	970	58	8	78	19.0	100.5	10.9	196
2	1155	58	2	79	34.0	98.5	19.9	299
3	1538	59	1	80	25.0	9.2		316
1	1109	38	1	79	113.0	97.0	18.8	216
1	1354	30	8	79	57.0	105.0	12.9	155
1	949	42	8	78	78.0	118.0	15.5	212
2	1066	43	11	78	73.0	104.0	20.6	201
1	1168	29	3	79	69.0	111.0	16.0	175

Table 38.1 *(cont.)*

Obs. No.	Hospital ID	Age	Date M	Date Y	Measurements * CK	H	PK	LD
2	1447	30	10	79	177.0	103.5	19.8	241
1	911	35	6	78	48.0	98.0	16.4	233
2	951	35	7	78	34.0	96.5	10.4	122
3	1009	35	9	78	42.0	100.1	17.1	184
1	1358	44	9	79	109.0	81.0	25.3	227
1	1115	35	9	79	925.0	81.0	62.9	279
2	1203	35	4	79	1288.	82.0	51.6	368
3	1381	36	9	79	325.0	76.3	33.9	413
1	929	53	6	78	59.0	93.0	22.2	240
2	1236	54	4	79	69.0	92.6	20.9	243
1	1202	30	4	79	363.0	91.3	36.0	325
1	1050	35	11	78	37.0	84.0	12.8	156
1	1289	53	6	79	101.0	77.5	11.7	280
1	1173	41	3	79	99.0	93.2	18.6	156
2	1008	40	9	78	125.0	90.5	19.4	438
3	1328	42	8	79	52.0	93.3	11.2	272
1	1303	59	6	79	560.0	106.0	21.0	345
1	956	31	8	78	85.0	94.0	20.1	198
2	1253	32	5	79	79.0	9.0		137
3	1302	32	6	79	72.0	88.0	8.3	166
1	953	52	6	78	197.0	91.5	25.2	236
2	1163	52	3	78	242.0	85.5	16.6	168
3	1334	53	8	79	245.0	89.5	22.7	269
1	1030	39	10	78	154.0	103.5	21.3	296
2	1306	39	6	79	228.0	104.0	10.2	236
3	1493	40	12	79	123.0	25.4		275
1	1323	43	8	79	80.0	90.5	12.1	269
1	902	44	6	78	28.0	104.0	22.0	142
2	1296	45	6	79	35.0	86.3	14.4	184
1	1249	33	5	79	57.0	88.0	8.9	190
1	955	26	11	78	326.0	98.0	27.1	358
2	1307	26	6	79	700.0	90.0	49.1	343
1	984	61	9	78	100.0	101.0	11.8	301
2	1141	61	2	79	80.0	97.5	15.1	262
1	1305	48	6	79	115.0	79.0	14.2	258

* CK, creatine kinase
H, hemopexin
PK, pyruvate kinase
LD, lactate dehydroginase

39. Lengths of Remissions for Children with Acute Leukemia

Source Freireich, E.J., Gehan, E.A., Frei, E., Shroeder, L.R., Wolman, I.J., Anbari, R., Burgert, E.D., Mills, S.N., Pinkel, D., Selawry, O.S., Moon, J.H., Gendel, B.R., Spurr, C.L., Storrs, R., Haurani, F., Hoogstraten, B. and Lee, S. (1963). The effect of 6-mercaptopurine on the duration of steroid-induced remissions in acute leukemia: A model for evaluation of other potentially useful therapy. *Blood* **21**, 699-716.

Contributor E.A. Gehan University of Texas System Cancer Center

Prior to the publication of Freireich *et al* (1963), the existence of an effective therapy for acute leukemia had helped to hamper the evaluation of new and potentially more effective therapeutic agents. The data, given in Table 39.1, on preferences for 6-mercaptopurine (6-MP) in 21 pairs of patients, established the usefulness of this drug in prolonging the duration of complete remissions in childhood acute leukemia. The design of study involved a paired comparison of 6-MP *versus* placebo treatment, both administered in randomized and blinded fashion, the pairing being based on the institution at which the patient was treated and completeness of remission, complete or partial. The observation recorded for each patient was length of remission and the analysis was based on preferences among the pairs of patients, as determined by the difference in the lengths of remission between patients in each pair. The study was a sequential one designed according to an Armitage restricted sequential procedure (Armitage, 1957) and the study was stopped after 21 pairs of patients, 18 preferences favouring 6-MP and 3 favouring the placebo.

The data are of interest to statisticians for a variety of reasons: as an application of a restricted sequential procedure for binomial observations, or exponentially or Weibull-distributed remission times when the times are used in analysis; an application of a paired two-sample test or simply a two-sample test (Gehan,1965), ignoring the pairing, when there is a mixture of uncensored and censored observations; and as an application of procedures for testing regression models, such as a proportional hazard model for goodness of fit to the data (Cox, 1972).

Finally, the data are of interest in the field of cancer clinical trials since they were the first to demonstrate the value of treatment, 6-MP, when the patient was in an apparently well state, remission. Prior to this time, the usual clinical practice was to observe the patient for signs of relapse, effectively equivalent to placebo treatment, and treat the patient at that time.

This study was the first prospective, randomized, double-blind, placebo-controlled, sequential comparison of treatments for maintaining remissions in acute leukemia.

References

Armitage, P. (1957). Restricted sequential procedures. *Biometrika* **44**, 9-26.

Cox, D.R. (1972). Regression models and life tables (with discussion). *J.R. Statist. Soc.* **34**, 187-220.

Gehan, E.A. (1965). A generalized Wilcoxon test for comparing arbitrarily single-censored samples. *Biometrika* **53**, 203-223.

Table 39.1
Preferences for 6-MP and Placebo in Sequential Phase

Pair	Remission status	Drug	Length of remission, wks.	Preference
1	partial	placebo	1	6-MP
	partial	6-MP	10	
2	complete	placebo	22	placebo
	complete	6-MP	7	
3	complete	placebo	3	6-MP
	complete	6-MP	32+	
4	complete	placebo	12	6-MP
	complete	6-MP	23	
5	complete	placebo	8	6-MP
	complete	6-MP	22	
6	partial	placebo	17	placebo
	partial	6-MP	6	
7	complete	placebo	2	6-MP
	complete	6-MP	16	
8	complete	placebo	11	6-MP
	complete	6-MP	34+	
9	complete	placebo	8	6-MP
	complete	6-MP	32+	
10	complete	placebo	12	6-MP
	complete	6-MP	25+	
11	complete	placebo	2	6-MP
	complete	6-MP	11+	
12	partial	placebo	5	6-MP
	partial	6-MP	20+	
13	complete	placebo	4	6-MP
	complete	6-MP	19+	
14	complete	placebo	15	placebo
	complete	6-MP	6	
15	complete	placebo	8	6-MP
	complete	6-MP	17+	
16	partial	placebo	23	6-MP
	partial	6-MP	35+	
17	partial	placebo	5	6-MP
	partial	6-MP	6	
18	complete	placebo	11	6-MP
	complete	6-MP	13	
19	complete	placebo	4	6-MP
	complete	6-MP	9+	
20	complete	placebo	1	6-MP
	complete	6-MP	6+	
21	complete	placebo	8	6-MP
	complete	6-MP	10+	

+ denotes patient still in remission at end of trial.

40. Testis Tumours in Japan

Source Lee, J.A.H., Hitosugi, M. and Peterson, G.R. (1973). Rise in mortality from tumors of the testis in Japan, 1947-70. *J. Nat. Cancer Inst.* **51**, 1485-1490.

Contributor J.A.H. Lee University of Washington

The death rate from malignant tumours of the testis in Japan has risen from 1.53 per million per year in 1947-49 to 3.81 in 1966-70. This rise has been most marked in young adults and children, where the death rate is now greater than in the U.S. white population. Fatal testicular tumours in Japanese boys occur at a younger age than in U.S. white boys. The increase in the Japanese mortality rate cannot be associated with particular years in the total period studied, but rather appears to be related to increased lifetime risks with successively later years of birth.

Table 40.1 gives the aggregate populations in thousands of persons and total deaths by age and period of observation for Japanese and United States (white). Table 40.2 gives death rates per million per year from malignant tumours of testis by birth cohorts, age, and period of observation; Japan 1947-49 to 1966-70.

Table 40.1
Aggregate Populations in Thousands of Persons and Total Deaths by Age and Period of Observation for Japanese and White Americans

Japan

Age	1947-49 Pop.	1947-49 Deaths	1951-55 Pop.	1951-55 Deaths	1956-60 Pop.	1956-60 Deaths	1961-65 Pop.	1961-65 Deaths	1966-70 Pop.	1966-70 Deaths
0	15501	17	26914	51	21027	65	20246	69	21596	74
5	14236	-	25380	6	26613	7	20885	8	20051	7
10	13270	-	23492	3	25324	3	26540	7	20718	11
15	12658	2	21881	6	23211	15	24931	25	26182	39
20	10696	5	20404	27	21263	39	22228	56	24033	83
25	7563	5	17242	40	19994	58	20606	97	21805	125
30	7074	7	12609	18	17128	54	19864	77	20750	129
35	7038	10	11712	13	12476	36	17001	70	19890	101
40	6418	9	11478	26	11450	32	12275	29	16794	67
45	5981	7	10274	16	11157	26	11147	34	11962	37
50	4944	7	9325	16	9828	27	10705	27	10741	29
55	3993	7	7562	17	8718	19	9206	32	10086	39
60	3098	6	5902	13	6796	21	7869	21	8399	31
65	2317	4	4244	12	4911	26	5728	29	6715	34
70	1513	7	2845	17	3197	22	3737	25	4448	33
75	688	5	1587	9	1812	10	2061	25	2482	31
80	264	2	583	6	787	6	904	14	1068	9
85	73	2	179	2	246	3	335	3	419	3

White Americans

Age	1951-55 Pop.	1951-55 Deaths	1956-60 Pop.	1956-60 Deaths	1961-65 Pop.	1961-65 Deaths	1966-69 Pop.	1966-69 Deaths
0	39318	29	43224	35	44633	46	32166	27
5	34301	14	39897	11	43664	8	36152	7
10	27966	8	34611	16	40038	24	35003	12
15	23706	102	27576	133	34636	168	31971	188
20	24569	334	23852	314	27804	365	27304	397
25	26680	519	24976	457	24283	458	22029	442
30	26875	504	26854	498	25044	434	19491	337
35	25680	396	27095	429	27022	440	20188	301
40	23951	292	25336	343	26815	313	21539	244
45	21873	189	23731	225	24867	210	20878	183
50	19124	134	20711	144	22638	171	19000	134
55	17282	146	18129	113	19533	108	16767	102
60	14959	116	15597	90	16310	103	13914	92
65	11991	108	13060	122	13248	102	10971	87
70	8360	102	9697	75	10445	106	8246	77
75	5150	84	5943	64	6860	61	5849	61
80	2556	57	2990	47	3488	50	3239	44
85	1351	37	1597	38	1778	26	1618	20

Table 40.2
Death Rates per Million per Year in Japan from Malignant Tumours of Testis by Birth Cohorts, Age, and Period of Observation

	Period of Observation				
Age	1947-49	1951-55	1956-60	1961-65	1966-70
0	1.10	1.89	3.09	3.41	3.43
5	-	0.24	0.26	0.38	0.35
10	-	0.13	0.12	0.26	0.53
15	0.16	0.27	0.64	1.00	1.46
20	0.47	1.32	1.83	2.52	3.45
25	0.66	2.32	2.90	4.71	5.73
30	0.99	1.43	3.15	3.88	6.22
35	1.42	1.11	2.89	4.12	5.08
40	1.40	2.27	2.79	2.36	3.99
45	1.17	1.56	2.33	3.05	3.09
50	1.42	1.72	2.75	2.52	2.70
55	1.75	2.25	2.18	3.48	3.87
60	1.94	2.20	3.09	2.67	3.69
65	1.73	2.83	5.29	5.06	5.06
70	4.63	5.98	6.88	6.69	7.42
75	7.27	5.67	5.19	12.13	12.49
80	7.58	10.29	7.62	15.49	8.43
85	27.40	11.17	12.20	8.96	7.16

41. Consecutive Measurements of Plasma Citrate Concentrations

Source Andersen, A.H., Jensen, E.B. and Schou, G. (1981). Two-way analysis of variance with correlated errors. *Internat. Statist. Rev.* **49**, 153-167.

Contributor E.B. Jensen University of Aarhus

In order to study the variation of plasma citrate concentrations during the day, an experiment including ten subjects was performed.

For each subject, the concentration of citrate in plasma was determined at 14 successive time points during the day. The measurements covered the period from 8 a.m. to 9 p.m., the time interval between consecutive measurements being one hour. Meals were given at 8 a.m., at noon and at 5 p.m. The data are given in Table 41.1.

In Andersen, Jensen and Schou (1981), the two-way table of log-transformed concentrations is analyzed. The usual two-way analysis of variance is modified in such a way that a correlation between adjacent measurements from the same subject is taken into account.

Table 41.1
Plasma Citrate Concentrations *

Sub-	Time points													
ject	1	2	3	4	5	6	7	8	9	10	11	12	13	14
1	93	109	114	121	101	109	112	107	97	117	89	132	121	124
2	116	116	111	135	107	115	114	106	92	98	116	105	135	93
3	125	166	180	137	142	114	119	121	95	105	152	154	102	110
4	144	157	161	173	158	138	148	147	133	124	122	133	122	130
5	105	134	128	119	136	126	125	125	103	91	98	112	133	124
6	109	121	100	83	87	110	109	100	93	80	98	100	104	97
7	89	109	107	95	101	96	88	83	85	91	95	109	116	86
8	116	138	138	128	102	116	122	100	123	107	117	120	119	99
9	151	165	156	149	136	142	141	128	130	126	154	148	138	127
10	137	155	145	139	150	141	125	109	118	109	112	102	107	107

* Measurements are in μmol per litre.

42. Time to Death and Type of Death in Mice Receiving Various Doses of Red Dye No. 40

Source Lagakos, S.W. and Mosteller, F. (1981). A case study of statistics in the regulatory process: The FD&C Red No. 40 experiments *J. Natl. Cancer Inst.* **66**, 197-212.

Contributor S.W. Lagakos and F. Mosteller Harvard University

A lifetime feeding experiment involving 400 mice was undertaken in 1976 to assess the carcinogenicity of FD&C Red No. 40, Red 40, a colour additive widely used in foods in the U.S. For each sex, fifty mice were allocated to each of four groups, a control group and three dose level groups of Red 40.

Table 42.1 gives the time to death and types of death observed. Type of death is classified according to the presence or absence of reticuloendothelial, RE, tumours, which can be detected only at death. The experiment was scheduled to run for two years, with all mice still alive at that time intentionally killed. The unscheduled killing of 142 mice at week 42 was precipitated by some unexpectedly early RE deaths in the exposed groups. No RE tumours were found among the 142 mice. Animals that survived to week 104 were killed as planned. Table 42.2 gives the number of mice sacrificed at 42 and 104 weeks and the number which were found to have tumours.

A primary objective of the experiment was to determine whether Red 40 has any effect on the development of RE tumours. It was generally believed that RE tumours kill their mouse hosts shortly after onset and hence that time to RE death approximates time to RE tumour onset.

It should be emphasized that a thorough evaluation of the safety of Red 40 involves a number of considerations other than these data. Some of these considerations are reviewed in Lagakos and Mosteller (1981) and the references therein.

Note in the case where the mice received a low dose one mouse of each sex was missing or autolyzed.

Table 42.1
Time to and Type of Natural Death of Mice Receiving Red Dye No. 40

Dosage	Sex	Time of death, weeks	Tumour code *
Control	Female	70	1
Control	Female	77	0
Control	Female	83	1
Control	Female	87	0
Control	Female	92	0
Control	Female	92	0
Control	Female	93	1
Control	Female	96	1
Control	Female	100	1
Control	Female	102	0
Control	Female	102	0
Control	Female	103	1
Control	Male	29	0
Control	Male	30	0
Control	Male	38	0
Control	Male	48	0
Control	Male	53	0
Control	Male	56	0
Control	Male	62	0
Control	Male	70	0
Control	Male	71	0
Control	Male	74	0
Control	Male	74	0
Control	Male	76	0
Control	Male	85	1
Control	Male	86	0
Control	Male	86	0
Control	Male	92	0
Control	Male	97	1
Control	Male	99	1
Control	Male	101	0
Control	Male	102	0
Control	Male	103	0
Low	Female	49	1
Low	Female	60	1
Low	Female	63	0
Low	Female	67	0
Low	Female	70	0
Low	Female	74	1

Table 42.1 *(cont.)*

Dosage	Sex	Time of death, weeks	Tumour code *
Low	Female	77	1
Low	Female	80	0
Low	Female	80	0
Low	Female	89	1
Low	Female	89	0
Low	Female	90	0
Low	Female	90	0
Low	Female	97	0
Low	Female	100	0
Low	Female	102	0
Low	Male	27	0
Low	Male	31	1
Low	Male	31	0
Low	Male	35	0
Low	Male	55	0
Low	Male	79	0
Low	Male	81	0
Low	Male	83	1
Low	Male	87	1
Low	Male	93	0
Low	Male	94	0
Low	Male	99	0
Low	Male	101	0
Low	Male	102	0
Medium	Female	30	0
Medium	Female	37	1
Medium	Female	56	1
Medium	Female	65	1
Medium	Female	76	0
Medium	Female	83	1
Medium	Female	87	1
Medium	Female	90	0
Medium	Female	94	0
Medium	Female	97	0
Medium	Female	97	0
Medium	Female	102	1
Medium	Male	5	0
Medium	Male	54	1
Medium	Male	67	0
Medium	Male	70	0
Medium	Male	79	0

Table 42.1 *(cont.)*

Dosage	Sex	Time of death, weeks	Tumour code *
Medium	Male	80	0
Medium	Male	83	1
Medium	Male	86	0
Medium	Male	88	0
Medium	Male	91	0
Medium	Male	93	0
Medium	Male	94	0
Medium	Male	95	0
Medium	Male	97	0
Medium	Male	100	0
Medium	Male	100	0
Medium	Male	100	0
Medium	Male	103	0
Medium	Male	103	0
High	Female	34	1
High	Female	36	1
High	Female	48	1
High	Female	48	0
High	Female	65	1
High	Female	91	1
High	Female	91	0
High	Female	98	0
High	Female	102	0
High	Female	102	0
High	Female	103	1
High	Male	2	0
High	Male	16	0
High	Male	23	0
High	Male	34	0
High	Male	35	1
High	Male	35	1
High	Male	67	1
High	Male	77	0
High	Male	77	0
High	Male	79	1
High	Male	84	0
High	Male	89	0
High	Male	92	0
High	Male	96	0
High	Male	97	0
High	Male	98	0
High	Male	99	0

* 0, tumours absent;
1, tumours present.

Table 42.2
Types of Death of Sacrificed Mice Receiving Red Dye No. 40

Dosage	Sex	Number sacrificed at 42 weeks	Number sacrificed at 104 weeks	Number with tumours
Control	Female	20	18	2
Control	Male	17	12	0
Low	Female	20	13	3
Low	Male	16	19	0
Medium	Female	18	20	2
Medium	Male	19	12	1
High	Female	18	21	3
High	Male	14	19	0

43. Cariogenic Effects of Diet

Contributor M.B. Carroll General Food Corporation

One hundred and twenty rats were randomly assigned to one of eight diets to see if Treatments A and B would reduce the cariogenic effects of Diet 2.

After completing the feeding period, the rats were sacrificed and their teeth removed and stained. Twenty-eight occlusal surfaces in each rat were examined for caries and scored according to severity of decay.

Several rats died before completion of the study; i.e. in Diet 3, rat numbers 6 and 7 and in Diet 8, rat number 11. The data are given in Table 43.1.

Table 43.1
Results of Examinations for Caries in Twenty-eight Occlusal Surfaces in 120 Rats given Eight Diets

Rat no.	Diet *	Replicate number	Occlusal scores **
			Diet 1
131	1	1	1 1 0 1 1 1 0 1 1 1 1 1 1 0 0 0 0 0 0 0 1 0 0 1 0 0 0 0
40	1	2	0 0 0 1 1 1 1 1 1 1 1 0 0 0 0 0 2 1 1 1 1 1 1 1 1 0 0 0
51	1	3	0 0 1 1 1 1 1 1 2 2 1 1 0 0 0 0 1 1 1 0 1 1 1 1 1 1 1 1
96	1	4	0 0 1 1 1 1 1 1 1 1 1 1 1 1 0 0 2 1 2 2 2 2 2 2 2 2 1 0
104	1	5	0 0 2 2 2 1 1 0 1 1 0 1 1 0 0 0 1 1 1 1 1 1 1 1 1 1 0 0
79	1	6	1 0 1 1 1 1 1 1 1 1 0 0 0 0 0 0 1 1 1 1 1 1 1 1 0 0 1 0
4	1	7	0 0 0 1 1 1 1 1 1 1 0 0 0 0 2 1 2 2 2 1 2 2 2 2 1 1 1 0
65	1	8	1 1 2 2 2 0 1 2 2 2 0 1 1 0 1 1 1 1 2 1 2 1 2 2 1 1 1 1
143	1	9	0 0 1 2 2 2 1 1 2 2 1 0 0 0 0 0 1 1 1 2 2 2 2 2 1 2 1 0
86	1	10	0 0 1 1 1 1 1 1 1 2 0 1 0 1 0 0 1 1 1 1 1 1 1 2 1 1 1 1
127	1	11	0 0 1 1 1 1 1 0 1 1 1 1 1 0 0 0 1 1 1 1 1 1 1 1 0 1 1 0
120	1	12	0 0 2 2 2 0 1 1 2 2 2 2 1 0 0 0 1 1 1 1 1 1 1 2 1 0 0 1
29	1	13	0 0 1 1 1 1 2 2 2 2 1 0 1 1 0 0 2 2 2 2 2 2 2 2 1 1 1 1
16	1	14	0 0 0 1 1 1 0 1 1 1 0 1 0 1 0 0 0 1 1 1 1 1 1 1 1 1 1 0
48	1	15	1 1 1 1 1 1 1 2 2 2 0 1 0 0 0 0 1 1 2 1 2 1 1 2 1 1 0 1
			Diet 2
61	2	1	1 1 2 2 2 2 2 2 2 2 1 1 0 0 2 2 2 2 2 1 1 2 2 2 0 0 1 0
130	2	2	2 2 2 2 2 2 2 2 2 2 2 1 1 1 0 0 2 2 2 1 1 1 2 1 2 2 1 1
142	2	3	1 1 2 2 2 1 1 2 2 2 2 2 1 1 1 1 2 2 2 2 2 2 2 2 1 0 0 0
8	2	4	1 1 2 2 2 2 2 2 2 2 2 1 1 1 0 0 2 1 1 1 1 2 2 2 2 1 0 1
82	2	5	2 2 2 2 2 2 2 2 2 2 2 1 1 0 2 1 1 2 2 1 1 2 2 2 2 1 2 1
44	2	6	1 1 2 2 2 2 2 2 2 2 1 1 2 1 1 1 2 2 2 0 2 1 2 2 1 0 1 1
14	2	7	1 1 2 2 2 2 2 2 2 2 2 2 1 1 1 1 2 2 2 2 2 2 2 2 1 1 1 0
33	2	8	2 2 2 2 2 2 2 2 2 2 2 1 1 0 2 2 2 2 2 2 2 2 2 2 2 2 2 1
135	2	9	1 1 2 2 1 2 2 2 2 2 2 2 2 0 1 0 2 1 1 2 2 2 2 2 2 2 2 2
52	2	10	1 1 2 2 2 2 2 2 2 2 1 1 1 1 1 1 2 2 2 0 1 1 2 2 1 1 0 0
99	2	11	1 1 2 2 2 2 2 2 2 2 1 1 1 1 1 2 2 2 2 2 2 2 2 2 0 0 1 1
76	2	12	1 2 2 2 2 2 2 2 2 2 1 1 1 1 1 1 1 1 1 1 1 1 2 2 0 0 1 1
105	2	13	1 1 2 2 2 1 1 2 2 2 0 0 0 1 1 1 2 2 1 1 2 2 2 2 1 0 1 0
111	2	14	1 1 2 2 2 1 1 2 2 2 0 0 2 2 1 1 2 2 2 1 1 2 2 2 0 1 1 0
21	2	15	1 1 2 2 2 1 1 2 2 2 0 1 2 2 2 1 2 2 2 0 1 2 2 2 1 1 1 1
			Diet 3
12	3	1	1 1 2 2 2 2 2 2 2 2 1 1 1 0 1 1 2 2 2 2 2 2 2 2 2 2 2 2
108	3	2	2 2 2 2 2 2 2 2 2 2 1 1 0 1 1 1 2 2 2 2 2 2 2 2 0 1 1 1
72	3	3	1 1 2 2 2 2 2 2 2 2 2 2 1 1 1 1 2 2 2 2 2 2 2 2 2 1 2 2
89	3	4	1 1 2 2 2 0 0 2 2 2 1 1 1 2 1 1 2 2 2 0 1 2 2 2 1 0 0 0
66	3	5	2 1 2 2 2 1 2 2 2 2 1 1 1 1 1 1 2 2 2 1 2 2 2 2 1 0 0 0
91	3	8	1 1 1 1 1 1 2 2 2 2 0 1 0 0 1 1 2 1 1 1 1 2 2 2 1 1 1 1
126	3	9	2 2 2 2 2 1 2 2 2 2 0 1 1 0 1 1 2 2 2 1 1 2 2 2 1 1 2 0
50	3	10	1 2 2 2 2 2 2 2 2 2 2 1 2 1 1 1 2 2 2 2 2 2 2 2 2 2 0 0
34	3	11	1 2 2 2 2 2 3 2 2 2 1 0 1 1 1 1 2 2 2 2 2 2 2 2 1 1 1 1
118	3	12	0 1 2 2 2 2 2 2 2 2 1 0 0 0 0 0 2 2 2 1 2 2 2 2 1 0 1 1
150	3	13	1 1 1 1 1 1 1 1 1 1 0 0 0 1 1 1 1 1 2 1 1 1 1 1 1 0 0 0
136	3	14	1 1 2 2 2 2 2 2 2 2 0 1 2 1 1 1 2 2 2 1 2 2 2 2 1 1 1 1
1	3	15	1 1 2 2 2 2 2 2 2 2 2 2 2 2 1 1 2 2 2 2 2 2 2 2 0 1 1 0

Table 43.1 *(cont.)*

Rat no.	Diet *	Replicate Number	Occlusal scores **
			Diet 4
64	4	1	0 0 2 2 2 1 1 1 1 2 0 1 1 1 1 1 1 1 1 2 1 2 2 2 0 1 2 1
41	4	2	0 0 1 1 2 1 2 1 1 1 1 0 0 0 0 0 2 2 1 1 1 1 2 2 1 0 1 0
88	4	3	0 0 2 2 2 2 2 2 2 2 2 2 0 0 1 1 2 2 2 2 2 1 1 1 1 1 0 0
141	4	4	1 1 2 2 2 2 2 2 2 1 2 1 1 1 1 1 2 2 2 0 1 2 2 2 1 1 1 1
107	4	5	1 1 2 2 2 2 2 2 2 2 1 1 0 1 1 1 2 2 2 1 2 2 2 2 1 1 1 1
32	4	6	1 1 2 2 2 1 1 2 2 2 1 2 1 0 1 1 2 2 2 1 2 2 2 2 2 2 1 0
55	4	7	1 1 2 2 2 1 1 1 2 2 0 1 0 1 0 0 1 1 2 1 1 1 1 2 1 1 0 1
24	4	8	1 1 2 2 2 2 2 2 2 2 2 1 1 1 1 1 2 2 2 2 2 2 2 2 0 1 2 2
78	4	9	2 2 2 2 2 2 2 2 2 2 1 0 0 0 1 2 2 2 2 2 2 2 2 2 1 1 1 0
20	4	10	1 0 2 2 2 1 1 1 2 2 1 0 1 0 1 0 2 2 2 1 0 2 2 2 0 1 0 0
137	4	11	0 0 2 2 2 2 2 2 2 2 2 2 2 1 1 2 2 2 2 2 2 2 2 2 1 1 1 1
112	4	12	1 2 2 2 2 2 2 2 2 2 1 1 1 1 0 0 2 2 2 2 2 2 1 1 0 1 1 1
7	4	13	1 1 2 2 2 2 2 2 2 2 1 1 0 1 1 0 2 2 1 1 1 2 2 2 1 1 1 0
124	4	14	1 1 1 1 1 1 1 2 2 2 1 0 1 0 1 1 1 1 1 1 1 1 1 1 0 1 0 1
100	4	15	1 1 2 2 2 2 2 2 2 2 2 2 1 1 1 1 2 2 2 2 3 2 2 2 2 2 1 1
			Diet 5
80	5	1	1 0 2 2 2 2 2 2 2 2 0 0 0 0 1 1 2 2 2 2 2 2 2 2 1 0 1 1
144	5	2	0 0 1 1 1 1 1 1 1 1 1 0 0 0 0 0 1 1 1 1 0 1 1 1 1 0 0 0
110	5	3	0 0 1 1 1 0 1 1 1 1 0 0 0 0 0 0 1 1 1 0 1 1 1 1 0 0 0 0
60	5	4	1 2 2 2 2 1 1 2 2 2 1 1 0 0 0 0 2 2 2 1 2 2 2 2 1 1 1 0
42	5	5	1 1 2 2 2 2 2 2 2 2 1 0 0 0 1 1 2 2 2 1 2 2 1 2 1 1 0 0
84	5	6	0 0 2 2 2 1 2 2 2 2 1 0 0 0 0 0 2 1 1 1 1 1 1 1 1 0 0 0
70	5	7	1 1 2 2 2 2 2 2 2 2 0 0 1 0 1 1 2 2 2 1 2 2 2 2 1 0 1 0
37	5	8	1 1 2 2 2 1 2 2 2 2 0 0 0 0 0 0 2 1 1 1 1 1 2 1 1 0 1 1
3	5	9	2 1 2 2 2 2 2 2 2 2 0 0 2 1 1 1 2 2 2 1 2 2 2 2 1 1 0 0
121	5	10	1 2 2 2 2 2 2 2 2 2 2 2 1 1 1 2 2 2 2 2 2 2 2 2 2 2 1 0
23	5	11	1 1 2 2 2 1 2 2 2 2 1 1 0 1 1 0 2 2 2 1 1 1 1 1 1 1 1 1
113	5	12	1 1 2 1 1 2 2 2 2 2 0 0 1 0 0 0 2 2 2 1 1 1 1 1 1 1 0 0
92	5	13	1 1 2 2 2 1 2 1 1 1 1 1 0 0 1 1 2 2 2 1 2 2 2 2 1 1 0 1
18	5	14	1 0 0 1 1 0 0 1 1 1 0 1 0 1 0 0 1 1 2 0 0 1 2 2 0 0 1 0
138	5	15	1 0 2 2 2 1 1 1 2 2 1 0 1 1 0 0 2 2 2 1 1 2 2 2 0 1 1 0
			Diet 6
115	6	1	1 1 2 2 2 2 2 2 2 2 2 2 2 2 0 0 2 2 2 2 2 2 2 2 2 1 0 0
9	6	2	1 1 2 2 2 2 2 2 2 2 2 1 2 1 1 1 2 2 2 2 2 2 2 2 1 1 1 0
128	6	3	1 1 2 2 2 2 2 2 2 2 0 0 1 0 1 1 2 2 1 1 2 1 2 2 1 1 1 0
46	6	4	1 1 2 2 2 1 1 2 2 2 1 0 1 1 1 1 2 1 1 1 1 1 2 2 0 0 1 1
13	6	5	1 1 2 2 2 2 2 2 2 2 2 2 1 1 1 1 2 2 1 1 0 2 2 2 1 1 2 1
39	6	6	1 1 2 1 1 2 2 2 2 2 1 1 1 1 1 1 2 2 1 2 2 2 2 2 2 2 1 1
101	6	7	2 2 2 2 2 2 2 2 2 2 1 1 1 0 2 2 2 2 2 2 2 2 2 2 2 2 1 1
146	6	8	1 1 2 2 2 1 2 2 2 2 2 2 1 1 0 0 2 2 2 1 1 2 2 2 1 1 2 1
57	6	9	1 1 2 2 2 2 2 2 2 2 0 1 1 0 1 1 2 2 2 2 2 2 2 1 1 1 1 1
74	6	10	0 1 2 2 2 1 2 2 2 2 0 0 1 0 1 0 1 2 1 0 0 2 2 2 0 1 0 0
27	6	11	1 1 2 2 2 2 2 2 2 1 1 1 1 1 1 0 2 2 2 2 2 2 2 2 2 2 1 1
139	6	12	1 0 1 1 1 1 1 1 2 2 2 2 1 1 1 1 2 2 2 2 2 2 2 2 2 1 1 1
67	6	13	1 1 2 2 2 1 2 2 2 2 2 2 2 2 1 0 2 2 2 0 0 2 2 2 2 1 0 0
94	6	14	1 1 2 2 2 1 1 2 2 2 0 1 1 0 1 1 2 2 2 1 1 1 2 2 0 0 1 1
83	6	15	0 0 2 1 2 1 2 2 2 2 0 1 1 0 1 1 2 1 1 1 2 2 2 2 1 1 0 0

Table 43.1 *(cont.)*

Rat no.	Diet *	Replicate Number	Occlusal scores **
			Diet 7
109	7	1	1 1 2 2 2 2 2 2 2 2 0 0 0 0 2 1 2 2 1 2 2 2 2 2 2 2 2 1
31	7	2	2 2 2 2 2 2 2 2 2 2 2 2 1 0 2 2 2 2 2 1 2 2 2 2 0 1 1 1
122	7	3	0 0 2 1 1 1 1 2 2 2 0 0 1 1 1 1 1 1 1 0 0 1 1 1 0 1 0 1
63	7	4	1 2 1 2 2 2 2 1 2 2 0 0 2 0 1 1 2 2 2 1 1 1 2 2 0 0 1 0
95	7	5	1 1 2 2 2 2 2 2 2 2 1 0 1 1 1 0 2 2 2 2 2 2 2 2 1 0 1 0
49	7	6	1 1 2 2 1 2 2 2 2 2 2 2 1 2 2 1 2 2 2 2 2 2 2 2 1 0 0 0
148	7	7	1 1 2 2 2 1 2 2 2 2 2 2 2 2 1 1 2 2 2 2 2 2 2 2 2 2 2 1
117	7	8	1 1 2 2 2 2 2 2 2 2 1 1 2 1 1 1 2 2 2 1 1 2 2 2 1 1 0 0
26	7	9	2 2 2 2 2 2 2 2 2 2 1 1 1 1 0 0 2 2 2 1 1 2 2 2 1 0 0 0
75	7	10	1 1 2 2 2 2 2 2 2 2 0 0 1 0 1 1 2 2 2 2 2 2 2 2 1 0 0 0
85	7	11	1 1 2 2 2 1 1 2 2 2 0 1 1 0 1 0 2 2 2 1 2 2 2 2 1 0 0 0
17	7	12	2 2 2 2 2 2 3 2 2 2 2 2 1 1 2 2 2 2 2 2 2 2 2 2 2 2 1 1
134	7	13	0 0 2 1 1 1 1 2 2 2 1 1 1 1 1 1 2 2 2 1 2 2 2 2 1 1 0 1
58	7	14	1 1 2 2 2 1 1 1 2 2 1 1 1 1 2 1 2 2 2 1 1 2 2 2 1 1 2 2
5	7	15	1 1 1 2 2 1 2 2 2 2 1 1 0 1 0 0 0 0 0 0 0 1 1 1 0 0 0 0
			Diet 8
25	8	1	1 1 1 2 2 2 2 1 2 2 2 2 1 1 1 1 2 2 2 1 2 2 2 2 0 0 0 0
129	8	2	2 2 2 2 2 2 2 2 2 2 1 1 1 1 1 2 2 2 2 2 2 2 2 2 0 1 1 1
19	8	3	1 1 2 2 2 0 1 1 2 2 0 1 2 1 1 1 2 2 2 1 1 2 2 2 1 1 1 0
36	8	4	1 0 2 2 2 2 2 2 2 2 2 1 1 0 0 0 2 1 1 2 2 1 2 2 2 1 1 0
149	8	5	1 1 2 2 2 2 2 2 2 2 1 1 1 1 1 1 2 2 2 2 2 2 2 2 2 2 1 0
59	8	6	0 0 1 2 2 1 2 1 2 2 0 0 0 0 1 1 2 2 2 1 2 1 2 2 0 0 0 0
119	8	7	1 1 2 2 2 2 2 2 2 2 1 0 2 1 1 1 2 2 2 1 0 2 2 2 1 0 1 0
71	8	8	1 1 2 2 2 2 2 2 2 2 2 2 2 0 2 2 2 2 2 2 2 2 2 2 0 1 0 0
87	8	9	1 1 2 2 2 2 2 2 2 2 1 0 0 0 1 1 2 2 2 2 2 2 2 2 0 1 2 0
6	8	10	0 0 2 2 2 1 1 2 2 2 0 1 0 1 0 1 1 1 1 1 1 1 1 1 1 0 1 0
133	8	12	2 2 2 2 2 2 2 2 2 2 1 1 1 1 2 2 2 2 2 2 3 2 2 2 1 1 0 1
103	8	13	0 1 2 2 2 1 1 2 2 2 0 0 1 1 1 1 1 1 1 1 1 1 1 1 0 0 1 1
98	8	14	1 1 1 2 1 1 1 2 2 2 1 1 2 1 1 1 2 2 2 2 1 2 2 2 2 1 2 1
47	8	15	1 1 2 2 2 2 2 2 2 2 2 2 2 1 1 1 1 2 2 2 2 2 2 2 2 2 1 0

* Diet
1, noncariogenic control (NC)
2, cariogenic control (CC)
3, CC + 0.25% of treatment A
4, CC + 0.50% of treatment A
5, CC + 1.00% of treatment A
6, CC + 0.25% of treatment B
7, CC + 0.50% of treatment B
8, CC + 1.00% of treatment B

** Severity of decay
0, no decay
1, decay into enamel
2, decay into dentin

44. Physical Characteristics of Urines With and Without Crystals

Source The data were obtained from the laboratory of James S. Elliot M.D. of the Urology Section, Veteran's Administration Medical Center, Palo Alto and the Division of Urology, Stanford University School of Medicine, Stanford.

Contributor D.P. Byar National Cancer Institute

The 79 urine specimens, given in Table 44.1, were analyzed in an effort to determine if certain physical characteristics of the urine might be related to the formation of calcium oxalate crystals.

The six physical characteristics of the urine are: (1) *specific gravity,* the density of the urine relative to water; (2) *pH,* the negative logarithm of the hydrogen ion; (3) *osmolarity* (mOsm), a unit used in biology and medicine but not in physical chemistry. Osmolarity is proportional to the concentration of molecules in solution; (4) *conductivity* (mMho milliMho). One Mho is one reciprocal Ohm. Conductivity is proportional to the concentration of charged ions in solution; (5) *urea concentration* in millimoles per litre; and (6) *calcium concentration* (CALC) in millimoles/litre. It may be anticipated that some of these characteristics are highly correlated. Another point worth noting is that the units of measurement vary by several orders of magnitude among the six characteristics.

Table 44.1
Physical Characteristics of Urines With and Without Crystals *

Patient number	Crystals**	Specific gravity	pH	mOsm	mMho	Urea	Calcium
1	1	1.021	4.91	725	-	443	2.45
2	1	1.017	5.74	577	20.0	296	4.49
3	1	1.008	7.20	321	14.9	101	2.36
4	1	1.011	5.51	408	12.6	224	2.15
5	1	1.005	6.52	187	7.5	91	1.16
6	1	1.020	5.27	668	25.3	252	3.34
7	1	1.012	5.62	461	17.4	195	1.40
8	1	1.029	5.67	1107	35.9	550	8.48
9	1	1.015	5.41	543	21.9	170	1.16
10	1	1.021	6.13	779	25.7	382	2.21
11	1	1.011	6.19	345	11.5	152	1.93
12	1	1.025	5.53	907	28.4	448	1.27
13	1	1.006	7.12	242	11.3	64	1.03
14	1	1.007	5.35	283	9.9	147	1.47
15	1	1.011	5.21	450	17.9	161	1.53
16	1	1.018	4.90	684	26.1	284	5.09
17	1	1.007	6.63	253	8.4	133	1.05
18	1	1.025	6.81	947	32.6	395	2.03
19	1	1.008	6.88	395	26.1	95	7.68
20	1	1.014	6.14	565	23.6	214	1.45
21	1	1.024	6.30	874	29.9	380	5.16
22	1	1.019	5.47	760	33.8	199	0.81
23	1	1.014	7.38	577	30.1	87	1.32
24	1	1.020	5.96	631	11.2	422	1.55
25	1	1.023	5.68	749	29.0	239	1.52
26	1	1.017	6.76	455	8.8	270	0.77
27	1	1.017	7.61	527	25.8	75	2.17
28	1	1.010	6.61	225	9.8	72	0.17
29	1	1.008	5.87	241	5.1	159	0.83
30	1	1.020	5.44	781	29.0	349	3.04
31	1	1.017	7.92	680	25.3	282	1.06
32	1	1.019	5.98	579	15.5	297	3.93
33	1	1.017	6.56	559	15.8	317	5.38
34	1	1.008	5.94	256	8.1	130	3.53
35	1	1.023	5.85	970	38.0	362	4.54
36	1	1.020	5.66	702	23.6	330	3.98
37	1	1.008	6.40	341	14.6	125	1.02
38	1	1.020	6.35	704	24.5	260	3.46
39	1	1.009	6.37	325	12.2	97	1.19
40	1	1.018	6.18	694	23.3	311	5.64

Table 44.1 *(cont.)*

Patient number	Crystals**	Specific gravity	pH	mOsm	mMho	Urea	Calcium
41	1	1.021	5.33	815	26.0	385	2.66
42	1	1.009	5.64	386	17.7	104	1.22
43	1	1.015	6.79	541	20.9	187	2.64
44	1	1.010	5.97	343	13.4	126	2.31
45	1	1.020	5.68	876	35.8	308	4.49
46	2	1.021	5.94	774	27.9	325	6.96
47	2	1.024	5.77	698	19.5	354	13.00
48	2	1.024	5.60	866	29.5	360	5.54
49	2	1.021	5.53	775	31.2	302	6.19
50	2	1.024	5.36	853	27.6	364	7.31
51	2	1.026	5.16	822	26.0	301	14.34
52	2	1.013	5.86	531	21.4	197	4.74
53	2	1.010	6.27	371	11.2	188	2.50
54	2	1.011	7.01	443	21.4	124	1.27
55	2	1.022	6.21	-	20.6	398	4.18
56	2	1.011	6.13	364	10.9	159	3.10
57	2	1.031	5.73	874	17.4	516	3.01
58	2	1.020	7.94	567	19.7	212	6.81
59	2	1.040	6.28	838	14.3	486	8.28
60	2	1.021	5.56	658	23.6	224	2.33
61	2	1.025	5.71	854	27.0	385	7.18
62	2	1.026	6.19	956	27.6	473	5.67
63	2	1.034	5.24	1236	27.3	620	12.68
64	2	1.033	5.58	1032	29.1	430	8.94
65	2	1.015	5.98	487	14.8	198	3.16
66	2	1.013	5.58	516	20.8	184	3.30
67	2	1.014	5.90	456	17.8	164	6.99
68	2	1.012	6.75	251	5.1	141	0.65
69	2	1.025	6.90	945	33.6	396	4.18
70	2	1.026	6.29	833	22.2	457	4.45
71	2	1.028	4.76	312	12.4	10	0.27
72	2	1.027	5.40	840	24.5	395	7.64
73	2	1.018	5.14	703	29.0	272	6.63
74	2	1.022	5.09	736	19.8	418	8.53
75	2	1.025	7.90	721	23.6	301	9.04
76	2	1.017	4.81	410	13.3	195	0.58
77	2	1.024	5.40	803	21.8	394	7.82
78	2	1.016	6.81	594	21.4	255	12.20
79	2	1.015	6.03	416	12.8	178	9.39

* - : value unknown

** 1 : no crystals

2 : crystals

45. Multiple Tumour Recurrence Data for Patients with Bladder Cancer

Source and Contributor D.P. Byar National Cancer Institute

These data were obtained in a randomized clinical trial conducted by the Veterans Administration Co-operative Urological Research Group (VACURG). All patients had superficial bladder tumours when they entered the trial. These tumours were removed transurethrally and patients were assigned randomly to one of three treatments: placebo pills, pyridoxine (vitamin B_6) pills, or periodic instillation of a chemotherapeutic agent, thiotepa, into the bladder. The rationale for the latter two treatments is given in Byar, Blackard and VACURG (1977). At subsequent follow-up visits any tumours noticed were removed and the treatment was continued. The goal of the analysis should be to determine the effect of treatment on the frequency of tumour recurrence. For the purpose of this analysis, the word recurrence will refer to a visit at which one or more tumours are found in the bladder regardless of whether these are thought to be recurrences or new tumours. This term is not to be confused with the number of tumours present at any single visit because the tumours are often multiple. The data, Table 45.1, consist of the number of recurrences experienced by each of 118 patients, the number of tumours present initially at the time of randomization in the trial and the diameter of the largest of these, the months from the beginning of the study until each recurrence, the number of tumours present at each recurrence, and the diameter of the largest of these. An analysis of perhaps secondary importance would compare treatments with respect to numbers and size of tumours.

Two publications based on these data have appeared (Byar, Blackard and VACURG, 1977; and Byar 1980), but the data in these publications are not necessarily identical to those presented here.

References

Byar, D.P., Blackard, C. and the VACURG (1977). Comparisons of placebo, pyridoxine, and topical thiotepa in preventing recurrence of stage I bladder cancer. *Urology* **10**, 556-561.

Byar, D. P. (1980). The Veterans Administration study of chemoprophylaxis for recurrent stage I bladder tumors: comparisons of placebo, pyridoxine, and topical thiotepa. In *Bladder Tumors and Other Topics in Urological Oncology* edited by M. Pavone-Macaluso, P.H. Smith and F. Edsmyr. New York: Plenum, pp. 363-370.

Table 45.1 : Multiple Tumour Recurrence Data

Patient number	Treatment group	Follow-up time, months	Survival status	No. of recurrences	Initial Number	Initial Size
1	1	0	2	0	1	1
2	1	1	2	0	1	3
3	1	4	0	0	2	1
4	1	7	0	0	1	1
5	1	10	2	0	5	1
6	1	10	2	1	4	1
7	1	14	0	0	1	1
8	1	18	0	0	1	1
9	1	18	2	1	1	3
10	1	18	2	2	1	1
11	1	23	0	0	3	3
12	1	23	0	2	1	3
13	1	23	0	3	1	1
14	1	23	1	3	3	1
15	1	24	0	4	2	3
16	1	25	0	3	1	1
17	1	26	0	0	1	2
18	1	26	0	1	8	1
19	1	26	0	2	1	4
20	1	28	0	1	1	2
21	1	29	0	0	1	4
22	1	29	0	0	1	2
23	1	29	2	0	4	1
24	1	30	0	2	1	6
25	1	30	0	3	1	5
26	1	30	2	5	2	1
27	1	31	0	3	1	3
28	1	32	0	0	1	2
29	1	34	2	0	2	1
30	1	36	0	0	2	1
31	1	36	0	1	3	1
32	1	37	0	0	1	2
33	1	40	0	4	4	1
34	1	40	0	6	5	1
35	1	41	0	0	1	2
36	1	43	0	1	1	1
37	1	43	0	1	2	6
38	1	44	0	3	2	1
39	1	45	2	5	1	1
40	1	48	0	1	1	1
41	1	49	0	0	1	3
42	1	51	0	1	3	1
43	1	53	0	1	1	7
44	1	53	0	5	3	1
45	1	59	0	0	1	1

for Patients with Bladder Cancer *

	Recurrence								
	1	2	3	4	5	6	7	8	9
Patient number	M#S	M#S	M#S	M#S	M#S	M#S	M#S	M#S	M#S
1									
2									
3									
4									
5									
6	6 1 1								
7									
8									
9	5 2 4								
10	12 2 2	16 3 -							
11									
12	10 6 1	15 3 1							
13	3 8 1	16 8 -	23 8 -						
14	3 1 1	9 1 2	21 8 8						
15	7 8 2	10 7 1	16 5 3	24 7 -					
16	3 1 1	15 1 -	25 3 -						
17									
18	1 8 1								
19	2 4 1	26 8 -							
20	25 3 -								
21									
22									
23									
24	28 2 1	30 1 1							
25	2 4 1	17 2 1	22 4 -						
26	3 1 -	6 3 -	8 3 -	12 3 -	26 3 -				
27	12 2 -	15 3 1	24 1 -						
28									
29									
30									
31	29 8 1								
32									
33	9 8 1	17 2 1	22 5 -	24 1 -					
34	16 1 1	19 8 1	23 1 1	29 2 1	34 1 1	40 3 -			
35									
36	3 3 1								
37	6 1 1								
38	3 5 1	6 3 1	9 4 1						
39	9 1 1	11 3 1	20 1 1	26 4 1	30 3 1				
40	18 1 1								
41									
42	35 1 1								
43	17 1 1								
44	3 7 1	15 2 1	46 3 -	51 2 -	53 1 1				
45									

Table 45.1

Patient number	Treatment group	Follow-up time, months	Survival status	No. of recurrences	Initial Number	Initial Size
46	1	61	0	9	3	2
47	1	64	0	5	1	3
48	1	64	0	8	2	3
49	2	0	0	0	8	1
50	2	2	2	0	1	2
51	2	4	0	2	4	6
52	2	4	2	0	1	1
53	2	5	0	2	1	1
54	2	7	2	0	2	3
55	2	8	0	0	1	1
56	2	8	0	1	4	3
57	2	11	0	1	1	1
58	2	14	2	0	1	1
59	2	26	0	0	1	2
60	2	29	2	0	1	2
61	2	30	0	1	8	1
62	2	32	2	0	1	3
63	2	33	0	0	1	1
64	2	34	0	5	1	4
65	2	37	0	5	3	7
66	2	38	0	0	1	1
67	2	39	0	9	1	2
68	2	40	0	0	1	1
69	2	40	0	0	1	1
70	2	42	0	9	3	1
71	2	45	0	0	1	1
72	2	45	0	1	2	1
73	2	46	0	2	1	4
74	2	46	2	8	1	1
75	2	48	0	1	1	1
76	2	54	0	0	2	1
77	2	54	0	2	1	1
78	2	55	0	8	4	1
79	2	57	0	0	1	1
80	2	60	0	0	3	8
81	3	1	0	0	1	3
82	3	1	2	0	1	1
83	3	5	2	1	8	1
84	3	9	0	0	1	2
85	3	10	2	0	1	1
86	3	13	0	0	1	1
87	3	14	0	1	2	6
88	3	17	2	5	5	3
89	3	18	2	0	5	1
90	3	18	2	1	1	3

(cont.)

	Recurrence								
	1	2	3	4	5	6	7	8	9
Patient number	M#S	M#S	M#S	M#S	M#S	M#S	M#S	M#S	M#S
46	2 1 3	15 3 1	24 4 1	30 3 2	34 4 1	39 1 -	43 1 -	49 1 -	52 1 -
47	5 3 1	14 4 1	19 2 1	27 5 1	41 - -				
48	2 1 1	8 3 1	12 6 1	13 2 1	17 2 1	21 1 1	33 1 1	49 1 -	
49									
50									
51	3 1 8	4 3 -							
52									
53	2 1 2	3 1 2							
54									
55									
56	4 1 -								
57	3 1 1								
58									
59									
60									
61	5 1 1								
62									
63									
64	3 1 1	10 3 1	22 4 1	26 8 -	34 8 -				
65	3 2 1	9 1 1	15 - 1	19 3 5	25 3 -				
66									
67	3 3 1	7 2 1	12 4 1	16 1 1	19 2 1	28 1 -	34 8 -	36 8 -	39 8 -
68									
69									
70	2 2 1	6 1 1	10 1 1	16 5 1	23 4 1	27 1 1	36 3 -	39 5 -	42 2 -
71									
72	10 2 1								
73	6 1 1	20 2 1							
74	8 1 1	15 5 1	18 4 1	20 6 1	22 4 1	25 8 1	38 1 1	40 3 1	
75	42 1 -								
76									
77	44 4 -	47 8 -							
78	8 1 1	14 1 1	20 2 1	25 8 1	29 1 3	33 5 1	48 8 -	49 8 -	
79									
80									
81									
82									
83	5 8 1								
84									
85									
86									
87	3 1 1								
88	1 5 2	3 2 1	5 5 1	7 2 1	10 2 1				
89									
90	17 2 1								

Table 45.1

Patient number	Treatment group	Follow-up time, months	Survival status	No. of recurrences	Initial Number	Initial Size
91	3	19	1	1	5	1
92	3	21	2	2	1	1
93	3	22	0	0	1	1
94	3	25	0	0	1	3
95	3	25	0	0	1	5
96	3	25	0	0	1	1
97	3	26	0	3	1	1
98	3	27	0	1	1	1
99	3	29	0	1	2	1
100	3	36	0	2	8	3
101	3	38	0	0	1	1
102	3	39	0	4	1	1
103	3	39	2	7	6	1
104	3	40	0	4	3	1
105	3	41	0	0	3	2
106	3	41	2	0	1	1
107	3	43	0	2	1	1
108	3	44	0	0	1	1
109	3	44	0	5	6	1
110	3	45	0	0	1	2
111	3	46	0	1	1	4
112	3	46	2	0	1	4
113	3	49	0	0	3	3
114	3	50	0	0	1	1
115	3	50	0	3	4	1
116	3	54	0	0	3	4
117	3	54	0	1	2	1
118	3	59	2	0	1	3

* Treatment:	1, placebo 2, pyridoxine 3, thiotepa
Survival Status:	0, alive 1, dead from bladder cancer 2, dead from other unknown cause
Number:	8 denotes eight or more initial tumours
Size:	measured in centimetres
Recurrence:	- denotes missing observation M, months of follow-up when recurred #, number of tumours at recurrence 8 denotes eight or more S, size of largest tumour

(cont.)

Patient number	Recurrence 1	2	3	4	5	6	7	8	9
	M#S	M#S	M#S	M#S	M#S	M#S	M#S	M#S	M#S
91	2 2 1								
92	17 1 1	19 1 1							
93									
94									
95									
96									
97	6 2 1	12 3 1	13 1 -						
98	6 1 1								
99	2 2 -								
100	26 3 -	35 3 -							
101									
102	22 2 1	23 1 1	27 2 1	32 3 -					
103	4 1 1	16 3 1	23 3 1	27 3 1	33 8 -	36 - -	37 8 1		
104	24 3 1	26 2 -	29 1 -	40 2 -					
105									
106									
107	1 1 2	27 1 -							
108									
109	2 2 1	20 1 1	23 2 1	27 1 1	38 8 -				
110									
111	2 1 1								
112									
113									
114									
115	4 1 1	24 1 1	47 1 -						
116									
117	38 2 1								
118									

46. Prognostic Variables for Survival in a Randomized Comparison of Treatments for Prostatic Cancer

Source and Contributor D.P. Byar National Cancer Institute

These data were obtained from a randomized clinical trial comparing four treatments for patients with prostatic cancer in Stages 3 and 4. Stage 3 represents local extension of the disease without evidence of distant metastasis and Stage 4 represents distant metastasis as evidenced by elevated acid phosphatase, x-ray evidence, or both. The trial was double-blinded and the treatments were placebo pill, 0.2 mg diethylstilbestrol, DES, 1.0 mg of DES, or 5.0 mg of DES, all drugs administered daily by mouth. Patients were not required to remain indefinitely on their assigned treatment, but were allowed to have their treatment changed at the discretion of the physician if signs of tumour progression or symptoms appeared. This provision was required for ethical reasons. Patients were followed according to a standard protocol at 6 month intervals or more frequently if required.

The data for 506 patients given in Table 46.1 consist of an identification number, stage of tumour, a code for the treatment to which the patient was assigned, the date of randomization, the total months of follow-up since randomization, an indicator for the survival status or cause of death, and the values of twelve pretreatment covariates.

The goal of an analysis should be to compare the treatments with respect to survival of the patients. Since this was a randomized study it would ordinarily not be necessary to adjust for the values of the pretreatment covariates. However, in such studies it is advisable to examine the prognostic significance of the covariates and to confirm that they are balanced across treatment groups. In addition, the analyst should look for important treatment-covariate interactions which might lead to the definition of subsets of patients in which treatment differences were significantly more marked or even reversed.

These data have been analyzed at an earlier stage in their collection in Byar and Corle (1977) and have been re-analyzed in their present form in Byar and Green (1980).

References

Byar, D.P. and Corle, D.K. (1977). Selecting optimal treatment in clinical trials using covariate information. *Chronic Diseases* **30**, 445-469.

Byar, D.P. and Green, S.B. (1980). The choice of treatment for cancer patients based on covariate information: application to prostate cancer. *Bull. Cancer*, Paris **67**, 477-488.

Table 46.1
Prognostic Variables for Survival in a Randomized Comparison of Treatments for Prostatic Cancer *

						Pretreatment covariates											
Pat no.	Stage	Rx	Date on-study	Mos FU	Surv stat	Age, yrs	Wt	PF	HX	SBP	DBP	EKG	HG	SZ	SG	AP	BM
1	3	2	8 10 67	72	0	75	76	0	0	15	9	4	138	2	8	3	0
2	3	2	9 21 67	1	5	54	116	0	0	13	7	3	146	42	-	7	0
3	3	4	1 12 68	40	3	69	102	0	1	14	8	4	134	3	9	3	0
4	3	2	3 18 68	20	3	75	94	1	1	14	7	1	176	4	8	9	0
5	3	1	3 21 68	65	0	67	99	0	0	17	10	0	134	34	8	5	0
6	3	2	6 13 68	24	1	71	98	0	0	19	10	0	151	10	11	6	0
7	3	1	6 26 68	46	2	75	100	0	0	14	10	1	130	13	9	8	0
8	3	1	6 27 68	62	0	73	114	0	1	17	11	4	126	3	9	6	0
9	3	3	7 31 68	61	0	60	110	0	0	12	8	0	146	4	10	7	0
10	3	3	8 6 68	60	0	78	107	0	1	13	8	5	130	21	6	4	0
11	3	3	8 7 68	60	0	77	89	0	0	15	8	0	156	3	8	6	0
12	3	4	9 11 68	59	0	74	105	0	1	18	14	0	136	6	8	4	0
13	3	2	9 18 68	59	0	74	107	0	0	14	9	5	144	6	9	3	0
14	3	1	9 18 68	49	1	55	112	0	1	16	9	4	139	4	9	10	0
15	3	4	10 1 68	20	3	73	88	0	0	19	10	4	120	15	10	6	0
16	3	4	10 1 68	3	2	87	81	1	1	17	12	2	134	3	9	4	0
17	3	1	10 23 68	58	0	64	90	0	0	14	8	0	162	6	9	7	0
18	3	3	11 4 68	29	4	79	104	0	0	13	8	1	150	5	8	5	0
19	3	4	4 9 69	26	2	62	90	0	1	13	8	1	144	2	9	7	0
20	3	4	4 23 69	52	0	74	107	0	0	13	8	4	164	10	9	7	0
21	3	4	4 23 69	20	2	83	72	1	1	13	9	4	150	1	10	7	0
22	3	3	5 21 69	51	0	65	113	0	0	16	10	0	141	21	9	7	0
23	3	1	5 26 69	51	0	61	97	0	0	11	8	0	135	8	8	6	0
24	4	4	8 23 67	31	1	56	95	0	0	15	10	2	148	26	13	357	1
25	4	3	10 2 67	42	3	73	103	0	0	16	10	0	146	6	11	39	0
26	4	2	11 1 67	38	2	69	75	0	0	16	8	4	107	25	11	62	0
27	4	1	11 9 67	69	0	76	82	0	0	12	8	5	117	12	10	25	0
28	4	1	11 13 67	39	1	71	103	0	0	12	7	0	144	11	11	120	0
29	4	3	11 24 67	33	5	66	126	0	0	17	10	4	140	32	13	242	0
30	4	2	12 5 67	8	1	76	90	0	0	15	9	4	128	40	13	23	0
31	4	2	12 15 67	13	1	74	77	2	0	16	9	4	105	17	13	3160	1
32	4	4	1 3 68	33	7	71	150	0	1	16	10	5	150	9	11	19	0
33	4	3	1 11 68	9	1	71	88	0	1	16	9	4	123	38	13	354	0
34	4	1	1 19 68	21	1	79	89	0	0	16	9	4	128	26	12	121	0
35	4	4	2 1 68	66	1	75	101	0	0	15	9	0	146	8	14	38	0
36	4	4	2 1 68	5	1	76	86	0	1	16	6	4	82	10	11	992	1
37	4	1	1 12 68	27	7	72	93	0	0	16	10	4	130	2	11	13	0
38	4	2	2 27 68	2	2	69	84	0	1	12	8	1	104	15	13	289	1
39	4	2	5 8 68	20	1	53	99	0	0	12	9	1	142	26	13	56	1
40	4	3	8 7 68	19	1	76	85	0	0	14	9	4	130	6	11	14	1
41	4	1	8 28 68	60	0	72	95	0	1	17	8	4	136	12	12	23	0
42	4	3	10 7 68	58	0	-	76	0	0	15	8	4	128	35	13	300	1
43	4	4	10 9 68	32	1	82	89	0	0	14	8	0	130	27	13	223	1
44	4	1	3 4 69	53	0	74	124	0	0	16	8	3	94	26	15	13	0
45	4	3	3 26 69	53	0	75	93	0	0	13	9	2	153	16	13	33	0

Table 46.1 *(cont.)*

						Pretreatment covariates											
Pat no.	Stage	Rx	Date on-study	Mos FU	Surv stat	Age, yrs	Wt	PF	HX	SBP	DBP	EKG	HG	SZ	SG	AP	BM
46	4	4	4 18 69	43	1	74	81	0	0	17	10	4	134	28	11	200	1
47	4	2	5 26 69	21	1	60	96	0	0	12	8	0	143	27	13	70	0
48	3	4	6 23 67	30	9	59	116	1	0	12	8	4	152	-	9	7	0
49	3	4	6 28 67	74	0	49	99	0	0	13	7	0	148	14	8	5	0
50	3	1	8 11 67	72	0	74	96	1	1	11	6	5	104	20	11	6	0
51	3	1	12 15 67	14	1	73	101	0	0	16	7	0	124	18	11	6	0
52	3	4	3 14 68	33	2	76	98	0	1	17	10	5	171	13	8	5	0
53	3	2	3 20 68	8	3	72	84	0	1	12	8	4	148	15	8	6	0
54	3	3	4 10 68	10	2	73	82	0	1	16	8	4	141	20	9	5	0
55	3	1	4 11 68	51	1	78	108	2	1	18	8	5	148	24	9	3	0
56	3	2	5 6 68	7	2	71	112	0	1	15	9	0	132	14	11	4	0
57	3	3	5 17 68	63	0	75	92	0	0	13	8	4	152	41	-	8	0
58	3	2	10 24 68	33	1	64	130	0	0	13	9	0	142	23	11	2	0
59	3	2	3 1 69	3	3	78	118	0	1	16	9	4	124	3	9	4	0
60	3	1	3 29 69	43	5	60	99	0	0	12	9	0	136	7	9	6	0
61	4	3	6 5 67	37	1	68	104	0	0	14	9	4	144	32	13	73	1
62	4	1	10 27 67	5	3	77	93	1	0	21	9	5	129	24	13	117	0
63	4	2	10 27 67	6	1	73	102	0	1	14	9	4	120	51	15	87	0
64	4	1	11 3 67	2	1	70	76	2	0	12	8	0	78	26	11	20	1
65	4	2	3 1 68	5	1	53	106	0	0	14	8	4	143	10	11	7	0
66	4	4	5 1 68	64	0	55	116	0	0	13	9	4	148	26	15	46	1
67	4	4	12 3 68	1	8	80	91	1	1	14	8	5	117	15	13	13	0
68	4	3	1 11 69	21	1	59	71	0	0	8	4	4	87	27	11	1751	1
69	4	1	2 27 69	16	7	80	111	0	0	16	9	0	113	34	15	11	1
70	3	3	4 14 67	21	3	68	97	1	1	14	8	5	121	18	8	6	0
71	3	3	5 10 67	12	2	77	103	0	1	13	7	2	144	5	9	7	0
72	3	3	5 10 67	75	0	74	110	0	0	16	8	1	136	7	8	5	0
73	3	1	6 7 67	74	0	70	94	0	0	15	7	1	137	24	7	8	0
74	3	2	6 7 67	74	0	72	142	1	1	14	13	0	126	-	9	5	0
75	3	2	6 16 67	54	1	65	90	0	0	14	8	0	140	47	9	3	0
76	3	4	7 17 67	73	0	74	97	0	1	13	7	0	129	6	8	2	0
77	3	1	7 26 67	36	2	72	110	0	1	22	13	5	147	5	10	6	0
78	3	1	7 26 67	59	7	69	97	0	1	13	7	0	143	4	9	3	0
79	3	4	8 4 67	6	3	72	115	0	1	16	10	0	120	5	8	3	0
80	3	2	8 21 67	58	2	71	87	0	1	13	10	2	134	13	9	8	0
81	3	3	8 30 67	16	6	73	93	0	0	11	7	0	138	8	9	3	0
82	3	4	9 6 67	19	2	79	106	0	1	16	7	5	136	13	9	2	0
83	3	4	10 10 67	39	2	63	86	0	1	16	8	5	160	7	9	4	0
84	3	2	10 27 67	26	5	74	101	0	1	14	9	2	106	14	8	4	0
85	3	1	10 27 67	60	6	72	118	0	1	17	11	0	156	9	9	6	0
86	3	2	11 2 67	43	7	78	117	0	1	10	7	4	136	16	8	4	0
87	3	4	12 13 67	7	2	76	92	0	1	12	7	5	136	12	9	5	0
88	3	3	12 13 67	9	5	70	72	0	1	16	9	3	120	17	8	1	0
89	3	3	1 17 68	67	0	69	120	0	1	16	10	4	136	21	10	7	0
90	3	4	1 30 68	62	5	76	102	0	1	12	8	0	132	13	8	5	0
91	3	1	5 7 68	17	6	84	104	0	1	15	10	1	134	8	6	4	0
92	3	2	5 7 68	25	5	80	98	0	1	15	8	0	123	8	9	5	0

Table 46.1 *(cont.)*

						Pretreatment covariates											
Pat no.	Stage	Rx	Date on-study	Mos FU	Surv stat	Age, yrs	Wt	PF	HX	SBP	DBP	EKG	HG	SZ	SG	AP	BM
93	3	3	6 14 68	000	6	81	97	0	0	10	7	0	134	11	9	7	0
94	3	4	6 21 68	5	2	81	76	0	1	13	7	4	149	28	9	5	0
95	3	4	7 24 68	34	2	71	100	0	1	12	10	4	144	11	6	3	0
96	3	1	9 17 68	59	0	77	109	0	0	14	7	0	120	7	10	3	0
97	3	2	10 2 68	31	2	70	109	0	1	18	8	4	120	28	9	5	0
98	3	1	10 9 68	58	0	76	108	0	1	12	7	5	120	3	9	3	0
99	3	3	10 21 68	30	3	76	102	0	1	16	9	5	182	52	11	6	0
100	3	2	11 6 68	57	0	71	97	0	0	13	7	3	134	19	9	6	0
101	3	2	11 13 68	5	2	76	96	0	1	17	8	0	137	33	9	6	0
102	3	4	1 8 69	55	0	64	100	0	1	16	10	0	137	6	9	2	0
103	3	3	1 30 69	55	0	74	113	0	0	12	7	0	143	7	10	8	0
104	3	1	4 17 69	52	0	76	88	0	1	16	9	0	123	3	8	6	0
105	3	4	4 29 69	000	4	82	82	0	1	13	9	2	116	5	10	6	0
106	3	3	5 14 69	51	0	65	87	0	1	16	8	0	145	2	9	8	0
107	4	2	4 14 67	17	2	70	83	1	0	14	6	0	127	14	11	97	1
108	4	4	4 14 67	70	5	76	92	0	0	13	6	4	97	6	10	62	1
109	4	1	6 7 67	74	0	77	93	0	0	10	6	1	110	00	12	49	0
110	4	2	6 16 67	74	0	77	99	0	1	17	9	0	128	15	12	24	0
111	4	3	8 18 67	20	2	70	103	0	1	12	7	0	135	23	11	102	0
112	4	1	11 7 67	2	1	73	80	0	1	13	8	4	129	61	12	397	0
113	4	2	1 15 68	8	1	63	119	0	1	13	8	4	128	14	15	417	1
114	4	4	2 14 68	27	7	71	91	0	0	13	6	4	120	1	10	12	0
115	4	3	3 13 68	23	3	78	103	1	1	14	8	0	100	38	12	18	1
116	4	1	1 7 69	1	2	74	100	0	1	13	8	5	144	1	11	11	0
117	4	4	4 4 69	52	0	70	110	0	1	16	8	5	159	4	10	20	0
118	4	3	5 7 69	51	0	73	106	0	1	11	8	1	112	10	13	12	1
119	4	4	5 20 69	6	1	74	98	0	1	13	6	2	140	41	12	784	1
120	4	3	5 27 69	000	1	77	91	0	1	12	7	4	143	6	11	17	1
121	4	4	5 27 69	26	2	85	85	0	1	15	7	0	140	12	12	259	0
122	3	1	9 14 67	65	6	72	79	0	1	13	7	0	123	16	9	8	0
123	3	1	9 14 67	52	8	74	106	0	0	14	6	5	138	17	-	2	0
124	3	2	10 23 67	26	1	78	83	0	0	13	7	0	103	25	11	5	0
125	3	4	10 24 67	70	0	60	110	0	0	14	10	4	143	13	-	5	0
126	3	1	11 8 67	6	5	72	76	0	0	11	9	1	100	16	10	7	0
127	3	2	11 8 67	6	4	68	124	0	0	13	7	2	145	17	9	18	0
128	3	3	12 12 67	68	0	71	93	0	0	15	9	0	145	19	9	3	0
129	3	4	1 3 68	33	2	65	125	0	0	17	9	5	150	19	8	8	0
130	3	2	2 2 68	67	0	67	96	0	0	13	7	5	157	7	9	5	0
131	3	2	3 4 68	58	2	76	94	0	0	12	7	0	142	-	7	3	0
132	3	4	3 8 68	23	1	74	98	0	0	15	9	3	142	3	12	41	0
133	3	4	4 9 68	6	5	78	91	0	0	11	7	0	110	4	9	3	0
134	3	3	5 10 68	11	2	74	85	0	0	14	8	1	140	16	9	9	0
135	3	1	6 6 68	6	1	71	92	0	0	13	7	3	140	9	9	8	0
136	3	3	11 14 68	57	0	74	113	0	0	14	7	2	105	20	9	5	0
137	3	3	11 20 68	57	0	72	110	0	0	17	10	4	145	20	8	4	0
138	3	1	1 24 69	29	2	57	97	0	0	14	8	4	147	20	9	7	0

Table 46.1 *(cont.)*

						Pretreatment covariates											
Pat no.	Stage	Rx	Date on-study	Mos FU	Surv stat	Age, yrs	Wt	PF	HX	SBP	DBP	EKG	HG	SZ	SG	AP	BM
139	3	2	2 7 69	9	5	74	104	0	0	14	7	0	138	25	8	8	0
140	3	3	3 26 69	53	0	72	105	0	0	12	6	2	136	5	9	7	0
141	3	3	5 16 69	51	0	65	99	0	0	13	8	0	144	10	9	6	0
142	3	4	5 21 69	8	4	83	115	0	0	17	9	4	167	25	9	6	0
143	4	1	7 31 67	000	7	71	98	2	1	18	9	4	123	34	13	429	1
144	4	4	8 22 67	21	2	73	103	0	0	12	8	0	128	21	11	28	0
145	4	2	11 29 67	27	1	71	102	0	0	14	7	0	105	30	13	3670	1
146	4	3	11 29 67	22	6	75	124	0	0	18	11	4	156	22	11	70	0
147	4	3	12 5 67	12	1	71	77	0	0	10	6	2	70	17	13	36	1
148	4	3	12 12 67	68	0	56	69	0	0	13	8	0	107	11	13	14	1
149	4	3	12 20 67	21	7	83	88	0	0	13	7	0	133	15	13	217	0
150	4	2	12 26 67	68	0	71	116	0	0	15	8	4	150	46	10	32	0
151	4	4	1 29 68	000	7	73	98	1	1	15	8	2	102	23	13	1278	0
152	4	4	4 9 68	2	1	78	74	0	0	12	9	4	140	32	11	47	1
153	4	1	5 27 68	14	1	50	109	0	0	12	8	0	93	17	11	2252	1
154	4	1	6 26 68	18	2	73	101	0	0	16	11	4	140	9	11	27	0
155	4	4	8 15 68	5	1	67	111	0	0	12	9	5	96	17	11	70	1
156	4	3	8 21 68	60	0	72	96	0	0	15	9	0	153	2	15	436	1
157	4	1	8 22 68	40	3	72	96	0	0	14	10	1	105	26	12	299	1
158	4	1	3 19 69	31	1	68	98	0	0	18	10	5	150	-	13	17	0
159	4	2	5 20 69	51	0	78	116	0	0	15	10	4	157	6	13	11	0
160	3	1	12 6 67	50	9	72	91	0	1	19	9	4	120	16	8	6	0
161	3	2	1 4 68	46	9	76	118	0	1	18	8	4	145	4	8	7	0
162	3	2	3 28 68	65	0	77	111	0	0	19	9	3	125	20	9	8	0
163	3	4	5 9 68	9	4	70	127	0	0	13	9	2	118	14	8	2	0
164	3	2	6 24 68	21	5	74	96	0	1	10	7	0	102	18	9	9	0
165	3	2	9 16 68	27	9	71	92	0	1	15	8	5	106	5	8	4	0
166	3	3	2 28 69	54	0	69	89	0	1	12	7	0	135	4	8	5	0
167	3	3	4 7 69	52	0	70	111	0	1	18	8	2	82	22	9	8	0
168	4	3	5 15 67	75	0	58	99	0	0	19	12	4	131	24	11	16	0
169	4	4	5 9 67	75	0	78	99	0	0	14	7	0	126	27	-	46	1
170	4	3	7 7 67	30	2	73	105	0	1	14	6	5	138	9	15	51	0
171	4	4	10 23 67	70	0	69	108	0	0	12	8	0	122	23	11	20	0
172	4	3	11 3 67	14	5	73	94	0	0	12	7	5	134	31	13	34	0
173	4	1	7 3 68	000	7	73	93	0	0	14	8	0	78	23	15	12	0
174	4	4	7 9 68	42	2	68	100	0	0	17	10	4	117	32	13	264	0
175	4	3	7 15 68	12	2	75	88	0	1	14	6	0	134	31	13	216	1
176	4	1	7 19 68	40	2	68	104	0	1	17	10	5	150	25	10	9	1
177	4	1	7 22 68	2	1	76	88	0	1	16	7	0	112	32	13	262	0
178	4	3	10 30 68	12	1	76	87	1	1	14	5	4	117	28	14	428	1
179	4	2	3 9 69	53	0	75	89	0	1	14	8	0	126	8	11	20	0
180	4	2	3 12 69	11	1	59	116	0	1	10	6	0	128	6	11	24	0
181	3	4	6 27 67	18	2	72	96	0	0	11	8	0	103	9	11	33	1
182	3	3	9 6 67	13	3	80	91	0	0	15	8	2	123	4	8	6	0
183	3	2	10 10 67	45	7	78	112	0	1	16	8	0	150	7	9	3	0
184	3	3	10 12 67	8	9	52	90	0	1	13	8	4	148	4	9	3	0

Table 46.1 *(cont.)*

						Pretreatment covariates											
Pat no.	Stage	Rx	Date on-study	Mos FU	Surv stat	Age, yrs	Wt	PF	HX	SBP	DBP	EKG	HG	SZ	SG	AP	BM
185	3	4	12 27 67	28	1	75	87	0	0	9	5	4	127	8	8	10	0
186	3	4	1 4 68	67	0	67	109	0	1	14	8	0	140	3	9	8	0
187	3	2	1 16 68	67	0	74	103	0	0	16	6	4	130	4	9	4	0
188	3	1	1 24 68	24	2	79	107	1	1	23	10	4	116	2	9	2	0
189	3	2	4 10 68	64	0	73	96	0	0	15	10	2	120	4	9	5	0
190	3	3	4 24 68	26	2	75	113	2	0	12	7	4	135	1	13	7	0
191	4	2	8 28 67	57	1	72	88	0	0	14	7	4	120	7	12	594	0
192	4	2	9 11 67	32	1	55	96	0	1	15	9	4	103	12	11	2	0
193	4	1	9 13 67	71	0	68	-	1	1	16	9	0	146	18	11	2	0
194	4	2	9 13 67	42	3	48	100	1	1	14	10	4	123	3	11	15	0
195	4	1	9 18 67	20	2	75	88	0	1	9	5	4	140	1	11	48	0
196	4	3	9 19 67	71	0	56	94	0	0	12	8	0	144	1	10	11	0
197	4	4	4 18 68	54	6	70	86	0	0	14	9	4	144	4	11	12	0
198	4	4	4 23 68	64	0	54	89	0	0	15	6	2	148	8	13	12	1
199	4	4	4 23 68	51	2	87	82	0	0	14	7	0	110	3	15	21	1
200	3	4	5 15 67	1	2	73	94	0	1	14	9	4	133	22	8	8	0
201	3	1	6 1 67	2	7	76	94	0	0	16	10	0	107	8	9	6	0
202	3	4	6 27 67	000	1	68	106	0	0	15	9	0	107	69	13	9	0
203	3	1	6 27 67	54	3	70	97	0	1	17	10	2	138	10	9	6	0
204	3	2	11 17 67	69	0	65	83	0	0	12	6	0	160	8	9	6	0
205	3	1	12 4 67	68	0	79	95	0	0	14	8	4	134	11	9	4	0
206	3	1	1 10 68	67	0	70	120	0	0	19	11	0	147	7	9	5	0
207	3	3	2 26 68	66	0	70	99	0	0	12	8	0	160	8	9	6	0
208	3	1	4 10 68	64	0	59	90	0	1	14	8	0	104	2	9	3	0
209	3	2	5 23 68	63	0	72	73	2	0	13	8	0	142	11	8	6	0
210	3	3	6 10 68	35	3	76	88	0	1	14	9	0	152	4	5	6	0
211	3	4	6 20 68	62	0	78	123	0	0	13	8	1	168	00	10	9	0
212	3	3	7 16 68	7	7	76	113	0	1	14	6	4	128	3	8	7	0
213	3	4	7 31 68	35	7	66	87	0	0	14	9	0	143	3	8	4	0
214	3	3	10 23 68	24	4	69	91	0	1	16	10	3	150	11	8	3	0
215	3	4	11 22 68	57	0	71	88	0	0	16	8	4	154	3	9	5	0
216	3	2	12 31 68	1	5	75	120	0	0	15	7	4	151	4	8	6	0
217	3	2	2 12 69	4	2	74	99	0	1	17	9	2	157	7	7	6	0
218	4	2	5 23 67	74	5	72	103	0	0	15	9	3	165	4	10	10	0
219	4	2	5 24 67	000	1	78	96	2	0	16	9	4	101	54	12	46	1
220	4	2	6 16 67	51	2	71	110	0	0	14	9	5	134	6	10	7	1
221	4	4	7 17 67	36	2	75	101	0	1	16	10	4	110	15	10	16	0
222	4	2	8 2 67	16	1	73	89	2	0	15	8	4	72	14	11	204	1
223	4	3	10 3 67	32	1	79	79	2	0	15	7	1	133	16	13	289	1
224	4	4	12 7 67	68	0	70	92	0	0	16	8	0	135	22	14	27	0
225	4	2	12 15 67	17	1	73	76	0	0	17	8	0	134	50	13	89	0
226	4	1	2 21 68	66	0	75	98	0	1	14	8	0	147	16	13	28	0
227	4	3	3 14 68	65	0	76	83	0	0	12	6	2	114	13	11	5960	1
228	4	1	3 19 68	19	1	71	77	1	0	17	7	5	118	61	15	43	1
229	4	3	3 22 68	41	1	76	91	0	0	17	8	0	151	24	11	82	0
230	4	1	4 5 68	28	2	73	106	0	1	16	9	0	147	11	11	369	0

Table 46.1 *(cont.)*

						Pretreatment covariates											
Pat no.	Stage	Rx	Date on-study	Mos FU	Surv stat	Age, yrs	Wt	PF	HX	SBP	DBP	EKG	HG	SZ	SG	AP	BM
231	4	1	4 11 68	000	2	70	80	0	1	10	8	4	122	5	11	43	0
232	4	4	4 26 68	46	1	72	100	0	1	12	7	5	160	62	11	548	1
233	4	4	5 24 68	63	0	74	101	0	0	14	9	5	148	25	10	24	0
234	4	3	8 5 68	60	0	62	98	0	0	12	8	4	156	15	13	45	0
235	4	3	9 19 68	59	0	53	116	0	0	13	8	0	164	7	12	10	0
236	3	4	5 1 67	46	2	54	91	0	0	13	10	0	130	15	9	7	0
237	3	3	5 1 67	63	1	58	94	0	1	16	10	3	138	12	9	2	0
238	3	4	6 13 67	8	2	78	99	0	1	17	9	5	120	9	8	1	0
239	3	3	6 15 67	74	0	77	100	0	1	10	7	3	124	25	9	4	0
240	3	1	6 21 67	18	2	61	152	0	1	15	10	4	141	18	7	2	0
241	3	1	9 12 67	33	7	78	98	1	1	13	7	4	136	9	9	5	0
242	3	1	9 25 67	12	2	68	88	1	1	24	12	4	144	12	9	2	0
243	3	2	9 27 67	1	7	62	76	1	0	10	6	-	116	13	9	3	0
244	3	3	2 29 68	66	0	70	109	0	0	15	10	0	145	15	11	6	0
245	3	3	6 17 68	28	5	75	81	0	0	16	10	0	137	19	10	5	0
246	3	2	9 5 68	30	2	80	91	0	0	12	8	4	108	22	11	9	0
247	3	2	9 18 68	57	7	76	99	0	1	13	8	5	112	20	8	5	0
248	3	3	9 20 68	50	1	68	102	0	0	16	8	0	120	20	11	7	0
249	3	1	10 18 68	39	2	73	81	0	1	19	4	4	125	18	11	3	0
250	3	1	10 18 68	40	2	79	96	0	0	15	10	4	140	16	6	3	0
251	3	2	4 4 69	24	3	78	81	0	1	14	7	5	110	4	9	4	0
252	3	4	6 3 69	6	2	79	94	0	1	11	8	5	122	4	6	4	0
253	4	2	8 14 67	30	2	72	90	1	1	22	12	4	123	15	14	5	1
254	4	2	9 8 67	58	1	52	106	0	1	11	7	0	150	13	14	53	0
255	4	3	12 18 67	68	0	76	97	0	1	15	10	4	146	19	11	3535	0
256	4	2	1 3 68	24	1	74	85	0	0	16	10	2	100	19	13	139	1
257	4	1	1 12 68	33	8	71	86	0	0	14	8	0	102	8	11	4	1
258	4	4	3 25 68	65	0	72	96	0	0	14	10	2	123	17	12	200	0
259	4	4	3 25 68	2	4	77	113	0	0	15	8	0	138	22	12	55	0
260	4	2	4 11 68	64	0	73	100	0	1	16	10	4	138	23	11	97	0
261	4	1	6 4 68	19	1	75	83	0	0	12	7	2	118	17	13	23	0
262	4	1	10 9 68	58	0	73	-	0	0	13	7	5	133	20	13	11	0
263	4	4	3 18 69	15	1	71	98	0	0	15	10	4	147	12	11	431	1
264	4	2	3 21 69	10	1	76	94	0	1	16	9	0	105	20	12	336	0
265	4	4	3 21 69	12	1	49	90	1	0	14	8	4	141	21	15	56	1
266	4	4	3 27 69	53	0	74	91	2	1	13	7	2	134	17	11	2783	1
267	4	3	5 5 69	51	0	59	97	0	0	13	6	0	141	17	13	22	0
268	4	3	5 29 69	1	2	79	116	0	1	15	9	2	126	23	11	38	0
269	3	3	10 20 67	40	1	78	113	0	0	20	10	-	132	36	11	6	0
270	3	1	2 1 68	40	1	78	110	0	0	14	8	-	148	15	9	3	0
271	3	3	2 8 68	66	0	79	96	0	0	14	8	-	148	19	8	4	0
272	3	2	2 19 68	2	4	72	82	0	0	13	7	4	94	33	11	6	0
273	3	2	4 8 68	21	1	58	136	1	1	14	8	1	105	30	12	7	0
274	3	4	4 11 68	35	1	73	101	0	0	12	6	0	128	46	11	6	0
275	3	1	9 23 68	29	1	75	113	0	1	19	8	5	109	21	11	4	0
276	3	4	12 9 68	54	1	78	107	0	0	14	9	0	128	42	11	297	0

Table 46.1 *(cont.)*

						Pretreatment covariates											
Pat no.	Stage	Rx	Date on-study	Mos FU	Surv stat	Age, yrs	Wt	PF	HX	SBP	DBP	EKG	HG	SZ	SG	AP	BM
277	3	4	1 20 69	37	2	60	94	0	0	15	9	0	130	10	9	4	0
278	3	2	3 7 69	10	5	72	83	0	0	11	7	0	138	15	9	4	0
279	4	4	4 27 67	29	1	59	99	3	0	13	8	0	127	17	13	9999	1
280	4	3	9 27 67	71	0	84	107	0	0	14	7	-	130	15	11	379	1
281	4	1	9 27 67	12	1	74	107	0	1	15	8	0	146	18	11	84	0
282	4	2	10 19 67	13	1	76	97	0	0	14	8	0	137	36	15	155	0
283	4	1	12 4 67	27	1	80	135	0	1	11	7	5	129	17	11	38	0
284	4	2	11 8 67	66	1	72	123	0	1	15	8	2	158	00	13	18	0
285	4	1	2 29 68	28	1	59	100	0	0	14	8	-	158	21	11	53	0
286	4	3	4 3 68	31	1	74	111	0	0	13	8	0	152	25	13	42	0
287	4	3	5 21 68	12	1	65	102	0	0	15	9	0	155	19	11	22	0
288	4	1	5 21 68	9	2	71	119	0	0	19	10	4	136	24	13	256	1
289	4	2	10 17 68	21	1	76	94	0	0	14	6	0	106	24	11	109	1
290	4	4	11 27 68	19	1	74	74	3	0	14	5	3	112	2	11	37	1
291	4	3	5 23 69	51	0	75	114	0	0	18	7	5	138	22	11	23	0
292	3	1	4 6 67	71	1	73	134	1	1	22	10	4	140	22	11	6	0
293	3	3	4 12 67	76	0	74	99	0	1	17	7	0	111	12	11	4	0
294	3	4	4 21 67	46	1	72	112	1	1	16	8	0	130	30	11	6	0
295	3	2	5 5 67	47	5	75	107	0	1	15	9	4	162	7	9	11	0
296	3	2	5 5 67	75	0	70	99	1	0	12	7	0	139	7	9	9	0
297	3	4	5 5 67	75	0	61	94	0	0	12	7	4	120	4	9	8	0
298	3	3	5 5 67	49	2	71	94	0	1	16	7	4	156	9	9	3	0
299	3	4	5 10 67	75	0	57	84	0	0	12	8	0	139	6	8	5	0
300	3	4	5 15 67	75	0	65	111	0	0	14	8	5	137	15	9	8	0
301	3	1	5 31 67	31	2	78	96	1	1	17	11	0	149	2	10	2	0
302	3	1	6 5 67	69	1	60	122	0	0	13	9	0	161	26	9	5	0
303	3	2	6 23 67	74	0	72	102	0	0	16	8	0	140	23	7	3	0
304	3	3	7 10 67	25	7	81	102	1	0	15	8	5	128	4	8	5	0
305	3	4	8 15 67	72	0	70	87	0	0	15	8	0	146	5	10	3	0
306	3	2	8 21 67	28	7	62	88	0	1	12	7	6	135	10	8	3	0
307	3	3	8 25 67	36	7	68	98	0	0	16	7	2	133	40	9	4	0
308	3	1	8 31 67	8	1	74	97	0	0	14	8	4	136	14	8	5	0
309	3	2	9 28 67	71	0	65	117	0	1	15	8	0	165	9	9	8	0
310	3	1	10 10 67	60	3	74	95	0	1	18	9	0	98	2	8	7	0
311	3	3	10 30 67	14	4	76	99	0	1	15	9	4	120	9	11	8	0
312	3	2	11 29 67	22	7	70	78	0	0	14	7	2	149	9	11	2	0
313	3	4	12 4 67	67	8	73	105	0	0	13	6	2	149	39	9	5	0
314	3	1	12 15 67	62	2	71	107	0	1	18	10	4	145	9	8	3	0
315	3	4	12 15 67	68	0	73	118	0	0	17	9	3	134	4	9	7	0
316	3	1	12 20 67	68	0	71	108	0	0	16	12	4	141	7	9	4	0
317	3	3	12 22 67	27	3	75	145	0	0	13	8	1	154	4	9	5	0
318	3	2	12 29 67	68	0	70	98	0	0	11	7	4	142	19	10	5	0
319	3	3	12 29 67	68	0	77	98	0	0	12	6	5	155	9	9	4	0
320	3	1	1 4 68	67	0	71	87	0	0	15	8	0	156	8	8	6	0
321	3	1	1 4 68	67	0	75	101	0	0	14	7	3	135	18	7	5	0
322	3	2	1 4 68	3	2	74	105	0	1	16	7	4	141	10	11	3	0

Table 46.1 *(cont.)*

						Pretreatment Covariates											
Pat no.	Stage	Rx	Date on-study	Mos FU	Surv stat	Age, yrs	Wt	PF	HX	SBP	DBP	EKG	HG	SZ	SG	AP	BM
323	3	4	1 15 68	49	2	72	100	0	0	11	6	4	150	8	11	3	0
324	3	4	1 16 68	12	6	89	92	0	1	13	6	5	136	8	9	7	0
325	3	2	1 16 68	30	6	75	96	0	0	12	6	4	143	10	12	7	0
326	3	3	1 26 68	67	0	71	101	0	0	14	8	4	147	3	11	5	0
327	3	3	1 31 68	5	5	80	95	0	1	16	10	4	145	7	9	9	0
328	3	3	2 12 68	65	5	51	96	0	0	13	8	0	118	2	6	6	0
329	3	1	2 13 68	24	7	71	81	0	0	9	4	0	137	10	9	4	0
330	3	2	2 19 68	66	0	72	95	0	1	10	5	0	140	8	8	7	0
331	3	4	2 22 68	36	2	70	91	0	1	16	9	3	159	7	9	10	0
332	3	1	2 26 68	66	0	55	103	0	0	16	10	5	167	2	8	8	0
333	3	2	2 26 68	40	3	76	104	0	1	15	7	4	146	3	9	5	0
334	3	2	2 26 68	13	7	68	97	0	0	17	9	4	95	11	8	5	0
335	3	4	3 20 68	40	3	75	81	0	1	12	8	5	155	14	8	4	0
336	3	2	3 26 68	65	0	52	101	0	1	15	10	4	140	4	-	6	0
337	3	2	4 1 68	000	7	75	91	0	0	20	10	4	118	3	11	2	0
338	3	3	4 11 68	14	2	75	94	0	1	11	7	4	175	15	9	8	0
339	3	3	4 11 68	45	3	72	77	0	0	13	7	0	110	4	8	5	0
340	3	3	4 11 68	64	0	74	118	0	0	14	8	0	142	4	6	6	0
341	3	3	4 11 68	13	6	75	91	0	0	13	7	2	121	5	8	5	0
342	3	1	4 18 68	24	5	75	79	0	1	13	9	5	109	8	8	4	0
343	3	1	5 6 68	15	2	71	96	0	1	12	8	4	173	5	11	8	0
344	3	1	5 9 68	26	2	76	102	0	1	15	9	4	112	10	8	5	0
345	3	3	5 9 68	5	1	70	83	0	0	14	7	5	117	33	13	3	0
346	3	2	5 13 68	63	0	73	129	0	1	21	11	4	167	10	8	7	0
347	3	4	7 8 68	35	2	70	107	0	1	13	8	5	161	7	11	8	0
348	3	1	7 8 68	61	0	75	93	0	0	13	8	0	137	10	12	1	0
349	3	4	7 19 68	61	0	70	127	0	0	14	7	0	162	8	9	6	0
350	3	4	7 24 68	50	9	74	97	0	1	13	7	0	123	1	8	4	0
351	3	4	7 31 68	57	2	72	109	0	0	16	10	0	175	8	9	7	0
352	3	4	8 6 68	21	3	62	109	0	1	16	8	4	164	15	9	7	0
353	3	1	8 6 68	26	1	72	111	0	0	14	8	0	153	37	11	5	0
354	3	1	8 6 68	11	6	77	74	0	0	12	6	2	124	45	11	5	0
355	3	3	9 23 68	59	0	52	90	0	0	12	7	4	119	5	8	5	0
356	3	1	9 25 68	42	1	57	89	0	0	12	6	0	139	24	12	5	0
357	3	3	9 30 68	27	3	80	90	0	1	14	8	4	143	9	8	6	0
358	3	2	11 27 68	50	2	62	85	0	1	11	7	0	137	6	9	3	0
359	3	3	11 27 68	57	0	72	101	0	0	13	8	0	146	8	10	5	0
360	3	3	11 27 68	57	0	72	116	0	0	14	8	4	152	11	9	8	0
361	3	1	12 16 68	000	2	77	91	0	1	14	10	5	145	5	9	5	0
362	3	2	12 19 68	1	2	72	134	0	1	15	8	4	148	23	9	7	0
363	3	3	2 28 69	54	0	73	100	0	0	18	10	5	117	8	9	7	0
364	3	2	3 12 69	53	1	74	98	0	1	16	8	5	138	11	8	10	0
365	3	2	3 13 69	23	2	74	82	0	1	17	4	5	109	8	8	4	0
366	3	4	3 24 69	8	2	72	110	0	0	15	8	0	131	4	9	8	0
367	3	4	3 28 69	51	4	76	85	0	1	14	8	5	159	25	9	9	0
368	3	1	4 3 69	27	7	78	111	0	1	15	7	4	123	7	9	5	0

Table 46.1 *(cont.)*

						Pretreatment Covariates											
Pat no.	Stage	Rx	Date on-study	Mos FU	Surv stat	Age, yrs	Wt	PF	HX	SBP	DBP	EKG	HG	SZ	SG	AP	BM
369	3	4	4 7 69	52	0	83	88	0	1	16	8	5	100	1	7	6	0
370	3	2	4 25 69	52	0	80	99	0	1	13	7	4	130	6	9	5	0
371	3	4	4 30 69	52	0	76	90	0	0	15	8	3	133	52	13	4	0
372	3	1	4 30 69	45	3	79	126	0	0	13	8	5	154	7	11	7	0
373	3	2	5 16 69	48	7	76	116	0	1	13	8	4	169	2	9	4	0
374	3	1	5 20 69	18	1	71	110	0	1	14	8	4	130	7	11	3	0
375	4	1	4 6 67	2	7	61	95	0	0	12	8	4	115	37	12	256	1
376	4	4	4 24 67	2	2	77	96	1	1	14	7	4	119	36	10	9	1
377	4	3	5 10 67	35	3	76	82	0	0	10	6	4	103	10	13	93	1
378	4	3	5 19 67	75	0	71	106	0	0	15	8	0	141	26	13	100	0
379	4	1	5 31 67	75	0	77	87	0	0	11	6	1	125	5	11	22	0
380	4	1	6 22 67	9	4	78	89	0	1	18	8	2	143	39	13	253	1
381	4	3	8 4 67	39	1	58	92	0	0	12	8	4	137	5	11	108	1
382	4	4	8 4 67	000	2	74	113	0	1	11	8	4	142	41	13	23	0
383	4	1	10 6 67	26	3	75	116	0	1	18	9	2	160	26	11	5	1
384	4	1	10 25 67	17	4	62	102	0	0	18	11	5	146	16	13	101	0
385	4	2	11 27 67	43	6	80	92	0	0	15	9	4	131	00	13	11	0
386	4	4	12 4 67	53	1	61	79	1	0	13	8	0	103	14	11	40	0
387	4	3	12 29 67	47	3	76	98	0	1	17	9	5	134	19	11	38	0
388	4	2	3 28 68	11	1	79	99	1	0	16	8	4	91	38	13	314	1
389	4	2	3 28 68	65	0	66	116	0	0	12	8	0	149	18	12	581	0
390	4	3	3 29 68	16	1	61	83	0	1	12	6	2	136	4	15	385	1
391	4	4	3 29 68	21	8	79	99	0	0	17	9	4	161	4	13	61	0
392	4	2	4 23 68	64	0	70	99	0	1	13	7	0	142	11	11	11	0
393	4	2	7 8 68	7	1	73	105	0	1	15	7	3	156	10	9	11	0
394	4	4	7 12 68	38	1	72	90	0	0	11	8	4	148	11	15	15	0
395	4	3	6 24 68	55	1	81	98	0	1	13	6	0	112	38	12	87	1
396	4	1	6 24 68	5	6	77	92	0	0	15	9	4	115	6	10	6	0
397	4	2	8 9 68	28	1	76	76	0	0	11	6	4	148	29	13	18	0
398	4	1	8 16 68	38	1	75	113	0	1	10	6	2	133	9	13	4	1
399	4	2	8 16 68	20	1	72	79	0	0	14	7	0	122	24	11	72	1
400	4	4	8 21 68	24	1	74	108	0	0	16	9	0	126	12	13	11	0
401	4	4	8 23 68	27	1	70	85	0	0	16	8	2	164	9	13	18	0
402	4	3	9 30 68	31	1	61	91	0	1	14	8	1	107	25	13	1522	1
403	4	1	10 1 68	9	1	77	116	0	1	14	8	2	152	7	13	470	0
404	4	3	10 9 68	9	1	60	73	0	0	11	8	-	141	46	13	170	1
405	4	3	11 12 68	11	8	72	86	1	0	12	7	2	140	50	15	11	0
406	4	4	11 27 68	56	1	64	104	0	0	17	7	0	120	43	13	24	1
407	4	1	12 5 68	36	1	78	95	0	0	15	9	4	138	55	10	32	0
408	4	2	12 16 68	56	0	74	108	0	1	14	8	5	106	4	15	11	0
409	4	4	12 16 68	15	5	73	83	0	1	12	7	4	112	00	11	42	0
410	4	4	5 22 69	51	0	71	104	0	0	15	7	3	160	26	11	8	1
411	3	1	9 25 67	33	3	75	113	0	1	17	9	4	138	9	9	7	0
412	3	3	10 9 67	70	0	72	105	0	0	13	8	0	138	3	9	8	0
413	3	4	12 5 67	32	2	68	89	0	1	19	10	4	138	3	9	6	0
414	3	3	2 13 68	5	5	74	91	0	0	15	8	0	151	3	9	4	0

Table 46.1 *(cont.)*

						Pretreatment Covariates											
Pat no.	Stage	Rx	Date on-study	Mos FU	Surv stat	Age, yrs	Wt	PF	HX	SBP	DBP	EKG	HG	SZ	SG	AP	BM
415	3	4	3 19 68	23	2	72	93	0	1	16	8	4	166	4	9	2	0
416	3	1	5 14 68	63	0	79	110	0	0	15	9	5	169	2	9	7	0
417	3	2	6 5 68	30	1	62	104	0	1	12	8	4	160	32	11	4	0
418	3	2	6 10 68	12	1	69	92	0	1	17	7	5	155	22	-	9	0
419	3	1	7 10 68	53	2	62	119	0	0	19	11	4	164	14	9	5	0
420	3	1	9 11 68	12	2	72	128	0	0	15	8	4	164	15	9	6	0
421	3	1	10 8 68	58	0	72	109	0	0	13	8	4	133	12	9	6	0
422	3	3	2 4 69	54	0	51	112	0	0	13	9	0	158	7	8	8	0
423	3	1	2 19 69	36	1	72	104	0	0	18	10	0	164	4	9	6	0
424	3	2	3 18 69	20	6	81	104	0	0	15	8	0	169	16	13	8	0
425	4	4	6 7 67	74	0	71	81	0	0	11	7	5	131	29	11	118	0
426	4	1	6 19 67	34	1	77	103	0	0	14	7	2	143	35	11	13	0
427	4	2	10 3 67	70	0	73	114	0	1	14	8	0	147	26	11	5	0
428	4	1	2 21 68	25	1	56	100	0	0	17	8	4	138	30	11	506	1
429	4	2	3 6 68	65	0	77	107	0	0	14	9	4	147	1	13	16	0
430	4	1	4 10 68	4	1	78	104	0	0	16	9	3	127	27	11	226	0
431	4	3	7 16 68	10	1	76	96	0	0	12	7	5	138	34	11	360	1
432	4	3	10 10 68	58	0	68	79	0	0	18	9	2	134	9	11	12	0
433	4	1	10 28 68	37	1	73	90	0	0	14	9	1	94	21	13	51	0
434	4	4	12 23 68	56	0	56	135	0	0	16	9	0	147	10	10	32	0
435	4	3	4 15 69	6	1	74	112	0	1	20	10	4	131	9	11	213	1
436	3	4	6 13 67	54	1	76	84	0	1	15	8	0	126	19	-	4	0
437	3	4	9 14 67	000	2	69	113	0	1	13	8	0	113	1	9	1	0
438	3	3	10 24 67	70	0	73	84	0	1	10	6	1	156	1	9	5	0
439	3	3	10 30 67	70	0	71	105	0	0	12	8	0	136	2	10	2	0
440	3	2	11 15 67	28	2	75	97	0	1	14	8	5	108	2	9	2	0
441	3	3	12 27 67	40	2	69	85	0	0	15	7	5	59	2	9	3	0
442	3	3	1 22 68	67	0	73	95	0	0	11	7	0	138	7	8	4	0
443	3	2	6 3 68	36	6	59	81	0	0	15	10	0	160	2	9	4	0
444	3	1	6 3 68	23	6	68	93	0	0	13	7	0	125	2	8	9	0
445	3	1	6 3 68	23	2	78	115	0	1	14	7	3	130	2	8	1	0
446	3	1	6 3 68	62	0	63	95	0	0	12	9	0	132	3	8	3	0
447	3	2	6 14 68	62	0	74	85	0	0	12	9	3	141	2	9	4	0
448	3	1	6 27 68	62	0	72	103	0	0	16	7	4	143	00	5	4	0
449	3	4	8 6 68	2	2	78	103	0	0	11	7	5	112	00	6	6	0
450	3	2	9 5 68	34	1	70	85	0	0	14	9	1	149	2	12	6	0
451	3	3	10 10 68	4	2	72	101	0	0	14	10	2	164	1	9	8	0
452	3	2	10 18 68	12	1	76	103	0	1	13	7	5	125	4	8	5	0
453	3	1	12 4 68	56	0	66	106	0	0	12	9	5	155	4	9	5	0
454	3	4	12 19 68	1	2	88	99	0	1	13	6	5	141	17	9	7	0
455	3	4	5 6 69	7	7	78	88	0	1	11	6	5	109	5	9	3	0
456	4	1	7 3 67	4	1	79	105	0	0	12	7	0	122	5	12	30	0
457	4	1	10 12 67	70	0	77	106	0	1	15	8	2	160	1	9	22	0
458	4	4	3 25 68	65	0	71	101	0	0	12	7	0	127	2	10	15	0
459	4	2	6 27 68	7	1	74	91	0	1	15	8	0	126	8	13	113	0
460	4	3	7 29 68	30	2	75	124	0	0	15	10	2	132	3	13	17	0

Table 46.1 *(cont.)*

						Pretreatment Covariates											
Pat no.	Stage	Rx	Date on-study	Mos FU	Surv stat	Age, yrs	Wt	PF	HX	SBP	DBP	EKG	HG	SZ	SG	AP	BM
461	4	2	9 11 68	40	2	77	90	0	1	10	6	3	147	2	12	82	0
462	4	1	12 19 68	13	2	75	74	0	1	15	9	0	158	4	11	11	0
463	4	2	2 14 69	22	1	64	110	0	0	16	10	2	93	6	11	63	1
464	4	3	2 14 69	000	2	79	89	0	0	17	10	3	103	19	13	20	1
465	4	1	3 21 69	18	1	60	134	2	1	16	8	5	141	9	13	12	0
466	3	2	9 7 67	64	9	70	107	0	0	14	9	4	141	8	9	2	0
467	3	4	10 27 67	13	2	73	111	0	1	15	8	4	170	4	9	7	0
468	3	1	12 8 67	26	1	61	109	0	0	14	9	2	143	3	11	4	0
469	3	4	1 22 68	1	4	79	97	1	1	16	8	4	136	3	9	6	0
470	3	1	3 4 68	16	2	78	88	0	0	12	7	-	114	4	9	4	0
471	3	3	10 3 68	38	7	66	-	-	-	-	-	-	-	1	9	11	9
472	3	2	12 11 68	33	2	72	106	0	1	13	8	5	165	17	11	6	0
473	3	3	2 18 69	14	6	76	-	-	-	-	-	-	-	29	9	12	9
474	3	1	2 18 69	16	2	76	90	0	0	13	8	2	120	5	-	11	0
475	3	1	3 20 69	53	0	76	-	-	-	-	-	-	-	-	-	6	9
476	4	2	5 22 67	22	1	51	100	0	0	16	9	0	146	5	13	6	0
477	4	2	6 6 67	30	1	59	87	0	1	15	9	0	151	32	13	30	0
478	4	4	7 24 67	53	3	73	91	0	1	16	9	2	146	4	10	32	0
479	4	2	9 7 67	53	1	74	90	0	1	13	7	2	118	5	13	13	0
480	4	1	9 7 67	7	1	77	93	2	1	14	9	5	118	43	11	430	1
481	4	2	10 12 67	61	1	74	95	0	0	15	7	2	135	5	11	244	0
482	4	1	10 23 67	40	1	60	125	0	0	11	7	0	140	9	11	13	0
483	4	1	1 22 68	9	2	72	83	0	1	14	8	2	129	18	13	14	1
484	4	3	3 21 68	14	1	79	88	0	1	12	7	5	86	20	-	14	0
485	4	3	9 4 68	59	0	50	117	0	0	12	8	0	153	-	13	103	0
486	4	3	9 27 68	59	0	73	71	0	0	16	8	0	134	3	11	7	1
487	4	4	10 25 68	7	2	72	110	0	0	14	6	0	137	4	11	21	0
488	4	2	12 11 68	56	0	60	-	-	-	-	-	-	-	-	11	140	9
489	4	4	5 28 69	12	1	80	105	2	1	14	7	5	85	6	5	1600	1
490	4	4	6 27 69	46	3	69	71	1	1	11	8	5	137	5	13	14	1
491	4	3	6 27 69	50	0	75	116	0	0	14	8	0	160	1	11	45	0
492	4	4	6 27 69	000	4	71	99	1	1	14	8	3	141	1	11	17	0
493	3	4	4 10 69	52	0	73	114	0	1	14	8	4	136	8	8	5	0
494	3	3	4 30 69	19	5	76	99	0	1	15	9	3	160	7	9	2	0
495	3	1	4 30 69	52	0	73	114	0	0	15	7	0	117	5	9	3	0
496	3	3	6 2 69	51	0	76	114	0	1	30	18	4	116	6	9	5	0
497	3	4	6 2 69	51	0	68	98	0	0	14	8	0	134	6	9	7	0
498	4	2	4 18 69	14	1	64	100	1	0	14	7	0	140	23	11	6	0
499	4	2	5 23 69	27	1	75	102	0	0	16	8	0	142	5	11	296	0
500	4	3	5 22 69	51	0	60	123	0	1	14	10	4	145	6	8	5	1
501	4	4	5 26 69	5	2	73	100	0	1	19	10	0	168	2	11	29	0
502	3	3	1 10 69	000	5	78	108	0	0	11	6	0	158	9	13	6	0
503	3	2	3 3 69	41	8	78	127	0	1	16	10	4	158	5	9	5	0
504	3	3	3 11 69	53	0	77	93	0	0	17	10	3	160	11	9	8	0
505	4	4	1 24 69	19	1	82	96	0	1	12	6	4	124	32	13	84	0
506	4	3	5 21 69	4	1	74	97	1	0	15	6	0	212	33	-	222	1

*

-	denotes unknown value
R_x:Treatment:	1, placebo 2, 0.2 mg. estrogen 3, 1.0 mg. estrogen 4, 5.0 mg. estrogen
Date on-study:	Date of randomization given as month-day-year
Mos FU:	Complete months of follow-up
Survival Status:	Survival status at last visit: 0, alive 1, dead from prostatic cancer 2, dead from heart or vascular disease 3, dead from cerebrovascular accident 4, dead from pulmonary embolus 5, dead from other cancer 6, dead from respiratory disease 7, dead from other specific non-cancer cause 8, dead from unspecified non-cancer cause 9, dead from unknown cause
Age:	89 denotes 89 years or more - denotes age unknown
Wt:	Weight index: Weight in kg. - Height in cm. + 200
PF:	Performance Rating: 0, normal activity 1, in bed less than 50% of daytime 2, in bed more than 50% of daytime 3, confined to bed
HX:	History of cardiovascular disease: 0, no; 1, yes
SBP:	Systolic blood pressure: only hundreds and tens digits are recorded, eg. 95 coded as 9, 122 coded as 12
DBP:	Diastolic blood pressure: coding same as SBP
EKG:	Electrocardiogram code: 0, normal 1, benign 2, rhythmic disturbances and electrolyte changes 3, heart blocks or conduction defects 4, heart strain 5, old myocardial infarct 6, recent myocardial infarct
HG:	Serum haemogoblin in g/100 ml.: coded as a percentage with one implied decimal place. eg., 13.8 coded as 138
SZ:	Size of primary tumour estimated in cm^2 from rectal examination: 00, no palpable tumour
SG:	Combined index of tumour stage and histologic grade
AP:	Serum prostatic acid phosphatase in King-Armstrong units: Coded with one implied decimal place
BM:	Bone metastases: 0, no; 1, yes

47. Visual Pattern Recognition

Source Foster, D.H. (1978). Visual comparison of random-dot patterns: evidence concerning a fixed visual association between features and feature-relations. *Quarterly J. Exp. Psychol.* **30**, 637-654.

Contributor D.H. Foster University of Keele

One of the major problems in the understanding of visual perception concerns how spatially structured stimuli are visually encoded and processed by the nervous system. Over the last decade two main theories of form perception and pattern recognition have emerged. The one supposes that spatially structured stimuli are represented internally by the visual system as unstructured, approximately point-for-point "images" of the stimulus. Judgements on the similarity of stimuli are made by subjecting these internal pattern representations to families of internal transformations that serve to bring the representations as closely as possible into coincidence. The amount of *overlap* between the transformed representations determines the judged similarity of the patterns. The other theory supposes that spatially structured stimuli are internally represented in an essentially symbolic fashion, listing local features of the stimulus pattern, for example, *blobs, edges* and *corners*, and the spatial relations that exist between these local features, for example, *left of, joined to* and *far from* . Judgements on the similarity of stimuli are achieved by an internal matching of their symbolic descriptions; see, for example, Barlow, Narasimhan and Rosenfeld (1972), Sutherland (1973) and Kahn and Foster (1981).

From recent experimental work, it appears that both theories may in some sense be correct, and merely hold at different levels of visual processing in different judgemental operations, although the second, feature-relation theory, is applicable in a greater variety of visual tasks. Experimental evidence is available suggesting the identity of some of the components making up these putative feature-relation representations, but remarkably little is known about how local features and their spatial relations may be tied together in an internal representation. The sets of binary data presented here were produced in an experimental study designed to resolve this problem.

The principle of the experiment was this. Subjects made *same-different* judgements on pairs of patterns according to one of two criteria. The first criterion was designed to involve the use of only local features of the hypothesized internal representation; the second criterion, which provided a control, was designed to involve the use of both local features and spatial relations. In each case, *same* responses were measured as a function of variations in the relational structures of the patterns. If local features of separate patterns could be compared without their relations, then, for patterns with the same local features, there should be a high level of *same* responses with the first criterion, and, in contrast to performance with the second criterion, this level should remain constant with changes in the relational structures of the patterns.

Briefly, the methods of the experiment were as follows. The stimulus patterns consisted of collections of dots distributed randomly in the plane as illustrated in Fig. 47.1. There is evidence suggesting that small groupings of closely spaced dots would have constituted the local features of the internal representations of the patterns. It was not necessary, however, that the local features actually selected by the visual system were known, merely that under certain conditions the relations between the local features could be changed. These changes were achieved by rotating the patterns in the plane. As the angle of clockwise rotation increased, an "oriented" spatial relation such as *left of* would become *above*, then *right of*, and then *below*. The two criteria by which *same-different* judgements on pairs of patterns were made were equality of dot-number and equality of shape. Given the dot-number criterion, if local features could be accessed independently of spatial relations in internal pattern representations, then reliable comparisons based solely on local features, for example dot-clusters of a certain density and area, could occur even though spatial relations might alter with pattern rotation. In contrast, given the shape criterion, then whether or not local features could be accessed independently of relations, reliable comparisons would necessarily entail the use of oriented spatial relations. Visual recognition of same-shape patterns is known to depend critically on relative pattern orientation (Hake, 1966).

The experiment was performed with forty-eight subjects who were each presented sequentially with 96 pairs, $s(m)$ $(m = 1,2,\ldots,96)$ of random-dot patterns. Each pattern pair (Fig. 47.1) was classified in one of the following four ways:

(i) same shape, same dot-number;
(ii) same shape, except that one pattern had fewer dots than the other;
(iii) different shape, same dot-number;
(iv) different shape, different dot-number.

Each class contained 24 pattern pairs and in each class the orientation θ between two patterns in each pair varied over the pairs in that class: $\theta = 0°\ (15°) \ldots 345°$. For classes (iii) and (iv), the relative orientation was defined arbitrarily. Each subject was instructed to indicate by a forced choice after each stimulus presentation whether the two patterns were the same according to a fixed criterion: either

(1) equality of dot-number, or
(2) equality of shape.

The same set of 96 pattern pairs was used for each sequence. The ordering of the 96 pairs within each sequence was adjusted to form a cross-over design without reference to the classification (i) - (iv), partially balanced for residual (carry-over) effects, the same design and patterns being used for the two criterion groups. The algorithm for generating the design was as follows:

(a) A random ordering $s(k_m)$ ($m = 1,2,...,96$) of the pattern pairs $s(m)$ was constructed.
(b) For the ith trial of the jth run, the pattern pair $s(k_m)$ was presented where $m' = i.j$ (mod 97).

Only the first 24 rows of this 96 × 96 Latin square were used. Although balancing was thus not complete, it was at least true that, over all runs, any given pattern pair s appeared no more than once in position m ($m = 1,2,...,96$); and, over all runs, pattern pair s′ succeeded pattern pair s by n places ($1 \leq n < 96$) no more than once. For a discussion of the construction of cross-over designs, see, for example, Finney (1960) and Wright (1965). Different permutations of the angles θ between the patterns of each pair were used for different subjects. Each subject was tested on a 96 pattern-pair sequence in one experimental session lasting about 20 minutes. The subjects were 48 male students in the Department of Physics, Imperial College of Science and Technology. Their ages ranged from 18 to 27 years. Half of the subjects were given the dot-number criterion for pattern judgement and half the shape criterion. All were unaware of the purpose of the experiment.

The data are summarized in Tables 47.1 and 47.2 for the two criterion groups, *same* responses indicated by 1, *different* responses by 0. Data for each of the conditions (i) to (iv) are shown separately. In each table and condition, each row gives responses for a single subject, each column responses for a fixed rotation angle θ and each leading diagonal and parallel off-diagonal, including wrap-around, responses for a fixed pattern pair.

To determine whether subjects were using the judgement criteria effectively, comparisons among classes were made on data that had been collapsed over both rotation angles θ and subjects. Simple normal-theory-approximation tests for differences in proportions showed (Foster, 1978) that there was reliable discrimination under the dot-number criterion of patterns differing solely in dot-number (Table 47.1, conditions (iii) and (iv)), and, under the shape criterion, reliable discrimination of patterns differing solely in shape (Table 47.2, conditions (i) and (iii)). For further analysis, see Foster (1978).

The main result of the study (Foster, 1978) followed from an analysis of the angular dependence of the responses to patterns in class (i); specifically, for judgements of dot-number equality (Table 47.1), is there evidence of a θ-effect, and if there is such an effect is it the same as that for judgements of shape equality (Table 47.2)? Differences between subjects were not of special interest here and data in Tables 47.1 and 47.2 were each collapsed over subjects; the θ-dependence of the data was smoothed by pooling over 45° intervals. Because of the relatively small sample sizes and the wide variation in proportion of *same* responses over θ, an analysis was performed using procedures described by Cox (1970) which make use of the logistic transform $\log\{p/(1-p)\}$ of the probability p.

It was expected that if the proportion of *same* responses in the dot-number task varied with orientation θ, then it would be maximum at $\theta = 0°$, and decrease as θ moves away from zero. The existence of such an elevation relative to the mean was tested for by means of a contrast (Cox, 1970, Section 3.4) and found to be significant ($P < 0.01$). An explanatory model of the θ -

dependence of the same-response data in Table 47.2(i) is given in Foster and Mason (1979).

To determine whether the θ-dependence of *same* responses differed in the two judgement tasks, the two sets of data were compared for evidence of a non-constant *versus* constant difference over θ (Cox, 1970, Section 6.3) and for evidence of a constant non-zero *vs* constant zero difference over θ (Cox, 1970, Section 5.3). Both comparisons yielded non-significant results ($P > 0.2$).

Although this outcome suggested the use of common processes by the visual system in the two judgement tasks, the main implication of the study concerned the θ-dependence of the dot-number data alone. It was concluded that in the encoding of random-dot patterns, the visual system seems to establish an association between local features and spatial relations such that local features cannot be accessed independently of the relations. For further discussion, see Foster (1978) and Kahn and Foster (1981).

References

Barlow, H.B., Narasimhan, R. and Rosenfeld, A. (1972). Visual pattern analysis in machines and animals. *Science* **177**, 567-575.

Cox, D.R. (1970). *The Analysis of Binary Data.* London: Methuen.

Finney, D.J. (1960). *An Introduction to the Theory of Experimental Design.* University of Chicago Press.

Foster, D.H. and Mason, R.J. (1979). Transformation and relational-structure schemes for visual pattern recognition. *Biol. Cybernet.* **32**, 85-93.

Hake, H.W. (1966). Form discrimination and the invariance of form. In *Pattern Recognition,* edited by L. Uhr. New York: Wiley, pp 142-173.

Kahn, J.I. and Foster, D.H. (1981). Visual comparison of rotated and reflected random-dot patterns as a function of their positional symmetry and separation in the field. *Quart. J. Exp. Psychol.* **33A**, 155-166.

Sutherland, N.S. (1973). Object recognition. In *Handbook of Perception,* Vol. 3, edited by E.C Carterette and M.P. Friedman. New York: Academic Press, pp. 157-185.

Wright, C.E. (1965). Field plans for a systematically designed polycross. *Rec. Agric. Res.* **14**, 31-41.

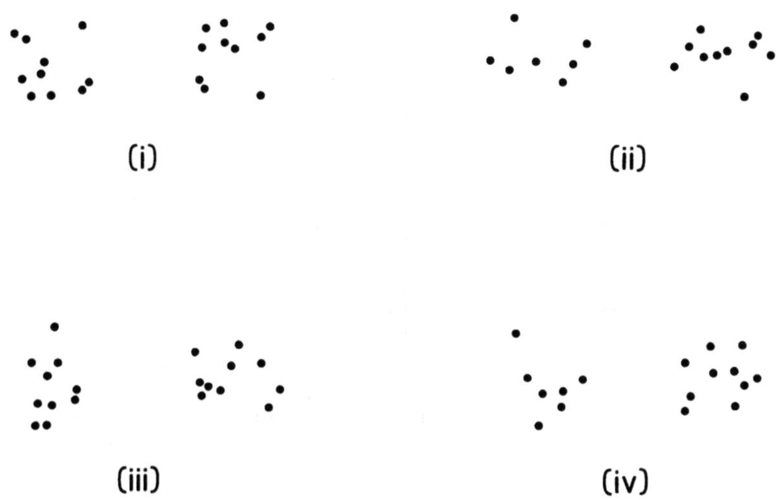

Figure 47.1
Illustrations of Typical Pattern Pairs

In (i) the patterns have the same shape and dot-number, and differ only in relative orientation, in (ii) they have part-same shape and different dot-number, in (iii) they have different shape but the same dot-number, and in (iv) they have different shape and different dot-number.

Table 47.1
Responses to Random-Dot Pattern Pairs - Equality of Dot-Number Criterion *

Condition	Subject	Angle of rotation of pairs, θ (180°(15°)0°(15°)165°)
		(i) Same shape, same dot-number pairs
i	1	1 1 1 1 0 1 1 1 1 1 1 1 1 1 1 1 0 1 1 1 1 1 1 0
i	2	1 1 1 0 1 1 1 0 1 0 1 1 1 1 1 1 1 1 1 0 1 1 0 1
i	3	1 1 1 1 0 0 1 1 1 1 1 0 1 1 1 1 0 1 1 1 0 0 1 1
i	4	0 1 0 1 0 1 0 0 1 0 0 0 1 1 1 1 1 0 1 0 0 0 0 1
i	5	1 1 1 0 1 1 1 1 0 1 1 1 1 1 1 1 1 0 0 1 1 0 0 1
i	6	0 1 1 1 1 0 1 0 1 0 0 1 1 1 1 0 1 1 1 0 0 1 1 0
i	7	0 1 0 1 0 1 0 1 1 1 1 1 1 1 0 0 1 1 1 0 0 1 0 0
i	8	1 1 1 1 0 0 0 1 1 1 0 1 1 1 1 1 1 1 0 1 1 1 1 0
i	9	1 1 1 0 0 1 1 1 1 1 1 1 1 1 1 1 1 0 1 1 1 1 1 1
i	10	1 1 1 0 1 0 1 0 1 1 1 1 1 0 1 1 1 0 1 0 1 1 1 1
i	11	0 1 1 0 1 1 1 1 1 0 0 1 1 1 1 1 0 1 0 1 1 0 0 1
i	12	0 0 1 1 1 1 1 1 0 1 1 1 1 0 1 0 1 1 1 1 1 1 0 1
i	13	1 1 1 1 0 1 1 1 1 0 1 1 1 1 1 1 0 1 1 1 1 0 1 1
i	14	1 1 1 0 1 1 0 0 0 1 0 1 1 1 1 0 0 1 1 1 1 1 0 1
i	15	0 1 1 1 0 0 0 0 1 0 1 1 0 0 0 1 0 1 0 1 1 1 0 0
i	16	0 0 0 1 1 0 1 1 0 1 0 1 1 0 0 0 0 1 0 1 0 0 1 0
i	17	0 1 1 1 0 1 0 0 1 0 0 1 1 0 1 1 1 1 0 0 1 1 0 0
i	18	1 1 1 0 1 1 0 1 0 1 0 1 1 1 1 1 0 1 1 1 0 0 0 1
i	19	1 1 1 0 1 0 0 0 1 0 0 1 1 1 0 1 0 1 1 1 1 1 0 1
i	20	1 0 1 1 0 1 0 0 0 0 1 1 1 1 1 0 0 1 1 1 1 0 0 1
i	21	1 1 0 1 1 1 0 1 0 1 0 1 1 1 0 1 0 0 1 1 1 1 1 0
i	22	0 1 0 1 1 1 1 0 1 1 1 1 1 1 1 0 1 0 1 0 0 0 0 1
i	23	0 1 0 1 1 0 0 0 0 0 1 1 1 1 1 0 0 1 0 1 0 1 0 1
i	24	1 1 1 1 1 1 1 1 1 1 1 1 1 1 1 1 0 1 1 1 1 1 1 1
		(ii) Same shape, different dot-number pairs
ii	1	1 0 0 1 0 1 0 1 1 1 1 1 0 1 1 1 1 0 1 1 0 0 1 1
ii	2	1 0 0 0 1 0 1 0 0 1 0 1 0 0 1 0 1 1 0 1 0 0 0 0
ii	3	0 1 1 1 0 0 1 1 1 0 1 0 0 1 0 1 0 1 0 0 1 0 0 0
ii	4	0 0 0 0 0 0 0 1 1 0 1 0 1 0 1 1 0 0 1 1 0 1 1 0
ii	5	1 0 0 0 0 1 0 1 1 1 1 1 1 0 0 0 1 0 1 0 1 0 0 1
ii	6	0 0 0 1 1 0 1 0 0 1 1 0 0 1 1 0 1 0 0 1 0 1 0 1
ii	7	0 1 0 0 1 1 1 1 0 1 1 0 0 1 0 0 1 0 0 0 0 0 0 0
ii	8	0 0 1 1 0 1 0 0 1 0 0 0 1 0 1 1 0 0 0 0 0 0 1 1
ii	9	1 0 0 1 0 0 0 1 0 0 0 1 1 1 0 1 0 0 1 0 0 0 1 0
ii	10	0 1 0 1 1 1 0 0 1 0 1 0 1 1 1 0 1 0 1 0 0 0 1 0
ii	11	1 1 1 0 1 1 0 0 1 1 0 1 0 1 0 1 0 0 1 0 1 1 0 0
ii	12	0 1 0 1 0 1 1 0 0 0 0 1 0 0 1 1 1 0 1 0 0 1 0 1
ii	13	0 1 1 1 1 0 1 1 0 1 1 1 0 0 0 1 1 1 1 0 0 0 1 0
ii	14	0 0 0 1 0 0 1 1 0 0 1 1 0 0 0 1 1 1 1 1 1 0 0 1
ii	15	0 0 0 1 1 1 0 0 1 0 0 0 1 0 0 1 0 0 0 1 1 1 1 0
ii	16	1 1 1 0 0 1 1 1 1 1 1 0 0 1 0 0 1 0 0 0 0 0 1 0
ii	17	1 1 0 1 0 0 1 1 1 0 0 1 1 1 0 0 1 0 0 0 1 0 1 1
ii	18	1 0 0 1 0 0 0 0 1 1 1 1 0 0 0 1 0 0 0 0 1 0 0 0
ii	19	0 0 0 0 0 0 0 0 0 1 1 0 1 0 0 0 0 0 0 0 1 1 1 1
ii	20	1 0 1 0 0 0 0 0 1 0 1 0 0 1 0 0 0 0 0 1 1 0 0 1
ii	21	1 0 0 1 1 0 0 0 0 1 0 1 1 1 1 0 0 0 1 0 0 1 0 1
ii	22	0 1 1 1 1 1 1 0 0 1 0 1 0 0 0 0 1 1 0 1 1 0 1 0
ii	23	0 0 0 0 0 0 0 0 0 1 0 0 0 0 1 0 0 0 0 0 0 0 1 0
ii	24	0 0 1 1 1 0 0 0 0 1 1 0 1 1 1 1 0 1 1 1 1 0 1 1

Table 47.1 (*cont.*)

Condition	Subject	Angle of rotation of pairs, θ (180°(15°)0°(15°)165°)
		(iii) Different shape, same dot-number pairs
iii	1	1 0 1 1 1 1 1 0 1 0 0 1 1 1 1 1 1 1 1 1 1 1 1 0
iii	2	0 0 1 0 0 0 1 0 0 0 1 0 1 1 1 0 0 0 1 0 0 0 0 1
iii	3	1 0 1 0 0 1 1 1 1 0 0 0 0 1 1 1 0 0 1 1 0 0 0 0
iii	4	0 1 0 0 0 1 0 0 0 1 0 0 0 0 0 0 1 0 1 1 1 0 0 0
iii	5	0 1 1 1 1 0 0 1 1 1 1 0 0 0 0 1 0 1 1 0 1 0 1 1
iii	6	0 0 1 1 0 1 0 1 1 1 0 1 0 1 1 0 1 1 0 0 0 1 1 0
iii	7	1 1 0 0 1 1 0 0 1 1 1 1 0 0 1 0 0 0 1 0 1 1 0 1
iii	8	1 1 1 0 0 0 1 1 1 0 1 1 1 1 0 0 0 0 0 1 1 0 1 1
iii	9	1 0 1 0 1 0 1 0 1 1 0 1 1 0 0 0 0 1 1 0 1 0 0 1
iii	10	1 0 1 0 1 1 0 0 1 1 0 0 0 1 1 1 0 0 0 0 1 0 0 0
iii	11	0 0 0 1 0 0 0 1 0 0 1 0 1 1 1 0 0 0 0 0 0 1 1 0
iii	12	1 0 1 1 0 1 1 1 1 0 1 1 0 1 0 1 1 1 0 1 1 0 1 1
iii	13	1 0 0 1 1 1 0 1 0 1 0 0 1 0 0 0 0 1 1 0 1 0 0 1
iii	14	0 1 1 0 0 0 0 0 0 1 1 1 0 0 0 0 1 1 0 1 0 0 0 1
iii	15	0 1 1 0 0 0 0 0 0 1 0 1 1 0 0 1 1 1 0 1 0 1 1 1
iii	16	1 0 0 0 1 1 0 1 0 1 0 1 1 1 1 0 0 0 1 0 0 1 0 1
iii	17	0 1 1 1 1 0 1 0 0 1 0 1 0 0 0 0 0 1 0 0 0 0 1 0
iii	18	0 0 0 0 0 1 1 1 1 0 0 0 1 0 0 1 1 0 1 0 0 1 1 1
iii	19	1 1 1 1 0 1 0 1 1 1 1 0 0 1 1 1 1 1 1 0 0 1 1 1
iii	20	0 1 0 1 0 1 0 1 1 0 0 0 1 0 0 1 1 0 1 0 1 0 1 0
iii	21	0 0 0 0 1 0 1 1 1 0 0 0 0 1 1 0 0 1 0 1 1 0 0 1
iii	22	0 1 1 1 0 1 0 0 0 0 0 0 0 0 0 1 0 1 0 1 1 0 1
iii	23	0 1 1 0 0 0 1 0 0 0 1 0 0 0 0 1 0 0 1 0 0 1 0 0
iii	24	0 1 1 1 1 1 0 1 1 0 1 1 1 1 1 0 1 1 0 0 1 1 1 1
		(iv) Different shape, different dot-number pairs
iv	1	1 0 1 0 1 0 1 1 0 1 0 1 1 1 1 1 0 1 1 1 1 0 1 1
iv	2	0 1 0 0 0 1 0 1 0 1 0 0 0 0 0 0 0 1 0 0 1 0 0 0
iv	3	1 0 0 0 0 0 1 1 1 1 1 0 0 1 0 0 0 0 0 0 0 1 0 0
iv	4	0 0 1 0 0 0 1 0 1 0 1 0 0 0 1 0 0 0 0 0 0 0 0 1
iv	5	1 0 0 1 1 1 0 0 1 1 1 1 0 1 0 1 0 0 0 1 0 0 0 0
iv	6	1 0 0 0 0 0 1 1 0 0 0 1 0 0 0 1 0 0 0 0 0 0 0 1
iv	7	0 0 1 0 0 0 0 0 0 0 1 0 0 1 0 0 1 0 0 0 0 0 1 0
iv	8	0 0 0 0 1 1 1 0 0 1 1 0 1 1 1 0 1 0 0 1 0 0 1 1
iv	9	0 0 0 0 0 1 0 1 0 0 0 0 1 1 1 1 0 0 1 0 0 0 0 0
iv	10	1 0 0 0 0 1 0 0 1 0 0 0 0 1 1 0 0 0 0 0 1 0 0 0
iv	11	1 0 0 0 0 1 1 0 0 0 0 1 0 1 0 0 1 1 1 0 1 1 0 0
iv	12	0 0 0 0 0 0 1 1 1 1 0 1 0 0 0 1 1 0 0 0 1 0 1 0
iv	13	1 0 0 0 0 0 1 0 1 0 1 0 0 1 1 1 1 0 1 1 1 0 0 0
iv	14	1 0 0 0 0 0 0 0 0 1 0 1 0 0 1 0 0 1 0 1 0 0 0 0
iv	15	0 0 0 1 0 0 0 0 0 0 0 0 0 0 0 0 0 0 1 1 1 0 0 1
iv	16	1 0 1 1 0 0 0 0 1 0 0 0 1 0 1 0 0 0 1 0 1 1 1 0
iv	17	0 0 0 0 0 1 0 1 1 0 1 1 0 0 0 0 0 0 1 0 0 0 0 1
iv	18	1 0 0 0 0 1 0 0 0 0 0 1 1 0 0 0 0 0 0 0 0 0 0 1
iv	19	0 0 0 1 1 0 0 0 0 1 0 1 1 0 0 1 0 1 0 0 0 0 1 0
iv	20	0 0 0 0 0 1 0 0 0 1 0 0 0 1 1 1 1 0 1 1 0 0 1 1
iv	21	0 1 0 1 1 0 0 1 0 0 1 1 0 0 0 1 1 0 1 1 0 1 0 1
iv	22	0 0 0 0 1 0 0 0 0 0 0 0 0 0 0 0 1 1 0 1 1 0 0 0
iv	23	0 0 0 0 0 0 0 0 0 0 0 0 0 0 0 0 1 1 0 0 0 0 0 0
iv	24	0 1 0 1 1 1 1 0 0 0 0 0 0 1 1 0 0 1 1 1 1 0 1 1

* 1, same; 0, different

Table 47.2
Responses to Random-Dot Pattern Pairs -
Equality of Shape Criterion *

Condition	Subject	Angle of rotation of pairs, θ (180°(15°)0°(15°)165°)
		(i) Same shape, same dot-number pairs
i	25	1 0 0 1 0 1 1 0 1 0 1 1 1 1 1 1 1 0 0 1 1 1 1 1
i	26	0 0 1 0 1 0 0 1 1 1 1 1 1 1 0 1 1 1 1 1 1 0 1 1
i	27	1 1 0 1 1 0 0 0 1 1 1 0 1 1 0 1 1 0 1 1 1 1 1 0
i	28	1 1 0 1 0 1 0 0 1 0 1 0 1 0 0 1 0 0 1 1 0 0 1 1
i	29	1 1 1 1 1 1 1 0 1 1 1 1 1 0 0 1 0 0 1 1 1 1 0 1
i	30	1 1 1 1 1 1 1 1 0 1 1 0 1 1 1 1 1 1 1 0 1 0 1 0
i	31	1 1 1 1 1 1 1 1 1 0 1 1 0 1 1 0 1 1 1 1 1 1 1 1
i	32	0 1 1 1 1 1 1 1 1 1 1 0 1 0 1 0 1 1 1 1 1 1 1 0
i	33	1 0 0 1 0 0 1 1 1 1 1 0 0 0 1 1 0 0 0 1 0 1 1 1
i	34	1 1 1 1 1 1 0 0 1 1 1 1 1 1 1 1 0 0 1 1 0 1 1 1
i	35	1 1 0 0 1 1 0 1 1 1 1 1 1 1 1 1 1 0 0 1 0 1 1 1
i	36	1 1 0 1 1 0 1 1 0 1 1 1 1 1 1 1 1 1 1 0 0 0 1 1
i	37	1 0 0 0 1 1 1 1 1 1 1 0 1 1 1 1 0 1 1 0 1 0 1 1
i	38	1 1 1 1 1 0 0 1 0 1 0 1 1 1 1 1 1 0 1 0 1 0 1 1
i	39	1 1 1 0 0 1 0 1 1 1 1 1 1 1 1 1 0 1 1 1 1 0 1 1
i	40	1 1 0 1 1 1 0 1 1 0 1 0 1 1 1 1 0 1 0 0 0 1 1 1
i	41	0 1 1 1 0 0 0 1 1 1 1 1 1 1 1 1 1 1 1 0 1 1 1 1
i	42	1 1 1 1 0 1 0 0 0 0 0 1 1 1 1 0 1 0 1 1 0 1 1 1
i	43	1 0 0 1 0 1 0 1 0 1 1 1 1 1 1 1 1 1 1 0 1 1 1 1
i	44	1 0 1 0 0 0 1 0 0 0 0 1 0 1 0 0 1 1 0 0 0 0 0 1
i	45	1 1 0 1 1 1 0 0 0 1 1 1 1 1 0 0 0 1 0 1 0 1 1 1
i	46	1 1 0 1 1 1 1 0 1 0 1 1 1 1 1 1 1 1 1 1 0 1 1 1
i	47	1 0 1 1 1 0 0 1 0 0 0 1 0 1 0 0 0 1 1 1 1 1 0 1
i	48	1 1 1 1 1 1 1 1 0 1 1 1 1 1 1 1 1 0 1 1 1 0 1 0
		(ii) Same shape, different dot-number pairs
ii	25	1 0 0 0 1 1 0 0 1 0 0 1 0 0 0 1 0 0 1 1 0 1 1 0
ii	26	0 0 0 0 1 1 0 0 0 1 0 1 1 0 1 0 0 0 0 1 1 0 1 0
ii	27	1 1 0 0 0 1 1 0 0 0 0 1 1 0 0 1 0 0 1 1 1 1 0 1
ii	28	1 0 1 1 0 0 1 0 1 0 1 1 0 0 1 1 0 0 1 1 0 0 1 0
ii	29	1 0 1 1 0 0 0 1 0 1 0 0 1 0 0 1 0 1 1 1 1 0 0 1
ii	30	1 0 1 0 0 0 0 0 1 0 1 0 0 1 1 0 1 0 0 1 1 1 0 0
ii	31	1 1 0 0 0 0 0 0 0 1 1 0 1 1 0 1 0 0 0 1 1 0 1 0
ii	32	0 1 1 0 1 1 1 0 0 0 0 0 1 0 1 0 0 1 1 0 0 1 1 1
ii	33	1 0 0 0 0 1 1 1 1 0 0 0 1 1 1 0 1 0 1 0 0 0 1 1
ii	34	1 1 0 0 1 1 1 1 1 1 0 0 1 1 0 0 1 1 0 0 0 1 0 1
ii	35	1 0 1 0 0 1 1 1 0 0 0 0 0 1 1 0 0 1 1 0 0 0 0 0
ii	36	1 1 1 1 1 1 1 0 1 0 1 0 1 1 0 1 1 0 0 1 1 0 1 1
ii	37	0 1 1 1 1 1 1 1 1 0 1 1 1 1 0 0 1 1 1 0 1 1 0 1
ii	38	1 0 1 0 1 0 1 1 1 0 1 1 1 1 1 0 0 0 1 0 0 0 0 1
ii	39	0 0 0 0 1 1 1 1 1 1 0 0 1 1 0 1 0 0 1 1 0 0 0 0
ii	40	1 1 1 0 1 1 0 1 1 1 1 1 1 1 1 1 0 1 1 1 1 0 1 1
ii	41	1 0 1 0 0 1 1 1 1 0 0 1 0 1 0 1 0 1 0 1 1 1 0 1
ii	42	1 1 1 0 1 0 0 1 0 1 0 1 1 0 1 0 0 0 0 0 1 0 1 0
ii	43	0 0 1 1 1 1 0 0 1 1 0 1 1 0 0 1 0 0 1 1 0 0 1 0
ii	44	0 0 0 0 0 0 0 0 0 1 0 0 0 0 0 0 1 1 0 0 0 0 0 1
ii	45	0 1 0 1 1 1 0 1 0 1 0 1 1 1 1 1 1 1 0 0 0 1 0 0
ii	46	1 1 1 0 1 0 0 1 0 0 1 1 1 1 0 0 0 0 0 0 1 0 1 0
ii	47	0 1 0 0 0 1 0 1 0 0 0 0 0 0 1 0 1 0 1 1 1 1 0 1
ii	48	0 0 1 0 1 0 1 0 0 1 1 0 1 0 1 0 1 0 1 0 1 0 1 1

Table 47.2 (*cont.*)

Condition	Subject	Angle of rotation of pairs, θ (180°(15°)0°(15°)165°)
		(iii) Different shape, same dot-number pairs
iii	25	0 0 0 1 0 0 1 0 1 1 0 0 0 0 0 0 0 1 0 0 0 0 1 1
iii	26	1 0 0 0 1 1 0 0 0 1 0 0 1 0 0 0 0 1 0 0 0 0 0 1
iii	27	1 1 1 0 0 1 0 0 0 0 0 0 0 1 0 0 1 1 1 0 0 0 1 0
iii	28	0 1 0 0 0 0 0 0 0 0 0 0 0 0 0 0 0 0 1 1 1 0 0 0
iii	29	0 1 0 0 1 0 0 1 1 1 0 0 0 0 0 0 0 0 1 0 1 1 0 1
iii	30	1 0 0 1 1 0 0 1 0 1 1 1 0 0 0 0 1 1 0 1 1 0 1 0
iii	31	0 0 0 0 0 1 0 0 0 1 0 0 1 0 1 0 0 1 0 0 0 1 1 1
iii	32	1 0 0 0 0 0 1 0 0 0 1 0 1 1 0 0 0 0 0 0 0 0 0 0
iii	33	1 1 0 1 0 0 1 0 0 0 0 1 1 0 0 0 0 1 0 1 0 0 0 0
iii	34	0 1 0 0 0 0 0 1 0 0 0 0 0 1 0 1 0 0 1 0 0 1 0 1
iii	35	1 1 0 0 0 1 0 1 0 0 1 0 1 0 0 0 0 0 0 0 1 0 0 0
iii	36	0 1 1 0 1 1 0 0 0 0 1 0 0 1 1 0 1 0 0 1 1 1 0 1
iii	37	1 1 0 0 1 1 0 1 0 1 1 0 1 0 1 0 0 1 1 0 0 0 0 0
iii	38	0 0 1 0 1 0 0 0 0 0 0 0 1 0 0 0 1 1 0 1 0 0 1 0
iii	39	0 0 0 1 0 0 1 0 0 0 0 0 1 1 0 0 1 1 0 0 0 0 0 1
iii	40	1 0 0 0 1 0 0 1 1 1 0 1 0 0 1 0 0 0 0 0 0 0 0 1
iii	41	0 0 0 1 0 0 1 1 1 1 1 0 0 0 1 1 1 0 0 0 0 1 0 0
iii	42	0 0 0 0 1 0 1 1 0 0 0 0 0 0 0 1 0 1 0 0 1 0 0 1
iii	43	1 0 0 0 0 0 0 0 0 1 0 0 0 1 1 1 0 1 0 0 0 0 0 0
iii	44	0 1 0 0 0 0 0 0 0 0 0 0 0 0 0 0 0 0 0 0 0 0 0 0
iii	45	1 1 1 0 0 0 0 0 0 1 0 0 0 0 1 0 0 0 0 0 1 0 1 1
iii	46	1 0 1 1 0 0 0 0 0 0 0 1 0 0 0 0 0 0 0 1 0 0 0 0
iii	47	0 0 0 0 1 0 0 0 0 0 0 0 0 0 0 0 1 1 0 0 1 0 0 0
iii	48	0 0 1 1 0 1 0 1 1 0 0 0 0 1 1 0 0 0 0 0 1 0 0 0
		(iv) Different shape, different dot-number pairs
iv	25	0 0 1 0 1 0 1 0 0 0 0 0 0 0 0 0 0 0 0 1 1 0 0 0
iv	26	0 0 0 1 0 0 1 0 0 1 0 0 1 0 0 0 0 0 0 1 0 0 1 0
iv	27	0 1 0 0 1 1 0 0 0 1 0 0 0 1 0 0 1 0 0 0 1 0 0 0
iv	28	0 0 0 0 0 0 0 1 1 0 0 1 0 0 0 1 0 1 0 0 1 0 1 0
iv	29	0 0 0 1 1 1 0 1 0 0 0 0 1 1 0 0 0 0 0 1 0 0 0 1
iv	30	1 1 1 0 1 0 0 0 0 1 0 1 1 1 0 1 0 0 0 1 0 0 1 0
iv	31	1 1 0 0 1 0 0 0 0 1 1 0 1 0 1 1 0 0 0 0 1 1 1 0
iv	32	0 0 1 0 0 0 1 0 0 0 0 0 0 0 1 0 0 0 0 0 0 1 1 0
iv	33	0 0 0 0 0 0 0 1 0 0 0 0 0 0 1 0 0 0 0 0 0 0 0 1
iv	34	0 0 0 0 0 0 0 0 0 0 1 1 0 1 0 1 1 0 0 0 0 1 0 0
iv	35	1 0 0 1 0 1 1 0 0 1 0 1 1 1 1 1 0 1 1 1 1 1 1 0
iv	36	1 0 0 1 0 1 0 0 0 0 0 0 1 1 1 1 0 1 1 0 1 0 1 0
iv	37	1 0 0 0 0 1 1 0 1 0 0 0 0 1 0 0 1 1 1 0 1 0 0 1
iv	38	1 0 0 0 1 0 1 0 0 0 0 0 0 0 0 0 0 1 0 0 0 0 1 0
iv	39	0 0 0 1 0 0 0 0 1 1 0 0 0 0 0 0 1 0 0 0 0 1 0 0
iv	40	0 0 0 1 0 0 0 1 0 1 1 1 1 0 1 0 0 1 1 1 0 1 0 1
iv	41	1 0 0 0 1 0 0 0 0 0 0 0 1 0 0 0 0 0 0 0 0 0 1 0
iv	42	1 1 0 0 1 1 0 1 0 0 0 1 1 0 0 0 1 0 1 0 0 1 0 0
iv	43	1 1 1 0 0 1 0 0 0 1 0 1 1 1 0 0 0 0 0 0 0 1 1 0
iv	44	0 0 0 0 0 0 0 0 0 1 0 0 0 0 0 0 0 0 0 0 0 0 1 0
iv	45	0 0 0 1 1 1 0 1 0 0 0 0 0 1 1 0 0 0 0 0 0 1 0 1
iv	46	1 0 0 0 0 1 0 0 0 0 0 1 0 0 0 0 0 0 0 0 1 0 0 0
iv	47	1 0 0 0 0 1 0 0 0 0 0 0 0 1 0 0 0 0 0 0 0 1 0 0
iv	48	0 0 0 0 0 1 1 0 0 0 0 0 0 0 0 0 0 0 0 0 0 0 1 0

* 1, same; 0, different

48. Circadian Rhythms

Source and Contributor M.D. Godfrey ICL

In a study of the effects of environment on two species of mammals, a large body of data was recorded at Princeton University in the late 1960's under the direction of Dr. C.S. Pittendrich. The data consist of temperature and activity recordings taken at either two or five minute intervals for up to 150 days. The animals were isolated in cages and were subjected to a variety of controlled photoperiods.

Tables 48.1 to 48.4 present body temperature recordings. The data for the four example channels comprise on the order of 100,000 numbers. Tables 48.1 to 48.4 each present only the first few observations, the rest being available in machine-readable form. The data are arranged so that time increases by columns and then by rows. The first number is a count of observations in the following block.

Four experiments were successfully completed during a period of several years. The experiments consisted of equipment to isolate each animal from its external environment as completely as possible, a telemetry system to transmit animal body temperature and an activity sensor in the floor of each cage. The telemetry system consisted of a small temperature transmitter implanted in each animal and a receiver and data logging system.

Three of the experiments ran for about three months each. The fourth ran for about five months. Each experiment used 18 cages with 17 containing animals and the eighteenth containing a temperature transmitter in an empty cage in order to monitor the environment.

Each cage contained a light source so that the photoperiod for each animal could be varied. A number of different photoperiods were applied as a function of animal and time. Generally, an animal was given a 24 hour photoperiod for an adjustment period and then given an extended period of total darkness (these were nocturnal animals), total light, or a photoperiod different from 24 hours.

Of course, various problems arose during the experiments:

1. Transient equipment failure. Data were time-stamped so that appropriate missing values could be filled in.

2. Permanent failure of individual channels, including death of animals.

Typically, about half of the channels survived to the end of the experiment.

For the temperature data, since the interest was periodicity and since the active temperature is substantially different from the rest temperature, no effort was made to precisely relate the telemeter frequency to actual temperature. The recordings are sampled telemeter frequency, counts, with higher counts indicating higher temperature.

The data logging equipment was imperfect, as it was based on a conversion to mechanical keypunch operation. Readings were punched in fields on standard 80 column cards. Some card punch problems, jams, misfeeds, were corrected during experiment operation. However, it was only after the experiments were completed and data analysis was under way that other malfunction syndromes became evident.

1. Simple mispunch (non-numeric or obviously out-of-range).
2. Column shift. From some column, all data shifted left by some number of columns.

During data analysis, "obvious" errors were corrected in order to retain as much information as possible. It is this form in which the data are presented. In subsequent spectrum and demodulation analysis trimming, drop-out interpolation and other techniques were applied.

Table 48.1
Activity Recordings for *Perognathus formosus*, frequency 2 min. under conditions: 8 days of 12 hours of light, 12 hours of dark 68 days constant darkness

465															
370	365	365	364	363	368	370	363	363	366	366	365	368	373	376	382
383	385	390	391	392	394	395	396	399	399	398	395	391	389	392	389
391	391	385	385	383	382	377	374	372	371	372	369	370	369	371	370
374	375	372	376	381	384	386	390	391	394	395	397	395	397	394	389
383	382	383	386	387	386	385	388	381	377	388	393	392	388	391	394
396	399	405	407	409	410	413	414	415	415	413	416	415	414	414	411
410	411	409	410	408	407	406	404	398	392	392	388	385	382	380	381
384	391	385	385	376	374	367	367	369	365	368	370	371	373	373	376
374	376	371	363	365	366	365	355	357	365	368	372	364	363	362	358
362	366	363	365	355	353	351	353	357	354	359	362	362	355	353	358
358	352	364	364	367	360	361	362	356	354	357	354	355	352	355	363
359	363	364	363	366	368	373	371	373	373	373	374	369	367	369	367
367	371	366	366	365	364	374	371	380	367	364	368	370	367	366	371
364	359	358	363	367	367	368	363	361	358	360	361	360	359	361	361
362	361	361	362	368	368	368	362	370	366	362	361	361	360	359	354
355	359	356	356	367	369	368	365	361	362	364	364	362	361	360	358
355	350	349	347	352	352	348	341	341	348	351	352	347	353	352	348
346	344	341	340	341	341	342	346	350	349	349	349	347	352	354	347
342	342	343	342	342	346	346	346	346	343	351	361	355	367	369	369
367	365	362	359	350	355	359	360	361	364	368	370	369	372	370	373
375	381	381	378	374	371	365	362	361	357	355	351	348	344	342	338
338	342	344	344	345	349	354	355	354	354	354	353	348	347	347	350
350	348	348	349	351	349	344	346	350	352	352	353	358	360	361	363
360	359	-1	357	353	350	350	349	347	350	350	352	350	350	349	347
345	342	342	345	344	342	339	342	348	354	357	356	354	347	347	345
344	341	342	345	346	349	352	373	387	393	399	400	401	396	388	381
375	366	362	358	355	350	347	344	341	337	336	335	335	336	334	334
334	333	335	338	337	341	341	342	346	342	342	340	345	348	352	351
350	349	346	346	346	346	348	346	345	344	346	349	350	352	354	352
370															

Table 48.2
Activity Recordings for *Peromyscus bardii*, frequency 5 min. under conditions:
20 days 12 hour light, 12 hours dark
10 days constant dark
12 hours light
26 days constant dark

465															
1	0	1	0	2	0	0	0	0	0	0	2	0	0	0	0
0	0	0	24	0	0	0	0	0	2	0	0	0	1	0	1
0	0	0	0	0	0	0	0	0	0	0	0	0	0	1	1
1	2	0	0	0	0	0	0	0	0	0	0	1	0	0	0
0	0	0	0	0	0	0	3	0	0	0	0	0	0	1	0
0	1	0	1	0	0	0	0	1	0	3	1	0	0	0	2
0	0	0	0	1	2	4	8	0	1	1	3	2	2	0	0
0	0	0	0	0	1	0	0	2	2	0	0	2	1	0	4
2	0	0	1	4	0	2	2	2	2	0	22	0	8	0	1
5	4	5	4	7	0	0	2	0	7	7	3	1	5	5	5
3	7	5	0	2	6	4	0	9	5	0	0	7	0	0	6
0	3	5	4	0	9	11	4	7	9	0	0	7	3	1	0
6	0	4	5	8	8	0	14	6	4	1	5	11	8	6	7
9	5	4	0	10	0	0	0	4	5	6	7	0	8	1	5
5	0	4	0	8	0	0	0	1	3	3	0	0	0	0	1
0	1	0	0	3	0	0	3	4	0	4	0	1	0	0	4
4	20	3	0	0	0	0	0	0	2	0	0	0	0	0	0
0	0	1	6	0	0	0	3	4	0	0	0	0	0	6	2
6	0	3	1	0	3	0	0	0	0	0	0	0	0	0	0
0	0	0	0	0	0	0	0	0	1	1	0	0	0	0	0
0	0	0	0	1	0	1	0	0	0	0	0	0	0	2	0
0	0	0	0	0	1	0	0	0	0	0	0	0	0	0	1
0	0	1	0	0	0	0	0	0	0	0	0	0	0	0	1
0	0	0	0	0	0	0	4	0	0	0	0	0	0	0	3
9	0	0	0	5	1	0	0	0	7	0	0	0	0	0	0
5	4	2	0	3	32	1	0	27	1	0	3	1	0	0	0
0	3	1	0	4	0	0	0	0	5	3	0	0	29	8	7
0	5	13	0	6	13	4	8	8	0	7	0	12	4	14	9
0	0	0	0	0	0	0	0	0	5	0	0	0	0	0	0
1															

Table 48.3
Activity Recordings for *Mesocricetus auratus*, frequency 5 min. under conditions:
18 days of 12 hours light, 12 hours dark
4 days 14 hours light, 10 hours dark
76 days constant light

465															
0	0	0	0	0	0	0	0	0	0	0	17	0	1	0	0
0	0	0	0	1	0	3	0	1	0	1	0	1	0	0	0
0	0	1	0	0	0	0	0	2	0	0	2	2	1	0	2
0	1	0	0	0	0	1	1	0	0	0	0	0	0	0	0
0	0	0	0	0	0	0	0	0	0	0	0	0	0	0	0
0	0	0	0	0	0	0	0	0	0	0	0	0	0	0	0
0	0	0	0	0	4	0	0	0	2	0	0	0	7	3	0
5	0	3	1	3	2	2	0	1	0	0	0	2	0	0	3
0	0	2	4	2	0	0	0	0	0	0	5	0	3	0	6
2	5	5	0	6	0	0	3	0	6	0	4	4	6	4	5
4	4	5	0	5	0	5	0	4	7	3	0	6	0	0	15
4	2	0	5	0	37	4	5	8	8	0	9	6	0	0	2
7	0	4	4	8	0	3	10	9	8	5	7	12	7	0	7
15	0	5	1	4	0	0	1	4	4	5	4	0	4	0	7
4	0	6	0	0	2	0	0	1	1	0	0	0	0	0	0
1	0	2	0	4	0	0	0	0	2	0	0	0	0	0	21
0	10	0	0	0	0	32	0	0	0	0	0	2	0	0	0
0	0	0	1	0	1	0	4	3	0	0	0	0	0	0	0
8	3	4	0	0	0	0	0	1	0	0	0	0	0	0	0
0	0	0	0	0	0	0	0	0	0	0	0	0	0	0	0
0	0	0	0	0	3	0	1	0	0	0	0	0	0	0	0
0	3	0	0	0	0	0	0	0	0	0	0	2	1	0	0
0	0	0	0	0	0	0	0	0	0	1	1	0	0	0	0
0	0	0	0	1	0	0	0	1	0	0	2	0	0	1	2
0	0	0	0	0	0	0	0	0	7	8	0	3	0	3	0
7	6	0	2	8	10	0	6	7	0	0	4	0	1	0	7
0	5	34	0	8	0	0	11	0	6	6	0	0	0	11	10
5	9	4	2	6	8	4	4	12	0	7	14	8	0	10	9
0	0	0	0	0	0	0	6	5	0	0	0	2	0	0	2
0															

Table 48.4
Activity Recordings for *Perognathus formosus*, frequency 5 min. under conditions: 37 days 12 hours light, 12 hours dark 48 days 12.5 hours light, 12 hours dark 15 days constant dark

465															
502	507	507	515	506	501	503	512	510	506	493	511	509	509	514	512
507	512	510	498	499	506	517	517	497	498	512	515	515	518	508	508
500	499	488	497	504	506	500	505	503	503	502	500	508	515	513	519
515	512	510	501	513	509	515	510	512	506	507	517	513	505	514	510
506	507	508	509	496	510	516	518	519	514	508	498	496	507	515	508
498	485	499	506	500	491	507	514	516	523	504	506	508	520	519	504
501	499	500	505	500	493	489	507	533	533	522	526	522	514	506	502
502	499	513	516	507	510	507	507	508	503	499	496	495	486	489	504
508	498	512	508	507	512	496	483	482	494	492	497	499	504	502	495
498	503	493	491	494	494	493	489	491	492	496	500	502	500	500	492
494	493	493	493	491	494	495	493	501	502	502	501	501	503	497	501
499	493	488	492	491	489	488	490	494	490	489	498	499	500	501	503
498	494	490	496	499	503	503	499	492	494	498	503	499	498	498	504
509	506	504	498	498	496	497	495	494	497	497	499	497	492	495	498
499	498	499	501	504	500	507	512	507	503	499	494	493	493	512	508
501	495	493	500	508	506	496	501	506	507	513	510	506	510	506	512
515	511	501	506	513	511	509	507	498	512	517	523	517	519	514	510
500	505	503	501	498	498	495	503	517	504	501	507	505	503	495	486
484	482	502	495	502	504	503	505	495	506	519	505	496	495	497	497
504	506	501	500	504	508	504	508	507	516	512	506	496	495	503	515
512	508	514	502	503	511	513	502	507	511	525	524	517	515	516	524
512	521	526	512	509	506	500	505	512	502	505	501	502	503	501	503
513	509	502	497	496	508	520	522	517	513	520	509	513	496	497	497
506	514	526	523	514	515	511	509	511	515	510	513	502	499	498	503
511	520	516	509	510	507	507	500	501	497	496	511	521	526	513	511
519	508	512	515	509	500	503	496	500	499	502	497	497	510	500	490
496	500	495	487	495	503	499	499	510	509	508	501	501	501	488	487
485	485	487	486	487	491	490	489	488	486	484	484	478	482	495	497
493	494	494	493	491	492	496	496	496	494	496	499	501	506	507	504
502															

49. Number of Species in the Galápagos Islands

Source Johnson, M.P. and Raven, P.H. (1973). Species number and endemism: The Galápagos Archipelago revisited. *Science* **179**, 893-895.

Contributors S. M. Stigler University of Chicago and H. Thurnauer Boulder, Colorado

Wiggins and Porter (1971) state:

> The Galápagos Islands are a territory of the Republic of Ecuador. And though they are officially listed by that government as El Archipiélago de Colón, most local inhabitants and many of the Ecuadorian nationals who visit the islands persist in calling them Las Islas de los Galápagos, the Islands of the Tortoises. Elsewhere in the world, of course, the name Galápagos has long been current.
>
> The archipelago lies astride the Equator roughly 600 miles from the port of Guayaquil, or about 500 miles west of Punta Arena (La Puntilla), the westernmost point of continental Ecuador. It extends from approximately 89°14′ to 92°01′ West Longitude and from 1°25′ South to 1°40′ North Latitude; its span is thus about 190 statute miles (304 km) east to west, or 213 miles (341 km) northwest to southeast, i.e. from Isla Darwin to the southeastern tip of Isla Espanola. Though the islands appear to have no close geologic affinities with the mainland of North or South America, their native fauna and flora derive in the main from the American tropics.

The identification of factors which influence the development and survival of different species of life is of considerable interest. The Galápagos Islands provide a rich source of data bearing on this problem. Hamilton, Rubinoff, Barth and Bush (1963) studied the dependence of the number of plant species on the area of each of the Galápagos Islands. Johnson and Raven (1973) suggested that curvilinear regression may be more appropriate and that this leads to different conclusions. Table 49.1 presents the number of species and related variables for the thirty islands shown in Figure 49.1 obtained from Johnson and Raven (1973) and Grant, Price and Snell (1980).

Note that one very large island has a dominating effect on most simple regressions and that data ranging over several orders of magnitude often benefit from transformations.

References

Grant, P.R, Price, T.D. and Snell, H. (1980). The exploration of Isla Daphne Minor. *Noticias de Galápagos* **31**, 22-27.

Hamilton, T.H., Rubinoff, I., Barth, R.H. Jr. and Bush, G.L. (1963). Species abundance: natural regulation of insular variation. *Science* **142**, 1575-1577.

Wiggins, I.L. and Porter, D.M. (1971). *Flora of the Galápagos Islands.* Stanford University Press.

Table 49.1
Galápagos Islands
Species and Geography

Island	Observed species		Area km^2	Elevation, m	Distance, km		Area of adjacent island, km^2
	Number	Native			From nearest island	From Santa Cruz	
Baltra	58	23	25.09	-	0.6	0.6	1.84
Bartolomé	31	21	1.24	109	0.6	26.3	572.33
Caldwell	3	3	0.21	114	2.8	58.7	0.78
Champion	25	9	0.10	46	1.9	47.4	0.18
Coamaño	2	1	0.05	-	1.9	1.9	903.82
Daphne Major	18	11	0.34	119	8.0	8.0	1.84
Daphne Minor	24	-	0.08	93	6.0	12.0	0.34
Darwin	10	7	2.33	168	34.1	290.2	2.85
Eden	8	4	0.03	-	0.4	0.4	17.95
Enderby	2	2	0.18	112	2.6	50.2	0.10
Española	97	26	58.27	198	1.1	88.3	0.57
Fernandina	93	35	634.49	1494	4.3	95.3	4669.32
Gardner*	58	17	0.57	49	1.1	93.1	58.27
Gardner†	5	4	0.78	227	4.6	62.2	0.21
Genovesa	40	19	17.35	76	47.4	92.2	129.49
Isabela	347	89	4669.32	1707	0.7	28.1	634.49
Marchena	51	23	129.49	343	29.1	85.9	59.56
Onslow	2	2	0.01	25	3.3	45.9	0.10
Pinta	104	37	59.56	777	29.1	119.6	129.49
Pinzón	108	33	17.95	458	10.7	10.7	0.03
Las Plazas	12	9	0.23	-	0.5	0.6	25.09
Rábida	70	30	4.89	367	4.4	24.4	572.33
San Cristóbal	280	65	551.62	716	45.2	66.6	0.57
San Salvador	237	81	572.33	906	0.2	19.8	4.89
Santa Cruz	444	95	903.82	864	0.6	0.0	0.52
Santa Fé	62	28	24.08	259	16.5	16.5	0.52
Santa María	285	73	170.92	640	2.6	49.2	0.10
Seymour	44	16	1.84	-	0.6	9.6	25.09
Tortuga	16	8	1.24	186	6.8	50.9	17.95
Wolf	21	12	2.85	253	34.1	254.7	2.33

* Near Española. † Near Santa María.
The values marked - are not known.

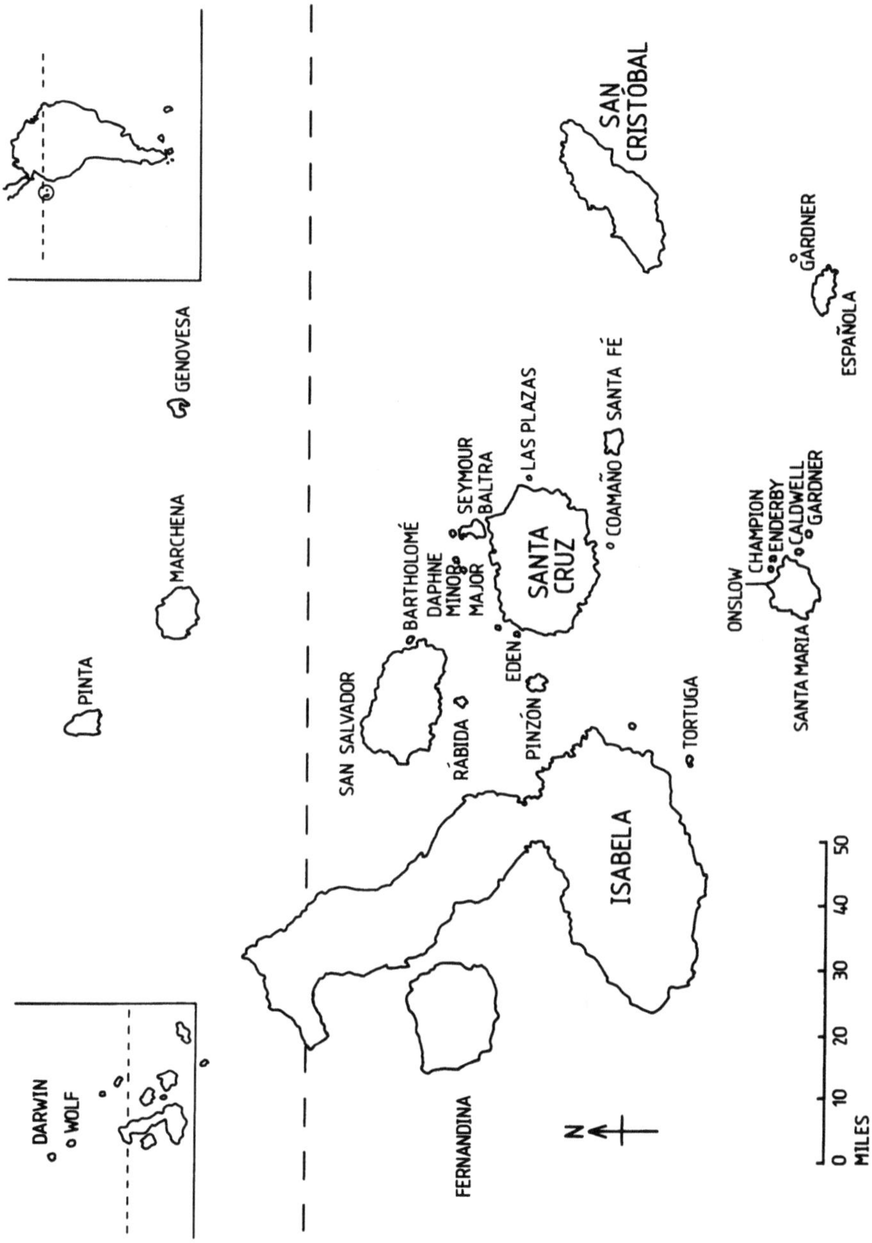

Figure 49.1
The Galápagos Islands

50. The Spatio-Temporal Spread of Fox Rabies

Source Sayers, B.McA., Mansourian, B.G., Phan Tan, T. and Bögel, K. (1977). A pattern analysis study of a wild-life rabies epizootic. *Medical Informatics* **2**, 11-34.

Contributor B.McA. Sayers Imperial College of Science and Technology

A retrospective study was conducted on the features of a wild life epizootic that spread into a study area in South Germany during 1963, spread southeast during the next few years and started to move out of the region in 1971.

The data are in the form of monthly case occurrences of fox rabies presented according to location in a 32 $\times$32 field covering an area 133 $\times$133 km. The purpose of the analysis is to determine the main features of the physical propagation of case occurrences in relation to discernible characteristics of the geographical region in which the epizootic spreads. The tactical steps are to reduce the temporal, i.e. month-by-month, and spatial variability, by longitudinal and spatial smoothing, then to generate a suitable display of the say, quarterly estimates of spatial case occurrence density; the longitudinal evolution of the major features in the spatial occurrence density pattern can then be analyzed with regard to the geographical characteristics of the region and to seasonal factors. It is deduced that a focal propagation occurs in this region, because of the exhibiting of a small number of trajectories of foci of inferred "infectivity". The role of the terrain is well illustrated by the influence of rivers and bridges on the trajectories, but the apparent effect of roadways is also evident.

Longitudinal averaging by aggregating cases in each grid square over three-monthly periods was followed by a spatial convolution with a two-dimensional Gaussian function, σ = 8 km., for example, in each direction, and contour mapping of the usually normalized result. This spatial smoothing has two objectives: reduction of the statistical sampling variability of the data, and minimization of the effect of missed cases due to the failure to locate some fox carcasses.

Several alternative steps are then possible. The successive locations of the maximum case density can be identified in sequential epochs and these are found to form a small number of trajectories. Alternatively, a wavefront approach can be made by studying the advancing front of a pre-selected contour level, at say, 50% of the maximum, and the global effect of the movement of trajectories across river courses can be seen. A different approach is to separate the translation and evolution components of case occurrence densities by calculating the local contour pattern of cases collected with respect to the focal location that locally applies. In this way, it is possible to study seasonal or other factors that affect the time-evolution of case occurrences, i.e. the absolute case density in a given area, and the pattern and extent of the spatial spread. Table 50.1 presents the aggregated data from January 1963 to December 1970. Table 50.2 gives the data for this period month by month. Only the data for January 1963 is printed here.

Table 50.1: Case Occurrences by
January 1963 to

Loc-ation	Location 1 - 8								Location 9 - 16							
1	0	0	0	0	0	0	0	0	0	1	5	4	6	3	1	2
2	0	0	0	1	0	1	0	2	0	3	5	2	4	6	1	2
3	0	0	0	0	0	0	2	0	0	1	1	2	5	3	2	1
4	0	1	0	0	0	0	0	2	0	1	5	3	1	8	5	2
5	0	1	0	0	1	0	1	1	0	0	6	6	5	10	5	3
6	0	0	0	0	0	0	0	0	2	0	1	8	5	5	5	4
7	0	0	0	0	0	0	0	0	1	3	13	15	5	2	8	1
8	0	0	0	0	0	0	0	0	1	8	8	8	7	8	8	1
9	0	0	0	1	1	0	0	0	0	0	7	5	2	2	4	1
10	0	1	1	3	1	0	0	0	0	0	20	4	8	5	4	4
11	0	0	2	3	0	0	0	0	0	5	6	10	2	3	4	4
12	0	1	1	6	0	0	0	0	0	0	7	3	5	2	1	3
13	0	0	0	1	1	0	0	0	1	5	2	2	1	6	2	3
14	1	0	1	2	1	0	0	0	0	5	9	1	9	7	2	3
15	0	2	0	1	1	0	1	0	0	1	0	2	2	3	5	3
16	0	1	2	5	0	0	0	0	0	1	6	0	4	1	8	4
17	0	1	0	2	0	0	3	0	1	2	2	0	3	4	6	1
18	0	0	0	1	0	0	1	1	0	4	1	0	1	4	5	8
19	0	0	0	0	0	1	0	2	0	3	4	2	2	6	2	8
20	0	0	0	0	0	1	0	0	0	4	3	4	1	7	4	3
21	0	0	0	0	1	0	1	0	0	2	3	2	1	5	1	4
22	0	0	1	0	0	0	1	0	0	1	9	2	0	3	2	2
23	0	0	0	0	0	0	0	2	0	1	10	1	1	5	1	5
24	0	1	0	0	0	0	1	0	0	8	9	4	2	2	3	1
25	0	0	0	0	0	0	0	0	1	3	3	4	3	5	1	3
26	0	0	0	0	0	0	0	0	1	1	1	3	4	4	2	7
27	0	0	0	0	0	0	0	1	0	2	4	3	4	5	5	9
28	0	1	0	1	1	0	0	0	0	4	3	7	5	8	2	2
29	0	0	0	0	0	0	0	0	0	4	1	3	1	6	5	4
30	0	0	0	0	1	0	1	0	0	1	3	0	3	1	1	1
31	0	0	0	0	0	0	0	0	0	1	2	1	0	1	3	3
32	0	0	0	0	0	0	0	0	0	1	2	1	0	1	3	3

Location in the Study-Area, December 1970

Location	Location 17 - 24								Location 25 - 32							
1	2	4	4	3	1	1	7	3	4	5	2	5	3	4	0	2
2	4	12	4	7	3	5	2	8	5	4	3	4	7	3	2	2
3	12	10	4	7	3	5	1	4	6	7	10	8	5	3	4	1
4	7	4	2	1	7	2	5	0	5	3	2	12	4	6	3	0
5	1	6	1	3	6	4	1	7	2	1	3	13	9	4	5	4
6	10	2	2	5	3	4	3	3	4	0	1	6	4	8	7	4
7	3	2	1	5	2	3	1	1	0	0	3	9	4	6	2	5
8	6	3	3	3	2	5	1	4	3	1	0	4	5	12	8	4
9	2	1	3	7	12	3	0	1	0	1	2	3	3	6	14	3
10	2	2	5	6	13	6	6	0	2	2	1	2	6	6	9	1
11	4	1	1	10	5	4	4	0	0	1	0	2	5	1	8	1
12	4	1	4	1	10	4	4	1	1	0	2	0	3	3	8	2
13	2	1	2	4	13	10	5	0	1	0	0	2	2	2	3	1
14	6	4	1	0	12	3	5	1	1	0	3	1	1	1	5	0
15	6	3	1	3	3	5	2	1	1	6	1	1	3	0	6	0
16	5	6	0	1	2	1	0	0	2	0	0	2	2	1	1	1
17	4	0	0	1	1	3	2	0	0	1	1	0	4	2	0	0
18	2	5	1	5	0	3	2	0	0	2	1	0	0	2	1	0
19	4	1	1	5	4	0	0	0	0	0	5	1	0	0	0	0
20	0	2	2	1	0	3	2	1	1	2	6	5	0	0	0	0
21	0	1	0	1	1	3	7	2	2	3	5	4	2	2	0	0
22	0	4	0	3	0	4	6	8	3	6	9	7	0	3	2	0
23	3	1	2	4	1	11	8	4	8	12	5	6	0	1	0	1
24	2	5	2	5	3	8	7	7	5	15	7	9	1	1	1	0
25	1	5	4	2	7	5	10	6	9	6	7	5	2	1	4	1
26	9	2	7	4	3	4	10	10	8	8	3	0	2	2	1	0
27	13	3	7	6	7	4	6	8	3	4	6	2	1	1	1	1
28	8	4	2	10	6	5	8	7	7	3	3	1	3	3	0	1
29	6	3	2	10	6	2	4	17	8	1	2	4	1	0	1	2
30	8	9	1	4	4	5	3	11	14	5	8	4	2	1	1	0
31	6	2	6	3	5	5	4	6	6	8	3	4	0	4	1	2
32	6	2	6	3	5	5	4	6	6	8	3	4	0	4	1	2

Table 50.2
Monthly Case Occurrences
by Location in the Study-Area
January 1963 *

Loc-ation	Location 1 - 8	9 - 16	17 - 24	25 - 32
1	0 0 0 0 0 0 0 0	0 0 0 0 0 0 0 0	0 0 0 0 0 0 0 0	0 0 0 0 0 0 0 0
2	0 0 0 0 0 0 0 0	0 0 0 0 0 0 0 0	0 0 0 0 0 0 0 0	0 0 0 0 0 0 0 0
3	0 0 0 0 0 0 0 0	0 0 0 0 0 0 0 0	0 0 0 0 0 0 0 0	0 0 0 0 0 0 0 0
4	0 0 0 0 0 0 0 0	0 0 0 0 0 0 0 0	0 0 0 0 0 0 0 0	0 0 0 0 0 0 0 0
5	0 1 0 0 0 0 0 0	0 0 0 0 0 0 0 0	0 0 0 0 0 0 0 0	0 0 0 0 0 0 0 0
6	0 0 0 0 0 0 0 0	0 0 0 0 0 0 0 0	0 0 0 0 0 0 0 0	0 0 0 0 0 0 0 0
7	0 0 0 0 0 0 0 0	0 0 0 0 0 0 0 0	0 0 0 0 0 0 0 0	0 0 0 0 0 0 0 0
8	0 0 0 0 0 0 0 0	0 0 0 0 0 0 0 0	0 0 0 0 0 0 0 0	0 0 0 0 0 0 0 0
9	0 0 0 0 0 0 0 0	0 0 0 0 0 0 0 0	0 0 0 0 0 0 0 0	0 0 0 0 0 0 0 0
10	0 0 0 0 0 0 0 0	0 0 0 0 0 0 0 0	0 0 0 0 0 0 0 0	0 0 0 0 0 0 0 0
11	0 0 0 0 0 0 0 0	0 0 0 0 0 0 0 0	0 0 0 0 0 0 0 0	0 0 0 0 0 0 0 0
12	0 0 0 0 0 0 0 0	0 0 0 0 0 0 0 0	0 0 0 0 0 0 0 0	0 0 0 0 0 0 0 0
13	0 0 0 0 0 0 0 0	0 0 0 0 0 0 0 0	0 0 0 0 0 0 0 0	0 0 0 0 0 0 0 0
14	0 0 0 0 0 0 0 0	0 0 0 0 0 0 0 0	0 0 0 0 0 0 0 0	0 0 0 0 0 0 0 0
15	0 0 0 0 0 0 0 0	0 0 0 0 0 0 0 0	0 0 0 0 0 0 0 0	0 0 0 0 0 0 0 0
16	0 0 0 0 0 0 0 0	0 0 0 0 0 0 0 0	0 0 0 0 0 0 0 0	0 0 0 0 0 0 0 0
17	0 0 0 0 0 0 0 0	0 0 0 0 0 0 0 0	0 0 0 0 0 0 0 0	0 0 0 0 0 0 0 0
18	0 0 0 0 0 0 0 0	0 0 0 0 0 0 0 0	0 0 0 0 0 0 0 0	0 0 0 0 0 0 0 0
19	0 0 0 0 0 0 0 0	0 0 0 0 0 0 0 0	0 0 0 0 0 0 0 0	0 0 0 0 0 0 0 0
20	0 0 0 0 0 0 0 0	0 0 0 0 0 0 0 0	0 0 0 0 0 0 0 0	0 0 0 0 0 0 0 0
21	0 0 0 0 0 0 0 0	0 0 0 0 0 0 0 0	0 0 0 0 0 0 0 0	0 0 0 0 0 0 0 0
22	0 0 0 0 0 0 0 0	0 0 0 0 0 0 0 0	0 0 0 0 0 0 0 0	0 0 0 0 0 0 0 0
23	0 0 0 0 0 0 0 0	0 0 0 0 0 0 0 0	0 0 0 0 0 0 0 0	0 0 0 0 0 0 0 0
24	0 0 0 0 0 0 0 0	0 0 0 0 0 0 0 0	0 0 0 0 0 0 0 0	0 0 0 0 0 0 0 0
25	0 0 0 0 0 0 0 0	0 0 0 0 0 0 0 0	0 0 0 0 0 0 0 0	0 0 0 0 0 0 0 0
26	0 0 0 0 0 0 0 0	0 0 0 0 0 0 0 0	0 0 0 0 0 0 0 0	0 0 0 0 0 0 0 0
27	0 0 0 0 0 0 0 0	0 0 0 0 0 0 0 0	0 0 0 0 0 0 0 0	0 0 0 0 0 0 0 0
28	0 0 0 0 0 0 0 0	0 0 0 0 0 0 0 0	0 0 0 0 0 0 0 0	0 0 0 0 0 0 0 0
29	0 0 0 0 0 0 0 0	0 0 0 0 0 0 0 0	0 0 0 0 0 0 0 0	0 0 0 0 0 0 0 0
30	0 0 0 0 0 0 0 0	0 0 0 0 0 0 0 0	0 0 0 0 0 0 0 0	0 0 0 0 0 0 0 0
31	0 0 0 0 0 0 0 0	0 0 0 0 0 0 0 0	0 0 0 0 0 0 0 0	0 0 0 0 0 0 0 0
32	0 0 0 0 0 0 0 0	0 0 0 0 0 0 0 0	0 0 0 0 0 0 0 0	0 0 0 0 0 0 0 0

* The remaining months are available on the tape.

51. Measuring the Avoidance of Super-Parasitism: Are Balls Scattered Non-Randomly into Boxes?

Source Daley, D.J. and Maindonald, J.H. (1984). A unified view of models describing the avoidance of super-parasitism. To appear.

Contributor D.J. Daley Australian National University

Various species of parasites which reproduce by laying eggs in hosts in which at most one egg per host can develop to the adult stage, exhibit avoidance of super-parasitism, that is they tend to avoid laying more than one egg in a host, thereby reducing egg wastage. Laboratory experiments can be devised (i) to demonstrate that the phenomenon is indeed a real one, and, when so, (ii) to measure the extent of avoidance. While the former question is akin to showing that eggs, balls, are not distributed randomly in the hosts, boxes, the latter question requires some modelling of the behaviour of the ovipository events in the experiment.

Both a variety of models, and a range of estimation techniques for the simplest non-trivial case of these models, have been reviewed in Daley and Maindonald (1984) from which the data given in Table 51.1 have been obtained. The data list the results of seventy experiments in which P parasites, wasps *Spalangia endius,* were confined with H hosts, pupae of the housefly *Musca domestica,* for a given time, where X_i is the number of pupae experiencing i ovipositions when the total number of ovipositions is N. The durations of the experiments are fixed for given (P,H) values, but may otherwise vary. Many more data are in Markwick's (1974) thesis from which the present data set originates. A more application-oriented discussion is given in Markwick and Maindonald (1983).

References

Markwick, N. (1974). *A Comparative Study of Four Housefly Parasites Hymenoptera: chalcidoidea.* Ph.D. thesis, Victoria University of Wellington.

Maindonald, J.H. and Markwick, N. (1983). An application of models describing the avoidance of super-parasitism to experimental results for four species of parasitic wasp. Technical Report, New Zealand DSIR, Applied Mathematics Division.

Table 51.1
The Number of Pupae of the Housefly, *Musca domestica*, Experiencing Ovipositions When Being Confined with Wasps, *Spalangia endius* *

			Trial Number																			
P	H		1	2	3	4	5	6	7	8	9	10	11	12	13	14	15	16	17	18	19	20
1	5	X_0	5	2	3	5	1	2	1	2	1	0	3	0	2	0	0	1	5	2	4	3
		X_1	0	3	2	0	3	1	3	3	4	5	2	4	3	5	5	4	0	2	1	2
		X_2	0	0	0	0	1	2	1	0	0	0	0	1	0	0	0	0	0	0	0	0
		N	0	3	2	0	5	5	5	3	4	5	2	6	3	5	5	4	0	5	1	2
1	15	X_0	11	9	8	8	9	10	7	10	11	7	13	7	8	9	8	8	7	10	11	10
		X_1	4	6	5	7	6	5	8	5	4	7	2	8	6	6	6	7	8	5	4	5
		X_2	0	0	2	0	0	0	0	0	0	1	0	0	1	0	1	0	0	0	0	0
		N	4	6	9	7	6	5	8	5	4	9	2	8	8	6	8	7	8	5	4	5
1	30	X_0	20	18	19	19	18	18	23	19	19	17	23	13	12	24	17	17	14	24	19	13
		X_1	10	12	10	11	12	12	7	10	11	13	7	17	17	6	12	11	16	10	11	16
		X_2	0	0	1	0	0	0	0	1	0	0	0	0	1	0	1	2	0	0	0	1
		N	10	12	12	11	12	12	7	12	11	13	7	17	19	6	14	15	16	10	11	18
10	50	X_0	3	13	2	5	7	1	5	11	10	4										
		X_1	19	20	22	25	32	24	27	29	25	19										
		X_2	20	9	18	14	9	20	16	10	14	22										
		N	86	65	83	71	56	80	65	49	56	80										

* P, number of wasps
H, number of pupae of the housefly
X_i , number of pupae experiencing i ovipositions (i = 0,1,2)
N, total number of ovipositions

52. Effect of Chemicals on Earthworm Populations

Contributors R.P. Blackshaw Agricultural and Food Science Centre, Belfast
and P.J. Diggle University of Newcastle upon Tyne

In an experiment to investigate the effects of several chemicals on earthworm populations, a five by eight array of two metre square plots with one metre buffers was laid out in a field of winter wheat according to the systematic layout given in Fig. 52.1. The four treatments were: A: water only (control); B : 0.5 kg/ha, Benlate; C: 0.6 kg/ha, Bevistin; D: 1.4 kg/ha, Cercobin; all diluted to 1000 *l*/ha for application. Samples were collected on three occasions during the period April to September 1977. Temporal variation in the earthworm population over this period is known to be considerable but data-collection for each sample was completed within two days. The first sample was collected prior to the application of any treatments. The sampling technique was to apply an irritant solution of formaldehyde, which causes worms to rise to the surface, to 50 cm square sub-plots as indicated in Fig. 52.2.

The data are presented in Table 52.1 as the observed numbers of worms for each of the nine occurring species and the total biomass/m^2. The nine species are: *Lumbricus terrestris, Lumbricus rubellus, Lumbricus festivus, Dendrobaena mammalis, Allolobophora rosea, Allolobophora chlorotica, Allolobophora longa, Allolobophora limicola, Allolobophora caliginosa.*

The data could be analyzed in terms of a classical linear model with the usual assumption of independent, constant variance errors, possibly after transformation. An alternative approach would be to incorporate dependencies between neighbouring plots, as discussed by Bartlett (1978); see, also, Besag (1975).

References

Bartlett, M.S. (1978). Nearest neighbour models in the analysis of field experiments (with discussion). *J.R. Statist. Soc.* **40**, 147-174.

Besag, J.E. (1975). Statistical analysis of non-lattice data. *The Statistician* **24**, 179-195.

Table 52.1
Data for Nine Species of Earthworms

			Species *									
Plot **	Sub-plot ***	Treat-ment ****	1	2	3	4	5	6	7	8	9	Total, biomass/m²
1	1	A	1	0	0	0	0	1	0	0	0	17.01
2	1	B	1	2	0	0	0	4	3	0	2	32.68
3	1	C	1	0	0	0	1	2	6	2	7	35.40
4	1	D	0	4	0	0	1	4	2	7	3	43.70
5	1	A	1	3	0	0	1	2	0	1	1	7.73
6	1	B	1	3	0	0	0	1	4	0	0	44.36
7	1	C	0	0	0	0	0	1	2	1	1	5.60
8	1	D	0	3	0	0	1	1	1	1	2	6.44
9	1	A	2	0	0	0	0	0	3	0	0	17.55
10	1	B	0	0	0	0	1	0	2	1	3	4.62
11	1	C	0	0	0	0	0	4	0	1	0	8.15
12	1	D	0	1	0	0	0	2	2	0	2	8.09
13	1	A	2	0	0	0	2	3	0	0	0	28.59
14	1	B	0	0	0	0	1	1	0	0	1	4.17
15	1	C	1	0	0	0	4	2	0	0	0	20.43
16	1	D	0	0	0	0	0	2	1	0	0	8.40
17	1	A	1	2	0	0	1	2	4	1	0	10.61
18	1	B	1	0	0	0	2	1	6	0	5	64.46
19	1	C	0	2	0	0	3	2	0	0	2	3.44
20	1	D	0	1	0	0	4	10	3	1	9	42.24
21	1	A	0	0	0	0	0	5	6	0	5	46.42
22	1	B	1	4	0	0	1	10	7	0	0	44.22
23	1	C	0	0	0	0	1	6	8	0	1	65.61
24	1	D	1	0	0	0	0	8	10	3	6	44.80
25	1	A	1	4	0	0	1	10	8	2	5	81.32
26	1	B	0	0	0	0	0	6	5	0	1	16.53
27	1	C	1	5	0	0	2	8	10	2	4	77.43
28	1	D	0	0	0	0	4	2	4	3	4	19.42
29	1	A	1	0	0	1	7	2	6	1	3	26.98
30	1	B	0	0	0	0	5	4	0	1	3	13.22
31	1	C	2	0	0	2	1	9	5	0	4	33.95
32	1	D	2	0	0	2	1	2	6	1	7	71.73
33	1	A	2	1	0	0	1	4	1	0	4	8.07
34	1	B	2	1	0	1	2	5	3	1	7	25.90
35	1	C	1	0	0	0	0	9	0	3	6	15.67
36	1	D	1	0	0	0	0	3	4	0	6	13.54
37	1	A	2	0	0	0	9	12	9	0	4	51.96
38	1	B	0	0	0	0	2	6	2	0	2	35.52
39	1	C	2	0	0	0	1	5	3	0	7	51.06
40	1	D	0	0	0	1	0	7	3	0	6	27.44
					Sample 2							
1	2	A	0	0	0	0	1	1	0	0	0	0.76
2	2	B	0	0	0	0	0	0	0	0	0	0.00
3	2	C	0	1	0	0	1	4	0	0	3	4.29
4	2	D	0	2	0	0	0	2	0	0	1	3.07
5	2	A	1	0	0	0	0	4	0	0	2	4.06

Table 52.1 *(cont.)*

Plot **	Sub-plot ***	Treat-ment ****	Species * 1	2	3	4	5	6	7	8	9	Total, biomass/m^2
6	2	B	0	0	0	0	0	1	0	0	0	0.84
7	2	C	0	0	0	0	0	2	1	0	2	23.68
8	2	D	2	0	0	0	1	1	0	0	2	55.68
9	2	A	0	1	0	0	2	2	0	0	0	1.82
10	2	B	2	1	0	1	0	2	0	0	0	53.37
11	2	C	0	0	0	0	0	4	1	0	0	3.79
12	2	D	1	2	0	0	0	1	0	0	4	38.17
13	2	A	1	0	0	4	0	2	1	0	0	12.45
14	2	B	0	2	0	0	0	1	0	0	6	3.24
15	2	C	0	1	0	0	0	1	1	0	3	6.43
16	2	D	1	0	0	1	0	6	0	0	3	20.74
17	2	A	1	1	0	1	1	2	1	0	1	37.29
18	2	B	0	1	0	0	1	1	0	0	1	2.71
19	2	C	0	3	0	0	0	0	0	0	2	2.42
20	2	D	1	5	1	1	0	3	0	0	2	54.75
21	2	A	0	1	0	0	1	11	0	0	2	4.73
22	2	B	1	2	0	0	1	3	0	0	0	27.11
23	2	C	0	8	0	0	1	4	0	0	1	13.24
24	2	D	0	4	0	0	7	2	1	0	0	7.06
25	2	A	0	3	0	0	1	0	0	0	3	4.71
26	2	B	0	0	0	0	1	2	0	0	1	2.23
27	2	C	2	3	0	0	1	1	0	0	2	39.66
28	2	D	0	3	0	0	2	3	0	0	1	11.24
29	2	A	0	3	0	0	1	4	0	0	1	4.93
30	2	B	0	2	0	0	0	0	0	0	1	2.90
31	2	C	0	0	0	0	0	3	1	0	3	5.90
32	2	D	0	7	0	0	0	0	0	0	0	4.01
33	2	A	0	0	0	2	2	2	0	0	0	5.20
34	2	B	0	2	0	0	1	1	0	0	3	2.66
35	2	C	2	1	0	0	0	4	0	0	0	45.68
36	2	D	0	4	0	0	0	1	1	0	0	6.22
37	2	A	2	0	0	0	0	2	0	0	1	39.57
38	2	B	1	1	0	0	1	2	1	0	0	28.10
39	2	C	0	1	0	0	0	1	0	0	0	0.91
40	2	D	0	2	0	0	1	1	0	0	8	4.52
					Sample 3							
1	3	A	0	0	0	0	5	8	0	1	5	17.61
2	3	B	1	0	0	0	2	8	0	1	17	72.61
3	3	C	1	0	0	0	0	5	3	2	6	57.10
4	3	D	0	0	0	0	0	0	4	0	11	32.34
5	3	A	0	2	0	0	1	1	1	0	5	21.19
6	3	B	0	0	0	0	1	2	3	0	9	24.47
7	3	C	1	0	0	0	0	3	5	0	4	74.06
8	3	D	0	1	0	0	0	0	4	0	0	22.17
9	3	A	0	0	0	0	0	2	2	0	9	19.34
10	3	B	0	1	0	0	0	1	1	0	4	9.38

Table 52.1 *(cont.)*

Plot **	Sub-plot ***	Treat-ment ****	Species * 1	2	3	4	5	6	7	8	9	Total, biomass/m^2
11	3	C	0	0	0	0	1	2	2	0	10	23.74
12	3	D	0	0	0	0	1	4	2	0	4	26.20
13	3	A	0	0	0	0	0	1	4	0	2	33.11
14	3	B	1	2	0	0	1	4	6	0	10	63.90
15	3	C	0	1	0	0	0	2	4	0	4	28.40
16	3	D	0	0	0	0	2	7	6	1	13	59.82
17	3	A	1	0	0	0	0	4	1	0	2	26.63
18	3	B	0	2	0	0	1	3	6	0	4	36.10
19	3	C	1	2	0	0	0	3	6	0	3	32.31
20	3	D	0	0	0	0	0	3	4	0	1	26.90
21	3	A	0	2	0	0	1	4	2	0	3	24.49
22	3	B	0	4	0	0	0	0	5	0	3	28.38
23	3	C	0	1	0	0	1	6	5	0	3	32.15
24	3	D	0	3	0	0	0	8	10	1	9	70.68
25	3	A	2	1	0	0	0	1	1	0	2	39.12
26	3	B	0	1	0	0	0	3	5	0	1	18.91
27	3	C	0	3	0	0	0	3	9	0	2	78.15
28	3	D	0	1	0	0	1	2	9	0	3	63.01
29	3	A	0	1	0	0	0	0	2	0	5	16.40
30	3	B	0	3	0	0	0	2	4	0	7	36.77
31	3	C	0	1	0	0	0	2	3	0	7	23.20
32	3	D	0	2	0	0	0	3	10	0	10	55.54
33	3	A	0	0	0	0	0	1	10	0	6	53.32
34	3	B	0	0	0	0	0	2	1	0	9	10.65
35	3	C	0	1	0	0	0	5	5	0	4	21.63
36	3	D	0	1	0	0	0	4	10	0	8	49.26
37	3	A	1	2	0	0	0	2	1	0	7	39.26
38	3	B	0	1	0	0	2	12	8	1	16	49.58
39	3	C	1	0	0	0	1	4	8	0	10	68.21
40	3	D	0	2	0	0	1	5	13	0	8	78.62

* 1. *Lumbricus terrestris*
2. *Lumbricus rubellus*
3. *Lumbricus festivus*
4. *Dendrobaena mammalis*
5. *Allolobophora rosea*
6. *Allolobophora chlorotica*
7. *Allolobophora longa*
8. *Allolobophora limicola*
9. *Allolobophora caliginosa.*

** Plot number: see Fig. 52.1.

*** Subplot number 1. pre-treatment, 2 and 3; see Fig. 52.2.

**** A: water only (control),
B: 0.5kg/ha, Benlate,
C: 0.6kg/ha, Bevistin,
D: 1.4kg/ha, Cercobin.

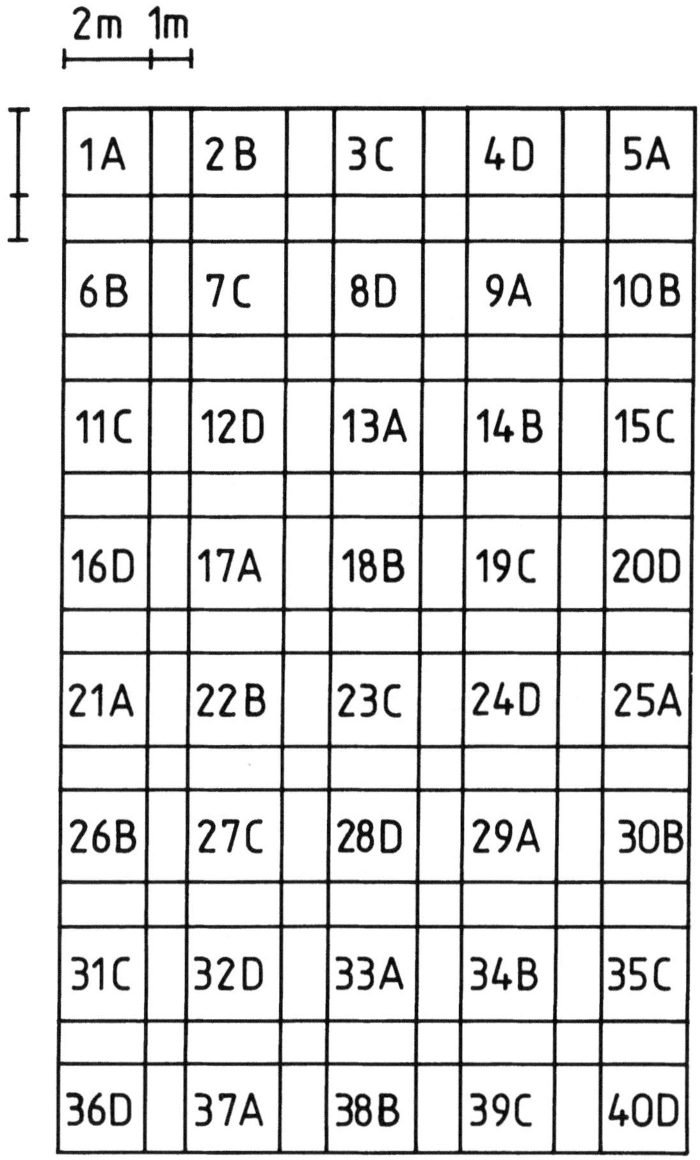

Figure 52.1
Experimental Layout

A five by eight array of two metre square plots with one metre buffers, 1,. . .,40 denote plots, A,B,C,D denote treatments.

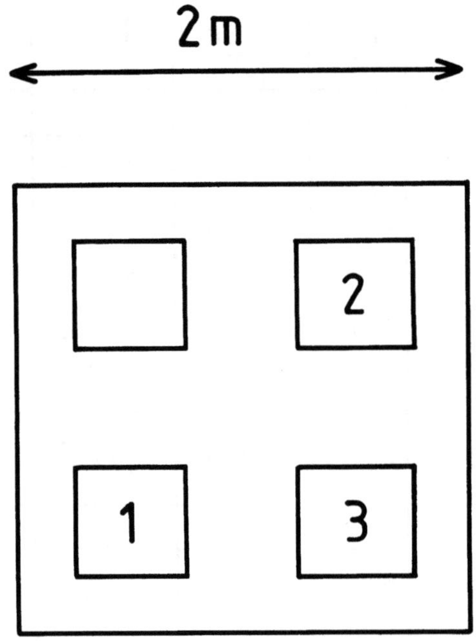

Figure 52.2
Fifty Centimetre Square Subplots of Two Metre Square Plots

53. The Classification of Three Historical Specimens of Grey Kangaroos

Source Data contributed by W. Poole CSIRO Lyneham

Contributor S.M. Carpenter CSIRO Canberra

In 1803 a French research vessel captured 19 live specimens of *Macropus fuliginosus* from Kangaroo Island. Despite a long and arduous voyage, and a diet consisting sometimes of rum and damper, some live specimens reached France, including the type specimen still held at Paris. During the nineteenth century, only three preserved specimens of *fuliginosus* were extant in Europe, and a great deal of taxonomic confusion arose between *fuliginosus* and a large male Tasmanian specimen of the eastern grey kangaroo, *Macropus giganteus*, held at the British Museum.

Skull measurements are available on modern populations of grey kangaroos and on three historical skulls, the large old male type specimen of *Macropus fuliginosus*, a small young female *Macropus fuliginosus*, and the large male *Macropus giganteus tasmaniensis*.

Canonical variate analysis can be used to separate modern populations using untransformed linear combinations of skull and molar measurements. To successfully allocate the historical skulls, however, various transformations of the variables are needed. In particular, the female *fuliginosus* skull is very difficult to successfully allocate, presumably because it had been captured as a young animal, and suffered skull and molar deformation during its critical growth period.

Details of the history of the three historical specimens are given by Poole (1976). Details of the analysis are given by Poole, Carpenter and Simms (1980) and Poole and Carpenter (1980).

The data given are skull measurements on 148 reference animals of known sex and species, and on the three historical specimens. The reference animals are from *M. giganteus* (25 male, 25 female), *M.f. melanops* (23 male, 25 female) and *M.f. fuliginosus* (25 male, 25 female), in that order. The species *M.f. melanops* has been included because it is the mainland form of the western grey kangaroo, and was sometimes confused with the insular form from Kangaroo Island, *M.f. fuliginosus*

Table 53.1 gives the data from eighteen skull measurements from each animal. Although most of these skull measurements discriminate significantly between species, or sexes, or both, not all the variables do so equally, and a subset is probably adequate to separate the groups. Basilar, occipitonasal and nasal length, nasal, zygomatic, crest and mandible width, and mandible depth and ascending ramus height were chosen in the analyses described in the references. Choosing a subset is desirable to identify and reduce the number of necessary variables for discrimination, but the correlations between the variables and the large number available make this subset selection difficult.

Kangaroos, and their skulls, continue to grow very slightly throughout their life. The number, b, given at the end of Table 53.1 is an index which can be used to determine an age adjusted measurement. This adjustment is based on measurements of teeth not presented here.

Missing values can be estimated as the group mean or, for identification of the historical specimens, by the overall mean.

References

Poole, W.E. (1976). Breeding biology and current status of the grey kangaroo, *Macropus fuliginosus fuliginosus*, of Kangaroo Island, South Australia. *Aust. J. Zool.* **24**, 169-187.

Poole, W.E. and Carpenter, S.M. (1980). Species determination by multivariate analysis of measurements from the skulls of historic specimens of grey kangaroos, attributed to *Macropus fuliginosus fuliginosus* (Desmarest) from Kangaroo Island, South Australia. *Aust. J. Zool.* **28**, 607-613.

Poole, W.E., Carpenter, S.M. and Simms, N.W. (1980). Multivariate analysis of skull morphometrics from the two species of grey kangaroos, *Macropus giganteus*, Shaw, and *M. fuliginosus* (Desmarest). *Aust. J. Zool.* **28**, 591-605.

For Table 53.1, see pages 310–317.

Table 53.1: Skull Measurements for

Animal number	Sex[a]	Species[b]	Measurements *								
			1	2	3	4	5	6	7	8	9
1	1	0	1312	1445	882	-	609	241	180	394	782
2	1	0	1439	1503	985	230	629	222	150	416	824
3	1	0	1378	1464	934	-	620	233	135	403	778
4	1	0	1315	1367	895	230	564	207	158	394	801
5	1	0	1413	1500	969	-	645	247	161	426	823
6	1	0	1090	1195	740	-	493	189	122	350	673
7	1	0	1294	1421	872	239	606	226	155	396	780
8	1	0	1377	1504	954	248	660	240	159	417	812
9	1	0	1296	1439	878	208	630	215	-	387	759
10	1	0	1470	1563	987	236	672	231	185	429	856
11	1	0	1612	1699	1119	281	778	263	185	441	921
12	1	0	1388	1500	936	227	616	220	150	412	805
13	1	0	1575	1655	1100	295	727	271	178	461	905
14	1	0	1717	1821	1184	307	810	284	185	490	960
15	1	0	1587	1711	1115	293	778	279	184	461	910
16	1	0	1604	1770	1132	268	823	272	173	474	880
17	1	0	1630	1703	1122	294	755	268	190	467	902
18	1	0	1490	1599	1007	289	710	278	179	483	897
19	1	0	1552	1540	1031	234	701	238	192	447	852
20	1	0	1595	1709	1092	267	803	255	204	455	904
21	1	0	1840	1907	1279	297	855	308	225	536	984
22	1	0	1740	1817	1225	291	838	281	185	493	977
23	1	0	1846	1893	1300	-	830	288	213	527	1013
24	1	0	1702	1860	1270	332	864	306	221	513	947
25	1	0	1768	1890	1173	-	837	285	252	512	968
1	2	0	1112	1225	755	-	525	175	141	354	702
2	2	0	1262	1367	855	211	575	200	145	386	760
3	2	0	1423	1490	971	229	626	226	159	441	839
4	2	0	1373	1494	952	247	664	238	171	428	828
5	2	0	1414	1517	971	256	659	229	197	433	853
6	2	0	1415	1523	973	257	671	247	181	443	859
7	2	0	1374	1463	927	241	629	205	169	393	833
8	2	0	1427	1487	978	221	646	210	173	402	823
9	2	0	1382	1494	942	229	674	221	170	407	803
10	2	0	1440	1557	985	233	687	225	175	413	832
11	2	0	1575	1667	1105	285	756	249	185	467	903
12	2	0	1462	1516	1016	255	687	223	190	432	873
13	2	0	1559	1650	1087	282	734	245	182	462	920
14	2	0	1507	1590	1043	-	700	255	175	446	-
15	2	0	1546	1621	1072	275	729	238	180	445	914
16	2	0	1512	1567	1052	287	695	251	183	439	885
17	2	0	1400	1501	1041	267	731	242	167	440	878

108 Known and Three Unknown Specimens

Animal number	Sex[a]	Species[b]	Measurements * 10	11	12	13	14	15	16	17	18
1	1	0	249	227	531	153	88	1086	131	179	591
2	1	0	233	248	632	141	100	1158	148	181	643
3	1	0	244	240	575	144	107	1131	116	169	610
4	1	0	224	242	568	116	79	1090	132	189	594
5	1	0	241	252	607	120	99	1175	131	197	654
6	1	0	234	185	462	188	90	901	101	138	476
7	1	0	237	238	577	149	101	1084	124	168	578
8	1	0	240	245	614	128	91	1149	129	175	628
9	1	0	248	219	584	151	117	1069	121	159	578
10	1	0	227	268	659	103	94	1240	132	196	683
11	1	0	251	284	699	86	79	1345	148	232	772
12	1	0	236	249	628	107	85	1179	132	180	652
13	1	0	251	281	692	82	125	1309	145	210	712
14	1	0	258	288	764	104	81	1443	157	222	731
15	1	0	234	282	673	81	75	1339	135	207	692
16	1	0	249	258	702	57	109	1361	140	208	713
17	1	0	261	314	710	81	104	1377	137	206	754
18	1	0	250	299	674	115	107	1253	142	194	688
19	1	0	246	287	691	82	89	1293	137	213	722
20	1	0	243	306	695	83	130	-	138	183	701
21	1	0	277	350	780	84	116	1539	163	238	795
22	1	0	273	265	736	121	83	-	159	227	770
23	1	0	252	330	784	21	89	1530	152	232	829
24	1	0	239	315	765	39	137	1526	138	218	776
25	1	0	233	347	798	41	100	1547	140	243	842
1	2	0	221	201	497	203	82	933	106	154	533
2	2	0	226	220	569	164	72	1059	113	164	578
3	2	0	256	234	613	148	106	1197	129	173	651
4	2	0	249	228	605	147	99	1131	126	167	621
5	2	0	233	264	664	131	94	1215	139	200	660
6	2	0	240	288	665	125	104	1202	147	192	676
7	2	0	235	247	615	129	100	1176	128	171	638
8	2	0	238	242	589	125	99	1198	128	169	649
9	2	0	210	222	599	101	106	1151	122	176	629
10	2	0	229	265	646	121	88	1214	130	195	668
11	2	0	233	301	722	104	117	1324	147	198	731
12	2	0	222	286	663	135	105	1258	131	205	688
13	2	0	248	258	699	103	70	1355	149	193	724
14	2	0	236	264	644	112	91	1257	139	215	681
15	2	0	251	272	672	80	93	1344	137	204	735
16	2	0	235	271	646	82	67	1297	135	203	709
17	2	0	234	259	688	123	130	1265	137	175	684

Table 53.1

Animal number	Sex[a]	Species[b]	Measurements *								
			1	2	3	4	5	6	7	8	9
18	2	0	1464	1573	1037	286	717	258	182	450	848
19	2	0	1491	1578	1028	259	682	253	177	455	875
20	2	0	1530	1576	1037	276	704	241	191	449	910
21	2	0	1570	1618	1064	294	725	261	160	455	894
22	2	0	1607	1647	1118	276	741	261	184	475	911
23	2	0	1558	1646	1087	-	740	232	177	478	908
24	2	0	1589	1750	1128	311	811	289	187	479	911
25	2	0	1548	1663	1079	265	784	251	206	457	907
1	1	1	-	1493	980	-	635	236	161	394	798
2	1	1	1299	1345	895	219	565	204	163	385	764
3	1	1	1337	1395	910	234	562	216	145	404	794
4	1	1	1372	1456	920	-	580	225	156	420	814
5	1	1	1336	1441	903	185	596	220	151	414	788
6	1	1	1301	1387	888	225	579	219	157	411	787
7	1	1	1360	1467	915	-	636	201	158	406	813
8	1	1	1276	1351	824	-	559	213	134	408	766
9	1	1	1351	-	921	-	615	228	180	424	833
10	1	1	1613	1726	1136	269	740	234	180	496	883
11	1	1	1542	1628	1020	213	677	237	187	455	885
12	1	1	1440	1580	972	248	675	217	158	399	815
13	1	1	1474	1555	989	239	629	211	170	433	888
14	1	1	1503	1603	1031	259	692	238	152	454	825
15	1	1	1597	1653	1060	-	710	221	178	470	908
16	1	1	1671	1689	1130	248	730	281	191	495	892
17	1	1	1673	1720	1140	-	763	292	299	497	946
18	1	1	1458	1588	969	-	686	251	183	435	836
19	1	1	1568	1689	1086	290	717	231	183	450	900
20	1	1	1650	1707	1173	319	737	275	200	523	943
21	1	1	1774	1838	1210	281	816	275	198	547	994
22	1	1	1893	1945	1315	267	893	260	213	499	994
23	1	1	1765	1781	1240	295	766	261	192	476	978
1	2	1	1030	1121	665	174	454	141	124	303	640
2	2	1	1389	1486	931	245	625	203	152	411	801
3	2	1	1263	1372	857	225	587	203	146	400	782
4	2	1	1067	1167	702	-	474	151	121	320	683
5	2	1	1379	1500	947	-	676	222	160	427	801
6	2	1	1335	1407	900	-	586	189	136	397	899
7	2	1	1282	-	835	172	-	186	165	385	785
8	2	1	1413	1490	962	244	630	235	170	430	834
9	2	1	1470	1583	1000	211	690	242	166	451	855
10	2	1	1377	1479	923	-	630	239	162	452	874
11	2	1	1464	1539	971	246	663	196	165	411	821

(cont.)

Animal number	Sex[a]	Species[b]	Measurements *								
			10	11	12	13	14	15	16	17	18
18	2	0	205	262	625	70	118	1235	124	180	659
19	2	0	244	288	647	114	111	1290	148	194	706
20	2	0	243	271	-	193	101	1321	143	218	741
21	2	0	212	302	684	74	103	1355	141	206	706
22	2	0	222	306	699	95	97	1370	139	218	733
23	2	0	205	320	704	28	93	1394	131	182	733
24	2	0	238	326	716	122	90	1366	129	174	715
25	2	0	244	282	687	100	119	1314	126	198	706
1	1	1	242	269	589	135	85	1160	125	170	659
2	1	1	235	215	542	153	93	1054	118	156	556
3	1	1	272	236	577	154	102	1094	122	158	625
4	1	1	230	250	589	124	88	1110	136	179	636
5	1	1	245	240	574	156	95	1083	133	178	623
6	1	1	231	229	582	113	78	1115	123	164	616
7	1	1	226	261	570	138	76	1102	123	171	603
8	1	1	212	235	533	129	91	-	112	159	608
9	1	1	231	275	640	136	100	1117	135	194	654
10	1	1	246	302	-	75	103	1365	148	184	745
11	1	1	242	247	-	94	100	1249	148	190	709
12	1	1	227	234	647	129	84	1213	131	186	634
13	1	1	251	253	662	134	91	1225	144	205	716
14	1	1	220	268	638	83	95	1264	141	203	712
15	1	1	243	294	-	104	107	-	146	194	761
16	1	1	236	306	-	62	78	1357	138	208	770
17	1	1	270	318	709	107	120	1361	153	196	755
18	1	1	239	243	624	115	105	1181	128	192	676
19	1	1	252	318	694	18	98	1337	140	194	759
20	1	1	268	-	744	72	90	1395	146	184	768
21	1	1	275	368	753	56	80	1482	146	227	794
22	1	1	234	345	754	13	92	1568	163	216	824
23	1	1	220	313	726	38	95	1503	159	211	775
1	2	1	202	-	435	216	96	856	101	132	473
2	2	1	245	242	587	125	114	1164	123	166	645
3	2	1	232	237	554	170	113	1055	120	166	580
4	2	1	215	193	475	209	76	894	103	134	511
5	2	1	236	248	607	129	107	1125	130	171	631
6	2	1	233	-	603	159	72	1132	129	197	619
7	2	1	233	235	-	157	60	1070	132	182	612
8	2	1	240	271	626	88	96	1198	132	191	665
9	2	1	249	271	634	129	80	1209	138	210	708
10	2	1	283	256	-	148	95	1136	129	190	679
11	2	1	236	237	645	138	83	1219	143	188	702

Table 53.1

Animal number	Sex[a]	Species[b]	Measurements * 1	2	3	4	5	6	7	8	9
12	2	1	1452	1592	973	236	694	236	176	461	855
13	2	1	1444	1552	972	217	667	261	156	467	869
14	2	1	1487	1586	1002	242	681	240	171	439	822
15	2	1	1515	1617	1036	-	712	217	153	440	854
16	2	1	1536	1620	1070	255	716	242	165	455	894
17	2	1	1499	1606	1028	275	699	254	177	439	835
18	2	1	1505	1569	1042	220	703	207	164	433	843
19	2	1	1519	1594	1066	259	704	238	184	430	865
20	2	1	1566	1606	1046	227	669	238	198	450	882
21	2	1	1509	1601	1023	256	693	228	183	434	828
22	2	1	1553	1607	1059	235	675	259	203	484	882
23	2	1	1660	1701	1171	300	769	253	188	482	960
24	2	1	1719	1757	1190	275	800	245	194	492	939
25	2	1	1688	1720	1188	285	798	235	183	459	893
1	1	2	1382	1493	928	230	573	231	169	427	861
2	1	2	1438	1481	976	250	566	218	173	428	912
3	1	2	1304	1408	959	261	525	200	176	413	889
4	1	2	1391	1430	950	256	551	202	176	421	886
5	1	2	1716	1688	1158	249	698	247	192	471	972
6	1	2	1530	1578	1046	259	638	208	185	452	929
7	1	2	1625	1630	1132	274	658	234	193	470	939
8	1	2	1559	1567	1069	243	630	216	196	461	955
9	1	2	1578	1566	1094	263	628	237	203	472	953
10	1	2	1562	1580	1076	295	638	240	177	455	925
11	1	2	1656	1640	1146	280	624	226	194	500	969
12	1	2	1477	1486	1001	244	590	206	155	402	867
13	1	2	1568	1601	1095	262	628	250	205	504	932
14	1	2	1656	1646	1142	297	656	232	198	471	947
15	1	2	1619	1678	1106	256	719	253	193	473	946
16	1	2	1687	1692	1153	260	687	263	204	491	986
17	1	2	1748	1731	1212	282	700	262	243	494	1000
18	1	2	1783	1788	1243	303	746	275	239	503	972
19	1	2	1745	1738	1235	304	715	246	237	513	1070
20	1	2	1680	1668	1151	293	685	255	224	479	1032
21	1	2	1653	1701	1151	299	734	239	215	490	1004
22	1	2	1732	1702	1200	328	708	255	230	496	990
23	1	2	1729	1694	1199	303	691	259	229	494	1015
24	1	2	1711	1695	1191	291	699	233	221	461	989
25	1	2	1834	1823	1247	320	737	278	280	535	1090
1	2	2	1271	1334	859	211	503	171	152	381	824
2	2	2	1329	1375	885	229	549	197	172	396	838
3	2	2	1248	1286	827	236	494	196	154	380	804

(cont.)

Animal number	Sex[a]	Species[b]	Measurements *								
			10	11	12	13	14	15	16	17	18
12	2	1	261	240	656	140	87	1210	143	194	695
13	2	1	262	251	650	176	105	1194	138	197	704
14	2	1	245	290	656	112	108	1237	127	187	679
15	2	1	245	271	641	101	90	1245	132	196	729
16	2	1	222	326	625	87	85	1287	135	196	692
17	2	1	254	264	650	112	91	1243	125	200	668
18	2	1	230	266	629	127	88	1228	136	194	709
19	2	1	238	273	645	70	121	1275	131	192	689
20	2	1	268	267	656	135	114	1296	142	219	699
21	2	1	240	256	638	120	107	1248	129	198	692
22	2	1	239	294	698	140	117	1314	135	212	722
23	2	1	247	320	695	86	126	1430	154	205	770
24	2	1	270	333	705	65	97	1450	154	240	813
25	2	1	247	312	-	73	95	1415	138	220	760
1	1	2	261	250	618	151	98	1162	133	186	640
2	1	2	249	261	619	172	86	1220	147	193	655
3	1	2	253	250	-	138	96	1187	151	196	700
4	1	2	260	252	646	140	106	1187	146	202	688
5	1	2	290	300	687	189	88	1450	160	236	799
6	1	2	243	265	667	134	104	1315	144	201	750
7	1	2	240	283	685	144	102	-	151	200	734
8	1	2	230	280	678	110	95	-	156	212	760
9	1	2	254	294	687	172	93	-	146	207	772
10	1	2	243	272	675	118	80	1320	154	217	753
11	1	2	235	314	738	114	92	1422	166	227	805
12	1	2	242	268	639	160	88	1245	147	185	672
13	1	2	265	290	719	115	96	1367	159	223	777
14	1	2	217	309	721	104	75	-	149	207	778
15	1	2	242	276	689	119	94	1369	159	215	765
16	1	2	240	288	735	111	104	1412	157	227	823
17	1	2	236	340	735	129	98	-	166	239	859
18	1	2	207	330	764	80	105	-	159	226	841
19	1	2	258	328	744	149	96	1502	167	239	853
20	1	2	213	337	728	78	76	1466	158	214	835
21	1	2	239	319	742	126	94	1408	169	209	836
22	1	2	233	312	703	106	87	1474	158	239	822
23	1	2	222	322	737	125	117	-	161	222	799
24	1	2	210	308	751	60	73	1452	154	215	815
25	1	2	262	371	770	151	100	1555	156	271	880
1	2	2	240	216	563	199	70	1078	130	181	634
2	2	2	214	239	587	153	79	1103	133	182	666
3	2	2	230	227	574	169	72	1046	131	170	616

Table 53.1

Animal number	Sex[a]	Species[b]	Measurements *								
			1	2	3	4	5	6	7	8	9
4	2	2	1271	1355	855	212	522	190	138	374	799
5	2	2	1317	1311	891	198	497	167	152	390	807
6	2	2	1356	1387	926	214	554	195	160	392	837
7	2	2	1250	1327	859	249	532	232	176	421	830
8	2	2	1048	1145	693	-	434	167	131	337	725
9	2	2	1363	1430	926	227	578	217	167	408	844
10	2	2	1389	1415	944	230	574	212	172	405	822
11	2	2	1370	1424	934	248	571	205	171	408	859
12	2	2	1405	1426	956	212	571	199	182	407	868
13	2	2	1428	1430	977	252	570	214	162	437	880
14	2	2	1456	1489	992	245	602	219	188	427	878
15	2	2	1395	1426	942	240	571	222	182	430	853
16	2	2	1385	1399	945	215	553	191	156	386	827
17	2	2	1441	1449	984	-	568	221	166	435	879
18	2	2	1464	1496	1008	226	636	230	183	436	897
19	2	2	1486	1483	1019	287	603	213	180	442	927
20	2	2	1499	1511	-	255	699	212	190	435	936
21	2	2	1485	1500	1016	277	552	205	203	454	919
22	2	2	1468	1536	996	264	667	222	190	431	951
23	2	2	1510	1546	1043	264	656	218	197	423	891
24	2	2	1526	1512	1052	281	625	250	201	470	934
25	2	2	1570	1583	987	285	646	244	198	482	984
A[+]	1	0	1899	1925	1327	306	905	310	265	560	1067
B	1	2	1848	-	1276	-	751	287	257	-	1045
C	2	2	1115	-	748	182	-	-	178	311	756
b			38.6	34.9	28.6	5.2	19.3	5.7	5.5	9.7	16.8

* Measurements:
1, basilar length
2, occipitonasal length
3, palatilar length
4, palate width
5, nasal length
6, nasal width
7, squamosal depth
8, inter-lacrymal width
9, zygomatic width
10, post orbital width
11, rostral width
12, supra-occipital - paroccipital depth
13, crest width
14, incisive foramina length
15, mandible length
16, mandible width
17, mandible depth
18, ascending ramus height

(cont.)

Animal number	Sex[a]	Species[b]	Measurements *								
			10	11	12	13	14	15	16	17	1
4	2	2	247	215	585	167	90	1058	132	179	62
5	2	2	226	237	609	155	93	1073	130	178	64
6	2	2	220	245	611	181	85	1140	135	188	65
7	2	2	274	266	-	214	80	1067	127	176	63
8	2	2	217	173	481	198	61	880	108	152	51
9	2	2	234	249	603	153	73	1163	129	194	65
10	2	2	241	241	612	170	73	1163	139	191	64
11	2	2	246	235	623	154	98	1152	143	194	68
12	2	2	235	250	-	167	107	1176	134	194	68
13	2	2	248	263	620	170	83	1179	140	189	65
14	2	2	235	253	657	118	78	-	141	193	71
15	2	2	228	259	650	138	79	1199	148	202	70
16	2	2	217	247	585	137	98	1156	134	190	64
17	2	2	231	257	635	160	82	1233	142	185	70
18	2	2	234	264	646	172	78	1263	150	216	71
19	2	2	230	274	649	139	97	1291	157	217	73
20	2	2	256	278	642	200	118	1291	150	200	74
21	2	2	225	278	676	122	74	1260	148	194	75
22	2	2	217	305	650	178	82	1287	141	199	73
23	2	2	190	270	651	78	87	1337	158	210	74
24	2	2	236	289	680	145	106	1334	153	211	73
25	2	2	253	291	699	188	103	1354	153	223	80
A[+]	1	0	-	-	770	46	108	1648	174	257	88
B	1	2	266	-	-	-	128	1583	-	-	84
C	2	2	226	-	-	-	48	1009	-	204	59
b			0.4	8.3	14.7	-7.1	1.0	34.2	2.4	4.8	1

- denotes missing value

All measurements are in millimetres × 10.

[a] 1, male
2, female

[b] 0, *M. giganteus*
1, *M. f. melanops*
2, *M. f. fuliginosus*

[+] A, British Museum of Natural History, London, male *M. giganteus*
B, Muséum Nationale d'Histoire Naturelle, Paris, type-specimen of *M. f. fuliginosus*
C, Rijksmuseum van Natuurlijkee, Leiden, female *M. f. fuliginosus*

54. Social Grooming in North American River Otters

Source Beckel, A.L. (1982). Behavior of free-ranging and captive river otters in northcentral Wisconsin. PhD. Thesis, University of Minnesota, Minneapolis, 191 pp.

Contributor A.L. Beckel University of Wisconsin

Social grooming, i.e. one animal grooming another, is a common interaction among many group-living animals. The behaviour is often regarded, rather uncritically, as the "social cement" of animal groups. Although thought to play an important role in the bonding of group members, social grooming has been the subject of few detailed studies. The data presented in Table 54.1, on social grooming in North American river otters, *Lutra canadensis*, were taken from a larger study of the species' social behaviour and were obtained from observations of five groups of captive otters. All animals within a group were observed simultaneously.

The questions of interest are:

(1) Do individuals within a group engage in social grooming at equal rates, or are some animals consistently "groomers" while others are consistently "recipients?"

(2) In groups containing more than two members (Groups A and H), do individuals exhibit preferences in whom they groom?

(3) Do females groom males more frequently than males groom females?

(4) Do rates of grooming, in general, increase during the breeding season?

  (a) Do rates of grooming, males to females, increase during the breeding season?

  (b) Do rates of grooming, females to males, increase during the breeding season?

  (c) Do rates of grooming between individual pairs of females and males increase during the breeding season?

Table 54.1
Frequency of Social Grooming Among Otters

Group *	Season **	Time observed, min.	Groomer	Recipient	Frequency of grooming
A	N	127	F1	M2	6
A	N	127	F1	M3	1
A	N	127	F1	M4	2
A	N	127	M2	F1	0
A	N	127	M2	M3	0
A	N	127	M2	M4	0
A	N	127	M3	F1	1
A	N	127	M3	M2	1
A	N	127	M3	M4	0
A	N	127	M4	F1	0
A	N	127	M4	M2	0
A	N	127	M4	M3	0
A	N	111	F1	M2	8
A	N	111	F1	M3	1
A	N	111	F1	M4	12
A	N	111	M2	F1	0
A	N	111	M2	M3	0
A	N	111	M2	M4	1
A	N	111	M3	F1	1
A	N	111	M3	M2	2
A	N	111	M3	M4	3
A	N	111	M4	F1	0
A	N	111	M4	M2	0
A	N	111	M4	M3	0
A	N	63	F1	M2	1
A	N	63	F1	M3	0
A	N	63	F1	M4	1
A	N	63	M2	F1	0
A	N	63	M2	M3	0
A	N	63	M2	M4	0
A	N	63	M3	F1	1
A	N	63	M3	M2	2
A	N	63	M3	M4	3
A	N	63	M4	F1	0
A	N	63	M4	M2	2
A	N	63	M4	M3	1
A	N	274	F1	M2	16
A	N	274	F1	M3	2
A	N	274	F1	M4	8
A	N	274	M2	F1	3
A	N	274	M2	M3	2
A	N	274	M2	M4	2
A	N	274	M3	F1	4

Table 54.1 *(cont.)*

Group *	Season **	Time observed, min.	Groomer	Recipient	Frequency of grooming
A	N	274	M3	M2	3
A	N	274	M3	M4	8
A	N	274	M4	F1	0
A	N	274	M4	M2	1
A	N	274	M4	M3	2
A	N	62	F1	M2	3
A	N	62	F1	M3	0
A	N	62	F1	M4	2
A	N	62	M2	F1	0
A	N	62	M2	M3	0
A	N	62	M2	M4	0
A	N	62	M3	F1	0
A	N	62	M3	M2	0
A	N	62	M3	M4	0
A	N	62	M4	F1	0
A	N	62	M4	M2	0
A	N	62	M4	M3	0
A	N	103	F1	M2	0
A	N	103	F1	M3	5
A	N	103	F1	M4	2
A	N	103	M2	F1	3
A	N	103	M2	M3	2
A	N	103	M2	M4	3
A	N	103	M3	F1	4
A	N	103	M3	M2	4
A	N	103	M3	M4	2
A	N	103	M4	F1	0
A	N	103	M4	M2	5
A	N	103	M4	M3	1
A	B	101	F1	M2	3
A	B	101	F1	M3	1
A	B	101	F1	M4	6
A	B	101	M2	F1	1
A	B	101	M2	M3	0
A	B	101	M2	M4	3
A	B	101	M3	F1	5
A	B	101	M3	M2	0
A	B	101	M3	M4	4
A	B	101	M4	F1	0
A	B	101	M4	M2	1
A	B	101	M4	M3	0
A	B	115	F1	M2	11
A	B	115	F1	M3	2
A	B	115	F1	M4	2
A	B	115	M2	F1	3

Table 54.1 *(cont.)*

Group *	Season **	Time observed, min.	Groomer	Recipient	Frequency of grooming
A	B	115	M2	M3	0
A	B	115	M2	M4	6
A	B	115	M3	F1	1
A	B	115	M3	M2	2
A	B	115	M3	M4	2
A	B	115	M4	F1	3
A	B	115	M4	M2	3
A	B	115	M4	M3	0
A	B	235	F1	M2	36
A	B	235	F1	M3	2
A	B	235	F1	M4	4
A	B	235	M2	F1	6
A	B	235	M2	M3	2
A	B	235	M2	M4	2
A	B	235	M3	F1	6
A	B	235	M3	M2	4
A	B	235	M3	M4	3
A	B	235	M4	F1	7
A	B	235	M4	M2	1
A	B	235	M4	M3	8
A	B	163	F1	M2	5
A	B	163	F1	M3	0
A	B	163	F1	M4	0
A	B	163	M2	F1	3
A	B	163	M2	M3	0
A	B	163	M2	M4	0
A	B	163	M3	F1	3
A	B	163	M3	M2	1
A	B	163	M3	M4	1
A	B	163	M4	F1	3
A	B	163	M4	M2	1
A	B	163	M4	M3	0
A	B	66	F1	M2	11
A	B	66	F1	M3	1
A	B	66	F1	M4	0
A	B	66	M2	F1	5
A	B	66	M2	M3	0
A	B	66	M2	M4	1
A	B	66	M3	F1	0
A	B	66	M3	M2	3
A	B	66	M3	M4	1
A	B	66	M4	F1	1
A	B	66	M4	M2	0
A	B	66	M4	M3	0
A	B	360	F1	M2	34

Table 54.1 *(cont.)*

Group *	Season **	Time observed, min.	Groomer	Recipient	Frequency of grooming
A	B	360	F1	M3	8
A	B	360	F1	M4	8
A	B	360	M2	F1	28
A	B	360	M2	M3	0
A	B	360	M2	M4	6
A	B	360	M3	F1	26
A	B	360	M3	M2	7
A	B	360	M3	M4	6
A	B	360	M4	F1	12
A	B	360	M4	M2	5
A	B	360	M4	M3	6
A	N	173	F1	M2	29
A	N	173	F1	M3	5
A	N	173	F1	M4	5
A	N	173	M2	F1	13
A	N	173	M2	M3	3
A	N	173	M2	M4	2
A	N	173	M3	F1	8
A	N	173	M3	M2	2
A	N	173	M3	M4	6
A	N	173	M4	F1	2
A	N	173	M4	M2	2
A	N	173	M4	M3	4
A	N	451	F1	M2	27
A	N	451	F1	M3	8
A	N	451	F1	M4	6
A	N	451	M2	F1	0
A	N	451	M2	M3	1
A	N	451	M2	M4	7
A	N	451	M3	F1	4
A	N	451	M3	M2	8
A	N	451	M3	M4	6
A	N	451	M4	F1	1
A	N	451	M4	M2	3
A	N	451	M4	M3	4
A	N	339	F1	M2	17
A	N	339	F1	M3	5
A	N	339	F1	M4	2
A	N	339	M2	F1	2
A	N	339	M2	M3	4
A	N	339	M2	M4	3
A	N	339	M3	F1	2
A	N	339	M3	M2	4
A	N	339	M3	M4	4

Table 54.1 *(cont.)*

Group *	Season **	Time observed, min.	Groomer	Recipient	Frequency of grooming
A	N	339	M4	F1	0
A	N	339	M4	M2	2
A	N	339	M4	M3	0
H	N	61	F21	F22	1
H	N	61	F21	M23	0
H	N	61	F21	M24	0
H	N	61	F22	F21	0
H	N	61	F22	M23	0
H	N	61	F22	M24	0
H	N	61	M23	F21	2
H	N	61	M23	F22	0
H	N	61	M23	M24	0
H	N	61	M24	F21	0
H	N	61	M24	F22	0
H	N	61	M24	M23	0
H	N	230	F21	F22	5
H	N	230	F21	M23	1
H	N	230	F21	M24	10
H	N	230	F22	F21	3
H	N	230	F22	M23	3
H	N	230	F22	M24	1
H	N	230	M23	F21	0
H	N	230	M23	F22	1
H	N	230	M23	M24	0
H	N	230	M24	F21	3
H	N	230	M24	F22	1
H	N	230	M24	M23	0
H	N	109	F21	F22	2
H	N	109	F21	M23	2
H	N	109	F21	M24	0
H	N	109	F22	F21	1
H	N	109	F22	M23	2
H	N	109	F22	M24	0
H	N	109	M23	F21	0
H	N	109	M23	F22	0
H	N	109	M23	M24	1
H	N	109	M24	F21	0
H	N	109	M24	F22	1
H	N	109	M24	M23	0
H	N	215	F21	F22	10
H	N	215	F21	M23	8
H	N	215	F21	M24	9
H	N	215	F22	F21	9
H	N	215	F22	M23	6
H	N	215	F22	M24	5

Table 54.1 *(cont.)*

Group *	Season **	Time observed, min.	Groomer	Recipient	Frequency of grooming
H	N	215	M23	F21	6
H	N	215	M23	F22	5
H	N	215	M23	M24	2
H	N	215	M24	F21	5
H	N	215	M24	F22	7
H	N	215	M24	M23	2
H	N	113	F21	F22	6
H	N	113	F21	M23	1
H	N	113	F21	M24	1
H	N	113	F22	F21	1
H	N	113	F22	M23	0
H	N	113	F22	M24	2
H	N	113	M23	F21	5
H	N	113	M23	F22	1
H	N	113	M23	M24	2
H	N	113	M24	F21	0
H	N	113	M24	F22	0
H	N	113	M24	M23	0
H	B	174	F21	F22	17
H	B	174	F21	M23	14
H	B	174	F21	M24	7
H	B	174	F22	F21	1
H	B	174	F22	M23	2
H	B	174	F22	M24	1
H	B	174	M23	F21	7
H	B	174	M23	F22	3
H	B	174	M23	M24	1
H	B	174	M24	F21	1
H	B	174	M24	F22	3
H	B	174	M24	M23	4
H	B	117	F21	F22	2
H	B	117	F21	M23	5
H	B	117	F21	M24	1
H	B	117	F22	F21	1
H	B	117	F22	M23	6
H	B	117	F22	M24	2
H	B	117	M23	F21	2
H	B	117	M23	F22	5
H	B	117	M23	M24	0
H	B	117	M24	F21	1
H	B	117	M24	F22	0
H	B	117	M24	M23	3
H	B	89	F21	F22	4
H	B	89	F21	M23	2
H	B	89	F21	M24	2

Table 54.1 *(cont.)*

Group *	Season **	Time observed, min.	Groomer	Recipient	Frequency of grooming
H	B	89	F22	F21	3
H	B	89	F22	M23	1
H	B	89	F22	M24	2
H	B	89	M23	F21	0
H	B	89	M23	F22	0
H	B	89	M23	M24	0
H	B	89	M24	F21	7
H	B	89	M24	F22	3
H	B	89	M24	M23	2
H	B	181	F21	F22	8
H	B	181	F21	M23	7
H	B	181	F21	M24	5
H	B	181	F22	F21	7
H	B	181	F22	M23	3
H	B	181	F22	M24	6
H	B	181	M23	F21	6
H	B	181	M23	F22	2
H	B	181	M23	M24	3
H	B	181	M24	F21	7
H	B	181	M24	F22	8
H	B	181	M24	M23	6
H	B	94	F21	F22	14
H	B	94	F21	M23	3
H	B	94	F21	M24	6
H	B	94	F22	F21	0
H	B	94	F22	M23	2
H	B	94	F22	M24	5
H	B	94	M23	F21	3
H	B	94	M23	F22	2
H	B	94	M23	M24	15
H	B	94	M24	F21	2
H	B	94	M24	F22	7
H	B	94	M24	M23	0
H	N	90	F21	F22	6
H	N	90	F21	M23	0
H	N	90	F21	M24	3
H	N	90	F22	F21	2
H	N	90	F22	M23	1
H	N	90	F22	M24	1
H	N	90	M23	F21	8
H	N	90	M23	F22	2
H	N	90	M23	M24	3
H	N	90	M24	F21	1
H	N	90	M24	F22	1
H	N	90	M24	M23	0
H	N	119	F21	F22	22

Table 54.1 *(cont.)*

Group *	Season **	Time observed, min.	Groomer	Recipient	Frequency of grooming
H	N	119	F21	M23	16
H	N	119	F21	M24	18
H	N	119	F22	F21	0
H	N	119	F22	M23	4
H	N	119	F22	M24	5
H	N	119	M23	F21	2
H	N	119	M23	F22	0
H	N	119	M23	M24	5
H	N	119	M24	F21	0
H	N	119	M24	F22	0
H	N	119	M24	M23	2
B	N	61	F7	M8	0
B	N	61	M8	F7	0
B	N	83	F7	M8	8
B	N	83	M8	F7	4
B	N	80	F7	M8	5
B	N	80	M8	F7	10
B	N	126	F7	M8	10
B	N	126	M8	F7	7
B	N	116	F7	M8	8
B	N	116	M8	F7	3
B	B	122	F7	M8	15
B	B	122	M8	F7	6
B	B	79	F7	M8	5
B	B	79	M8	F7	2
B	B	100	F7	M8	29
B	B	100	M8	F7	8
B	B	228	F7	M8	13
B	B	228	M8	F7	15
B	B	197	F7	M8	15
B	B	197	M8	F7	11
B	N	131	F7	M8	4
B	N	131	M8	F7	3
B	N	177	F7	M8	5
B	N	177	M8	F7	6
C	N	132	F9	M15	2
C	N	132	M15	F9	0
C	N	133	F9	M15	1
C	N	133	M15	F9	7
C	N	70	F9	M15	1
C	N	70	M15	F9	2
C	N	91	F9	M15	5
C	N	91	M15	F9	5
C	N	92	F9	M15	6
C	N	92	M15	F9	0
C	B	125	F9	M15	8
C	B	125	M15	F9	4

Table 54.1 *(cont.)*

Group *	Season **	Time observed, min.	Groomer	Recipient	Frequency of grooming
C	B	121	F9	M15	4
C	B	121	M15	F9	12
C	B	79	F9	M15	0
C	B	79	M15	F9	6
C	B	102	F9	M15	0
C	B	102	M15	F9	4
C	N	123	F9	M15	1
C	N	123	M15	F9	6
D	N	70	F5	M6	9
D	N	70	M6	F5	21
D	N	68	F5	M6	13
D	N	68	M6	F5	9
D	N	144	F5	M6	27
D	N	144	M6	F5	13
D	N	87	F5	M6	17
D	N	87	M6	F5	5
D	N	60	F5	M6	4
D	N	60	M6	F5	2
D	B	84	F5	M6	19
D	B	84	M6	F5	8
D	B	109	F5	M6	34
D	B	109	M6	F5	11
D	B	107	F5	M6	25
D	B	107	M6	F5	5
D	B	95	F5	M6	28
D	B	95	M6	F5	6
D	B	213	F5	M6	118
D	B	213	M6	F5	45
D	N	123	F5	M6	18
D	N	123	M6	F5	12
D	N	224	F5	M6	17
D	N	224	M6	F5	2
D	N	208	F5	M6	22
D	N	208	M6	F5	9

* Group A - F1 (adult female)
M2, M3, M4 (adult males)

Group B - F7 (adult female)
M8 (adult male)

Group C - F9 (adult female)
M15 (adult male)

Group D - F5 (adult female)
M6 (adult male), siblings

Group H - F21 (subadult female)
F22 (young adult female)
M23 (subadult male)
M24 (young adult male)

** B denotes observations conducted during breeding season, N nonbreeding season.

55. Species Composition in a Complex of Woodlands

Source The data were collected by final year Applied Biology students at the University of Bradford.

Contributor A.W. Kemp University of Bradford

The Northcliffe/Heaton complex of woodlands lies two miles northwest of Bradford city centre in two deep ice-cut valleys which run approximately west to east and join at their eastern extremities. They form a continuous, roughly semi-circular arc of woodland approximately three kilometers in length.

The area was planted with deciduous trees around 1870. Multiple ownership in the last half-century has resulted in differing management practices, including virtual neglect; these in turn have affected the extent of regeneration in different parts of the complex, and hence the species composition.

The data were obtained from Bitterlich counts of trees, saplings excluded, at several sites within each of the component woodlands. These areas of woodland are presented in the order in which they occur in the arc of woodland, going from northwest clockwise to southwest. The data are given in Table 55.1.

The purpose of the investigation was to discover whether similarity in species composition between the component woodlands can be related to management history rather than to contiguity in space.

There are many ways of constructing similarity coefficients; particularly appropriate for discrete multivariate data are the generalized distance measures between groups of Balakrishnan and Sanghvi (1968) and Kurczynski (1970). Everitt (1974, 1978) gives a variety of methods of cluster analysis, including graphical methods, which can then be applied.

References

Balakrishnan, V. and Sanghvi, L.D. (1968). Distance between populations on the basis of attribute data. *Biometrics* **24**, 859-865.

Everitt, B. (1974). *Cluster Analysis.* London: Heinemann.

Everitt, B. (1978). *Graphical Techniques for Multivariate Data.* London: Heinemann.

Kurczynski, T. W. (1970). Generalized distance and discrete variables. *Biometrics* **26**, 525-534.

Table 55.1
Tree Counts in a Continuous Complex of Woodlands

		Species						
Woodland	Site	Oak	Sycamore	Birch	Beech	Ash	Elm	Horse chestnut
			1. Dungoon (DU)					
DU	1	4	7	2	1	0	0	0
DU	2	1	10	0	0	0	0	2
DU	3	1	12	0	0	0	0	3
DU	4	7	6	0	3	0	0	0
			2. Northcliffe West (NW)					
NW	1	2	4	0	4	0	0	0
NW	2	1	5	4	1	0	0	0
NW	3	9	0	0	1	0	0	0
NW	4	12	0	0	2	0	0	0
			3. Northcliffe Middle (NM)					
NM	1	2	4	0	0	1	0	0
NM	2	4	1	0	2	0	0	0
NM	3	0	1	0	1	7	0	0
NM	4	5	5	0	0	1	0	0
NM	5	0	2	0	1	2	0	0
			4. Northcliffe East (NE)					
NE	1	13	2	0	0	0	0	0
NE	2	11	0	0	0	0	0	0
NE	3	13	0	0	0	0	0	0
NE	4	13	0	0	0	0	0	0
NE	5	9	2	0	0	0	0	0
			5. Low Wood (LW)					
LW	1	2	2	5	0	0	0	0
LW	2	7	0	2	0	0	0	0
LW	3	2	1	5	0	0	0	0
LW	4	7	0	3	0	0	0	0
LW	5	3	2	5	0	0	0	0
LW	6	2	9	0	0	0	0	0
			6. Dixon's Wood (DW)					
DW	1	0	3	4	0	0	0	0
DW	2	0	0	8	0	0	0	0
DW	3	0	4	1	0	0	1	2
DW	4	6	0	0	0	0	0	0
DW	5	5	2	0	0	0	0	0
DW	6	2	8	0	0	0	0	0
			7. Royd's Cliffe (RC)					
RC	1	4	0	0	2	0	1	0
RC	2	4	0	0	7	0	2	0
RC	3	3	0	0	0	6	2	0
RC	4	2	0	0	7	0	3	0
RC	5	2	3	0	2	0	2	0
			8. Weather Royd's (WR)					
WR	1	5	1	0	1	0	0	0
WR	2	10	3	0	0	0	0	0
WR	3	10	3	0	0	0	0	0
WR	4	4	2	0	0	0	1	0
WR	5	3	4	1	1	0	3	1

56. Distribution Patterns of Plant Species

Source Evans, F.C. (1952). The influence of size of quadrat on the distributional patterns of plant populations. *Contrib. Lab. Vert. Biol.* University of Michigan, No. **54**.

Contributor J.B. Douglas University of New South Wales

Cain and Evans (1952) mapped in detail an old-field grasslands community in southeastern Michigan, plotting the occurrence of three plant species: *Lespedeza capitata, Liatris aspera* and *Solidago rigida.* From these, Evans (1952) prepared quadrat coverages of 16, 8, 4, 2, 1, 1/2, 1/4, 1/8 and 1/16 square metres, recording the frequencies with which each of the species appeared in the quadrats. For *Solidago rigida,* golden rod, the frequency distributions for the three largest quadrat sizes are given in Table 56.1.

Table 56.1
Frequency Distribution of *Solidago rigida*

Quadrat coverage	Frequency												
	0	1	2	3	4	5	6	7	8	9	10	14	15+
16 sq.m.	245	94	36	31	8	10	0	2	0	1	1	1	0
8 sq.m.	615	162	48	20	5	4	1	1	1	1	0	0	0
4 sq.m.	1425	222	51	13	2	2	0	1	0	0	0	0	0

By visual inspection of the field, it was clear that there was some degree of clustering in the occurrence of the plants, and an attempt to fit Poisson distributions to the observed distributions bears this out: the fit, corresponding to random placement of individuals over the field, is very poor.

Perhaps the next simplest assumption is that the plants occur in clusters, of relatively negligible area, which are randomly placed over the field, and that the numbers of plants per cluster is Poisson distributed with constant mean. An approximation to this is given by the Neyman Type A distribution, but carrying out fitting for this also leads to extremely poor agreement between the observed and fitted frequencies.

The assumptions that the distribution of the numbers of clusters per quadrat can be approximated by a continuous gamma distribution, and that the number of plants per cluster is Poisson distributed leads to very satisfactory fits, with a negative binomial distribution. Unfortunately, when moving from smaller to larger quadrats, the estimates of the mean numbers of clusters per quadrat are substantially constant instead of approximately doubling, while the estimates of plants per cluster almost double. This model is thus not tenable either.

However, the assumption that the distribution of clusters per quadrat is Poisson, and that the numbers of individuals per cluster is logarithmic leads to a negative binomial distribution: the fits given are as good as before, while now estimates of clusters per quadrat go up by a factor of 1.7 to 1.8, close to the "ideal" 2. Similarly, estimates of individuals per cluster, though increasing, perhaps because greater proportions of some clusters are in the larger quadrats by a factor of 1.1 to 1.2, are much more nearly constant.

This last model, of the four mentioned, is thus at any rate not clearly rejected by the data set exhibited, but should of course be investigated in more detail, for example by endeavouring to identify clusters in the field and then examining the size distribution directly within clusters.

References

Cain, S.A. and Evans, F.C. (1952). The distribution patterns of three plant species in an old field community in southeastern Michigan. *Contrib. Lab. Vert. Biol.* Univ. Michigan, No. **52**.

57. The Garrison Bay Project, Stock Assessment and Dynamics of the Littleneck Clam, *Protothaca staminea*

Contributor V.F. Gallucci University of Washington

Garrison Bay is a small bay in Washington State, U.S.A. The marine fauna is diversified, with especially large numbers of soft-substrate benthic organisms such as polychaetes and bivalves (Scherba and Gallucci, 1976). One of the popular recreational activities in the bay is clam digging. During the past five years this harvest has been monitored and data collected on the species harvested, the total weight of each digger's harvest, the size distribution of each species and the time needed to dig each catch. In addition, a periodic survey by stratified random sampling determines the abundance and size distributions of the unharvested standing stock of each species. These latter sampling data are presented for the littleneck clam, *Protothaca staminea*, the species most commonly taken.

Classical theory predicts that, under intensive harvest pressure, size and age distributions of harvested and standing stocks will move to the left since the larger animals are usually caught first. If this occurs, there are many biological and management consequences. For example, a stock which has been reduced to include relatively fewer fully mature individuals and more sub-adult organisms is more prone to catastrophe and perhaps extinction since the reproductive success is frequently lower for juveniles and subadults. Thus, a series of recruitment failures is more likely to devastate the standing crop.

The data presented were collected in a stratified sample in 1976 as part of the stock assessment in Garrison Bay. Garrison Bay is divided into four study sections, south, north, west and central, and a representative area within each study section is sampled by a stratified random sampling design. Each sample unit is a square quadrat measuring 0.375 metres on a side. The areas are stratified by tide height because for most species definite density gradients exist perpendicular to the water line. Each stratum is 100 metres long by 20 feet wide. Three strata are above the 0.0 tide line while two are below.

Only north and south strata are given. Stratum, location in the stratum, date and clam dimensions are presented in Table 57.1 for the first portion of the data. The remainder of the data is available in machine readable form. In addition, a gonad grade was assigned to a randomly chosen subsample as a preliminary assessment of the ripeness of the gonad. This grade may be somewhat subjective because it is an imprecise measurement.

Animals grown in different environments, sections, strata, etc. may have statistically different dimensions or patterns of growth because of differences in food availability, exposure to warmth, sediment differences, etc. See Scherba and Gallucci (1976) for details on differences in sediment; see Hylleberg and Gallucci (1975) for related work on the clam *Macoma nasuta*.

Are there relationships between the dimensions of the animals taken in one stratum or area and do these change significantly from one to another area or stratum?

Growth models are frequently fitted to linear dimensional data which encompasses most size groups. The comparison of the resulting nonlinear curves is a multivariate problem which could be approached by the simultaneous comparison of two or more parameters and their standard errors (Gallucci and Quinn, 1979). Bernard (1981) used basic multivariate methods.

There is, of course, a relationship between size and age, but it is not obvious. An age size key may be constructed to estimate ages from field data (Kimura, 1977; Westrheim and Ricker, 1978).

These data may be analyzed from the point of view of spatial dispersion since individual sample units are identified by their locations. Possible spatial dispersion techniques that could be applied are quadrat analysis (Elliott, 1971) and spatial autocorrelation analysis (Jumars, Thistle and Jones, 1977). Since data on linear dimensions are included for each clam it is possible to divide the clams into size classes for the analysis of spatial patterns, and spatial patterns may be compared among strata, within study areas and among study areas, as well. Since the data were collected over an entire year, it is important to decide whether it is appropriate to pool the samples from all months for spatial pattern analysis.

Gonad grade might be expected to relate to organism size, time of year, and sample unit placement, but variances may be large because grades are visual, probably with some subjectivity and personal biases.

These data are part of a larger set being used for modeling the dynamics of the beach. If age specific mortality, i.e. loss, and biomass addition rates are guessed from the literature, the distribution of ages in the data is a basis for Leslie age structure modelling. Many different of model may be constructed with these data, as well as other estimators of population-level or individual condition.

This research was done with financial support from the Washington Sea Grant Program, National Oceanic & Atmospheric Administration, Grant No. NA 81-AA-D-0030 as part of the Garrison Bay Studies Project. Assistance was given by Z. Zachwieja, F. Hastings, R. Westley and A. O. D. Willows.

References

Bernard, D.R. (1981). Multivariate analysis as a means of comparing growth in fish. *Canadian J. Fish. Aquat. Sci.* **38**, 233-236.

Elliott, J.M. (1971). Some methods for the statistical analysis of samples of benthic invertebrates. *Freshwater Biological Association, Scientific Publication No.* **25**.

Gallucci, V.F. and Quinn, T. II (1979). Reparameterizing, fitting and testing a simple growth model. *Trans. Amer. Fish. Soc.* **108**, 14-25.

Hylleberg, J. and Gallucci, V. (1975). Selectivity in feeding by the deposit-feeding bivalve. *Macoma nasuta. Mar. Biol.* **32**, 167-178.

Jumars, P.,Thistle, D. and Jones, M.L. (1977). Detecting two-dimensional spatial structure in biological data. *Oecologia* **28**, 109-123.

Kimura D.K. (1977). Statistical assessment of the age-length key. *J. Fish. Res. Board Can.* **34**, 317-324.

Scherba, S. and Gallucci, V.F. (1976). The application of systematic sampling to a study of infaunal variation in a soft substrate intertidal environment. *Fishery Bulletin* **74**, 937-948.

Westrheim, S.J. and Ricker, W.E. (1978). Bias in using an age-length key to estimate age-frequency distributions. *J. Fish. Res. Board Can.* **35**, 184-189.

Table 57.1:
the littleneck clam, *Protothaca staminea,*

Clam no.	Length mm.	Width mm.	Breadth mm.	Gonad grade[a]
1	530	494	337	3
2	517	477	334	3
3	505	471	338	3
4	512	413	302	2
5	487	407	286	2
6	481	427	315	3
7	485	408	298	3
8	479	430	314	2
9	452	395	282	3
10	468	417	272	3
11	459	394	282	2
12	449	397	278	3
13	472	402	281	3
14	471	401	271	2
15	455	385	269	3
16	394	338	253	3
17	475	422	287	3
18	335	288	193	3
19	508	464	298	3
20	486	436	275	3
21	474	414	317	3
22	465	402	299	2
23	420	383	265	3
24	402	340	216	
25	410	349	253	3
26	393	333	209	2
27	389	356	249	3
28	330	268	188	3
29	305	264	172	
30	169	141	81	0
31	91	77	42	
32	537	498	345	3
33	519	456	312	3
34	509	433	284	3
35	511	447	285	3

* The rest of the data is on the tape.

a. Gonad grade: Measure of fullness of gonad with egg or sperm, 0,1,2,3.

b. Section sampled: 1, south; 2, north; 3, west; 4, central.

c. Stratum of sample:
 0, 20 ft. just above the 0.0 tide line;
 1, next of 20 ft. toward high water;
 2, next toward high water mark;
 4, 20 ft. just below 0.0 tide line;
 5, next of 20 ft. toward low water.

d. Location of sample in stratum, meters from a starting point parallel to water (in paces, 1pace/m).

e. Location of sample in feet perpendicular to water.

Measurements on
made at Garrison Bay, Washington, U.S.A. *

			Location of Sample	
Date	Section sampled[b]	Stratum of sample[c]	Parallel distance,[d] m.	Perpendicular distance,[e] ft.
010176	2	4	8	6
010176	2	4	8	6
010176	2	4	8	6
010176	2	4	8	6
010176	2	4	8	6
010176	2	4	8	6
010176	2	4	8	6
010176	2	4	8	6
010176	2	4	8	6
010176	2	4	8	6
010176	2	4	8	6
010176	2	4	8	6
010176	2	4	8	6
010176	2	4	8	6
010176	2	4	8	6
010176	2	4	8	6
010176	2	4	44	16
010176	2	4	44	16
010176	2	0	45	17
010176	2	0	45	17
010176	2	0	45	17
010176	2	0	45	17
010176	2	0	45	17
010176	2	0	45	17
010176	2	0	45	17
010176	2	0	45	17
010176	2	0	45	17
010176	2	0	45	17
010176	2	0	45	17
010176	2	0	45	17
010176	2	0	45	17
010276	2	1	12	1
010276	2	1	12	1
010276	2	1	12	1
010276	2	1	12	1

58. Maize Fertilizer Experiments on the Islands of St. Vincent and Antigua

Source These data were collected under the auspices of the former Regional Field Experimental Programme and are printed with the permission of the Executive Director of the Caribbean Agricultural Research and Development Institute, Trinidad.

Contributor S.C. Pearce University of Kent at Canterbury

A series of experiments was designed to study the effect of concentration of three components of fertilizer on the growth of maize. All experiments contained 36 plots in four blocks of nine plots each. Half of the blocks contained the treatments

000 022 202 220 111(twice) 113 131 311

and the other half

002 020 200 222 111(twice) 113 131 311

where the three digits represent levels of nitrogen, phosphorus and potassium, respectively, in the fertilizer applications. The treatments 000, 002, 020, 022, 200, 202, 220, 222 and 111 were included to search the usual range of fertilizer applications for an optimum; 113, 131 and 311 were added to provide clues in case the optimum lay outside the expected range (Springer, 1972). If this design is regarded as an incomplete block design, there is no great difficulty obtaining a reasonable partition of the treatment sum of squares.

Plots measured 16 feet by 18 feet, of which a central area of 12 feet square was recorded. Table 58.1 gives the treatment, the number of good ears harvested and their weight in kilograms for Antigua; Table 58.2 gives the treatment and the weight of good ears in kilograms for St. Vincent.

Antigua is a coral island in the semi-arid zone of the Caribbean. It has, therefore, adequate level land on which to place experiments and there is little risk of damaging rain. Experiment 3 (TEAN), however, was damaged by goats, in particular Plots 27 and 35, Plot 36 being completely destroyed. Also, experiment 2 (LFAN) suffered early damage from cattle down one boundary, i.e. Plots 9, 18, 27 and 36, but this appears to have done little lasting harm. Plant development in Experiments 5 (WLAN) and 6 (NSAN) was considered to be disappointing.

The maize experiments on Antigua were part of a scheme extending over eleven territories in the West Indies, one of which was the island of St. Vincent, which is very different from Antigua. On the one hand, it is volcanic so areas of level, even land are hard to find; on the other hand, it is in the wet zone and at some seasons of the year rainfall can be heavy, so much so that at some sites storm drains had to be dug to protect the experimental plots. The data, therefore, present a contrast with those from Antigua.

References

Springer, B.G.F. (1972). Experimental design and analysis under limited resources. *Proc. Caribbean Food Corps Soc., 10th Annual Meeting,* Puerto Rico, 147-152.

Table 58.1
Treatment, Number of Ears and Weight of Ears of Maize Harvested on the Island of Antigua *

Location	Block	Plot	Treatment	Number of ears	Weight of harvest
			1. Dunbar, Friars Hill (DBAN)		
DBAN	1	1	111	42	4.96
DBAN	1	2	000	41	3.94
DBAN	1	3	311	49	6.35
DBAN	1	4	202	48	5.56
DBAN	1	5	111	45	5.36
DBAN	1	6	220	46	6.18
DBAN	1	7	113	42	4.71
DBAN	1	8	131	44	6.03
DBAN	1	9	022	42	2.88
DBAN	2	10	222	44	5.68
DBAN	2	11	311	42	5.80
DBAN	2	12	020	42	4.16
DBAN	2	13	200	46	4.90
DBAN	2	14	111	44	5.25
DBAN	2	15	131	48	5.80
DBAN	2	16	002	46	2.18
DBAN	2	17	111	48	4.36
DBAN	2	18	113	45	4.36
DBAN	3	19	131	43	4.81
DBAN	3	20	111	47	5.19
DBAN	3	21	311	47	5.96
DBAN	3	22	000	45	3.88
DBAN	3	23	220	50	5.72
DBAN	3	24	022	44	2.51
DBAN	3	25	111	47	4.94
DBAN	3	26	113	44	3.70
DBAN	3	27	202	40	2.77
DBAN	4	28	111	41	4.34
DBAN	4	29	200	45	4.33
DBAN	4	30	111	41	4.68
DBAN	4	31	113	46	4.76
DBAN	4	32	131	47	5.24
DBAN	4	33	020	42	1.86
DBAN	4	34	222	41	3.84
DBAN	4	35	002	41	1.36
DBAN	4	36	311	36	3.98
			2. Lower Friars Hill (LFAN)		
LFAN	1	1	111	37	2.92
LFAN	1	2	022	36	1.68
LFAN	1	3	113	42	4.50
LFAN	1	4	220	41	4.74
LFAN	1	5	202	40	4.44
LFAN	1	6	311	44	4.66
LFAN	1	7	111	44	2.94
LFAN	1	8	000	44	1.60
LFAN	1	9	131	38	1.82
LFAN	2	10	113	39	4.44

Table 58.1 *(cont.)*

Location	Block	Plot	Treatment	Number of ears	Weight of harvest
LFAN	2	11	222	40	5.78
LFAN	2	12	311	46	6.24
LFAN	2	13	111	47	5.40
LFAN	2	14	200	46	5.62
LFAN	2	15	002	35	1.72
LFAN	2	16	131	43	4.42
LFAN	2	17	111	46	4.14
LFAN	2	18	020	40	3.94
LFAN	3	19	220	41	5.00
LFAN	3	20	202	46	5.26
LFAN	3	21	000	36	2.90
LFAN	3	22	111	43	4.94
LFAN	3	23	113	45	4.43
LFAN	3	24	311	47	5.42
LFAN	3	25	022	34	1.22
LFAN	3	26	111	42	3.72
LFAN	3	27	131	41	3.70
LFAN	4	28	131	46	5.61
LFAN	4	29	111	45	6.02
LFAN	4	30	020	45	3.18
LFAN	4	31	311	47	6.07
LFAN	4	32	002	38	1.48
LFAN	4	33	222	47	5.94
LFAN	4	34	111	42	3.58
LFAN	4	35	200	39	3.95
LFAN	4	36	113	44	3.38
			3. Thibou's Estate (TEAN)		
TEAN	1	1	113	32	1.27
TEAN	1	2	111	29	2.10
TEAN	1	3	022	46	2.37
TEAN	1	4	202	28	2.08
TEAN	1	5	000	31	1.74
TEAN	1	6	131	39	3.16
TEAN	1	7	220	42	4.55
TEAN	1	8	111	36	3.20
TEAN	1	9	311	35	3.69
TEAN	2	10	200	28	1.52
TEAN	2	11	311	30	2.02
TEAN	2	12	222	29	1.97
TEAN	2	13	020	29	1.94
TEAN	2	14	131	30	2.01
TEAN	2	15	111	35	2.18
TEAN	2	16	002	36	3.24
TEAN	2	17	113	39	3.91
TEAN	2	18	111	48	4.20
TEAN	3	19	111	34	2.57
TEAN	3	20	202	37	2.88
TEAN	3	21	220	37	3.45
TEAN	3	22	000	42	2.89
TEAN	3	23	111	38	3.01

Table 58.1 *(cont.)*

Location	Block	Plot	Treatment	Number of ears	Weight of harvest
TEAN	3	24	113	39	2.87
TEAN	3	25	311	35	3.25
TEAN	3	26	131	41	3.92
TEAN	3	27	022	28	0.88
TEAN	4	28	311	25	1.14
TEAN	4	29	111	43	3.61
TEAN	4	30	113	42	3.05
TEAN	4	31	002	44	3.41
TEAN	4	32	222	48	4.46
TEAN	4	33	200	41	4.36
TEAN	4	34	111	39	3.42
TEAN	4	35	131	28	1.06
TEAN	4	36	020	-	-
			4. Woods Estate (WEAN)		
WEAN	1	1	000	48	4.02
WEAN	1	2	311	47	5.80
WEAN	1	3	202	32	2.16
WEAN	1	4	220	43	5.31
WEAN	1	5	111	42	5.12
WEAN	1	6	131	45	5.98
WEAN	1	7	113	46	5.46
WEAN	1	8	022	43	3.45
WEAN	1	9	111	45	4.96
WEAN	2	10	111	46	4.26
WEAN	2	11	131	48	6.35
WEAN	2	12	020	47	4.28
WEAN	2	13	111	41	4.94
WEAN	2	14	311	38	4.39
WEAN	2	15	002	44	3.92
WEAN	2	16	113	43	5.74
WEAN	2	17	200	46	4.98
WEAN	2	18	222	44	5.40
WEAN	3	19	111	45	6.10
WEAN	3	20	113	45	5.86
WEAN	3	21	022	47	4.59
WEAN	3	22	311	36	4.36
WEAN	3	23	000	41	4.16
WEAN	3	24	111	46	6.58
WEAN	3	25	220	38	5.11
WEAN	3	26	131	48	5.96
WEAN	3	27	202	36	3.45
WEAN	4	28	020	43	5.77
WEAN	4	29	111	46	6.60
WEAN	4	30	200	43	3.88
WEAN	4	31	222	34	4.04
WEAN	4	32	002	32	2.76
WEAN	4	33	311	42	6.31
WEAN	4	34	131	38	5.28
WEAN	4	35	113	44	4.61
WEAN	4	36	111	46	5.65

Table 58.1 *(cont.)*

Location	Block	Plot	Treatment	Number of ears	Weight of harvest
			5. Wireless, Clare Hall (WLAN)		
WLAN	1	1	000	58	2.00
WLAN	1	2	113	56	2.39
WLAN	1	3	131	52	2.04
WLAN	1	4	202	44	1.64
WLAN	1	5	220	45	1.83
WLAN	1	6	111	47	2.08
WLAN	1	7	022	55	3.13
WLAN	1	8	111	53	1.96
WLAN	1	9	311	43	1.30
WLAN	2	10	311	68	2.16
WLAN	2	11	002	40	1.74
WLAN	2	12	111	58	2.46
WLAN	2	13	200	44	2.17
WLAN	2	14	020	42	2.72
WLAN	2	15	113	55	2.07
WLAN	2	16	111	54	2.87
WLAN	2	17	222	48	2.14
WLAN	2	18	131	38	1.45
WLAN	3	19	311	69	2.05
WLAN	3	20	022	60	2.44
WLAN	3	21	111	50	2.88
WLAN	3	22	000	40	2.28
WLAN	3	23	111	51	3.44
WLAN	3	24	113	54	2.56
WLAN	3	25	202	41	2.41
WLAN	3	26	131	53	3.24
WLAN	3	27	220	51	1.66
WLAN	4	28	200	57	1.76
WLAN	4	29	113	55	2.84
WLAN	4	30	002	39	2.25
WLAN	4	31	131	63	2.73
WLAN	4	32	111	48	3.55
WLAN	4	33	311	51	3.13
WLAN	4	34	020	39	2.41
WLAN	4	35	111	43	3.49
WLAN	4	36	222	56	1.96
			6. North Sound (NSAN)		
NSAN	1	1	202	25	2.43
NSAN	1	2	111	13	1.28
NSAN	1	3	311	10	0.83
NSAN	1	4	131	24	2.44
NSAN	1	5	022	30	1.30
NSAN	1	6	000	25	1.34
NSAN	1	7	113	25	2.06
NSAN	1	8	220	32	3.02
NSAN	1	9	111	27	2.18
NSAN	2	10	131	26	2.35
NSAN	2	11	002	28	2.15
NSAN	2	12	111	36	3.24
NSAN	2	13	020	34	2.78

Table 58.1 *(cont.)*

Location	Block	Plot	Treatment	Number of ears	Weight of harvest
NSAN	2	14	222	27	2.91
NSAN	2	15	111	32	3.10
NSAN	2	16	200	28	2.86
NSAN	2	17	113	32	3.33
NSAN	2	18	311	32	3.00
NSAN	3	19	022	32	2.75
NSAN	3	20	131	36	2.92
NSAN	3	21	000	28	1.70
NSAN	3	22	220	34	3.36
NSAN	3	23	111	30	1.58
NSAN	3	24	202	27	1.98
NSAN	3	25	311	39	2.90
NSAN	3	26	113	26	2.05
NSAN	3	27	111	24	1.40
NSAN	4	28	002	33	1.95
NSAN	4	29	111	29	2.34
NSAN	4	30	200	19	1.55
NSAN	4	31	222	22	1.73
NSAN	4	32	020	30	2.05
NSAN	4	33	131	33	3.44
NSAN	4	34	113	33	2.11
NSAN	4	35	311	35	3.22
NSAN	4	36	111	21	1.60
			7. Orange Valley (OVAN)		
OVAN	1	1	111	28	2.04
OVAN	1	2	113	35	3.88
OVAN	1	3	311	38	5.53
OVAN	1	4	131	42	6.61
OVAN	1	5	000	31	2.97
OVAN	1	6	111	35	4.47
OVAN	1	7	220	26	4.06
OVAN	1	8	022	37	5.02
OVAN	1	9	202	38	4.56
OVAN	2	10	111	41	3.50
OVAN	2	11	020	37	4.58
OVAN	2	12	131	30	4.42
OVAN	2	13	200	47	6.16
OVAN	2	14	311	45	6.09
OVAN	2	15	113	38	4.94
OVAN	2	16	222	37	5.83
OVAN	2	17	111	37	5.05
OVAN	2	18	002	30	3.90
OVAN	3	19	131	35	3.92
OVAN	3	20	022	42	4.78
OVAN	3	21	111	40	5.26
OVAN	3	22	220	40	5.52
OVAN	3	23	111	42	5.87
OVAN	3	24	311	43	6.19
OVAN	3	25	000	39	5.26
OVAN	3	26	202	30	3.08

Table 58.1 *(cont.)*

Location	Block	Plot	Treatment	Number of ears	Weight of harvest
OVAN	III	27	113	29	3.14
OVAN	IV	28	222	45	4.56
OVAN	IV	29	113	38	4.74
OVAN	IV	30	311	38	5.32
OVAN	IV	31	111	45	5.57
OVAN	IV	32	002	26	1.69
OVAN	IV	33	020	40	4.57
OVAN	IV	34	131	38	6.18
OVAN	IV	35	111	48	6.90
OVAN	IV	36	200	33	3.93
			8. Old Road (ORAN)		
ORAN	I	1	113	43	5.26
ORAN	I	2	111	44	6.87
ORAN	I	3	202	44	7.78
ORAN	I	4	311	41	6.75
ORAN	I	5	022	44	6.56
ORAN	I	6	131	42	6.47
ORAN	I	7	111	41	6.71
ORAN	I	8	220	41	6.74
ORAN	I	9	000	45	6.92
ORAN	II	10	002	36	3.75
ORAN	II	11	020	35	3.56
ORAN	II	12	113	38	5.45
ORAN	II	13	311	40	7.55
ORAN	II	14	111	40	7.20
ORAN	II	15	222	44	6.48
ORAN	II	16	200	41	7.36
ORAN	II	17	131	41	7.18
ORAN	II	18	111	48	7.53
ORAN	III	19	000	36	5.74
ORAN	III	20	311	44	7.52
ORAN	III	21	220	46	7.87
ORAN	III	22	111	43	6.39
ORAN	III	23	113	46	7.04
ORAN	III	24	111	40	6.48
ORAN	III	25	131	47	7.56
ORAN	III	26	202	46	6.87
ORAN	III	27	022	41	6.12
ORAN	IV	28	200	47	7.45
ORAN	IV	29	111	45	6.75
ORAN	IV	30	131	42	6.41
ORAN	IV	31	020	37	5.28
ORAN	IV	32	311	35	5.27
ORAN	IV	33	002	46	7.28
ORAN	IV	34	111	45	7.39
ORAN	IV	35	222	43	6.61
ORAN	IV	36	113	37	4.75

* Weight in harvest is in kilograms. Note: - denotes missing value.

Table 58.2
Treatment and Weight of Ears of Maize Harvested on the Island of St. Vincent *

Location	Block	Plot	Treatment	Weight of harvest
		1. Carapan (CPSV)		
CPSV	1	1	022	24.0
CPSV	1	2	131	27.1
CPSV	1	3	111	26.5
CPSV	1	4	000	23.1
CPSV	1	5	202	22.1
CPSV	1	6	220	24.1
CPSV	1	7	111	26.1
CPSV	1	8	113	22.5
CPSV	1	9	311	23.0
CPSV	2	10	111	26.4
CPSV	2	11	311	24.8
CPSV	2	12	111	23.8
CPSV	2	13	222	24.0
CPSV	2	14	200	22.8
CPSV	2	15	131	26.0
CPSV	2	16	002	20.5
CPSV	2	17	113	24.4
CPSV	2	18	020	26.0
CPSV	3	19	220	19.0
CPSV	3	20	111	21.0
CPSV	3	21	111	24.1
CPSV	3	22	311	23.1
CPSV	3	23	131	20.1
CPSV	3	24	022	23.2
CPSV	3	25	202	19.8
CPSV	3	26	113	19.0
CPSV	3	27	000	22.2
CPSV	4	28	131	25.2
CPSV	4	29	222	24.8
CPSV	4	30	200	21.2
CPSV	4	31	111	23.1
CPSV	4	32	002	20.9
CPSV	4	33	311	20.0
CPSV	4	34	020	20.9
CPSV	4	35	111	21.5
CPSV	4	36	113	13.4
		2. Mount Pleasant (MPSV)		
MPSV	1	1	311	16.7
MPSV	1	2	131	13.0
MPSV	1	3	111	17.8
MPSV	1	4	111	11.5
MPSV	1	5	022	12.4
MPSV	1	6	113	12.1
MPSV	1	7	220	13.5
MPSV	1	8	000	8.5
MPSV	1	9	202	13.0

Table 58.2 *(cont.)*

Location	Block	Plot	Treatment	Weight of harvest
MPSV	2	10	131	12.2
MPSV	2	11	222	15.1
MPSV	2	12	200	11.8
MPSV	2	13	111	12.8
MPSV	2	14	002	7.1
MPSV	2	15	311	17.0
MPSV	2	16	020	6.2
MPSV	2	17	111	12.5
MPSV	2	18	113	15.2
MPSV	3	19	000	8.1
MPSV	3	20	113	13.1
MPSV	3	21	202	14.2
MPSV	3	22	311	18.9
MPSV	3	23	022	12.5
MPSV **	3	24	131	17.0
MPSV	3	25	111	13.5
MPSV	3	26	220	12.0
MPSV	3	27	111	12.8
MPSV	4	28	113	14.0
MPSV	4	29	002	10.5
MPSV	4	30	020	14.4
MPSV	4	31	111	13.8
MPSV	4	32	311	12.8
MPSV	4	33	131	14.5
MPSV	4	34	222	13.1
MPSV	4	35	200	14.0
MPSV	4	36	111	13.5
		3. Argyl (AGSV)		
AGSV	1	1	220	10.4
AGSV	1	2	000	9.1
AGSV	1	3	131	9.2
AGSV	1	4	022	10.1
AGSV	1	5	311	15.2
AGSV	1	6	111	13.9
AGSV	1	7	113	15.1
AGSV	1	8	111	12.2
AGSV	1	9	202	9.8
AGSV	2	10	113	11.8
AGSV	2	11	002	10.1
AGSV	2	12	020	13.0
AGSV	2	13	111	13.2
AGSV	2	14	311	10.2
AGSV	2	15	131	12.0
AGSV	2	16	222	13.2
AGSV	2	17	200	10.9
AGSV	2	18	111	15.1
AGSV	3	19	220	9.9
AGSV	3	20	022	12.0

Table 58.2 *(cont.)*

Location	Block	Plot	Treatment	Weight of harvest
AGSV	3	21	113	12.5
AGSV	3	22	111	14.5
AGSV	3	23	131	12.2
AGSV	3	24	111	11.5
AGSV	3	25	202	11.5
AGSV	3	26	311	16.5
AGSV	3	27	000	7.8
AGSV	4	28	200	13.0
AGSV	4	29	311	7.0
AGSV	4	30	111	12.4
AGSV	4	31	222	14.0
AGSV	4	32	131	13.1
AGSV	4	33	111	13.0
AGSV	4	34	020	16.0
AGSV	4	35	113	14.5
AGSV	4	36	002	11.0
		4. Colonaire (CASV)		
CASV	1	1	311	22.0
CASV	1	2	113	18.2
CASV	1	3	022	14.4
CASV	1	4	111	22.4
CASV	1	5	220	22.3
CASV	1	6	111	20.0
CASV	1	7	000	16.0
CASV	1	8	202	17.2
CASV	1	9	131	16.1
CASV	2	10	311	22.1
CASV	2	11	113	20.3
CASV	2	12	020	16.0
CASV	2	13	111	23.8
CASV	2	14	131	20.2
CASV	2	15	111	24.5
CASV	2	16	222	26.8
CASV	2	17	200	19.9
CASV	2	18	002	13.7
CASV	3	19	202	9.1
CASV	3	20	113	12.2
CASV	3	21	000	11.1
CASV	3	22	131	10.5
CASV	3	23	311	21.8
CASV	3	24	111	17.2
CASV	3	25	111	20.5
CASV	3	26	220	20.0
CASV	3	27	022	16.1
CASV	4	28	002	19.1
CASV	4	29	111	22.2
CASV	4	30	111	24.2
CASV	4	31	131	20.2

Table 58.2 *(cont.)*

Location	Block	Plot	Treatment	Weight of harvest
CASV	4	32	311	21.1
CASV	4	33	200	22.8
CASV	4	34	020	20.1
CASV	4	35	113	20.3
CASV	4	36	222	22.0
		5. Sans Souci (SSSV)		
SSSV	1	1	022	9.5
SSSV	1	2	202	10.5
SSSV	1	3	000	14.1
SSSV	1	4	220	9.7
SSSV	1	5	131	11.2
SSSV	1	6	111	13.1
SSSV	1	7	311	9.2
SSSV	1	8	113	13.1
SSSV	1	9	111	11.5
SSSV	2	10	111	11.2
SSSV	2	11	020	12.1
SSSV	2	12	002	11.0
SSSV	2	13	111	16.1
SSSV	2	14	200	12.5
SSSV	2	15	131	9.9
SSSV	2	16	222	17.5
SSSV	2	17	131	11.8
SSSV	2	18	311	12.2
SSSV	3	19	202	8.7
SSSV	3	20	131	16.4
SSSV	3	21	113	16.1
SSSV	3	22	220	9.1
SSSV	3	23	022	12.2
SSSV	3	24	111	12.4
SSSV	3	25	111	10.0
SSSV	3	26	000	12.3
SSSV	3	27	311	9.0
SSSV	4	28	111	15.2
SSSV	4	29	113	20.2
SSSV	4	30	020	13.7
SSSV	4	31	111	14.2
SSSV	4	32	200	14.5
SSSV	4	33	222	12.0
SSSV	4	34	002	11.5
SSSV	4	35	311	11.2
SSSV	4	36	131	10.2
		6. Union (UISV)		
UISV	1	1	202	15.4
UISV	1	2	000	12.2
UISV	1	3	220	26.0
UISV	1	4	111	20.1
UISV	1	5	022	14.1

Table 58.2 *(cont.)*

Location	Block	Plot	Treatment	Weight of harvest
UISV	1	6	131	21.8
UISV	1	7	113	20.1
UISV	1	8	311	28.1
UISV	1	9	111	23.1
UISV	2	10	311	22.4
UISV	2	11	113	28.9
UISV	2	12	111	19.0
UISV	2	13	131	21.4
UISV	2	14	111	20.5
UISV	2	15	020	14.0
UISV	2	16	200	29.5
UISV	2	17	002	13.3
UISV	2	18	222	23.7
UISV	3	19	202	25.1
UISV	3	20	220	30.1
UISV	3	21	111	23.1
UISV	3	22	113	24.5
UISV	3	23	131	17.8
UISV	3	24	022	15.1
UISV	3	25	311	25.8
UISV	3	26	111	23.2
UISV	3	27	000	17.2
UISV	4	28	113	20.4
UISV	4	29	111	22.5
UISV	4	30	200	22.2
UISV	4	31	311	26.8
UISV	4	32	131	23.0
UISV	4	33	002	11.4
UISV	4	34	020	19.5
UISV	4	35	222	25.5
UISV	4	36	111	19.8
		7. Orange Hill (OOSV)		
OOSV	1	1	111	16.0
OOSV	1	2	000	12.0
OOSV	1	3	113	21.1
OOSV	1	4	311	22.1
OOSV	1	5	111	20.0
OOSV	1	6	220	23.2
OOSV	1	7	202	20.6
OOSV	1	8	131	12.8
OOSV	1	9	022	9.0
OOSV	2	10	020	12.3
OOSV	2	11	002	11.0
OOSV	2	12	111	17.0
OOSV	2	13	113	24.0
OOSV	2	14	311	28.5
OOSV	2	15	200	25.7
OOSV	2	16	111	13.1

Table 58.2 *(cont.)*

Location	Block	Plot	Treatment	Weight of harvest
OOSV	2	17	222	16.3
OOSV	2	18	131	15.0
OOSV	3	19	131	13.1
OOSV	3	20	022	10.0
OOSV	3	21	202	14.0
OOSV	3	22	220	18.8
OOSV	3	23	111	14.0
OOSV	3	24	113	11.0
OOSV	3	25	111	18.1
OOSV	3	26	000	12.1
OOSV	3	27	311	28.0
OOSV	4	28	311	20.6
OOSV	4	29	113	10.1
OOSV	4	30	200	19.2
OOSV	4	31	002	7.6
OOSV	4	32	020	8.2
OOSV	4	33	222	23.1
OOSV	4	34	111	15.8
OOSV	4	35	111	18.6
OOSV	4	36	113	20.0
		8. Orange Hill (OTSV)		
OTSV	1	1	022	12.2
OTSV	1	2	113	19.8
OTSV	1	3	131	19.9
OTSV	1	4	311	25.3
OTSV	1	5	111	19.0
OTSV	1	6	111	20.5
OTSV	1	7	000	7.8
OTSV	1	8	202	17.8
OTSV	1	9	220	19.7
OTSV	2	10	020	13.1
OTSV	2	11	311	22.4
OTSV	2	12	113	23.2
OTSV	2	13	111	17.8
OTSV	2	14	200	20.0
OTSV	2	15	002	20.0
OTSV	2	16	131	15.6
OTSV	2	17	222	15.7
OTSV	2	18	111	29.1
OTSV	3	19	311	13.1
OTSV	3	20	022	20.4
OTSV	3	21	111	10.8
OTSV	3	22	000	18.4
OTSV	3	23	220	14.0
OTSV	3	24	131	14.2
OTSV	3	25	111	14.1
OTSV	3	26	113	18.4
OTSV	3	27	202	20.5

Table 58.2 *(cont.)*

Location	Block	Plot	Treatment	Weight of harvest
OTSV	4	28	113	13.0
OTSV	4	29	020	8.4
OTSV	4	30	002	10.0
OTSV	4	31	311	14.1
OTSV	4	32	200	18.1
OTSV	4	33	222	22.2
OTSV	4	34	131	14.2
OTSV	4	35	111	12.1
OTSV	4	36	111	13.0
		9. Lanley Park (LPSV)		
LPSV	1	1	311	14.2
LPSV	1	2	111	8.2
LPSV	1	3	220	3.5
LPSV	1	4	022	7.2
LPSV	1	5	131	9.2
LPSV	1	6	202	8.5
LPSV	1	7	111	6.7
LPSV	1	8	000	7.5
LPSV	1	9	113	14.2
LPSV	2	10	111	4.5
LPSV	2	11	200	8.0
LPSV	2	12	222	5.5
LPSV	2	13	131	9.1
LPSV	2	14	111	7.2
LPSV	2	15	311	8.7
LPSV	2	16	020	6.1
LPSV	2	17	113	16.3
LPSV	2	18	002	15.2
LPSV	3	19	111	7.1
LPSV	3	20	000	6.1
LPSV	3	21	220	16.0
LPSV	3	22	202	16.8
LPSV	3	23	131	13.5
LPSV	3	24	022	8.8
LPSV	3	25	111	18.2
LPSV	3	26	311	13.5
LPSV	3	27	113	15.9
LPSV	4	28	111	17.5
LPSV	4	29	113	11.0
LPSV	4	30	131	11.4
LPSV	4	31	111	14.9
LPSV	4	32	200	24.5
LPSV	4	33	002	9.0
LPSV	4	34	020	7.1
LPSV	4	35	222	12.9
LPSV	4	36	311	6.8

* The weight of harvest is in kilograms.

** The observation MPSV III 24 is described as uneven.

59. Disorder and Mineral Content in Apples

Source Ratkowsky, D.A. and Martin, D. (1974). The use of multivariate analysis in identifying relationships among disorder and mineral element content in apples. *Aust. J. Agric. Res.* **25**, 783-790.

Contributor D.A. Ratkowsky CSIRO, Hobart

Apple fruits are subjected to chemical analysis for their mineral element content and to examination for disorders such as bitter pit, breakdown, scald and fungal rots. The relationships among bitter pit incidence, calcium deficiency and mean fruit weight per tree is illustrated by using data obtained on Jonathan apples from potted trees.

Ratkowsky and Martin feel that although the analyses of variance and covariance are useful for indicating whether the data exhibit differences due to treatment, multivariate methods, for example principal component analyses, are required to study inter-relationships among variables.

The experiment consisted of 48 trees given four treatments, and divided into four blocks, the blocks comprising four plots of three trees each. The trees received one single-level and three double-level nitrogen treatments, the extra nitrogen being supplied respectively as (i) urea, (ii) calcium and potassium nitrate and (iii) ammonia and ammonium sulphate. Six of the 48 trees did not bear fruit, but the fruits of those that did were stored separately for 4 months at 33°F and examined for incidence of bitter pit and other disorders. The data are given in Table 59.1.

Table 59.1
Post-storage Samples from Jonathan Apples in Pot Culture, 1970: Mineral Content, Mean Fruit Weight and Incidence of Bitter Pit

Block no.	TN (ppm)	PN (ppm)	P (ppm)	K (ppm)	Ca (ppm)	Mg (ppm)	Mean fruit weight (g)	Bitter pit incidence (%)
Treatment A - Control								
1	3580	1790	932	8220	244	410	85.3	0.0
1	2880	1670	836	9840	142	367	113.8	3.2
1	3260	1530	740	8180	269	387	92.9	0.0
2	2870	1700	926	7550	272	332	48.9	0.0
2	3430	1800	899	9520	202	370	99.4	3.6
2	2930	1490	847	8310	272	413	79.1	0.0
3	3110	1700	770	8180	297	389	70.0	2.7
3	3300	1840	891	8970	225	362	86.9	1.8
4	3370	1780	899	9420	212	403	87.7	6.5
4	3290	1730	879	7240	206	330	67.3	4.3
Treatment B - Urea								
1	3040	1810	798	10760	138	414	117.5	47.0
1	4470	2020	886	9990	151	401	98.9	39.6
2	5810	2400	1037	11340	165	479	108.5	44.2
2	4610	2070	840	9070	151	351	104.4	19.0
2	4690	2070	914	9730	199	429	96.8	10.0
3	3010	1780	813	9830	159	403	94.5	18.5
3	6740	2310	1111	11150	158	458	90.6	7.3
3	4510	2320	912	10360	163	401	100.8	23.6
4	4890	2040	925	9550	239	397	96.0	6.5
4	4340	1990	915	10440	180	428	99.9	20.4
4	4130	1870	710	9040	199	363	84.6	0.0
Treatment C - Calcium and Potassium Nitrates								
1	4250	2040	932	11830	169	408	127.1	9.5
1	3710	1810	792	10530	210	392	108.5	3.9
1	4640	2340	883	11210	172	393	99.9	1.6
2	6950	2300	1202	12910	148	510	124.8	27.2
2	4880	1800	829	11210	219	411	94.5	2.0
3	4680	1940	850	11010	224	411	99.4	2.7
3	5170	2130	862	11750	152	419	117.5	13.9
3	5730	2560	1161	12440	160	454	135.0	50.0
4	5360	2000	898	10960	211	428	85.6	3.6
4	6310	2420	984	12210	178	428	102.5	14.3
4	4370	2080	874	12560	183	404	110.8	10.0
Treatment D - Ammonium and Sulphate								
1	4700	1990	938	8830	148	349	77.4	50.0
1	5930	2720	1211	11430	128	449	91.3	54.0
1	4840	2360	1038	10370	132	406	91.3	89.5
2	7230	3280	1233	10840	120	437	81.7	70.5
2	7650	2670	1289	10800	124	455	89.2	37.5
3	5760	2610	1137	9200	147	378	69.6	64.0
3	7140	2240	1074	9300	255	420	69.0	16.0
3	7950	2730	1200	10630	172	425	73.7	39.5
4	5040	2270	869	9120	140	334	75.1	36.1
4	3850	1880	823	8520	181	334	87.0	58.6

60. A Classical Apple Experiment

Source These data are printed with the permission of the Director of the East Malling Research Station

Contributor S.C. Pearce University of Kent at Canterbury

A commercial apple tree consists of two parts grafted together. The upper, the *scion*, determines the main characteristics of the fruit and leaves, while the lower, the *root-stock*, largely determines the size and development of the tree. At the beginning of the century it was generally accepted that a root-stock propagated asexually, for example by cuttings or from a stool-bed, gave a dwarf tree, whereas one propagated sexually, that is from seed, gave a large tree.

With increasing knowledge of genetics this was seen to be implausible and many research workers began to investigate the matter. When the East Malling Research Station was founded in 1913 one of its first activities was to collect asexually produced root-stocks from all over Europe and to study them botanically. It soon appeared that there were effectively only nine kinds, though known by a confusing range of names. These were numbered I to IX. Later seven root-stocks raised from seed in Germany were added and numbered X to XVI. A clone was raised from each of these sixteen sources, a *clone* being a set of plants raised asexually from a single parent. In the experiment considered here trees of the scion, Worcester Pearmain, were grafted on root-stocks from these clones. This classical experiment, planted in the winter of 1918-1919, went far to elucidate the effects of different kinds of root-stock.

In the winter of 1933-1934 a number of these trees were removed to make more room for the rest. The data presented here came from 104 trees, eight on each of thirteen kinds of root-stock. At that stage no trees on root-stocks VIII, XI and XIV were removed; therefore no data are available for those root-stocks.

Essentially a tree grows in two ways. Activity of the cambium, the meristematic tissue beneath the bark, may be measured by the girth, that is the circumference, of the trunk above the graft union. Activity of the meristematic tissue at the apex of shoots may be measured by the total extension of shoots, though that is a very laborious record to take. It is also indicated by the weight of the tree above ground level but that can be measured only when the tree is removed. Table 60.1 gives the measurements of the four variates representing cambial and apical activity recorded four and fifteen years after planting for the thirteen types of root-stock.

Table 60.1
Measurements of the Four Variates Representing Cambial and Apical Activity Recorded Four to Fifteen Years after Planting for 13 Types of Root-stock

	Four years		Fifteen years	
Root-stock	Trunk girth, mm.	Extension growth, cm.	Trunk girth, mm.	Weight of tree above ground, lbs.
		Root-stock I		
I	111	2569	358	760
I	119	2928	375	821
I	109	2865	393	928
I	125	3844	394	1009
I	111	3027	360	766
I	108	2336	351	726
I	111	3211	398	1209
I	116	3037	362	750
		Root-stock II		
II	105	2074	409	1036
II	117	2885	406	1094
II	111	3378	487	1635
II	125	3906	498	1517
II	117	2782	438	1197
II	115	3018	465	1244
II	117	3383	469	1495
II	119	3447	440	1026
		Root-stock III		
III	107	2505	376	912
III	99	2315	444	1398
III	106	2667	438	1197
III	102	2390	467	1613
III	115	3021	448	1476
III	120	3085	478	1571
III	120	3308	457	1506
III	117	3231	456	1458
		Root-stock IV		
IV	122	2838	389	944
IV	103	2351	405	1241
IV	114	3001	405	1023
IV	101	2439	392	1067
IV	99	2199	327	693
IV	111	3318	395	1085
IV	120	3601	427	1242
IV	108	3291	385	1017

Table 60.1 *(cont.)*

	Four years		Fifteen years	
Root-stock	Trunk girth, mm.	Extension growth, cm.	Trunk girth, mm.	Weight of tree above ground, lbs.
		Root-stock V		
V	91	1532	404	1084
V	115	2552	416	1151
V	114	3083	479	1381
V	105	2330	442	1242
V	99	2079	347	673
V	122	3366	441	1137
V	105	2416	464	1455
V	113	3100	457	1325
		Root-stock VI		
VI	111	2813	376	800
VI	75	840	314	606
VI	105	2199	375	790
VI	102	2132	399	853
VI	105	1949	334	610
VI	107	2251	321	562
VI	113	3064	363	707
VI	111	2469	395	952
		Root-stock VII		
VII	96	2091	266	414
VII	91	1583	241	335
VII	120	4099	380	885
VII	110	3383	401	1012
VII	102	2785	296	489
VII	105	2785	315	616
VII	110	3387	358	788
VII	108	3082	343	733
		Root-stock IX		
IX	83	1344	231	375
IX	90	2247	250	410
IX	87	1426	219	335
IX	94	2211	275	560
IX	69	877	205	251
IX	84	1431	213	272
IX	90	1863	266	478
IX	90	2001	226	278
		Root-stock X		
X	105	1964	299	506
X	117	2516	381	882
X	113	3016	362	737
X	113	3424	372	772
X	122	3174	369	827
X	109	2865	368	821
X	117	3634	408	1149
X	122	3393	410	1035

Table 60.1 *(cont.)*

	Four years		Fifteen years	
Root-stock	Trunk girth, mm.	Extension growth, cm.	Trunk girth, mm.	Weight of tree above ground, lbs.
		Root-stock XII		
XII	129	4387	431	1609
XII	135	4166	465	1658
XII	138	4595	484	1789
XII	142	5131	527	2375
XII	132	4041	463	1556
XII	123	3848	412	1418
XII	142	5471	514	2266
XII	144	5956	522	2508
		Root-stock XIII		
XIII	121	3705	387	1052
XIII	120	2886	414	1167
XIII	123	3856	387	981
XIII	109	2763	390	944
XIII	100	2223	327	737
XIII	116	2905	424	1392
XIII	117	3590	421	1326
XIII	105	2516	382	1052
		Root-stock XV		
XV	122	3484	448	1258
XV	116	2730	435	1304
XV	124	3924	451	1290
XV	122	3580	450	1288
XV	125	3355	428	1176
XV	122	3694	424	1177
XV	126	4698	482	1331
XV	119	3566	469	1490
		Root-stock XVI		
XVI	126	4299	452	1499
XVI	113	3432	412	1412
XVI	113	3357	425	1488
XVI	138	5475	460	1751
XVI	127	4482	464	1937
XVI	115	3333	457	1823
XVI	120	3960	463	1838
XVI	119	4040	473	1817

61. Mastitis Control by Penicillin and Novobiocin

Source Heald, C.W., Jones, G.M., Nickerson, S. and Bibb, T.L. (1977). Mastitis control by penicillin and novobiocin at drying-off. *Canadian Veterinary J.* **18**, 171-180.

Contributor G.G. Koch University of North Carolina Chapel Hill

In a study of the treatment of mastitis, cows from sixteen Southwestern Virginia Holstein dairy farms were used, including both milking parlour and stanchion barn herds. These herds also reflected a range of subjective management scores of poor, fair, good and excellent, as judged by the technicians who visited the farms on a weekly basis. Cows were assigned sequentially as they were identified, rather than randomly, to the treatments.

In addition, dairymen sometimes requested that certain cows be treated. These cows were given one of the eight active, non-control, treatments. The problems were:

(i) to optimize the dose of penicillin for the treatment of *Streptococcus agalactiae* at drying-off when the drug was used in combination with novobiocin for the treatment of *Staphylococcus aureus,*; and

(ii) to determine if a combination of penicillin and novobiocin was justified for treatment of both infections at drying-off.

The infection state of each quarter of the udder was measured both before and after treatment. Note that "cow" rather than "quarter of the udder" is the basic experimental unit. Therefore, the assumption of independence among the quarters of an udder is not required.

Koch, Grizzle, Semenya and Sen (1978) analyzed these categorical data by several methods, namely (a) assessing the extent of interaction among experimental factors and the implications to model fitting for describing relationships, (b) accounting for pre-treatment scores as a covariate and (c) undertaking multivariate analyses with respect to the four quarters simultaneously.

The data are given in Table 61.1.

References

Koch, G.G., Grizzle, J.E. Semenya, K. and Sen, P.K. (1978). Statistical methods for evaluation of mastitis treatment data. University of North Carolina, Chapel Hill, Institute of Statistics Mimeo Series No. 1156.

Table 61.1
Mastitis Treatment Data *

Cow identification number	Herd	Treatment	Q11	Q21	Q31	Q41	Q12	Q22	Q32	Q42	Herd management score	Herd milking type
8	1	1	0	0	0	7	0	3	0	0	3	0
3	1	1	7	1	1	1	0	0	1	1	3	0
124	1	1	0	0	0	1	0	0	0	0	3	0
123	1	1	2	2	7	2	0	0	2	0	3	0
120	1	2	0	2	0	7	0	0	0	0	3	0
2	1	2	0	1	7	7	2	1	2	0	3	0
275	1	2	0	0	0	0	0	0	0	0	3	0
122	1	3	0	0	0	0	0	0	0	0	3	0
4	1	3	0	0	0	2	0	0	0	0	3	0
125	1	4	1	1	2	1	0	0	2	0	3	0
6	1	4	0	0	0	0	2	0	0	2	3	0
119	1	5	7	7	7	0	0	2	2	0	3	0
187	1	5	2	2	0	1	2	0	0	0	3	0
5	1	5	0	0	0	0	2	2	2	0	3	0
121	1	6	7	7	7	7	2	7	0	2	3	0
10	1	6	7	0	0	0	0	0	0	0	3	0
9	1	6	0	0	0	0	2	2	2	1	3	0
11	1	7	7	0	0	0	0	0	0	0	3	0
118	1	7	0	0	0	0	2	2	1	2	3	0
117	1	7	7	7	7	7	7	0	0	0	3	0
1	1	8	0	0	0	0	0	0	0	0	3	0
188	1	8	2	1	1	2	1	2	2	1	3	0
126	1	8	2	0	0	7	2	0	0	0	3	0
7	1	8	7	7	1	7	2	2	0	0	3	0
183	1	9	0	7	0	0	6	7	7	2	3	0
182	1	9	1	0	1	1	2	2	0	0	3	0
181	1	9	0	0	0	0	7	0	2	0	3	0
16	2	1	7	0	0	0	0	0	7	0	3	0
15	2	2	7	7	2	7	0	0	0	0	3	0
14	2	3	0	0	0	0	0	0	0	0	3	0
13	2	4	0	0	0	0	0	0	0	0	3	0
12	2	5	7	7	7	7	0	0	0	0	3	0
189	2	6	0	0	0	0	0	0	0	0	3	0
19	2	6	0	0	2	0	0	0	0	0	3	0
17	2	8	0	2	0	0	0	0	0	0	3	0
190	2	8	0	0	0	0	0	0	0	0	3	0
96	2	9	7	0	0	0	0	7	0	0	3	0
194	3	1	0	0	0	0	6	6	6	6	3	0
26	3	1	0	0	0	0	0	0	2	0	3	0
193	3	2	0	2	0	2	0	0	0	0	3	0
25	3	2	0	0	0	0	0	0	0	0	3	0
191	3	3	0	0	0	0	0	0	0	0	3	0
20	3	3	1	0	0	1	0	0	0	0	3	0
27	3	4	7	7	7	7	0	0	0	0	3	0
23	3	5	0	2	0	0	0	0	0	0	3	0
21	3	6	7	7	7	7	0	0	0	0	3	0
192	3	6	0	0	0	0	0	0	0	0	3	0

Table 61.1 *(cont.)*

Cow identification number	Herd	Treatment	Q11	Q21	Q31	Q41	Q12	Q22	Q32	Q42	Herd management score	Herd milking type
22	3	7	0	0	0	0	0	0	0	0	3	0
24	3	8	2	1	0	0	1	1	0	0	3	0
184	3	9	0	0	6	0	0	0	0	0	3	0
138	4	1	0	0	1	0	0	0	0	1	2	1
205	4	1	3	0	0	3	0	0	2	0	2	1
213	4	1	1	0	1	0	0	7	1	0	2	1
212	4	1	2	0	1	7	0	0	1	7	2	1
31	4	1	0	0	0	0	0	0	0	0	2	1
142	4	1	7	7	7	7	0	0	0	0	2	1
133	4	2	0	0	0	0	0	0	0	0	2	1
40	4	2	0	3	3	0	0	0	0	0	2	1
217	4	2	0	1	0	0	0	0	0	0	2	1
208	4	2	0	0	0	1	0	0	0	0	2	1
206	4	2	0	0	0	0	0	0	0	0	2	1
200	4	3	3	0	0	3	0	2	2	7	2	1
35	4	3	2	2	7	2	7	0	7	7	2	1
32	4	3	0	0	0	0	0	0	0	0	2	1
26	4	3	7	7	7	2	0	0	0	0	2	1
135	4	3	0	0	0	7	0	0	0	0	2	1
203	4	3	7	2	0	0	0	0	0	0	2	1
34	4	4	0	0	7	0	0	0	7	0	2	1
41	4	4	0	3	0	3	2	0	0	0	2	1
198	4	4	0	0	0	0	0	0	6	6	2	1
204	4	4	3	0	7	3	0	0	0	0	2	1
210	4	4	2	7	7	7	0	0	0	0	2	1
29	4	4	0	0	0	0	0	0	0	0	2	1
209	4	5	0	0	0	0	0	0	0	0	2	1
207	4	5	0	0	0	0	0	0	0	0	2	1
36	4	5	0	6	0	0	0	7	0	0	2	1
33	4	5	0	0	0	0	0	0	0	0	2	1
214	4	5	7	2	0	0	0	0	0	0	2	1
199	4	5	2	2	2	2	0	0	0	0	2	1
215	4	6	1	0	0	1	0	0	0	0	2	1
211	4	6	0	0	0	7	0	0	0	2	2	1
140	4	6	0	7	0	0	2	0	2	0	2	1
201	4	6	0	0	0	0	0	0	0	0	2	1
216	4	6	0	1	1	1	0	0	0	2	2	1
30	4	7	2	0	0	0	0	6	0	0	2	1
37	4	7	0	0	0	2	0	2	0	6	2	1
202	4	8	0	0	0	3	7	0	0	0	2	1
141	4	8	7	0	7	2	0	0	0	0	2	1
139	4	8	0	7	7	0	0	0	7	0	2	1
137	4	8	1	1	7	0	1	1	0	0	2	1
136	4	8	0	7	2	0	0	0	0	0	2	1
134	4	8	2	2	2	2	0	0	0	0	2	1
143	4	9	2	3	3	1	0	3	3	1	2	1
97	4	9	7	7	7	2	0	7	1	2	2	1
98	4	9	7	0	2	0	0	0	2	0	2	1

Table 61.1 *(cont.)*

Cow identification number	Herd	Treatment	Q11	Q21	Q31	Q41	Q12	Q22	Q32	Q42	Herd management score	Herd milking type
39	4	9	0	0	0	0	0	0	7	7	2	1
38	4	9	0	0	0	0	0	6	6	2	2	1
185	4	9	2	3	3	1	0	3	3	1	2	1
46	5	1	0	0	0	0	0	7	0	2	3	0
43	5	1	3	0	3	3	0	2	0	0	3	0
230	5	1	7	7	7	0	0	0	0	0	3	0
226	5	2	0	0	6	6	7	0	0	7	3	0
225	5	2	0	0	0	7	0	0	0	2	3	0
49	5	2	2	2	3	3	0	0	0	0	3	0
47	5	3	0	0	0	0	0	0	0	6	3	0
224	5	3	0	0	0	7	0	0	0	0	3	0
219	5	3	7	7	7	7	0	0	0	0	3	0
144	5	4	7	0	0	7	0	0	0	0	3	0
42	5	4	1	3	1	1	6	0	0	0	3	0
44	5	5	7	0	0	7	0	2	2	7	3	0
223	5	5	3	0	0	0	0	0	2	2	3	0
229	5	5	0	3	3	0	0	2	2	0	3	0
228	5	6	7	7	0	0	0	0	0	0	3	0
221	5	6	0	2	0	0	0	2	0	0	3	0
48	5	6	1	1	3	2	0	0	0	0	3	0
220	5	7	1	1	1	0	0	0	0	0	3	0
45	5	8	3	0	0	0	0	0	0	0	3	0
218	5	8	7	7	3	7	0	0	0	0	3	0
222	5	8	0	0	0	7	0	0	0	0	3	0
227	5	8	0	0	0	0	0	0	0	2	3	0
99	5	9	7	2	7	3	0	3	7	3	3	0
186	5	9	7	0	1	7	0	0	1	0	3	0
100	5	9	2	7	2	0	0	7	0	0	3	0
52	6	1	7	7	7	7	0	0	0	0	2	1
50	6	1	0	0	0	0	0	0	0	0	2	1
151	6	2	0	2	0	0	2	7	0	0	2	1
146	6	2	7	7	0	0	0	0	0	0	2	1
236	6	2	7	3	3	3	0	0	7	7	2	1
234	6	3	0	0	0	3	0	2	0	0	2	1
235	6	3	0	2	0	7	0	0	0	0	2	1
150	6	4	7	0	0	0	0	0	0	0	2	1
231	6	4	0	2	0	0	0	0	0	0	2	1
147	6	4	3	3	3	3	0	0	0	0	2	1
152	6	5	0	0	0	0	0	0	0	0	2	1
51	6	6	7	7	0	1	0	0	0	0	2	1
154	6	6	0	0	0	0	0	0	0	0	2	1
233	6	6	3	0	0	7	0	0	1	0	2	1
232	6	7	0	7	0	0	0	0	0	0	2	1
148	6	7	0	0	0	0	0	0	0	0	2	1
149	6	8	7	0	7	0	0	0	0	0	2	1
153	6	8	2	7	7	7	0	0	0	7	2	1
145	6	8	3	3	3	3	0	7	0	0	2	1
101	6	9	7	0	0	0	2	0	7	7	2	1

Table 61.1 *(cont.)*

Cow identification number	Herd	Treatment	Q11	Q21	Q31	Q41	Q12	Q22	Q32	Q42	Herd management score	Herd milking type
102	6	9	0	0	2	2	0	0	7	7	2	1
58	7	1	7	0	0	0	0	0	0	0	4	1
244	7	2	7	0	0	0	0	7	0	0	4	1
156	7	2	0	0	0	0	0	0	0	2	4	1
59	7	3	0	0	0	0	0	0	0	0	4	1
157	7	3	0	0	0	0	0	0	0	0	4	1
245	7	4	7	0	0	0	2	0	7	0	4	1
60	7	4	0	0	0	0	0	0	0	0	4	1
155	7	5	3	2	0	3	0	0	0	0	4	1
54	7	6	0	0	0	0	0	0	0	0	4	1
56	7	6	2	0	2	3	2	0	0	0	4	1
246	7	7	0	0	0	0	0	0	0	0	4	1
53	7	8	0	0	0	0	0	0	0	2	4	1
247	7	8	0	0	0	0	0	0	0	0	4	1
55	7	8	7	0	0	0	0	0	2	0	4	1
61	7	8	0	0	0	0	0	0	0	0	4	1
103	7	9	0	0	0	0	1	0	0	0	4	1
57	7	9	0	0	0	0	7	0	0	7	4	1
158	8	1	0	0	0	0	0	0	0	0	4	0
63	8	2	0	0	0	0	0	0	0	0	4	1
160	8	3	0	0	7	7	0	7	7	7	4	0
161	8	4	0	0	0	0	0	0	0	0	4	0
62	8	5	0	0	0	0	0	0	0	0	4	0
159	8	6	7	0	0	7	0	0	0	0	4	0
64	8	7	7	0	0	7	0	0	0	0	4	0
69	9	1	2	3	3	3	2	0	0	0	2	0
165	9	1	3	3	3	3	2	2	2	0	2	0
163	9	2	2	1	7	7	1	1	0	1	2	0
65	9	2	3	2	3	3	2	2	2	1	2	0
250	9	3	7	7	7	7	0	1	1	2	2	0
67	9	4	0	1	6	3	0	0	0	0	2	0
251	9	5	7	7	3	7	1	1	1	1	2	0
68	9	5	1	3	7	1	2	2	2	0	2	0
252	9	6	7	7	7	7	0	2	1	1	2	0
164	9	6	2	1	1	1	0	0	0	0	2	0
66	9	7	0	3	3	3	0	0	0	0	2	0
249	9	8	0	7	7	0	0	0	0	0	2	0
70	9	8	0	7	1	0	0	2	2	0	2	0
71	9	9	0	0	0	0	0	0	0	0	2	0
73	10	1	0	0	0	0	0	0	7	0	2	1
253	10	1	0	0	0	0	0	0	2	0	2	1
254	10	2	0	0	0	0	0	0	0	7	2	1
167	10	2	2	2	2	2	0	0	0	0	2	1
74	10	2	3	0	0	0	0	0	0	0	2	1
169	10	3	7	3	7	3	0	2	7	0	2	1
77	10	3	2	2	0	2	0	2	0	2	2	1
72	10	4	1	0	0	0	0	0	0	0	2	1
265	10	4	7	3	7	7	0	0	2	0	2	1

Table 61.1 *(cont.)*

Cow identification number	Herd	Treatment	Q11	Q21	Q31	Q41	Q12	Q22	Q32	Q42	Herd management score	Herd milking type
255	10	5	2	2	0	0	0	7	6	0	2	1
76	10	5	0	0	0	2	0	0	2	0	2	1
75	10	6	0	0	0	0	0	0	0	0	2	1
267	10	6	7	0	0	7	0	0	0	0	2	1
168	10	6	7	0	0	1	0	0	0	0	2	1
166	10	6	0	0	3	0	0	0	3	0	2	1
80	10	7	0	0	0	2	0	7	0	0	2	1
79	10	7	1	0	3	7	0	0	0	0	2	1
170	10	8	0	0	0	0	0	0	0	0	2	1
264	10	8	7	7	7	3	0	0	0	0	2	1
256	10	8	0	0	0	0	0	3	0	0	2	1
78	10	8	6	7	2	0	6	7	0	7	2	1
266	10	9	7	7	7	7	0	7	7	7	2	1
269	11	1	7	0	7	7	1	0	2	0	1	1
81	11	1	7	7	7	2	0	0	0	0	1	1
89	11	1	0	0	0	0	2	0	0	0	1	1
87	11	2	0	0	0	7	0	0	0	0	1	1
270	11	2	0	0	0	0	0	0	3	0	1	1
178	11	2	0	7	7	1	0	0	0	0	1	1
84	11	3	0	0	0	7	0	0	0	2	1	1
273	11	4	0	0	0	0	0	0	0	3	1	1
85	11	4	7	7	2	7	0	0	0	0	1	1
82	11	5	7	7	7	7	0	0	0	0	1	1
274	11	5	1	7	7	7	0	2	2	1	1	1
86	11	6	7	7	7	7	0	0	3	0	1	1
268	11	6	7	0	0	0	0	0	0	0	1	1
179	11	6	3	7	3	3	1	1	3	1	1	1
88	11	6	7	0	2	7	0	0	0	0	1	1
271	11	7	7	0	1	7	0	0	0	0	1	1
272	11	8	7	7	7	7	0	0	0	7	1	1
180	11	8	3	0	0	3	0	0	2	0	1	1
177	11	8	7	0	7	0	0	0	7	0	1	1
83	11	8	7	0	0	2	0	0	0	0	1	1
104	11	9	7	7	0	7	7	7	0	7	1	1
92	12	1	0	0	2	1	1	0	0	1	1	1
162	12	2	0	2	0	2	0	0	0	0	1	1
94	12	3	7	7	7	7	2	0	0	0	1	1
91	12	6	2	2	2	0	0	0	0	0	1	1
90	12	7	7	1	7	7	0	1	0	2	1	1
93	12	8	7	0	7	0	0	0	0	0	1	1
248	12	8	0	0	7	1	0	0	0	1	1	1
95	12	9	2	1	0	0	1	1	0	0	1	1
258	13	1	0	3	0	0	0	0	0	0	3	1
172	13	1	7	1	0	3	0	7	0	0	3	1
261	13	2	7	0	0	3	0	0	0	0	3	1
260	13	2	7	7	7	1	6	0	0	0	3	1
262	13	2	0	1	0	2	0	0	0	0	3	1
173	13	3	7	7	0	7	0	0	0	0	3	1

Table 61.1 *(cont.)*

Cow identification number	Herd	Treatment	Q11	Q21	Q31	Q41	Q12	Q22	Q32	Q42	Herd management score	Herd milking type
263	13	4	0	2	7	7	0	0	0	0	3	1
105	13	5	2	0	0	0	0	0	0	0	3	1
176	13	6	7	0	0	0	0	0	0	0	3	1
175	13	6	7	7	0	3	0	0	0	0	3	1
257	13	7	0	0	0	0	0	0	0	0	3	1
259	13	8	0	0	0	0	0	0	2	0	3	1
171	13	8	1	0	0	1	0	0	1	1	3	1
174	13	9	0	0	0	3	0	0	0	0	3	1
113	14	1	7	7	0	7	0	0	1	0	3	1
108	14	1	0	0	0	0	0	0	0	0	3	1
128	14	2	7	7	0	7	0	0	0	0	3	1
111	14	3	7	7	7	7	0	0	0	0	3	1
110	14	5	7	7	1	7	2	0	0	0	3	1
109	14	6	7	7	7	7	0	7	7	7	3	1
127	14	7	3	0	0	0	0	0	0	7	3	1
107	14	7	0	0	0	0	0	0	0	0	3	1
106	14	8	0	0	0	7	0	0	0	0	3	1
112	14	9	0	0	0	0	0	0	0	0	3	1
131	15	1	0	2	0	0	0	0	0	0	3	0
197	15	1	0	0	0	0	0	0	0	0	3	0
195	15	2	7	7	7	0	0	2	0	2	3	0
114	15	2	0	0	0	0	0	2	0	0	3	0
132	15	4	0	0	0	0	0	0	0	0	3	0
196	15	5	7	7	7	7	0	0	0	0	3	0
116	15	6	0	0	0	1	0	0	0	2	3	0
115	15	8	0	2	2	2	0	2	0	2	3	0
130	15	8	0	2	0	0	0	0	0	0	3	0
129	15	9	0	0	0	0	2	0	0	0	3	0
241	16	3	1	3	7	7	2	2	2	2	2	1
243	16	4	0	2	0	2	2	0	2	2	2	1
239	16	5	0	0	0	0	0	0	0	0	2	1
237	16	6	0	0	0	0	0	0	0	0	2	1
240	16	7	0	1	0	0	0	0	0	2	2	1
242	16	8	7	0	7	0	0	0	0	0	2	1
238	16	9	0	7	0	0	7	0	0	0	2	1

* Code	Treatment
1	200,000 I. U. penicillin
2	400,000 I. U. penicillin
3	200,000 I. U. penicillin + 400 mg. novobiocin
4	400,000 I. U. penicillin + 400 mg. novobiocin
5	100,000 I. U. penicillin + 400 mg. novobiocin
6	400 mg. novobiocin
7	600 mg. novobiocin
8	100,000 I. U. penicillin
9	No treatment (control)

Q11 : Pre-treatment infection state of the left front quarter of the cow's udder.
Q21 : Pre-treatment infection state of the left rear quarter of the cow's udder.
Q31 : Pre-treatment infection state of the right rear quarter of the cow's udder.
Q41 : Pre-treatment infection state of the right front quarter of the cow's udder.
Q12 : Post-treatment infection state of the left front quarter of the cow's udder.
Q22 : Post-treatment infection state of the left rear quarter of the cow's udder.
Q32 : Post-treatment infection state of the right quarter of the cow's udder.
Q42 : Post-treatment infection state of the right front quarter of the cow's udder.

Code	Organisms
0	None (infection free)
1	*Staphylococcus aureus*
2	*Staphylococcus epidermis*
3	*Streptococcus agalactiae*
6	Coliform
7	Other

Herd management score (1, Poor; 2, Fair; 3, Good; 4, Excellent).

Herd milking type (0, Stanchion; 1, Milking parlour).

62. United Kingdom Pig Production 1967 - 1978

Source Ministry of Agriculture, Fisheries and Food.

Contributor G. Tunnicliffe Wilson University of Lancaster

A description of the state of the pig production is given by five indicators, measured quarterly over the twelve year period, 1967-1978. An approximate three-year cycle is evident in much of the data, and interest lies partly in explaining this by models which represent the interaction between the series. The quality of the data leaves something to be desired, being based on sampling schemes which change slightly over the twelve years, but is typical of what could reasonably be expected in many econometric situations.

Table 62.1 presents the data of five series.

Series 1 is a measure of the number of gilts in pig, taken in March, June, September and December of each year. These are sows in pig for the first time, so the series indicates the intake into the breeding herd. Series 2 is an index of the profit on the sale of pigs, being the ratio of an all-pig price to an index of pig fattener feed price.

Series 3 is the ratio of Sow and Boar slaughter numbers in the following quarter, to the total pig breeding herd size at the beginning of the quarter, starting from March 1967. It, therefore, measures the removal of pigs from the breeding herd.

Series 4 is a measure of the number of clean pigs, reared for meat as opposed to being culled from the breeding herd, which are slaughtered during the quarter, it is the main measure of pig production.

Series 5 is a measure of the actual breeding herd size.

Relevant facts of life about pigs are a gestational period of 16 weeks, with weaning at about 8 weeks, allowing two litters per year. Clean pigs are typically slaughtered from 4 to 6 months of age and the lifetime of a sow in the breeding herd is typically 4 or 5 years.

Table 62.1
United Kingdom Pig Production
1967-1978

Year	Quarter	Gilts in pig	Profit index	Sow and boar slaughter	Clean pig slaughter	Herd size
1967	1	105	8.075	10.80	2645	703
1967	2	119	7.819	9.16	2540	722
1967	3	119	7.366	9.38	2565	738
1967	4	109	8.113	10.39	2776	747
1968	1	117	7.380	9.44	2725	755
1968	2	135	7.134	8.69	2623	780
1968	3	126	7.222	9.60	2722	806
1968	4	112	7.768	11.28	3004	807
1969	1	116	7.386	11.20	2952	805
1969	2	122	6.965	9.94	2968	801
1969	3	115	6.478	11.21	2961	821
1969	4	115	8.105	11.69	3243	809
1970	1	122	8.060	9.67	3027	797
1970	2	138	7.684	7.87	2902	831
1970	3	135	7.580	8.15	3057	867
1970	4	125	7.093	8.83	3331	862
1971	1	115	6.129	10.51	3266	871
1971	2	108	6.026	9.03	3290	864
1971	3	100	6.679	9.93	3223	854
1971	4	96	7.414	10.27	3501	846
1972	1	107	7.112	9.56	3402	854
1972	2	115	7.762	8.74	3278	851
1972	3	123	7.645	10.32	3258	876
1972	4	122	8.639	10.31	3400	876
1973	1	128	7.667	9.97	3303	888
1973	2	136	8.080	8.99	3228	903
1973	3	140	6.678	12.22	3269	922
1973	4	122	6.739	12.90	3396	902
1974	1	102	5.569	14.00	3396	820
1974	2	103	5.049	12.77	3386	819
1974	3	89	5.642	12.61	3385	797
1974	4	77	6.808	12.16	3262	751
1975	1	89	6.636	11.16	3113	743
1975	2	94	8.241	8.90	2851	744
1975	3	104	7.968	10.24	2752	747
1975	4	108	8.044	10.29	2919	764
1976	1	119	7.791	10.03	2842	759
1976	2	126	7.024	9.05	2834	807
1976	3	119	6.102	12.00	2957	798
1976	4	103	6.053	12.70	3305	811
1977	1	96	5.941	12.95	3256	752
1977	2	95	5.386	11.69	3151	761
1977	3	80	5.811	12.65	3141	719
1977	4	88	6.716	12.21	3266	741
1978	1	93	6.923	10.68	3061	745
1978	2	105	6.939	10.47	3018	764
1978	3	107	6.705	11.05	3085	764
1978	4	100	6.914	10.31	3242	786

63. Comparison of Family Sizes: A Problem from the 1941 Canadian Census

Source Keyfitz, N. (1952). Differential fertility in Ontario. An application of factorial design to a demographic problem. *Population Studies* **6**, 123-134.

Keyfitz, N. (1953). A factorial arrangement of comparisons of family size. *Amer. J. Sociology* **53**, 470-480.

Contributor N. Keyfitz University of Toronto

Most of the numerous studies of differential fertility are based either on census or vital registration records which are substantially complete or on large portions of a population which are considered as samples. Before the nature of sampling error was understood it seemed dangerous to base conclusions on a small number of cases, but this need no longer be the case. Small samples may make possible investigations which avoid the limitations of census tables in which only a few variables can be cross-tabulated. In any serious attempt to study the effect of a particular variable, freed from the effects of other relevant variables, at least ten directions of simultaneous cross-classification are required rather than the three or four which are the maximum usually given in a census.

The variability of family size within cells is small enough for samples comprising a few hundred families to reveal all important differentials significantly. The work involved in hand tabulation is trifling; the entire project described required a small number of man-days of clerical work, including the searching of the Canadian census schedules for 1941 in the stacks where they were located and the calculation of all necessary tests of significance.

Two investigations were made: (i) French Catholic families in the Province of Quèbec, and (ii) English Protestant families in the Province of Ontario. It was of interest to find whether (a) the relationship between distance from cities and family size and (b) the relationship between education (and income) and family size were the same for both provinces. It was also of interest to ascertain the differences in family size between Quèbec and Ontario for groups similar in age at marriage, etc.

The purpose of the investigation of French Catholic families in the Province of Quèbec was to determine whether families living near cities had fewer children than those living farther away. The data consisted of 1056 families arranged in a 2^6 factorial design, i.e. one in which two levels could be compared for each of six factors. Special interest attaches to a by-product of the sample; both schooling and income showed a reversal of the usual inverse relationship to fertility. The data are given in Table 63.1.

The sample was confined to the Province of Quèbec. By measurement on a map the distance from the centre of each of the cities over 30,000 population in or near the province (Montreal, Quèbec, Hull-Ottawa, Trois Rivières, Sherbrooke) was found for the nearest point of each county.

The investigation was confined to 16 counties, chosen to show maximum contrast in respect to distance from cities. Income could not be controlled for individual families, since farm income is recorded in the Canadian census on a schedule filed separately from the population schedule, and the two cannot be matched without a great amount of effort; selection of counties was accordingly in income strata. The two income classes were (a) one in which the average annual income per farm family worker lay between $195 and $290, and (b) one in which the average was between $300 and $350. Exhibit 63.1 shows the counties which came into the study.

Exhibit 63.1
Counties Contrasting in Distance from Cities, in each of Two Income Classes for the Province of Quèbec

$195-$290 Average Net Income per Farm Family Worker	$305-$335 Average Net Income per Farm Family Worker
Distant from Cities	
Gaspé E.	Chicoutimi
Gaspé W.	Lac St. Jean E.
Madeleine Is.	Saguenay
Matane	
Matapedia	
Near Cities	
Chambly	Champlain
Deux Montagnes	Levis
Laprairie	Papineau
L'Assomption	
Napierville	

Examination of the map of Quèbec shows that in the lower-income group the five near counties are all clustered about Montreal, while the five distant ones are along the south shore of the St. Lawrence River and in the gulf. In the higher income group the three near counties are close to Trois Rivières, Quèbec City and Hull, and the distant ones are along the north shore of the St. Lawrence River.

In order to get a sufficient contrast of distance, it was necessary to make the income range rather broad for the low incomes; however, it turned out that the unweighted average of incomes per head in the distant places was $253 and in the near places $252; it was, therefore, considered that whatever benefit can be secured by the equalizing of the farm incomes of the counties had been secured.

Within the counties an effectively random sample, all those found in one-fifth of the enumeration areas, of all families meeting the qualifications described was selected. Though the sample was actually every fifth area taken systematically for convenience, there seemed little reason to fear that its departure from randomness would make the usual tests of significance inapplicable. There was no significant effect of clustering.

The conditions of selection of the individual families controlled extraneous factors far more effectively than is possible in published tabulations. The following conditions were imposed: (a) husband and wife both French-speaking and of French origin; (b) husband and wife both of Roman Catholic religion; (c) husband and wife both born on farm and in 1941 living on farm; (d) husband the operator of the farm on which the family lives; and (e) wife between 45 and 74 years of age in 1941, and married at 15 to 24 years of age. Confining the survey to this group excludes widows, French Protestants, and other combinations which, though not numerous, may confuse the results.

Further census information was used to sharpen comparisons without restricting the scope of the survey. The group of families selected was classified in the following way: (1) net farm income per farm worker in the county low ($195-$290) or high ($300-$350); (2) 1941 age of wife 45-54 years or 55-74 years; (3) age of wife at marriage 15-19 or 20-24; (4) schooling of wife less than seven years or seven or more; (5) farm in a district purely French or mixed, according as they contained fewer than 5 English families or 5 or more, and (6) farm near or far from cities of 30,000 and over population.

Because of the vastly simpler calculation needed to extract the information from a table involving only two classes for each variable and also because the problem was conceived essentially as one of finding out whether or not there was an effect due to cities, presence of English, etc., rather than how much effect, two values only of each of the independent variables are recognized.

The investigation for the families in the Province of Ontario followed similar lines to those of the one in the Province of Quèbec. The measure of income used was simply the ratio of gross farm income less farm expenses given in the 1941 Census of Agriculture, divided by the number of family workers. The forty-three counties of Ontario containing appreciable farm populations were arranged in order of this measure and the sixteen highest and sixteen lowest were selected. Within each income group the counties were arranged in order of their distance from the nearest city with a population of over 30,000 and the four nearest and four farthest counties from each income group were taken. The sixteen counties so chosen were supplemented by the addition of the next two in order of distance from cities in the "low income, far from cities" class, because of the small yield of qualifying families in this class. The counties specified by this procedure are given in Exhibit 63.2.

Exhibit 63.2
Counties Contrasting in Distance from Cities, in each of Two Income Classes for the Province of Ontario

Low Income	High Income
Distant from Cities	
Kenora	Victoria
Haliburton	Peterborough
Muskoka	Grey
Cochrane	Bruce
Rainy River	
Timiskaming	
Near Cities	
Norfolk	Wellington
Russell	Halton
Haldimand	Oxford
Lennox and Addington	Peel

Within the eighteen counties that constitute the universe for this study a systematic sample of one-tenth of the enumeration areas was selected, except for the six counties that were in the 'low income-far from cities' group, in which one-fifth of the areas were used. The selected enumeration areas were scanned line by line and the number of children ever born to the mother was recorded for each of the families within them that met the qualifications. The work sheet was divided into columns such as those of Table 63.2, providing for all combinations of (1) schooling under 7 years or 7 years and over, (2) age at marriage of mother 15-19 or 20-24 years, (3) present age of mother 45-54 or 55-74 years. For a family to qualify it was required that (1) both husband and wife be of Protestant religion, British origin and English mother tongue, (2) both be born on a farm, now living on a farm and living in the same municipality since childhood, (3) the husband to be a farm operator, either working by himself or employing labour. The procedure described yielded 475 families, the sizes of which are shown in Table 63.2.

A difficulty presented by Tables 63.1 and 63.2 is the unequal number of observations in the several cells. Yates (1934) showed an extremely simple method for dealing with unequal sub-class numbers which is applicable in the case of dichotomous variables.

References

Yates, F. (1934). The analysis of multiple classifications with unequal numbers in the different classes. *J. Amer. Statist. Ass.* **29**, 51-66.

Table 63.1
Quèbec Catholic Families, from 1941 Census, Showing for Each Cell Average Number of Children Ever Born and Number of Families on which Average is Based

Present age of mother	Age at marriage	Years of schooling	Income class	Language	Near city	Average number of children	Number of families
45-54	15-19	0-6	Low	French	Far	9.4	15
45-54	15-19	0-6	Low	French	Near	7.4	5
45-54	15-19	0-6	Low	mixed	Far	12.9	14
45-54	15-19	0-6	Low	mixed	Near	9.7	3
45-54	15-19	0-6	High	French	Far	10.9	35
45-54	15-19	0-6	High	French	Near	8.3	6
45-54	15-19	0-6	High	mixed	Far	12.8	9
45-54	15-19	0-6	High	mixed	Near	10.5	15
45-54	15-19	7+	Low	French	Far	10.7	14
45-54	15-19	7+	Low	French	Near	12.9	8
45-54	15-19	7+	Low	mixed	Far	10.9	11
45-54	15-19	7+	Low	mixed	Near	11.3	7
45-54	15-19	7+	High	French	Far	12.9	29
45-54	15-19	7+	High	French	Near	8.7	15
45-54	15-19	7+	High	mixed	Far	14.3	10
45-54	15-19	7+	High	mixed	Near	12.2	6
45-54	20-24	0-6	Low	French	Far	10.3	35
45-54	20-24	0-6	Low	French	Near	8.3	10
45-54	20-24	0-6	Low	mixed	Far	8.9	15
45-54	20-24	0-6	Low	mixed	Near	9.4	14
45-54	20-24	0-6	High	French	Far	10.6	24
45-54	20-24	0-6	High	French	Near	7.1	7
45-54	20-24	0-6	High	mixed	Far	9.4	14
45-54	20-24	0-6	High	mixed	Near	7.6	25
45-54	20-24	7+	Low	French	Far	9.8	20
45-54	20-24	7+	Low	French	Near	6.7	37
45-54	20-24	7+	Low	mixed	Far	9.8	21
45-54	20-24	7+	Low	mixed	Near	7.1	49
45-54	20-24	7+	High	French	Far	9.8	29
45-54	20-24	7+	High	French	Near	10.3	28
45-54	20-24	7+	High	mixed	Far	11.2	13
45-54	20-24	7+	High	mixed	Near	8.8	12

Table 63.1 *(cont.)*

Present age of mother	Age at marriage	Years of schooling	Income class	Language	Near city	Average number of children	Number of families
55-74	15-19	0-6	Low	French	Far	10.1	18
55-74	15-19	0-6	Low	French	Near	10	9
55-74	15-19	0-6	Low	mixed	Far	8.3	16
55-74	15-19	0-6	Low	mixed	Near	9	12
55-74	15-19	0-6	High	French	Far	12.1	31
55-74	15-19	0-6	High	French	Near	10.8	14
55-74	15-19	0-6	High	mixed	Far	10.6	14
55-74	15-19	0-6	High	mixed	Near	11	14
55-74	15-19	7+	Low	French	Far	14.5	6
55-74	15-19	7+	Low	French	Near	11	8
55-74	15-19	7+	Low	mixed	Far	12.8	9
55-74	15-19	7+	Low	mixed	Near	9.9	8
55-74	15-19	7+	High	French	Far	12.5	15
55-74	15-19	7+	High	French	Near	13.2	18
55-74	15-19	7+	High	mixed	Far	12	2
55-74	15-19	7+	High	mixed	Near	11	3
55-74	20-24	0-6	Low	French	Far	10.4	34
55-74	20-24	0-6	Low	French	Near	7.6	15
55-74	20-24	0-6	Low	mixed	Far	8.4	16
55-74	20-24	0-6	Low	mixed	Near	8.6	17
55-74	20-24	0-6	High	French	Far	9	22
55-74	20-24	0-6	High	French	Near	10.9	14
55-74	20-24	0-6	High	mixed	Far	9.9	9
55-74	20-24	0-6	High	mixed	Near	8.6	26
55-74	20-24	7+	Low	French	Far	9.8	12
55-74	20-24	7+	Low	French	Near	8.6	22
55-74	20-24	7+	Low	mixed	Far	9.6	17
55-74	20-24	7+	Low	mixed	Near	8.6	29
55-74	20-24	7+	High	French	Far	11.3	27
55-74	20-24	7+	High	French	Near	9.9	30
55-74	20-24	7+	High	mixed	Far	9	4
55-74	20-24	7+	High	mixed	Near	8.4	10

Table 63.2
Ontario Protestant Families, from 1941 Census, Showing Number of Children Ever Born

Present age of mother	Age at marriage	Years of schooling	Income and distance *	Number of children
45-54	15-19	0-6	A	14
45-54	15-19	0-6	A	13
45-54	15-19	0-6	A	4
45-54	15-19	0-6	B	14
45-54	15-19	0-6	B	10
45-54	15-19	0-6	B	2
45-54	15-19	0-6	B	16
45-54	15-19	0-6	B	13
45-54	15-19	0-6	C	5
45-54	15-19	0-6	C	0
45-54	15-19	0-6	C	0
45-54	15-19	0-6	C	13
45-54	15-19	0-6	D	3
45-54	15-19	0-6	D	9
45-54	15-19	0-6	D	2
45-54	15-19	0-6	D	10
45-54	15-19	0-6	D	11
45-54	15-19	0-6	D	13
45-54	15-19	0-6	D	5
45-54	15-19	0-6	D	14
45-54	15-19	7+	A	0
45-54	15-19	7+	A	4
45-54	15-19	7+	A	0
45-54	15-19	7+	A	2
45-54	15-19	7+	A	3
45-54	15-19	7+	A	3
45-54	15-19	7+	A	0
45-54	15-19	7+	A	4
45-54	15-19	7+	A	7
45-54	15-19	7+	A	1
45-54	15-19	7+	B	9
45-54	15-19	7+	B	4
45-54	15-19	7+	B	3
45-54	15-19	7+	C	3
45-54	15-19	7+	C	2
45-54	15-19	7+	C	16
45-54	15-19	7+	C	6
45-54	15-19	7+	C	0
45-54	15-19	7+	C	13
45-54	15-19	7+	C	2
45-54	15-19	7+	C	6
45-54	15-19	7+	C	6
45-54	15-19	7+	C	5
45-54	15-19	7+	D	9

Table 63.2 *(cont.)*

Present age of mother	Age at marriage	Years of schooling	Income and distance *	Number of children
45-54	15-19	7+	D	10
45-54	15-19	7+	D	5
45-54	15-19	7+	D	4
45-54	15-19	7+	D	3
45-54	15-19	7+	D	3
45-54	15-19	7+	D	5
45-54	15-19	7+	D	2
45-54	15-19	7+	D	3
45-54	15-19	7+	D	5
45-54	15-19	7+	D	15
45-54	15-19	7+	D	5
45-54	20-24	0-6	A	2
45-54	20-24	0-6	B	6
45-54	20-24	0-6	B	7
45-54	20-24	0-6	B	3
45-54	20-24	0-6	B	8
45-54	20-24	0-6	B	6
45-54	20-24	0-6	B	6
45-54	20-24	0-6	B	2
45-54	20-24	0-6	B	10
45-54	20-24	0-6	C	7
45-54	20-24	0-6	C	5
45-54	20-24	0-6	C	3
45-54	20-24	0-6	C	6
45-54	20-24	0-6	C	4
45-54	20-24	0-6	C	3
45-54	20-24	0-6	C	1
45-54	20-24	0-6	C	3
45-54	20-24	0-6	C	4
45-54	20-24	0-6	C	4
45-54	20-24	0-6	D	7
45-54	20-24	0-6	D	4
45-54	20-24	0-6	D	8
45-54	20-24	0-6	D	0
45-54	20-24	0-6	D	1
45-54	20-24	0-6	D	3
45-54	20-24	0-6	D	4
45-54	20-24	0-6	D	3
45-54	20-24	0-6	D	6
45-54	20-24	0-6	D	4
45-54	20-24	0-6	D	6
45-54	20-24	0-6	D	4
45-54	20-24	0-6	D	0
45-54	20-24	0-6	D	3

Table 63.2 *(cont.)*

Present age of mother	Age at marriage	Years of schooling	Income and distance *	Number of children
45-54	20-24	0-6	D	6
45-54	20-24	0-6	D	4
45-54	20-24	0-6	D	4
45-54	20-24	0-6	D	3
45-54	20-24	0-6	D	3
45-54	20-24	0-6	D	3
45-54	20-24	0-6	D	10
45-54	20-24	0-6	D	6
45-54	20-24	0-6	D	1
45-54	20-24	0-6	D	6
45-54	20-24	0-6	D	3
45-54	20-24	7+	A	3
45-54	20-24	7+	A	4
45-54	20-24	7+	A	3
45-54	20-24	7+	A	0
45-54	20-24	7+	A	3
45-54	20-24	7+	A	5
45-54	20-24	7+	A	2
45-54	20-24	7+	A	4
45-54	20-24	7+	A	3
45-54	20-24	7+	A	5
45-54	20-24	7+	A	4
45-54	20-24	7+	A	6
45-54	20-24	7+	A	3
45-54	20-24	7+	A	2
45-54	20-24	7+	A	2
45-54	20-24	7+	A	2
45-54	20-24	7+	A	2
45-54	20-24	7+	A	3
45-54	20-24	7+	A	2
45-54	20-24	7+	A	3
45-54	20-24	7+	B	1
45-54	20-24	7+	B	4
45-54	20-24	7+	B	3
45-54	20-24	7+	B	1
45-54	20-24	7+	B	6
45-54	20-24	7+	B	2
45-54	20-24	7+	B	0
45-54	20-24	7+	C	9
45-54	20-24	7+	C	3
45-54	20-24	7+	C	0
45-54	20-24	7+	C	2
45-54	20-24	7+	C	5
45-54	20-24	7+	C	4

Table 63.2 *(cont.)*

Present age of mother	Age at marriage	Years of schooling	Income and distance *	Number of children
45-54	20-24	7+	C	3
45-54	20-24	7+	C	3
45-54	20-24	7+	C	3
45-54	20-24	7+	C	1
45-54	20-24	7+	C	3
45-54	20-24	7+	C	9
45-54	20-24	7+	C	1
45-54	20-24	7+	C	7
45-54	20-24	7+	C	1
45-54	20-24	7+	C	3
45-54	20-24	7+	C	6
45-54	20-24	7+	C	12
45-54	20-24	7+	C	2
45-54	20-24	7+	C	1
45-54	20-24	7+	C	2
45-54	20-24	7+	C	7
45-54	20-24	7+	C	5
45-54	20-24	7+	C	5
45-54	20-24	7+	C	3
45-54	20-24	7+	C	6
45-54	20-24	7+	C	2
45-54	20-24	7+	C	5
45-54	20-24	7+	C	0
45-54	20-24	7+	C	4
45-54	20-24	7+	C	2
45-54	20-24	7+	C	5
45-54	20-24	7+	C	4
45-54	20-24	7+	C	1
45-54	20-24	7+	C	1
45-54	20-24	7+	C	8
45-54	20-24	7+	C	3
45-54	20-24	7+	C	6
45-54	20-24	7+	C	0
45-54	20-24	7+	C	4
45-54	20-24	7+	C	2
45-54	20-24	7+	C	6
45-54	20-24	7+	D	7
45-54	20-24	7+	D	3
45-54	20-24	7+	D	4
45-54	20-24	7+	D	5
45-54	20-24	7+	D	4
45-54	20-24	7+	D	3
45-54	20-24	7+	D	1
45-54	20-24	7+	D	3

Table 63.2 (cont.)

Present age of mother	Age at marriage	Years of schooling	Income and distance *	Number of children
45-54	20-24	7+	D	5
45-54	20-24	7+	D	7
45-54	20-24	7+	D	5
45-54	20-24	7+	D	6
45-54	20-24	7+	D	0
45-54	20-24	7+	D	1
45-54	20-24	7+	D	1
45-54	20-24	7+	D	2
45-54	20-24	7+	D	1
45-54	20-24	7+	D	10
45-54	20-24	7+	D	1
45-54	20-24	7+	D	2
45-54	20-24	7+	D	2
45-54	20-24	7+	D	7
45-54	20-24	7+	D	4
45-54	20-24	7+	D	4
45-54	20-24	7+	D	5
45-54	20-24	7+	D	3
45-54	20-24	7+	D	3
45-54	20-24	7+	D	5
45-54	20-24	7+	D	2
45-54	20-24	7+	D	3
45-54	20-24	7+	D	4
45-54	20-24	7+	D	2
45-54	20-24	7+	D	2
45-54	20-24	7+	D	2
45-54	20-24	7+	D	1
45-54	20-24	7+	D	4
45-54	20-24	7+	D	4
45-54	20-24	7+	D	2
45-54	20-24	7+	D	0
45-54	20-24	7+	D	3
45-54	20-24	7+	D	7
45-54	20-24	7+	D	9
45-54	20-24	7+	D	5
45-54	20-24	7+	D	1
45-54	20-24	7+	D	4
45-54	20-24	7+	D	1
45-54	20-24	7+	D	3
45-54	20-24	7+	D	3
45-54	20-24	7+	D	1
45-54	20-24	7+	D	0
45-54	20-24	7+	D	4
45-54	20-24	7+	D	3

Table 63.2 *(cont.)*

Present age of mother	Age at marriage	Years of schooling	Income and distance *	Number of children
55-74	15-19	0-6	A	6
55-74	15-19	0-6	A	5
55-74	15-19	0-6	A	7
55-74	15-19	0-6	B	4
55-74	15-19	0-6	B	9
55-74	15-19	0-6	B	7
55-74	15-19	0-6	B	7
55-74	15-19	0-6	B	9
55-74	15-19	0-6	B	4
55-74	15-19	0-6	B	9
55-74	15-19	0-6	B	7
55-74	15-19	0-6	B	14
55-74	15-19	0-6	C	2
55-74	15-19	0-6	D	3
55-74	15-19	0-6	D	5
55-74	15-19	0-6	D	2
55-74	15-19	0-6	D	2
55-74	15-19	0-6	D	7
55-74	15-19	0-6	D	5
55-74	15-19	0-6	D	10
55-74	15-19	0-6	D	2
55-74	15-19	7+	A	1
55-74	15-19	7+	A	1
55-74	15-19	7+	A	2
55-74	15-19	7+	A	6
55-74	15-19	7+	A	3
55-74	15-19	7+	A	6
55-74	15-19	7+	A	3
55-74	15-19	7+	A	4
55-74	15-19	7+	A	6
55-74	15-19	7+	A	5
55-74	15-19	7+	A	0
55-74	15-19	7+	B	3
55-74	15-19	7+	B	2
55-74	15-19	7+	B	4
55-74	15-19	7+	B	6
55-74	15-19	7+	C	2
55-74	15-19	7+	C	7
55-74	15-19	7+	C	5
55-74	15-19	7+	C	4
55-74	15-19	7+	C	5
55-74	15-19	7+	C	11

Table 63.2 *(cont.)*

Present age of mother	Age at marriage	Years of schooling	Income and distance *	Number of children
55-74	15-19	7+	C	4
55-74	15-19	7+	C	3
55-74	15-19	7+	D	6
55-74	15-19	7+	D	8
55-74	15-19	7+	D	10
55-74	15-19	7+	D	6
55-74	15-19	7+	D	6
55-74	15-19	7+	D	3
55-74	15-19	7+	D	4
55-74	15-19	7+	D	8
55-74	20-24	0-6	A	0
55-74	20-24	0-6	A	1
55-74	20-24	0-6	A	5
55-74	20-24	0-6	A	2
55-74	20-24	0-6	B	6
55-74	20-24	0-6	B	9
55-74	20-24	0-6	B	5
55-74	20-24	0-6	B	5
55-74	20-24	0-6	B	0
55-74	20-24	0-6	B	7
55-74	20-24	0-6	B	4
55-74	20-24	0-6	B	10
55-74	20-24	0-6	B	10
55-74	20-24	0-6	C	3
55-74	20-24	0-6	C	6
55-74	20-24	0-6	C	8
55-74	20-24	0-6	C	1
55-74	20-24	0-6	C	9
55-74	20-24	0-6	C	2
55-74	20-24	0-6	C	6
55-74	20-24	0-6	C	2
55-74	20-24	0-6	C	3
55-74	20-24	0-6	C	9
55-74	20-24	0-6	C	8
55-74	20-24	0-6	C	3
55-74	20-24	0-6	C	5
55-74	20-24	0-6	C	5
55-74	20-24	0-6	C	10
55-74	20-24	0-6	D	5
55-74	20-24	0-6	D	11
55-74	20-24	0-6	D	1
55-74	20-24	0-6	D	1
55-74	20-24	0-6	D	3

Table 63.2 *(cont.)*

Present age of mother	Age at marriage	Years of schooling	Income and distance *	Number of children
55-74	20-24	0-6	D	6
55-74	20-24	0-6	D	7
55-74	20-24	0-6	D	3
55-74	20-24	0-6	D	6
55-74	20-24	0-6	D	12
55-74	20-24	0-6	D	9
55-74	20-24	0-6	D	9
55-74	20-24	0-6	D	8
55-74	20-24	0-6	D	8
55-74	20-24	0-6	D	1
55-74	20-24	0-6	D	5
55-74	20-24	0-6	D	5
55-74	20-24	0-6	D	2
55-74	20-24	0-6	D	6
55-74	20-24	0-6	D	2
55-74	20-24	0-6	D	3
55-74	20-24	0-6	D	2
55-74	20-24	0-6	D	4
55-74	20-24	0-6	D	0
55-74	20-24	0-6	D	4
55-74	20-24	0-6	D	2
55-74	20-24	0-6	D	2
55-74	20-24	0-6	D	1
55-74	20-24	0-6	D	9
55-74	20-24	7+	A	10
55-74	20-24	7+	A	5
55-74	20-24	7+	A	5
55-74	20-24	7+	A	3
55-74	20-24	7+	A	2
55-74	20-24	7+	A	3
55-74	20-24	7+	A	3
55-74	20-24	7+	A	0
55-74	20-24	7+	A	4
55-74	20-24	7+	A	0
55-74	20-24	7+	A	14
55-74	20-24	7+	A	0
55-74	20-24	7+	A	3
55-74	20-24	7+	A	1
55-74	20-24	7+	A	1
55-74	20-24	7+	A	1
55-74	20-24	7+	A	2
55-74	20-24	7+	A	0
55-74	20-24	7+	A	8

Table 63.2 *(cont.)*

Present age of mother	Age at marriage	Years of schooling	Income and distance *	Number of children
55-74	20-24	7+	A	3
55-74	20-24	7+	A	6
55-74	20-24	7+	A	5
55-74	20-24	7+	A	9
55-74	20-24	7+	A	0
55-74	20-24	7+	A	1
55-74	20-24	7+	A	0
55-74	20-24	7+	A	9
55-74	20-24	7+	A	5
55-74	20-24	7+	A	4
55-74	20-24	7+	A	4
55-74	20-24	7+	A	4
55-74	20-24	7+	B	5
55-74	20-24	7+	B	9
55-74	20-24	7+	B	3
55-74	20-24	7+	B	4
55-74	20-24	7+	B	9
55-74	20-24	7+	B	2
55-74	20-24	7+	B	6
55-74	20-24	7+	B	8
55-74	20-24	7+	B	7
55-74	20-24	7+	B	3
55-74	20-24	7+	B	5
55-74	20-24	7+	B	6
55-74	20-24	7+	B	9
55-74	20-24	7+	B	7
55-74	20-24	7+	C	3
55-74	20-24	7+	C	0
55-74	20-24	7+	C	4
55-74	20-24	7+	C	2
55-74	20-24	7+	C	7
55-74	20-24	7+	C	3
55-74	20-24	7+	C	2
55-74	20-24	7+	C	0
55-74	20-24	7+	C	8
55-74	20-24	7+	C	2
55-74	20-24	7+	C	6
55-74	20-24	7+	C	3
55-74	20-24	7+	C	5
55-74	20-24	7+	C	6
55-74	20-24	7+	C	3
55-74	20-24	7+	C	2
55-74	20-24	7+	C	4

Table 63.2 *(cont.)*

Present age of mother	Age at marriage	Years of schooling	Income and distance *	Number of children
55-74	20-24	7+	C	2
55-74	20-24	7+	C	1
55-74	20-24	7+	C	5
55-74	20-24	7+	C	4
55-74	20-24	7+	C	2
55-74	20-24	7+	C	5
55-74	20-24	7+	C	1
55-74	20-24	7+	C	2
55-74	20-24	7+	C	0
55-74	20-24	7+	C	2
55-74	20-24	7+	C	6
55-74	20-24	7+	C	6
55-74	20-24	7+	C	5
55-74	20-24	7+	C	6
55-74	20-24	7+	C	0
55-74	20-24	7+	C	3
55-74	20-24	7+	C	6
55-74	20-24	7+	C	6
55-74	20-24	7+	C	10
55-74	20-24	7+	C	2
55-74	20-24	7+	C	1
55-74	20-24	7+	C	1
55-74	20-24	7+	C	2
55-74	20-24	7+	C	2
55-74	20-24	7+	D	1
55-74	20-24	7+	D	6
55-74	20-24	7+	D	1
55-74	20-24	7+	D	9
55-74	20-24	7+	D	1
55-74	20-24	7+	D	1
55-74	20-24	7+	D	11
55-74	20-24	7+	D	1
55-74	20-24	7+	D	5
55-74	20-24	7+	D	4
55-74	20-24	7+	D	1
55-74	20-24	7+	D	3
55-74	20-24	7+	D	2
55-74	20-24	7+	D	1
55-74	20-24	7+	D	10
55-74	20-24	7+	D	5
55-74	20-24	7+	D	7
55-74	20-24	7+	D	3
55-74	20-24	7+	D	8

Table 63.2 *(cont.)*

Present age of mother	Age at marriage	Years of schooling	Income and distance *	Number of children
55-74	20-24	7+	D	6
55-74	20-24	7+	D	1
55-74	20-24	7+	D	4
55-74	20-24	7+	D	5
55-74	20-24	7+	D	2
55-74	20-24	7+	D	8
55-74	20-24	7+	D	4
55-74	20-24	7+	D	5
55-74	20-24	7+	D	10
55-74	20-24	7+	D	2
55-74	20-24	7+	D	6
55-74	20-24	7+	D	1
55-74	20-24	7+	D	3
55-74	20-24	7+	D	5
55-74	20-24	7+	D	1
55-74	20-24	7+	D	7
55-74	20-24	7+	D	3
55-74	20-24	7+	D	6
55-74	20-24	7+	D	8
55-74	20-24	7+	D	8
55-74	20-24	7+	D	3
55-74	20-24	7+	D	8
55-74	20-24	7+	D	7
55-74	20-24	7+	D	3
55-74	20-24	7+	D	4
55-74	20-24	7+	D	4
55-74	20-24	7+	D	4
55-74	20-24	7+	D	4
55-74	20-24	7+	D	1
55-74	20-24	7+	D	12
55-74	20-24	7+	D	3
55-74	20-24	7+	D	4
55-74	20-24	7+	D	2
55-74	20-24	7+	D	6
55-74	20-24	7+	D	4
55-74	20-24	7+	D	1
55-74	20-24	7+	D	3
55-74	20-24	7+	D	5
55-74	20-24	7+	D	10
55-74	20-24	7+	D	2
55-74	20-24	7+	D	3

* A, Low income, near city;
B, Low income, far from city;
C, High income, near city;
D, High income, far from city.

64. Canadian Unemployment Data 1956 - 1975

Contributor P.B. Kenny U.K. Central Statistical Office

The problem of shifts in seasonal behaviour is one which turns up in many time-series, but particularly in the statistics of unemployment. Standard procedures for the seasonal adjustment of economic time-series assume that the amplitude of the seasonal variation either varies in proportion as the level of the series changes (multiplicative seasonality) or is independent of the level (additive seasonality). Economic theory provides little guidance, and some preliminary analysis of the data is usually necessary.

If an approximate trend line is fitted to the data, the sum of absolute deviations from the trend over twelve months gives a measure of seasonal amplitude. A plot of amplitude against average trend for each calendar year gives a guide as to whether amplitude is roughly proportional to trend level or independent of level. Difficulties arise if there is an abrupt change. Such a plot may be useful in two ways. To the economist, it may suggest a search for physical or institutional changes which could have produced the observed effects. To the statistician concerned with seasonal adjustment, it shows the need to look at separate portions of the series when choosing a model.

Table 64.1 gives the total monthly unemployment figures for Canada from 1956 to 1975. This gives a clear example of an abrupt change from predominantly multiplicative seasonality to purely additive. Similar changes occur in unemployment data from other countries at about the same times. Durbin and Murphy (1975) give formal tests for additive or multiplicative behaviour and develop models which are applicable to series exhibiting mixtures of additive and multiplicative seasonality.

References

Durbin, J. and Murphy, M.J. (1975). Seasonal adjustment based on a mixed additive-multiplicative model. *J.R. Statist. Soc. A* **138**, 385-410.

Table 64.1
Canadian Monthly
Total Unemployment Figures *
from 1956 - 1975

	Month											
Year	Jan	Feb	Mar	Apr	May	June	July	Aug	Sept	Oct	Nov	Dec
1956	315	340	321	273	175	127	112	116	116	110	149	211
1957	328	352	378	334	209	177	181	194	214	223	318	422
1958	579	600	638	553	388	339	310	317	284	328	378	466
1959	577	570	553	466	354	248	239	257	224	250	316	405
1960	546	597	607	550	417	313	328	350	325	366	426	525
1961	690	716	702	619	454	367	351	320	305	315	347	411
1962	543	582	559	484	335	300	307	279	259	282	342	414
1963	541	546	550	463	347	305	294	271	251	266	303	346
1964	466	467	456	403	293	282	265	246	217	257	257	284
1965	407	397	387	371	265	257	244	211	176	171	220	252
1966	359	356	341	298	247	230	244	228	205	195	238	266
1967	381	396	400	365	304	292	284	247	219	254	289	353
1968	464	482	488	436	366	395	371	319	262	288	338	373
1969	467	473	448	432	386	383	349	318	279	314	354	383
1970	485	526	542	544	513	529	518	448	398	419	476	538
1971	668	675	650	659	543	551	514	455	434	447	503	530
1972	665	627	642	592	552	568	543	503	459	483	524	584
1973	688	655	608	570	493	503	461	433	421	429	468	512
1974	637	635	599	568	524	469	465	447	431	430	493	597
1975	817	839	840	795	714	704	653	623	586	576	640	697

* Figures shown are in thousands.

65. United States of America Unemployment Data, 1948-1981

Contributor T.J. Plewes U.S. Bureau of Labor

Statistics on the labour force status of the United States civilian noninstitutional population are derived from a monthly sample survey of 60,000 households, the Current Population Survey, conducted by the Bureau of the Census for the Bureau of Labor Statistics. Respondents are interviewed to obtain information on the labor force status of each member of the household, 16 years of age and over. The inquiry relates to activity or status during the calendar week, which includes the 12th of the month.

Persons classified as unemployed are those who did not work during the survey week, who make special efforts to find a job within the past four weeks, and who were available for work during the survey week.

As might be expected, the United States unemployment series moves strongly over the course of the business cycle. The series also exhibit regularly recurring seasonal movements. Unemployment of young persons typically increases dramatically in the summer months, when schools are not in session, and falls when schools re-open. Adult male unemployment typically peaks in the early part of the year, and is lowest in the latter part of the Summer and early Fall. The seasonal pattern of adult female unemployment shows twin peaks during the year increasing in the early months, then declining through the Spring and early Summer, and rising again in the late Summer and early Fall, only to decline until after the Christmas season.

Each of these series, in addition to displaying its own pattern of movement over the course of the year, behaves differently in relationship to the overall trend-cycle. Youth unemployment reflects seasonality of an additive nature; that is, the level of seasonality of the series is fairly constant over time. Adult male unemployment is much more affected by the stage of the business cycle, and exhibits seasonality of a multiplicative nature: that is, the level of seasonality tends to be proportioned to the level of the series. Adult female unemployment shows mixed behaviour, neither clearly additive nor multiplicative.

These series are each independently seasonally adjusted by the Bureau of Labor Statistics. They are used in the computation of the overall seasonally adjusted unemployment level, which, in turn, is a component in computation of the seasonally adjusted unemployment rate.

Tables 65.1, 65.2, 65.3 and 65.4 give the monthly unemployment figures from 1948-1981 for men 16-19 years, men 20 years and over, women 16-19 years and women 20 years and over. The data presented here are not adjusted for seasonality.

Table 65.1
United States of America
Monthly Employment Figures
for Males Aged 16-19 Years
from 1948-1981 *

	Month											
Year	Jan	Feb	March	April	May	June	July	Aug	Sept	Oct	Nov	Dec
1948	278	310	298	225	185	385	356	274	197	166	189	203
1949	295	294	318	301	349	550	478	408	294	290	305	351
1950	476	448	337	291	328	500	361	271	228	187	177	213
1951	223	204	171	149	147	305	278	186	152	128	180	168
1952	250	244	198	166	188	324	266	178	176	122	162	183
1953	170	174	166	132	140	280	234	176	141	162	175	252
1954	294	374	324	277	314	385	427	346	279	188	239	276
1955	309	283	243	212	244	439	354	272	223	218	233	258
1956	275	290	227	219	279	494	359	241	178	178	262	226
1957	313	256	290	248	257	507	390	270	229	226	311	298
1958	361	391	393	377	415	634	580	414	383	335	342	372
1959	396	382	346	330	354	603	523	436	336	349	327	399
1960	381	330	424	367	388	749	548	456	329	370	365	400
1961	462	481	463	443	445	811	608	472	388	373	377	423
1962	401	425	396	355	416	699	469	380	338	307	380	328
1963	386	469	465	480	585	831	670	474	416	376	441	414
1964	442	425	475	490	527	872	569	483	394	327	424	417
1965	415	419	416	463	535	844	662	456	378	398	359	405
1966	391	359	411	397	496	777	576	382	352	317	342	388
1967	391	425	372	363	337	753	620	455	375	431	450	402
1968	385	417	400	320	292	778	627	396	339	368	385	410
1969	426	405	413	352	304	675	663	407	407	403	420	410
1970	479	500	475	483	440	947	807	586	581	603	620	669
1971	707	655	638	574	552	980	926	680	597	637	660	704
1972	758	835	747	617	554	929	815	702	640	588	669	675
1973	610	651	605	592	527	898	839	614	594	576	672	651
1974	714	715	672	588	567	1057	949	683	771	708	824	835
1975	980	969	931	892	828	1350	1218	977	863	838	866	877
1976	1007	951	906	911	812	1172	1101	900	841	853	922	886
1977	896	936	902	765	735	1234	1052	868	798	751	820	725
1978	821	895	851	734	636	994	990	750	727	754	792	817
1979	856	886	833	733	675	1004	956	777	761	709	777	771
1980	840	847	774	720	848	1240	1168	936	853	910	953	874
1981	1026	1030	946	860	856	1190	1038	883	843	857	1016	1003

* Figures shown are in thousands. The monthly data are not seasonally adjusted.

Table 65.2
United States of America
Monthly Employment Figures
for Males Aged 20 Years and Over
from 1948-1981 *

	Month											
Year	Jan	Feb	March	April	May	June	July	Aug	Sept	Oct	Nov	Dec
1948	1511	1694	1614	1482	1245	1108	1195	1177	1146	981	1163	1344
1949	1923	2292	2267	2081	2154	2163	2547	2293	2115	2419	2108	2270
1950	2882	3041	2802	2429	1930	1779	1826	1443	1328	1074	1186	1343
1951	1522	1450	1216	995	859	919	954	878	862	856	928	914
1952	1230	1312	1152	1022	862	930	1062	1008	894	702	752	835
1953	1334	1188	1034	1070	894	822	886	768	814	788	1099	1529
1954	2068	2386	2444	2329	2099	1978	2015	1924	1885	1720	1739	1831
1955	2245	2248	2123	2013	1485	1381	1332	1236	1095	1118	1253	1433
1956	1788	1863	1824	1504	1428	1384	1340	1199	1122	1028	1281	1543
1957	1809	1810	1618	1521	1353	1402	1327	1258	1307	1337	1690	2062
1958	2759	3198	3318	3073	2804	2752	2779	2601	2190	2072	2132	2497
1959	2850	2926	2577	1933	1689	1659	1691	1650	1662	1630	2021	1970
1960	2417	2306	2454	2025	1740	1787	1863	1890	1703	1806	2085	2638
1961	3211	3363	3190	2773	2522	2330	2383	2275	1973	1880	2008	2309
1962	2591	2548	2458	2115	1807	1847	1803	1886	1621	1536	1826	2151
1963	2663	2787	2499	2066	1770	1745	1723	1683	1454	1466	1769	2023
1964	2406	2348	2148	1786	1467	1609	1510	1501	1378	1394	1395	1676
1965	2033	2108	1835	1586	1320	1317	1271	1279	1094	1023	1108	1249
1966	1529	1512	1403	1103	941	1049	1035	1035	871	839	971	1149
1967	1312	1310	1262	1069	955	1062	989	986	839	893	968	1069
1968	1319	1391	1219	969	843	997	921	913	816	810	844	873
1969	1142	1134	1048	901	810	905	945	886	914	906	909	1052
1970	1456	1678	1606	1498	1404	1585	1669	1625	1566	1641	1821	2115
1971	2553	2590	2379	2079	1914	2004	1994	2001	1842	1769	1939	2100
1972	2502	2478	2350	2071	1858	1921	1868	1759	1625	1667	1599	1684
1973	2026	2093	1957	1725	1558	1579	1484	1457	1343	1281	1396	1582
1974	2094	2226	2003	1855	1624	1744	1733	1805	1728	1863	2146	2665
1975	3719	3950	4009	3731	3518	3501	3418	3184	3141	3060	3184	3297
1976	3750	3723	3473	3057	2761	3035	2981	2818	2696	2710	2978	3196
1977	3611	3811	3386	2855	2637	2641	2587	2532	2221	2371	2387	2485
1978	2996	2979	2834	2347	2147	2097	2150	2090	1952	1979	2054	2309
1979	2693	2719	2596	2263	2010	2090	2217	2209	2046	2152	2272	2429
1980	3206	3182	3228	3322	3410	3507	3675	3547	3303	3210	3308	3345
1981	4066	4043	3881	3403	3400	3392	3221	3290	3145	3421	3773	4343

* Figures shown are in thousands. The monthly data are not seasonally adjusted.

Table 65.3
United States of America
Monthly Employment Figures
for Females Aged 16-19 Years
from 1948-1981 *

	Month											
Year	Jan	Feb	March	April	May	June	July	Aug	Sept	Oct	Nov	Dec
1948	116	153	142	139	93	323	256	152	138	90	128	104
1949	149	180	169	171	193	388	356	230	271	190	192	191
1950	187	210	196	122	178	344	291	179	178	136	155	169
1951	125	130	142	113	106	264	218	154	152	100	136	98
1952	132	106	100	88	154	264	216	156	136	126	116	85
1953	108	118	90	108	98	206	158	110	108	116	110	145
1954	186	192	168	179	197	313	250	194	194	167	129	125
1955	133	144	136	138	180	297	215	189	178	137	207	157
1956	155	183	196	159	235	463	299	180	172	146	191	124
1957	143	173	146	149	227	391	293	200	161	131	175	173
1958	188	195	168	241	288	511	425	249	260	210	236	176
1959	166	153	195	249	274	508	345	276	226	215	262	205
1960	224	232	232	237	297	569	341	273	264	251	255	258
1961	271	277	291	260	334	686	519	384	357	291	323	196
1962	266	263	264	309	321	542	419	305	252	252	315	251
1963	278	295	280	300	454	762	547	327	346	349	376	278
1964	295	288	280	341	438	774	464	374	327	357	335	350
1965	340	339	325	410	419	745	483	325	359	325	381	294
1966	312	275	288	390	476	827	571	403	324	343	370	275
1967	276	346	276	260	269	710	584	446	394	397	409	320
1968	265	352	322	299	324	820	675	427	402	355	391	317
1969	276	323	322	322	319	761	587	458	435	434	388	325
1970	385	378	387	400	335	835	644	552	553	531	617	463
1971	463	474	507	460	433	902	805	596	597	528	589	468
1972	514	526	537	522	405	959	816	651	626	566	568	480
1973	448	558	507	524	469	905	742	570	625	552	585	501
1974	568	556	565	452	528	1010	914	623	720	704	701	638
1975	764	698	758	643	741	1101	976	863	826	779	732	738
1976	746	718	732	656	641	1095	928	911	781	734	733	690
1977	803	718	765	691	684	1157	929	808	833	750	742	589
1978	740	696	719	647	706	1071	964	817	809	699	711	653
1979	684	652	640	648	687	1061	875	791	777	763	685	658
1980	700	700	682	594	769	1067	969	814	762	706	698	599
1981	704	717	726	705	772	1068	933	799	832	846	824	678

* Figures shown are in thousands. The monthly data are not seasonally adjusted.

Table 65.4
United States of America
Monthly Employment Figures
for Females Aged 20 Years and Over
from 1948-1981 *

	Month											
Year	Jan	Feb	March	April	May	June	July	Aug	Sept	Oct	Nov	Dec
1948	446	650	592	561	491	592	604	635	580	510	553	554
1949	628	708	629	724	820	865	1007	1025	955	889	965	878
1950	1103	1092	978	823	827	928	838	720	756	658	838	684
1951	779	754	794	681	658	644	622	588	720	670	746	616
1952	646	678	552	560	578	514	541	576	522	530	564	442
1953	520	484	538	454	404	424	432	458	556	506	633	708
1954	1013	1031	1101	1061	1048	1005	987	1006	1075	854	1008	777
1955	982	894	795	799	781	776	761	839	842	811	843	753
1956	848	756	848	828	857	838	986	847	801	739	865	767
1957	941	846	768	709	798	831	833	798	806	771	951	799
1958	1156	1332	1276	1373	1325	1326	1314	1343	1225	1133	1075	1023
1959	1266	1237	1180	1046	1010	1010	1046	985	971	1037	1026	947
1960	1097	1018	1054	978	955	1067	1132	1092	1019	1110	1262	1174
1961	1391	1533	1479	1411	1370	1486	1451	1309	1316	1319	1233	1113
1962	1363	1245	1205	1084	1048	1131	1138	1271	1244	1139	1205	1030
1963	1300	1319	1198	1147	1140	1216	1200	1271	1254	1203	1272	1073
1964	1375	1400	1322	1214	1096	1198	1132	1193	1163	1120	1164	966
1965	1154	1306	1123	1033	940	1151	1013	1105	1011	963	1040	838
1966	1012	963	888	840	880	939	868	1001	956	966	896	843
1967	1180	1103	1044	972	897	1103	1056	1055	1287	1231	1076	929
1968	1105	1127	988	903	845	1020	994	1036	1050	977	956	818
1969	1031	1061	964	967	867	1058	987	1119	1202	1097	994	840
1970	1086	1238	1264	1171	1206	1303	1393	1463	1601	1495	1561	1404
1971	1705	1739	1667	1599	1516	1625	1629	1809	1831	1665	1659	1457
1972	1707	1607	1616	1522	1585	1657	1717	1789	1814	1698	1481	1330
1973	1646	1596	1496	1386	1302	1524	1547	1632	1668	1421	1475	1396
1974	1706	1715	1586	1477	1500	1648	1745	1856	2067	1856	2104	2061
1975	2809	2783	2748	2642	2628	2714	2699	2776	2795	2673	2558	2394
1976	2784	2751	2521	2372	2202	2469	2686	2815	2831	2661	2590	2383
1977	2670	2771	2628	2381	2224	2556	2512	2690	2726	2493	2544	2232
1978	2494	2315	2217	2100	2116	2319	2491	2432	2470	2191	2241	2117
1979	2370	2392	2255	2077	2047	2255	2233	2539	2394	2341	2231	2171
1980	2487	2449	2300	2387	2474	2667	2791	2904	2737	2849	2723	2613
1981	2950	2825	2717	2593	2703	2836	2938	2975	3064	3092	3063	2991

* Figures shown are in thousands. The monthly data are not seasonally adjusted.

66. Interorganizational Resource Links in Towertown, U.S.A.

Source E.O. Laumann University of Chicago

Contributors S.E. Fienberg Carnegie-Mellon and J. Galaskiewicz University of Minnesota

As part of a more elaborate study of social organization, sociologists from the University of Chicago gathered information on the formal organizations in a midwestern U.S. community of 32,000 persons, referred to here by its pseudonym, Towertown. This city lies 60 miles from a large metropolitan centre, and contains a large state university with an enrollment of almost 24,000 students. A total of 109 formal organizations were identified; for more details, see Galaskiewicz and Marsden (1978) and Galaskiewicz (1979). These organizations included all manufacturing firms having more than 20 employees, banks, savings and loans, law firms, business associations, service clubs, labour unions, city offices and departments, political organizations, mass media organizations, health institutions, public welfare institutions, educational institutions and churches.

The executive officers or principal agents of 73 of these 109 organizations were interviewed, and these are listed, in part using pseudonyms, in Table 66.1. Of those organizations not interviewed 10 were unions, mainly small ones, 22 churches, mainly small, a law firm refused, a cable television company, a medical association and a welfare association had disbanded.

In each of the 73 interviewed organizations, the officers or agents provided various types of background information. They were presented with a list of the other organizations in the community, and were asked the following six questions:

A. To which organizations on this list would (your organization) be likely to pass on important information concerning community affairs (or other matters that might affect them)?

B. And to which organizations on this list does (your organization) rely upon for information regarding community affairs (or other matters that might affect it)?

C. Now to which organizations on this list does (your organization) give substantial funds as payment for services rendered or goods received, loans, or donations?

D. And from which organizations on this list does (your organization) get substantial funds as payment for services rendered or goods provided, loans or donations?

E. Which organizations on this list does (your organization) feel a special duty to stand behind in time of trouble: that is, to which organizations would (your organization) give support?

F. Finally, which organizations on this list would be likely to come to (your organization's) support in time of trouble?

From the responses to these questions, three relations were defined for the group of 73 organizations by Galaskiewicz (1979), one each for information (Questions A and B), money (Questions C and D), and support (Questions E and F). An organization was determined to be "in relation to" another organization if the former organization answered yes to the first question in a pair, or the latter organization the second question in the pair. For each relation, a 73×73 sociomatrix, $\mathbf{X}$, was constructed with entries

$$x_{ij} = \begin{cases} 1 & \text{if organization } i \text{ relates to organization } j \\ 0 & \text{otherwise} \end{cases}$$

and with $x_{ii} = 0$. These three sociomatrices are included as Table 66.2, and have been analyzed in part by Fienberg and Wasserman (1981) and by Fienberg, Meyer and Wasserman (1981); see also Galaskiewicz and Marsden (1978).

References

Fienberg, S.E., Meyer, M.M. and Wasserman, S. (1981). Analyzing data from multivariate directed graphs: An application to social networks. In *Interpreting Multivariate Data*, edited by V. Barnett. Chichester: Wiley, pp. 289-306.

Fienberg, S.E. and Wasserman, S (1981). Categorical data analysis of single sociometric relations. In *Sociological Methodology 1981,* edited by S. Leinhardt, San-Francisco: Jossey-Bass, pp. 156-192.

Galaskiewicz, J. (1979). *Exchange Networks and Community Politics.* Beverly Hills: Sage.

Galaskiewicz, J. and Marsden, P.V. (1978). Interorganizational resource networks: Formal patterns of overlap. *Social Science Research* 7, 89-107.

Table 66.1
Towertown Organizations

1. Farm Bureau	39. The Towertown Newspaper
2. Farm Equip Co.	40. WTWR Radio Station
3. Clothing Mfg. Co.	41. Towertown Public Hospital Board
4. Farm Supply Co.	42. Towertown Public Hospital
5. Mechanical Co.	43. County Medical Society
6. Electric Equip. Co.	44. County Board of Mental Health
7. Metal Products Co.	45. County Board of Health
8. Music Equip. Co.	46. County Health Service Center (private)
9. Chamber of Commerce	47. State Highway Authority
10. Bankers' Association	48. Kiwanis Club - 1
11. 1st Towertown Bank	49. Kiwanis Club - 2
12. Towertown Savings and Loan	50. Rotary Club
13. Bank of Towertown	51. Lions Club
14. 2nd Towertown Bank	52. United Fund
15. Brinkman Law Firm	53. School Board
16. Cater Law Firm	54. Towertown High School
17. Lenhart Law Firm	55. Towertown Parent - Teacher Association
18. County Bar Association	56. Towertown Community College
19. Towertown Board of Realtors	57. State University
20. Towertown Small Business Association	58. Association of Churches - 1
21. Municipal Employees Union - 1	59. Association of Churches - 2
22. Municipal Employees Union - 2	60. St. Hilary's Catholic Church
23. Teacher's Union	61. 1st Baptist Church
24. Central Labor Union	62. 1st Church of the Light
25. City Council	63. 1st Congregational Church
26. City Manager's Office	64. 1st Methodist Church
27. County Board	65. Unity Lutheran
28. Fire Department	66. University Methodist Church
29. Human Relations Commission	67. State Department of Public Aid
30. Mayor's Office	68. County Housing Authority
31. Police Department	69. Family Services
32. Sanitary District	70. State Employment Services
33. Streets and Sanitation	71. YMCA
34. Park District	72. Mental Health Center
35. Zoning Board	73. Towertown Youth Services Bureau
36. Democratic Committee	
37. Republican Committee	
38. League of Women Voters	

Table 66.2
Information, Money and Support Flows for 73 Organizations *

Organ-ization	Organization 1 - 10	11 - 20	21 - 30	31 - 40	41 - 50	51 - 60	61 - 70	71 - 73
				1. Information				
1	0001000011	1110101101	0000111000	0000011011	0000110010	1000011100	0100000010	000
2	0011111110	0000000001	0100000010	0000000001	0000000000	0100001001	0000100010	000
3	0100000110	0000000000	0000000010	0000000001	0000000000	0001000000	0000000010	001
4	1100000010	1010101000	0000000010	0000000001	1000010001	0110000001	0110100010	001
5	0100000010	0000000000	0100000010	0000000001	0000000000	0000000000	0000000010	000
6	0100001010	1000000000	0100010010	0000000011	0000000001	0100000001	0000100010	000
7	0100010010	1000000000	0000011011	0000000001	0000000000	0000011000	0000000010	000
8	0110000010	1000000000	0000000010	0000000001	0000000000	0000000001	0010100010	000
9	1011111100	1111101001	1100111010	0010000001	1100001000	0110011010	1100000110	100
10	0000000000	0000000000	0000000000	0000000000	0000000000	0000000000	0100000000	000
11	1001011110	0010101010	0000111000	0000000011	0000000000	0011011011	1110000100	000
12	1000001010	1010111111	0000111101	1000100111	1101111000	1011011101	0100001111	110
13	1001000011	1000101110	0000000000	0000000011	0000000000	0011001000	0110000001	000
14	0000000010	0000000000	0000010001	0000000001	0000000000	0000000000	0100000000	000
15	1001001010	1110011100	0000000000	0000100000	1111111000	0000000000	0000000000	000
16	0000001010	1100100100	0000000000	0000000000	0100000000	0000000000	0001000101	001
17	1001000010	1110100100	0000010000	0100100011	0100001000	0010001000	0000001001	001
18	0000000010	1010000010	0000100000	0000000100	0000000000	0000000000	0000000000	000
19	0000000010	1110000001	0000110011	0001000011	0000001111	1100000000	0100000000	101
20	1000000010	0110000000	0100111011	1001000011	1000010000	0000111000	0100000000	000
21	0000000010	0000000000	0001110001	0000000011	0000000000	0000000000	0000000000	000
22	0100110010	0000000001	1011000110	1000011100	0000000000	0100001000	0000000000	000
23	0000000000	0000000000	0101000000	0000000011	0000000000	0010001000	0100000000	000
24	0000000000	0000000000	1110000000	0000000010	0000000000	0000000000	0000000000	000
25	1000000010	1100000001	0000011111	1001000111	1100000101	1000001011	1110000111	111
26	1010011010	1101001111	1000101111	1111000011	0100011100	1010101010	1111010111	011
27	1000001001	1100010001	0000110011	0000000101	0001100000	0010001010	1000000111	011
28	0000000000	0000000000	0100110010	1000000011	0100000000	0010100000	0000000100	000
29	0000000000	0000000001	0100110000	0000000111	0000000000	0000000010	1100001010	010
30	0100011010	1100000001	1000100000	0101000000	0100000000	0000000011	1100000000	011
31	0000000000	1000000001	0100110110	0010000001	0100000000	0010100000	1100000110	011
32	0000001000	0110001000	0000110000	0000000111	0000100000	0010000000	0000000001	000
33	0000000000	0000000000	0000110000	0100000011	0000000000	0000000000	0000000000	000
34	0000000000	1000000001	0000110001	0000000111	0000000111	1110000000	0000100000	101
35	0000000000	0110000001	0000100000	0000000001	0000000000	0000000000	0000000001	000
36	1000000000	0000000000	0100010000	0000000001	0000000000	0000000000	0000000000	000
37	1000000010	0000000000	0100010000	0000000001	0000000000	0000000000	0000000000	000

Table 66.2 *(cont.)*

Organ-ization	Organization 1 - 10	11 - 20	21 - 30	31 - 40	41 - 50	51 - 60	61 - 70	71 - 73
38	0000000000	1000000100	0100101010	0101000011	1000100001	0010000000	1001001100	011
39	1000010000	1110001011	1010110100	0111001101	1100011000	1011101101	1101111111	111
40	1111111110	1111001011	1010111110	1111111110	1111111111	1111111111	1111110111	110
41	0001000000	0000100001	0000110100	0000000111	0000010001	0000000010	0100001100	011
42	0000000000	0000111000	0000110001	0000000011	1001010000	0000011001	0100101110	010
43	0000000000	0000100000	0000000000	0000000001	0001100000	0000000000	0000000010	001
44	0000000000	0000100000	0000001000	0000000001	1100110000	0010000100	0000000010	011
45	1000000000	0000100000	0000011000	0000000111	0001010000	0000000000	0100001010	010
46	1001001010	1110100001	0000111001	0000000111	1111100001	0100001011	0000000010	111
47	0000000010	0100101000	0000110001	0000000011	0000000000	0000000000	0100000000	000
48	0000001000	1000000000	0000000000	0001000011	0000000000	0000000000	1000000011	100
49	1000001000	1000000000	0000100000	0001000001	0000000101	1010001000	0000100000	100
50	0101010000	1000000010	0000100000	0001000011	0100010000	0010000000	0000000000	100
51	0000000000	1100000000	0000110001	0001000011	1000000000	0010000000	0000000010	110
52	0001001000	1011000001	0100000000	0001000001	0100010000	0000001000	0100000010	100
53	1001000011	1110001011	0000110100	1001000111	0000100010	1001101010	1110000001	010
54	0100001010	0000000000	0000100010	0001000101	0000000110	1100111001	0000000011	110
55	0000000000	0000000001	0000010100	1000000011	0000000000	0011000000	0101000000	100
56	1100001010	1110000001	0000000000	0000000001	0100100000	0011001000	0000000111	100
57	1100001010	1110011101	0110111001	0000000011	0100110010	1111010001	1000000111	101
58	1000000000	0000000000	0000111000	0000000011	0000000000	0000000011	1111010010	100
59	0000000000	1010000000	0000111010	0000000000	1100000000	0010000101	1111110011	100
60	0000000000	0000000000	0000101000	0000000011	0100010000	0001001010	0000000010	000
61	0000000010	1000000000	0000111011	1000000111	0000000100	0010001010	0000000000	000
62	0000000011	1111000010	0000110000	0000000011	1000000000	0110100010	0000000000	000
63	1000000000	1010000000	0000010010	0000100011	1000000000	0000000111	0001001000	000
64	0000000000	0000000000	0000110000	0000000001	0000000000	0010000010	0000010000	000
65	0000000000	0000000000	0000000000	0000000011	0001000000	0000000010	0000000000	000
66	0000000000	0000000000	0000010000	0000000011	0000000000	0000000110	0001000000	000
67	0000000000	0110001000	0000000010	0000000100	0101100000	0110010000	0100000111	011
68	0000000010	1110010000	0000111100	1000000111	1100000000	0000011000	0100001011	001
69	1111111110	0010111111	0000111000	1000000111	1111110001	0101011111	0111001101	110
70	0000000000	0110011000	0000111000	0100100011	0000000100	0011011010	0000001110	111
71	0000001010	0100000000	0000110000	0001000011	0000010000	0100101000	0100000011	000
72	1001000000	0000010000	0000111011	1000000111	1111110001	0111000100	0100001011	101
73	0011000000	0000001010	0000111101	1001000110	0001010000	0010001000	0100001101	110

Table 66.2 *(cont.)*

Organization	1 - 10	11 - 20	21 - 30	31 - 40	41 - 50	51 - 60	61 - 70	71 - 73
				2. Money				
1	0000000000	1111000000	0000000000	0000000001	0000010000	0111010000	1000000010	000
2	0000000010	1000100000	0000000000	0000000011	0000000000	0100011000	0000000010	100
3	0000000010	1000001000	0000000000	0000000100	0000000000	0100000000	0100000010	100
4	1000000010	1110001000	0000000000	0000000000	1100010000	0111011000	0100000010	110
5	0000000010	1000000000	0000000000	0000000000	0100000000	0100000000	0000000010	000
6	0000000010	1100101000	0000000000	0000000011	0000010000	0100011000	0000000010	100
7	0000000010	1110100000	0000000000	0000000000	0000010000	0100011000	0000000010	100
8	1000000010	1100000000	0000000000	0000000000	0100010000	0000000000	0010000010	100
9	0000000100	1110000001	0000000000	0000000000	0000000000	0001000000	0000000000	000
10	0000000000	0100000000	0000000000	0000000000	1000000000	0000000000	0000000000	000
11	1101001111	0000111010	0000100000	0001000110	1100010010	1010011000	0100100110	100
12	1000000010	1000111010	0000000000	0000000100	1111010000	1000010000	0000000010	100
13	1001001011	0000101011	0000000000	0000000111	1100010000	1110011000	0100000010	100
14	0000000011	0000000000	0000000000	0000000111	0000000000	0000000000	0000000010	100
15	0000000000	1000000100	0000000000	0000011100	0100010000	0000000001	0001000010	100
16	0000000001	1010000111	0000001000	0000001100	1100101000	0000000001	0000001111	111
17	0001000010	1110000100	0000000000	0000000111	0000010001	0000101000	0001000010	100
18	0000000000	0000000000	0000000000	0000000000	0000000000	0000000000	0000000000	000
19	0000000000	1110000000	0000000000	0000000011	0000000000	0000000000	0000000000	000
20	0000000000	1100000000	0000000000	0000000000	0000000000	0000100000	0000000000	000
21	0000000000	0000000000	0001000000	0000000000	0000000000	0000000000	0000000010	000
22	0000000000	0010000000	0001000000	0000011000	0000000000	0100000000	0000000010	000
23	0000000000	0000000000	0101000000	0000000000	0000000000	0000000000	0000000010	000
24	0000000000	0010000000	0000000000	0000010000	0000000000	0000000000	0000000000	000
25	0000000000	1111000000	0000011111	1110100000	1100000000	0000001000	0000000000	001
26	0000000000	1111000001	0000000110	1010000000	1100100000	0000000000	0000000000	001
27	0000000000	1000000000	0000000000	0000000000	0001100000	0000000000	0000000100	010
28	0000000000	1100000000	0000000000	0000000000	0000000000	0000000000	0000000000	010
29	0000000000	1000000000	0000000000	0000000000	0000000000	0000000000	0000000000	000
30	0000000000	1000000000	0000100000	0000000000	0000000000	0000000000	0000000000	000
31	0000000000	1100000000	0000000000	0000000000	0000000000	0100000000	0000000000	000
32	0000000000	1110001000	0000110000	0000000000	0000000000	0000000000	0000000000	000
33	0000000000	1000000000	0000000000	0000000000	0000000000	0000000000	0000000000	000
34	0000000000	1110000000	0000000000	0000000000	0000000000	0000000000	0000000000	000
35	0000000000	1000000000	0000000000	0000000000	0000000000	0000000000	0000000000	000
36	0000000000	0010000000	0000000000	0000000011	0000000000	0000000000	0000000000	000
37	0000000000	0000000000	0000000000	0000000011	0000000000	0000000000	0000000000	000

Table 66.2 *(cont.)*

Organ-ization	Organization							
	1 - 10	11 - 20	21 - 30	31 - 40	41 - 50	51 - 60	61 - 70	71 - 73
38	0000000000	1100000000	0000000000	0000000000	0000000000	0000000000	0000000010	000
39	1111111110	1110101011	0000111111	1111111000	0000010000	0000001000	0000000010	100
40	0000000010	1110001010	0000000000	0000000000	0000010111	0100000000	0000000010	100
41	0000000000	1100010000	0000000000	0000000000	0000010000	0000000000	0000000000	000
42	0000000000	1110010000	0000010000	0000000010	0000010000	0000000000	0000000010	010
43	0000000000	0000000000	0000000000	0000000000	0000000000	0000000000	0000000010	000
44	0000000000	0010000000	0000000000	0000000000	0000000000	0000000000	0000000010	010
45	0000000000	0100000000	0000000000	0000000000	0000000000	0000000000	0000000000	000
46	0000000000	1110000000	0000000000	0000000000	0001000000	0000000000	0000000000	010
47	0000000000	0000111100	0000000000	0000000000	0000000000	0000000000	0000000000	000
48	0000000000	1100000000	0000000000	1001000010	0100010000	0101011000	0000000000	100
49	0000000000	1100000000	0000000000	0001000000	0000010000	0010000000	0000000000	100
50	0000000000	1110000000	0000000000	0000000000	0000010000	0100010000	0000000010	111
51	0000000000	0100000000	0000000000	1000000000	0000000000	0100010000	0000000010	110
52	0000000000	1100000000	0000000000	0000000000	0000000000	0000000000	0000000010	100
53	0000000000	1110001000	0000000000	0100000010	0000100000	0001010000	0000000010	000
54	0000000000	0000000000	0000000000	0000000000	0000000000	0000000000	0000000000	011
55	0000000000	0000000000	0000000000	0000000000	0000000000	0010000000	0000000000	100
56	0000000010	1000000000	0000000000	0000000011	0000000000	0000001000	0000000000	000
57	0000000000	1010001000	0000000100	0010000000	0000000000	0100000000	0000000000	100
58	0000000000	0000000000	0000000000	0000000010	0000010000	0000000000	0000000000	000
59	0000000000	1100000000	0000000000	0000000000	0000000000	0000000000	0000000000	000
60	0000000000	1100000000	0000000000	0000000000	0000000000	0111000000	0000000010	000
61	0000000000	1100000000	0000000000	0000000011	0000000000	0100000000	0000000010	000
62	0000000000	1010000000	0000001000	0000100011	0000000000	0000000000	0000000010	000
63	0000000000	1110000000	0000000000	0000000000	0000000000	0000000100	0000000010	000
64	0000000000	1100001000	0000000000	0000000001	0000000000	0000000100	0000000010	000
65	0000000000	1000000000	0000000000	0000000011	0100000000	0000001100	0000000010	000
66	0000000000	1000000000	0000000000	0000000000	0000000000	0000000100	0000000010	000
67	0000000000	0110000000	0000000100	0000000000	0100100000	0000011000	0000000000	010
68	0000000000	1000010000	0000000000	0000000000	0000000000	0000000000	0000000000	000
69	0000000000	1100000000	0000000000	0000000000	0000000000	0000000000	0000000000	100
70	0000000000	0100000000	0000000000	0000000000	0100000000	0000000000	0000000000	000
71	0000000000	1110000000	0000000000	0000000000	0000000000	0000000000	0000000010	000
72	0000000000	0010001000	0000000000	0000000011	0100100000	0000000000	0000000000	001
73	0000000000	1000000000	0000000000	0000000000	0000000000	0001000000	0000000000	010

Table 66.2 *(cont.)*

Organ-ization	Organization 1 - 10	11 - 20	21 - 30	31 - 40	41 - 50	51 - 60	61 - 70	71 - 73
				3. Support				
1	0001000010	0000000001	0000011000	0000000000	0100110000	0010011000	0000000010	010
2	0000000010	1000000000	0000000000	0000000000	0100000000	0000010000	0000000010	000
3	0000000010	1000000000	0000000000	0001000000	0100000000	0000010000	0000000010	000
4	1000000010	1010001000	0000000000	0000000000	1001010001	0110011000	0000001010	111
5	0000000010	1000000000	0000000000	0000000000	0000000000	0000010000	0000000010	000
6	0000000010	1000000000	0000010000	0000000000	0000010000	0100010000	0000000010	000
7	0000000010	1000000000	0000010001	0100000000	0000010000	0100011000	0000000010	000
8	0000000010	1000000000	0000000000	0000000000	0000000000	0000010000	0010000010	000
9	0111111100	1111001001	0000001000	0010000011	0000001000	1011011000	0000000010	000
10	0000000000	0000000001	0000000000	0000000000	0000000000	0000010000	0000000010	000
11	1001001010	0000001001	0000000000	0000000000	0000010000	1010011000	0010000110	000
12	1000000010	0000001001	0000111101	1101100111	1111111111	1110011010	0010001010	110
13	1001000010	0000001001	0000000001	0001000000	0000000001	1111111000	0110000010	010
14	0000000010	0000000001	0000000000	0000000000	0000000000	0000010000	0000000000	000
15	0000010000	1000011000	0000000000	0000000010	0000010000	0000000000	0000000010	010
16	0000000000	0000100000	0000000000	0000000000	0100000000	0000000000	0000000110	000
17	0000000010	0000100100	0000000000	0100000000	0000000001	1011001000	0001000010	110
18	0000000000	0000000000	0000000000	0000000100	0000000000	0000000000	0000000010	000
19	0000000000	1110000000	0000000000	0000000000	0000000000	0100000000	0000000000	000
20	1111111110	0000000000	0000111111	1111100000	1100011111	1111111000	0000000010	111
21	0000000000	0000000000	0111110110	1000000011	0000000000	0000000000	0000000000	000
22	0000000000	0000000000	1011000000	0000000000	0000000000	0000000000	0000000000	000
23	0000000000	0000000000	1101000000	0000000000	0000000000	0000000000	0000000000	000
24	0000000000	0000000000	1110000000	0000000000	0000000000	0000000000	0000000000	000
25	0000000000	0100000000	1000011111	1011100000	1000001000	0000111000	0000000010	011
26	0000001000	0100000000	1000101100	1011100000	0000001000	1000111000	0000000000	001
27	0000001000	0100000000	0000000000	0010000000	0001110000	0000011000	0000000010	010
28	0000000000	0000000000	0000010001	1000000000	0000000000	0000011000	0000000000	000
29	0000000000	0000000000	0000000001	0000000000	0000000000	0000010000	0000000000	011
30	0000001000	0100000000	1000010000	0001000000	0100000000	0000111000	0000000000	010
31	0000000000	0100000000	0000110101	0000000000	0000000000	0000011000	1000000010	010
32	0000001000	0000001000	0000000000	0000000000	0000000000	0000011000	0000000000	000
33	0000000000	0000000000	0000000000	0000000000	0000000000	0000010000	0000000000	000
34	0000000000	0000000000	0000110001	0000000000	0000000010	1010010000	0000000000	000
35	0000000000	0000001000	0000000000	0000000000	0000000000	0000000000	0000000000	000
36	0000000000	0000000000	0000000000	0000000000	0000000000	0000000000	0000000100	000
37	0000000000	0000000000	0000000000	0000000000	0000000000	0000000000	0000000100	000

Table 66.2 *(cont.)*

Organ-ization	Organization							
	1 - 20	11 - 20	21 - 30	31 - 40	41 - 50	51 - 60	61 - 70	71 - 73
38	0000000000	0000000000	0000011010	0000000000	0000100000	0011010000	0000000000	010
39	1000000010	0100001011	0000000010	0000000000	0100010001	0010011000	0000000010	011
40	1000000010	0000000011	0000010000	0000000100	0100010001	1010111000	0000100010	110
41	0001000000	0000000000	0000000000	0000000000	0001010000	0000011000	0000000010	011
42	1100000010	1000010000	0000000000	0000000000	0010000000	0000011100	0000000010	010
43	0000000000	0000000000	0000000000	0000000000	0001100000	0000011000	0000000010	010
44	0001000000	0000000000	0000000000	0000000000	1000110000	0000011100	0000000010	011
45	0000000000	0000000000	0000111111	1111100011	1111010000	0000011100	0000000000	010
46	1001000000	0100100000	0000000000	0000000000	1110100010	0000011100	0000000010	010
47	0000000000	0000000000	0000000000	0000000000	0000000000	0000000000	0000000000	000
48	0000000000	0000000000	0000000000	0001000000	0100010001	0001011000	0010000110	110
49	0000000000	0000000000	0000000000	0001000000	0000010001	0111011000	0010000100	100
50	0001010000	1110001000	0000000000	0000000100	1111110000	0011111000	0010000110	010
51	0000000000	0100001000	0000000000	0001000000	0000000000	0101011000	0010001111	110
52	0001000000	0000000000	0000000000	0000000000	0000000010	0000000000	0100000010	110
53	0001000000	0100001000	0000000000	0011000000	0000000010	0001111100	1000000010	001
54	0000000000	0000000000	0000110000	0000000010	0000000000	0010110100	0000000010	111
55	0000000000	0000000000	0000000000	0000000000	0000000000	0011010000	0000000000	001
56	1111111111	1111000001	0000111111	1111100011	1111110111	1111101111	1111111111	111
57	0000000010	0100000000	0000001110	1110000000	1111110010	0010010100	0000010010	000
58	0000000000	0000000000	0000000000	0000000000	0101110000	0011011011	1010001111	011
59	0000000000	0000000000	0000010010	0000000000	0000000000	0010011001	1010100010	010
60	0000000000	0000000000	0000101000	0000000000	0100010000	0001011110	0001000010	001
61	0000000000	0000000000	0000000000	1000000000	0000000000	0010010110	0000000010	000
62	0000000000	0000000000	0000000010	1001000000	0001010111	1100110010	1000101111	111
63	1000000010	0000000000	0000000000	0000000100	1101000000	0110011111	1101110010	111
64	0000000000	0000001000	0000010010	0000000000	0001100000	0010010110	0000011110	011
65	0000000000	0000000000	0000000000	0000000000	0000000000	0000010110	0000000010	000
66	0000000000	0000000000	0000010000	0000000000	0000000000	0000011100	0001000010	000
67	0001000000	0000000000	0000000010	0000000000	0100000000	0000011100	0000000111	011
68	0000000000	0000000000	0000000000	0000000000	0000000010	0000011100	0000000000	000
69	0001000000	0000000000	0000000000	0000000000	0001010000	0100011100	0001001100	110
70	1000000000	0000000000	0000000000	0000000000	0000000000	0001011100	0000000010	001
71	0001000000	0000000000	0000000000	0000000000	0000000010	0100011000	0000000010	010
72	1001000000	0000000000	0000000010	0000000000	1101110000	0101011110	0000000010	001
73	0001000000	0000000000	0000000010	0000000000	1001000000	0001111100	0000000001	010

1, organization *i* relates to organization *j* ; 0, otherwise.

67. Insurance Availability in Chicago

Source U.S. Commission on Civil Rights.

Contributor S.E. Fienberg Carnegie-Mellon University

In a study of insurance availability in Chicago, the U.S. Commission on Civil Rights attempted to examine charges by several community organizations that insurance companies were *redlining* their neighbourhoods, i.e. cancelling policies, refusing to insure or to renew, etc. Data were obtained from a variety of sources. First, the Illinois Department of Insurance provided the number of cancellations, nonrenewals, new policies, and renewals of homeowners and residential fire insurance policies by ZIP code for the months of December 1977 through February 1978. The companies that provided this information to the Department account for more than 70 percent of the homeowners insurance policies written in the city of Chicago. The Department also supplied the number of FAIR Plan policies written and renewed in Chicago, by ZIP code, for the months of December 1977 through May 1978. Since most FAIR Plan policyholders secure such coverage only after they have been rejected by the voluntary market, rather than as a result of a preference for that type of insurance, the distribution of FAIR Plan policies is another measure of insurance availability in the voluntary market.

Secondly, the Chicago Police Department provided crime data, by beat, on all thefts for the year 1975. Most insurance companies claim to base their underwriting activities on loss data from preceding years, i.e. a 2-3 year lag seems reasonable for analysis purposes. The Chicago Fire Department provided similar data on fires occurring during 1975. These theft and fire data were organized by ZIP code.

Finally, the U.S. Bureau of the Census supplied data on racial composition, income and the age and value of residential units for each ZIP code in Chicago. To adjust for differences in the population size associated with different ZIP code areas, the theft data were expressed as incidents per 1,000 population, and the fire and insurance data as incidents per 100 housing units. Table 67.1 gives the information for these factors and also median family income for all predominantly residential ZIP codes in Chicago. Figure 67.1 presents a map of the Chicago ZIP code areas.

The report on this study, U.S. Commission on Civil Rights (1979), gives the following objectives:

> Of particular interest is the extent to which the racial composition, income, and age of housing of a community affect current underwriting practices after controlling for those factors which directly cause compensable

losses, e.g. fire, theft. Second, underwriting activity is compared between neighborhoods containing a high proportion of minority residents and those which are predominantly white. Third, communities with similar theft and fire rates but which differ in minority composition, age of housing, and income level are identified and insurance practices between these areas are compared.

Note that any analyses of these data cannot indicate how many people are having insurance availability problems, as no individual information is provided. Also, the insurance data has been provided by companies actually underwriting insurance in Chicago, whereas much of the problem may lie with those companies that refuse to write insurance in Chicago at all.

Other factors clearly effect insurance availability, and a more comprehensive analysis of the data should include these. The data are organized to facilitate more elaborate analysis. A useful exercise involves the plotting of data, and residuals from various analyses, on a map of Chicago, organized by ZIP code.

References

U.S. Commission on Civil Rights (1979), *Insurance Redlining: Fact not Fiction.* A report prepared by the Illinois, Indiana, Michigan, Minnesota, Ohio, and Wisconsin Advisory Committees to the U.S. Commission on Civil Rights, Washington, D.C.

Table 67.1
Information on Predominantly Residential Areas in Chicago Based on ZIP Codes *

Zip code	Racial comp.	Fire	Theft	Age	Voluntary market activity	Involuntary market activity	Income
60626	10.0	6.2	29	60.4	5.3	0.0	11744
60640	22.2	9.5	44	76.5	3.1	0.1	9323
60613	19.6	10.5	36	73.5	4.8	1.2	9948
60657	17.3	7.7	37	66.9	5.7	0.5	10656
60614	24.5	8.6	53	81.4	5.9	0.7	9730
60610	54.0	34.1	68	52.6	4.0	0.3	8231
60611	4.9	11.0	75	42.6	7.9	0.0	21480
60625	7.1	6.9	18	78.5	6.9	0.0	11104
60618	5.3	7.3	31	90.1	7.6	0.4	10694
60647	21.5	15.1	25	89.8	3.1	1.1	9631
60622	43.1	29.1	34	82.7	1.3	1.9	7995
60631	1.1	2.2	14	40.2	14.3	0.0	13722
60646	1.0	5.7	11	27.9	12.1	0.0	16250
60656	1.7	2.0	11	7.7	10.9	0.0	13686
60630	1.6	2.5	22	63.8	10.7	0.0	12405
60634	1.5	3.0	17	51.2	13.8	0.0	12198
60641	1.8	5.4	27	85.1	8.9	0.0	11600
60635	1.0	2.2	9	44.4	11.5	0.0	12765
60639	2.5	7.2	29	84.2	8.5	0.2	11084
60651	13.4	15.1	30	89.8	5.2	0.8	10510
60644	59.8	16.5	40	72.7	2.7	0.8	9784
60624	94.4	18.4	32	72.9	1.2	1.8	7342
60612	86.2	36.2	41	63.1	0.8	1.8	6565
60607	50.2	39.7	147	83.0	5.2	0.9	7459
60623	74.2	18.5	22	78.3	1.8	1.9	8014
60608	55.5	23.3	29	79.0	2.1	1.5	8177
60616	62.3	12.2	46	48.0	3.4	0.6	8212
60632	4.4	5.6	23	71.5	8.0	0.3	11230
60609	46.2	21.8	4	73.1	2.6	1.3	8330
60653	99.7	21.6	31	65.0	0.5	0.9	5583
60615	73.5	9.0	39	75.4	2.7	0.4	8564
60638	10.7	3.6	15	20.8	9.1	0.0	12102
60629	1.5	5.0	32	61.8	11.6	0.0	11876
60636	48.8	28.6	27	78.1	4.0	1.4	9742
60621	98.9	17.4	32	68.6	1.7	2.2	7520
60637	90.6	11.3	34	73.4	1.9	0.8	7388
60652	1.4	3.4	17	2.0	12.9	0.0	13842
60620	71.2	11.9	46	57.0	4.8	0.9	11040
60619	94.1	10.5	42	55.9	6.6	0.9	10332
60649	66.1	10.7	43	67.5	3.1	0.4	10908
60617	36.4	10.8	34	58.0	7.8	0.9	11156
60655	1.0	4.8	19	15.2	13.0	0.0	13323
60643	42.5	10.4	25	40.8	10.2	0.5	12960
60628	35.1	15.6	28	57.8	7.5	1.0	11260
60627	47.4	7.0	3	11.4	7.7	0.2	10080
60633	34	7.1	23	49.2	11.6	0.3	11428
60645	3.1	4.9	27	46.6	10.9	0.0	13731

* Code	Explanation
Racial comp.	Racial composition in percent minority
Fire	Fires per 1000 housing units
Theft	Thefts per 1000 population
Age	Percent of housing units built in or before 1939
Voluntary market activity	New homeowner policies plus renewals, minus cancellations and non-renewals per 100 housing units
Involuntary market activity	New fair plan policies and renewals per 100 housing units
Income	Median family income

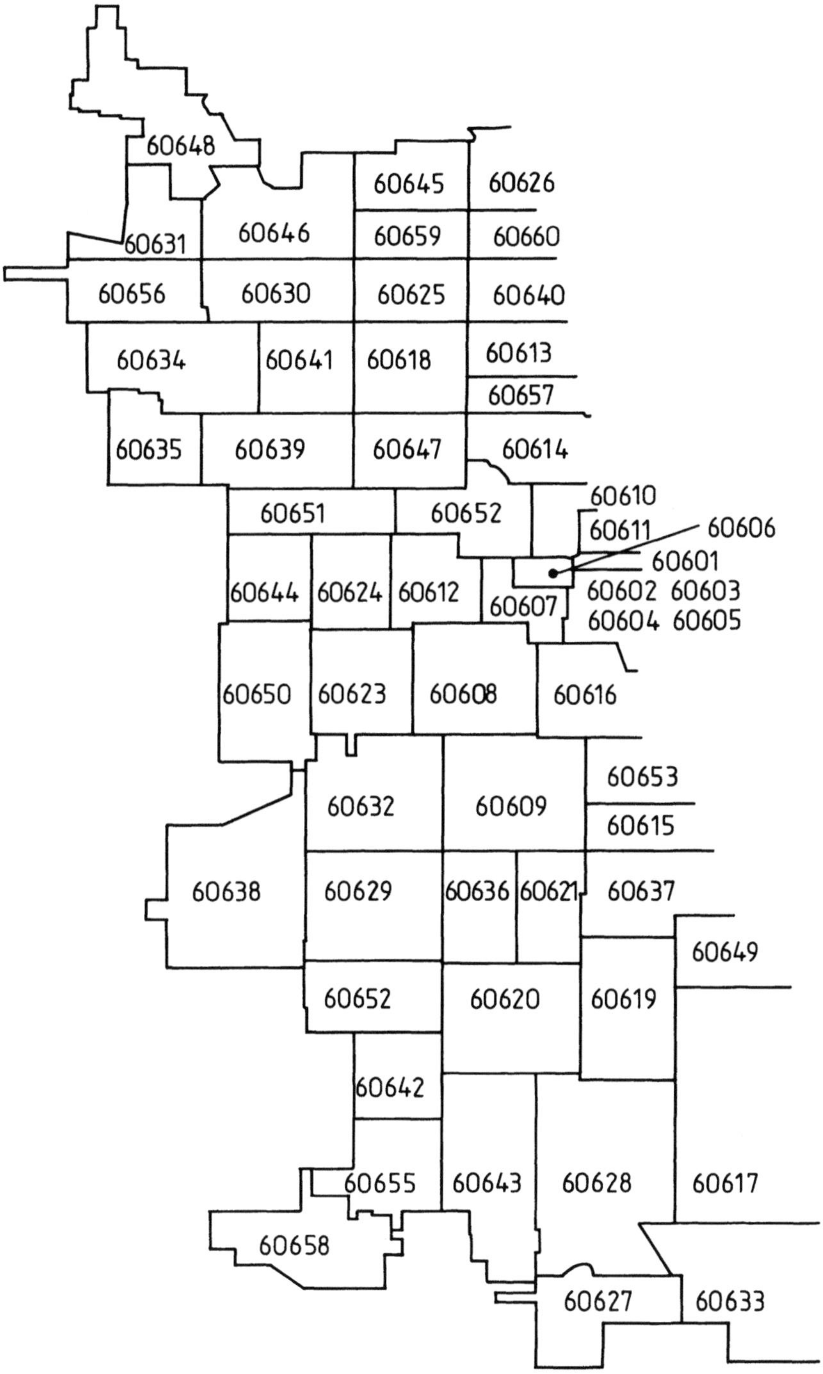

Figure 67.1
ZIP Code Areas of Chicago

68. Factors Influencing Motor Insurance Rates

Source Swedish Committee on the Analysis of Risk Premium in Motor Insurance.

Contributor J. Jung Försäkringsbranschens Serviceaktiebolag, Stockholm

In most countries, motor car insurance is obligatory. The problems concerning this type of insurance are the same, although the technical solution to these problems may vary. The data given in Table 68.1 present Swedish third party motor insurance for 1977 for one of seven geographical zones.*

In Sweden all motor insurance companies apply identical risk arguments to classify customers, and thus their portfolios and their claims statistics can be, and are, combined. The data were compiled by a Swedish Committee on the Analysis of Risk Premium in Motor Insurance. The Committee was asked to look into the problem of analyzing the real influence on claims of the risk arguments and to compare this structure with the actual tariff.

The influence of the tariff arguments represents different types of causal effects:

(i) Exposure is measured in kilometres per year, although the extreme class above 25000 kilometres per year is too wide.

(ii) The zones are given from a detailed investigation of 100 areas in 1972 and represent combinations of traffic intensity, state of roads, climatic differences, etc.

(iii) The bonus is a measure of individual claim history and thus some sort of Bayesian correction, but unfortunately, or rather fortunately, the accident frequency is too low, about 3% per year, to admit a good *a posterior* estimate.

The car models for private cars are classified into 10 premium classes, but in a special investigation for 1977 eight common pure models were chosen and the rest were put in a combined class for reference. A statistician who is not responsible for the tariff might cut out this class, i.e. Class 9. Originally the model year was included for comparison with the ordinary motor insurance. As no influence was shown it has been removed. The influence of the car make is interesting. It is associated not only with vehicle quality but also with the selection of the customers. For example, if one were to add some shiny decoration, put on the letters GT and add something to the price, the associated accident

* The six other geographical regions are available on the tape.

frequency is remarkably increased.

For a long time, the Swedish motor insurance companies have agreed upon a common structure, based on the identical classification of risks and of a multiplicative premium model, where the net premium was obtained from a basic premium, multiplied by a factor for each risk classification. In principle each company could use their own factors, although all companies agreed upon common bonus factors. When, a decade ago, the companies added the premium class factor, based on makes and models, it was in order to counteract a tendency by the auto industry to lower the selling prices of cars and to compensate themselves by increased repair prices, which hit the motor insurance companies. The result was good. However, the small companies could not classify the car makes and it was decided to make a common classification, based on the combined material. This was easy, due to common risk classes and to the multiplicative model, which made it possible to 'norm' each risk exposure for all factors but the make, and then to compare these normed claim statistics for all car makes. This method presumes that the multiplicative model is reasonably describing the risk variation, which ought to be studied by the committee. On the other side, the Supervisory Service asked if the companies had made certain that their tariff system was fair to the insured.

It is well-known, that in other countries, the tariff is "additive", i.e. that each risk classification gives constant additions to the premium. During the years, new risk classifications have been added, and the system is not simple to administrate. One of the problems is the subjective roll of the "mileage". Some customers change cars often, and it is very difficult to check the real distances driven except when a claim occurs. Now and then a committee is set up to see if the tariff could not be simplified and this happened in 1977. There is always a young man proposing that all factors except the bonus factor should be abolished, as the man driving short distances on the countryside should obtain a high bonus class. This might have been true if the average claim frequency had been one in the year.

These different reasons caused the Committee for Actuarial Research to appoint the Analysis of Risk group with the object to study the risk premium structure as a function of the risk classification; to see if this structure could reasonably well be described by some well known simple models, or if this were not the case, to look for better approximations; to check if any of the classifications could be abolished without creating systematic injustice to the policy holders.

It should be pointed out that the Analysis of Risk group is independent of the motor insurance companies. The report, unfortunately in Swedish, concluded that the multiplicative model is fairly good, that an additive model is worse but acceptable, that any model better than these two will be complicated to administrate, and that none of the factors used could be left out without biased premiums.

On the other hand the group concluded, that the premium classes based on statistics for separate makes were very uncertain for unusual makes, and might be replaced by classes based on motor strength and car weight.

This report was then handed over to the motor companies' own committees for their consideration.

Table 68.1
Third Party Motor Insurance for Sweden, 1977

Kilometers travelled *	Geographical zone **	No claims bonus †	Make ††	Number of insured, yrs × 100,000	Number of claims	Sum of payments, Skr
1	1	1	1	455.13	108	392491
1	1	1	2	69.17	19	46221
1	1	1	3	72.88	13	15694
1	1	1	4	1292.39	124	422201
1	1	1	5	191.01	40	119373
1	1	1	6	477.66	57	170913
1	1	1	7	105.58	23	56940
1	1	1	8	32.55	14	77487
1	1	1	9	9998.46	1704	6805992
1	1	2	1	314.58	45	214011
1	1	2	2	61.82	10	65303
1	1	2	3	47.06	5	20871
1	1	2	4	782.58	48	242894
1	1	2	5	115.43	11	23545
1	1	2	6	338.06	23	39598
1	1	2	7	70.44	7	48767
1	1	2	8	15.25	2	6560
1	1	2	9	6416.19	638	2873487
1	1	3	1	309.98	24	134931
1	1	3	2	49.18	6	50908
1	1	3	3	32.02	3	4399
1	1	3	4	497.20	23	112992
1	1	3	5	73.48	6	14788
1	1	3	6	278.01	9	48713
1	1	3	7	66.36	9	52076
1	1	3	8	17.86	3	13161
1	1	3	9	5063.15	408	1707680
1	1	4	1	318.48	29	103866
1	1	4	2	57.21	7	77588
1	1	4	3	35.33	4	11839
1	1	4	4	374.28	20	98140
1	1	4	5	85.18	7	27919
1	1	4	6	199.70	7	103910
1	1	4	7	60.46	4	38065
1	1	4	8	12.74	0	0
1	1	4	9	4263.09	300	1267678
1	1	5	1	444.37	25	69203
1	1	5	2	86.65	6	14620
1	1	5	3	53.81	5	40258
1	1	5	4	361.62	22	161455
1	1	5	5	117.91	3	20011
1	1	5	6	232.55	11	57214
1	1	5	7	81.27	3	4496
1	1	5	8	18.21	0	0
1	1	5	9	4761.37	301	1116208

Table 68.1 *(cont.)*

Kilometers travelled *	Geographical zone **	No claims bonus †	Make ††	Number of insured, yrs × 100,000	Number of claims	Sum of payments, Skr
1	1	6	1	1016.67	61	217617
1	1	6	2	150.56	12	58099
1	1	6	3	126.69	4	12268
1	1	6	4	517.31	16	59634
1	1	6	5	246.62	13	84966
1	1	6	6	482.96	19	137005
1	1	6	7	203.60	12	33767
1	1	6	8	25.88	3	6279
1	1	6	9	9197.99	522	1939894
1	1	7	1	5430.48	214	1048698
1	1	7	2	659.54	24	143915
1	1	7	3	657.34	22	153830
1	1	7	4	2795.72	60	202413
1	1	7	5	1119.12	41	180345
1	1	7	6	2861.69	92	484604
1	1	7	7	1111.00	37	152801
1	1	7	8	166.61	6	14084
1	1	7	9	48264.64	1875	8977527
2	1	1	1	638.51	148	578831
2	1	1	2	94.77	29	137355
2	1	1	3	126.87	29	137680
2	1	1	4	398.58	55	162678
2	1	1	5	167.88	41	73435
2	1	1	6	354.26	48	166416
2	1	1	7	95.57	11	21295
2	1	1	8	25.93	5	2059
2	1	1	9	6792.88	1410	6556534
2	1	2	1	605.01	86	484085
2	1	2	2	96.42	12	93305
2	1	2	3	94.57	12	62554
2	1	2	4	319.20	20	43734
2	1	2	5	152.42	24	109785
2	1	2	6	349.20	27	92572
2	1	2	7	94.39	8	76099
2	1	2	8	16.52	3	39939
2	1	2	9	5941.02	722	3249461
2	1	3	1	599.51	78	450864
2	1	3	2	121.49	17	142141
2	1	3	3	91.96	13	93038
2	1	3	4	234.48	13	65170
2	1	3	5	134.82	19	111059
2	1	3	6	318.69	21	162284
2	1	3	7	88.04	11	30153
2	1	3	8	22.47	5	47551
2	1	3	9	5162.92	484	2522341
2	1	4	1	597.21	55	213704
2	1	4	2	129.44	7	47589
2	1	4	3	94.80	9	47251
2	1	4	4	182.93	12	59369
2	1	4	5	111.64	11	20942

Table 68.1 *(cont.)*

Kilometers travelled *	Geographical zone **	No claims bonus †	Make ††	Number of insured, yrs × 100,000	Number of claims	Sum of payments, Skr
2	1	4	6	263.94	9	44772
2	1	4	7	107.37	9	61695
2	1	4	8	23.10	3	40634
2	1	4	9	4712.87	407	1631427
2	1	5	1	779.53	55	232521
2	1	5	2	157.99	13	31952
2	1	5	3	137.19	6	50067
2	1	5	4	156.34	11	18221
2	1	5	5	149.71	15	32164
2	1	5	6	295.12	13	59509
2	1	5	7	132.79	8	55603
2	1	5	8	29.02	2	2802
2	1	5	9	5373.06	363	1597528
2	1	6	1	1557.98	107	435444
2	1	6	2	329.74	29	133272
2	1	6	3	320.71	13	150031
2	1	6	4	210.91	4	15193
2	1	6	5	336.02	30	194477
2	1	6	6	576.63	24	137963
2	1	6	7	269.81	9	87411
2	1	6	8	56.96	9	144950
2	1	6	9	9805.53	618	2918359
2	1	7	1	8009.92	390	1992027
2	1	7	2	1407.85	67	332194
2	1	7	3	1538.74	49	397236
2	1	7	4	806.00	31	209824
2	1	7	5	1380.85	83	398103
2	1	7	6	2848.53	111	493340
2	1	7	7	1496.62	74	282000
2	1	7	8	267.80	14	95457
2	1	7	9	44511.41	2127	10315455
3	1	1	1	213.38	53	244561
3	1	1	2	76.97	26	187544
3	1	1	3	60.42	8	18624
3	1	1	4	60.89	10	81052
3	1	1	5	53.30	10	52745
3	1	1	6	104.58	13	168471
3	1	1	7	24.92	4	5100
3	1	1	8	12.91	3	42372
3	1	1	9	2375.34	456	1911772
3	1	2	1	276.13	28	106668
3	1	2	2	73.41	14	30232
3	1	2	3	69.26	14	61941
3	1	2	4	73.29	5	66851
3	1	2	5	52.25	7	57304
3	1	2	6	139.64	9	59193
3	1	2	7	42.93	3	5646
3	1	2	8	13.03	0	0
3	1	2	9	2472.22	297	1413344
3	1	3	1	299.10	36	196119
3	1	3	2	87.60	11	55966

Table 68.1 *(cont.)*

Kilometers travelled *	Geographical zone **	No claims bonus †	Make ††	Number of insured, yrs × 100,000	Number of claims	Sum of payments, Skr
3	1	3	3	67.91	6	13026
3	1	3	4	65.39	2	2186
3	1	3	5	60.44	8	23091
3	1	3	6	135.85	6	74018
3	1	3	7	48.67	2	3044
3	1	3	8	16.35	3	12589
3	1	3	9	2411.79	256	1266725
3	1	4	1	336.16	29	98220
3	1	4	2	119.03	9	61145
3	1	4	3	75.71	3	16285
3	1	4	4	44.78	0	0
3	1	4	5	69.52	11	67180
3	1	4	6	115.91	10	25912
3	1	4	7	59.11	2	3783
3	1	4	8	18.65	0	0
3	1	4	9	2419.59	209	905418
3	1	5	1	480.43	41	129476
3	1	5	2	164.22	16	133759
3	1	5	3	123.44	7	48811
3	1	5	4	50.82	2	4868
3	1	5	5	102.56	3	13127
3	1	5	6	159.79	15	44864
3	1	5	7	73.37	9	56128
3	1	5	8	32.92	2	1512
3	1	5	9	3214.83	274	1088795
3	1	6	1	1110.72	84	399470
3	1	6	2	327.30	32	100742
3	1	6	3	300.23	47	234344
3	1	6	4	52.79	2	6842
3	1	6	5	209.45	23	67428
3	1	6	6	303.79	20	70487
3	1	6	7	184.93	16	90830
3	1	6	8	62.38	2	62884
3	1	6	9	6151.48	625	3050824
3	1	7	1	4918.02	236	1326307
3	1	7	2	1398.19	72	342421
3	1	7	3	2349.67	45	142660
3	1	7	4	252.36	15	25307
3	1	7	5	881.09	52	195295
3	1	7	6	1520.51	58	307753
3	1	7	7	867.16	45	110181
3	1	7	8	306.00	7	68657
3	1	7	9	26905.35	1481	6783057
4	1	1	1	49.31	11	23218
4	1	1	2	22.93	6	14940
4	1	1	3	34.01	9	101091
4	1	1	4	6.08	0	0
4	1	1	5	17.77	4	4164
4	1	1	6	18.59	1	3044
4	1	1	7	3.89	0	0
4	1	1	8	4.31	1	400

Table 68.1 *(cont.)*

Kilometers travelled *	Geographical zone **	No claims bonus †	Make ††	Number of insured, yrs × 100,000	Number of claims	Sum of payments, Skr
4	1	1	9	601.75	103	381138
4	1	2	1	77.53	6	16812
4	1	2	2	29.09	5	9656
4	1	2	3	36.39	6	10543
4	1	2	4	11.32	1	8231
4	1	2	5	18.74	3	4571
4	1	2	6	32.02	3	38542
4	1	2	7	13.62	1	954
4	1	2	8	3.73	0	0
4	1	2	9	735.71	86	391512
4	1	3	1	85.77	4	14451
4	1	3	2	36.08	2	3784
4	1	3	3	34.31	6	8274
4	1	3	4	12.01	1	917
4	1	3	5	19.10	4	14439
4	1	3	6	32.80	4	12608
4	1	3	7	18.59	1	900
4	1	3	8	8.54	3	4956
4	1	3	9	729.62	88	599538
4	1	4	1	105.73	15	90111
4	1	4	2	35.14	6	38663
4	1	4	3	53.38	1	3000
4	1	4	4	6.89	1	4686
4	1	4	5	18.19	3	3372
4	1	4	6	34.63	2	3178
4	1	4	7	16.77	0	0
4	1	4	8	11.75	0	0
4	1	4	9	766.14	71	430373
4	1	5	1	166.10	7	40729
4	1	5	2	66.54	5	43356
4	1	5	3	48.99	7	17562
4	1	5	4	10.91	1	993
4	1	5	5	32.50	5	47591
4	1	5	6	53.90	4	15205
4	1	5	7	17.88	2	8643
4	1	5	8	17.04	2	3314
4	1	5	9	1064.66	96	403899
4	1	6	1	344.68	32	184118
4	1	6	2	125.36	15	34894
4	1	6	3	130.03	11	66353
4	1	6	4	15.00	0	0
4	1	6	5	73.29	10	122551
4	1	6	6	95.16	4	9655
4	1	6	7	51.85	3	33098
4	1	6	8	35.32	5	12104
4	1	6	9	2154.39	173	702215
4	1	7	1	1528.57	87	483084
4	1	7	2	567.56	35	209646
4	1	7	3	563.23	34	107767
4	1	7	4	65.95	3	12950
4	1	7	5	278.46	24	93032

Table 68.1 *(cont.)*

Kilometers travelled *	Geographical zone **	No claims bonus †	Make ††	Number of insured, yrs × 100,000	Number of claims	Sum of payments, Skr
4	1	7	6	395.47	10	27088
4	1	7	7	230.90	5	22495
4	1	7	8	163.53	8	27401
4	1	7	9	8738.13	520	2437429
5	1	1	1	39.38	8	19055
5	1	1	2	33.22	14	70516
5	1	1	3	25.94	6	12853
5	1	1	4	4.97	0	0
5	1	1	5	5.83	1	5315
5	1	1	6	10.55	2	14735
5	1	1	7	4.18	2	3699
5	1	1	8	8.91	4	2745
5	1	1	9	643.63	172	802056
5	1	2	1	40.03	7	53251
5	1	2	2	31.20	8	64325
5	1	2	3	40.44	6	51965
5	1	2	4	4.58	1	8390
5	1	2	5	10.55	0	0
5	1	2	6	20.21	2	8498
5	1	2	7	8.99	0	0
5	1	2	8	7.78	2	31442
5	1	2	9	697.77	122	505940
5	1	3	1	54.57	8	17149
5	1	3	2	36.47	8	46332
5	1	3	3	32.03	6	42364
5	1	3	4	4.03	0	0
5	1	3	5	11.06	0	0
5	1	3	6	18.96	1	1000
5	1	3	7	8.71	2	8327
5	1	3	8	14.50	5	39056
5	1	3	9	618.16	98	524453
5	1	4	1	61.37	3	67055
5	1	4	2	40.59	5	18027
5	1	4	3	32.00	0	0
5	1	4	4	2.43	0	0
5	1	4	5	16.54	1	1725
5	1	4	6	20.67	3	10781
5	1	4	7	9.94	1	3530
5	1	4	8	21.16	3	3762
5	1	4	9	617.39	81	449815
5	1	5	1	92.05	11	18385
5	1	5	2	65.64	10	47270
5	1	5	3	58.99	5	40553
5	1	5	4	1.57	0	0
5	1	5	5	18.46	2	8860
5	1	5	6	22.92	1	1845
5	1	5	7	11.36	3	36029
5	1	5	8	22.87	1	3801
5	1	5	9	820.04	111	426845
5	1	6	1	183.54	8	30510
5	1	6	2	116.62	15	66600

Table 68.1 *(cont.)*

Kilometers travelled *	Geographical zone **	No claims bonus †	Make ††	Number of insured, yrs × 100,000	Number of claims	Sum of payments, Skr
5	1	6	3	115.02	11	56739
5	1	6	4	4.61	0	0
5	1	6	5	42.25	4	6990
5	1	6	6	47.19	4	4778
5	1	6	7	23.47	1	2943
5	1	6	8	55.74	8	105275
5	1	6	9	1623.22	154	653118
5	1	7	1	717.63	47	208573
5	1	7	2	401.42	35	150942
5	1	7	3	439.61	30	266009
5	1	7	4	21.22	1	2000
5	1	7	5	145.24	9	53228
5	1	7	6	182.52	9	15099
5	1	7	7	102.40	5	8972
5	1	7	8	186.24	14	31626
5	1	7	9	5539.09	416	2249007

* Kilometers travelled
1: less than 1000 km. per year
2: 1000-15000 km. per year
3: 15000-20000 km. per year
4: 20000-25000 km. per year
5: more than 25000 km. per year

** Geographical zone
The car is classified according to the insured's home address.
1: Stockholm, Göteborg, Malmö with surroundings
2: Other bigger cities with surroundings
3: Smaller cities with surroundings in southern Sweden
4: Rural areas in southern Sweden
5: Smaller cities with surroundings in northern Sweden
6: Rural areas in northern Sweden
7: Gotland
Note: only data from zone 1 are presented.
The other zones are available on tape.

† No claims bonus
The insured starts in the class B = 1.
Every year there is no claim he is moved up one class.

†† Make
1-8: Different specified car models.

69. Disputed Authorship : The Federalist Papers

Source Mosteller, F. and Wallace, D.L. (1964). *Inference and Disputed Authorship: The Federalist.* Reading, Mass: Addison-Wesley, with the permission of the copyright owners.

Contributor F. Mosteller Harvard University and D.L. Wallace University of Chicago

The *Federalist* papers were written between 1787-1788 by Alexander Hamilton, John Jay and James Madison to persuade the citizens of the State of New York to ratify the American Constitution. As was common in those days, these short essays, about 900-3500 words in length, appeared in newspapers signed with a pseudonym, in this instance, 'Publius'. Seventy-seven essays first appeared in several different newspapers, and then Hamilton wrote an additional eight essays designed to complete the job.

It is of interest to discover something about the distribution of the words. It had been thought that they would have been distributed according to the Poisson distribution, but this was not the case. Thus there were considerable complications in the analysis of the data.

Table 69.1 gives the distribution of the occurrences of the *function* words for Hamilton and Madison. Function words are filler words such as a, an, by, to, that. For Alexander Hamilton the table gives the frequency distributions of the function words in 247 blocks of 200 words each of running text; for James Madison there are 262 blocks of about 200 words each. Here *about* means within one or two and usually exactly. These all came from writings known to be by these two authors. For a more detailed study, words with asterisks in the table might be analyzed separately.

Table 69.1
Distribution of Occurrences for Function Words

	Frequency										
Words	0	1	2	3	4	5	6	7	8	9	14
				Hamilton							
this	45	80	71	39	8	4					
an *	77	89	46	21	9	4	1				
or	86	69	49	21	9	6	6	1			
would	90	47	29	28	22	14	5	7	4	1	
from *	93	82	51	13	5	2	1				
will	105	60	31	26	12	8	3	2			
its	116	82	29	11	4	3	2				
their	118	66	34	16	9	2	1	1			
if	118	87	31	7	3	1					
any *	125	88	26	7		1					
may *	128	67	32	14	4	1	1				
upon *	129	83	20	9	5	1					
at	129	70	42	4	2						
all	132	67	32	13	2	1					
there	138	75	23	8	2	1					
been	138	65	24	14	3	3					
than	143	66	29	6	2	1					
on	145	67	27	7	1						
one	149	77	16	3	2						
more	152	70	15	7	2	1					
can *	157	60	20	5	2	2	1				
has	157	57	20	11	2						
should	161	58	26	2							
who	163	53	25	3	2	1					
no	167	60	11	8	1						
so	170	55	19	2	1						
such	173	56	13	5							
must	173	49	14	9	1		1				
into	183	50	12	2							
only	185	54	7	1							
every *	186	46	14	1							
what	188	44	11	3			1				
was	192	42	7	2	3	1					
were	194	44	7	1	1						
when	195	40	9	3							
had	200	35	8	4							
some	200	38	9								
even	204	39	4								
his *	192	18	17	7	3	2	4	1	2		1
our	212	23	9	1		1	1				
do *	228	16	2	1							
then	230	17									
up	231	14	2								
also	232	15									
now	234	13									
things	236	11									
down	240	6	1								
my *	241	6									
her	241	3	1	1						1	
unto	247										
your	247										

Table 69.1 *(cont.)*

	Frequency										
Words	0	1	2	3	4	5	6	7	8	9	14
				Madison							
on	63	80	55	32	20	8	4				
this	80	83	71	21	6	1					
from *	90	93	42	17	8	9	3				
or	103	74	51	16	12	3	2	1			
an *	122	77	40	14	8		1				
was	129	55	32	16	20	5	3		2		
been	132	74	35	12	4	5					
at	133	86	31	8	1	3					
any *	145	90	19	8							
all	146	68	28	16	4						
may *	156	63	29	8	4	1	1				
than	157	84	19	2							
its	158	66	29	7	2						
more	158	68	30	4	2						
their	159	71	17	12	2		1				
no	165	71	16	7	2	1					
has	167	64	19	9	1	1	1				
would	167	49	31	9	1	3		2			
such	170	66	22	3	1						
if	171	60	22	6	3						
one	172	66	15	5	3			1			
will	172	55	19	8	3	3		2			
into	178	71	11	2							
were	179	58	18	5	1	1					
only	182	65	11	4							
so	182	59	17	3	1						
had	185	52	16	6		2	1				
her	200	26	10	9	4	4	3	2	1	2	1
must	202	47	10	2	1						
some	205	53	4								
there	208	47	4	2	1						
every *	209	43	4	6							
can *	211	44	6	1							
his *	213	21	9	11	2	2	1	2	1		
should	218	32	12								
then	219	34	8	1							
what	220	33	8	1							
even	221	38	3								
also	222	36	4								
who	227	28	5	2							
now	236	24	2								
when	236	23	2	1							
do *	244	15	2	1							
our	245	13	4								
down	246	15	1								
up	254	8									
upon *	254	7	1								
things	254	7	1								
my *	259	3									
unto	262										
your	262										

70. Platonic Prose Rhythm

Source Wishart, D. and Leach, S.V. (1970). A multivariate analysis of Platonic prose. *Computer Studies of the Humanities and Verbal Behavior* **3**, 90-99.

Contributor D. Wishart Scottish Office, Edinburgh

Prose rhythm may be characterized by the occurrence of five-syllable sequences in passages of text. This characterization may be used to assess the similarity among passages. The data presented in Table 70.1 come from thirty-three passages representing ten Platonic texts. Each passage was divided into sentences, lists of words ending with a colon, question mark or period. Syllables within each sentence were classified as long or short. Each sequence of five syllables was identified as being one of the thirty-two possible groupings of five long or short syllables. Thus each sentence of length N greater than 5 contributed N-4 such identifications. Table 70.1 gives the percent of each of the thirty-two five-syllable sequences for each of ten books: *Timaeus, Critias, Laws, Republic, Phaedrus, Symposium, Sophistes, Philedus, Seventh Epistle* and *Politics*.

Table 70.1
Percentage Occurrences of the Thirty-two Five-syllable Groups

Syllable groups	Tim	Crit	Laws	Rep	Pha	Symp	Soph	Phil	Ept	Pol
11111	2.09	1.93	1.37	0.85	0.52	1.07	2.22	1.28	1.09	1.71
01111	2.77	2.79	2.10	1.64	1.11	1.68	2.80	2.32	2.50	3.01
10111	3.11	2.83	2.52	1.98	1.81	2.03	3.34	2.59	2.86	2.69
11011	3.13	3.00	2.00	2.36	2.18	2.50	3.07	2.09	2.46	2.41
11101	2.95	2.95	2.35	1.81	1.61	2.05	2.70	2.41	3.30	2.87
11110	2.75	2.83	2.02	1.60	1.08	1.76	2.87	2.30	2.39	2.90
00111	3.45	3.38	4.00	2.24	1.74	2.08	3.28	3.82	3.88	4.09
01011	2.99	2.57	2.23	2.60	3.33	2.64	3.24	2.09	2.72	2.45
01101	3.00	2.88	1.79	2.98	3.80	3.16	2.80	1.88	2.75	1.99
01110	3.71	3.40	4.46	2.53	2.49	2.51	3.55	4.02	4.27	3.81
10011	3.50	3.10	2.40	2.67	2.71	2.59	2.97	2.68	2.90	3.46
10101	2.57	1.84	1.81	2.79	3.80	2.71	2.32	1.91	2.54	1.47
10110	2.97	2.74	1.70	2.89	3.63	3.08	2.90	1.61	2.28	2.10
11001	3.55	2.85	2.66	2.58	2.81	2.57	3.17	2.83	2.83	3.36
11010	2.85	2.77	2.16	2.50	3.22	2.63	2.56	2.24	3.55	2.45
11100	3.50	3.24	4.00	2.35	1.95	2.25	3.65	3.90	3.40	4.09
00011	3.27	3.69	4.10	3.29	3.24	3.32	3.11	4.05	3.91	3.50
00101	3.20	2.75	2.79	3.69	3.94	3.39	3.48	3.01	2.72	2.94
00110	3.28	3.38	2.47	3.66	4.27	3.81	2.73	2.80	2.97	2.94
01001	3.55	3.32	3.19	3.58	3.71	3.12	3.55	3.07	3.33	3.50
01010	2.82	2.03	2.35	3.88	4.52	3.39	2.66	2.83	2.54	1.99
01100	3.24	3.27	2.41	3.47	4.13	3.76	2.90	2.56	2.54	3.08
11000	3.20	3.59	3.74	3.21	3.22	3.38	3.38	3.55	2.97	3.77
10100	3.05	2.94	2.74	3.50	3.98	3.21	3.00	3.22	3.51	2.90
10010	3.59	3.04	3.50	3.64	3.84	3.10	3.69	3.31	3.37	3.39
10001	3.28	3.48	2.94	3.70	4.20	3.52	3.11	3.19	3.11	2.97
10000	3.00	3.71	5.02	4.14	3.44	4.31	3.31	4.64	3.77	4.19
01000	3.08	3.57	4.15	4.64	4.45	4.46	3.04	4.44	3.84	3.53
00100	3.57	3.98	4.51	4.71	4.27	4.34	3.58	4.17	3.69	4.09
00010	3.18	3.62	4.05	4.63	4.45	4.55	3.24	3.96	3.19	3.74
00001	3.05	3.76	5.15	4.20	3.46	4.28	3.31	4.74	3.91	4.26
00000	2.75	4.66	7.32	5.68	3.07	6.50	4.47	6.47	4.93	4.37

71. Relationships between Birthday and Deathday

Contributor C. O'Brien Imperial College of Science and Technology, University of London

In Phillips (1972), an investigation is made of the relationship between date of birth and date of death in a sample of famous Americans. One conclusion is that famous people are less likely to die in the month preceding their birth month than at any other time. The data in Tables 71.1 and 71.2 allow comparisons to be made among various types of famous people and ordinary people.

The data have been classified in two ways: first by the number of full calender months elapsed since the last birthday on the day of death, Table 71.1, and secondly by month of birth and month of death, Table 71.2. The latter is the classification used by Phillips. The data may be examined for deviations from uniformity using contingency tables. Other techniques may be useful.

Notes on the Samples:

1. *Famous Scientists.* A random sample of 400 was taken from *Biographical Encyclopaedia of Science and Technology* by I. Asimov. Although the book covers many centuries the majority of data is from the period 1500-1900, since earlier information is often incomplete.

2. *Famous Writers.* The first 400 people, for whom complete information is available in *Everyman's Dictionary of Literary Biography, English and American*, were used. Since this book is arranged alphabetically, it was felt that the sample would be random in terms of dates.

3. *Ordinary People.* A random sample of 399 was taken from records at the Office of Population Censuses and Surveys of people who died in England and Wales in 1973. This information is stored on over three hundred tapes with approximately 2,000 entries per tape. A machine was available for reading the tapes and would find a specified entry on a tape. Random numbers were generated and the corresponding entries found on the assumption that there were 2,000 entries on each tape. This assumption, combined with the less than perfect accuracy of the counting device on the machine, means that the entries read did not correspond exactly with those indicated by the random numbers.

4. *Royal Family.* All the descendants of Queen Victoria who have died, excluding those killed in action, the members of the Russian Imperial family assassinated in 1918, and the Grand Ducal family of Hesse, killed in an airplane crash in 1937, make up this sample. The information was taken from *Whitakers Almanac, 1977.*

Table 71.3 gives the exact birth and death dates for the individuals in Example 4.

References

Asimov, I. (Ed.) (1972). *Biographical Encyclopedia of Science and Technology*, revised edition. London: George Allen and Unwin.

Browning, D.C. (Ed.) (1958). *Everyman's Dictionary of Literary Biography, English and American*. London: Dent.

Phillips, D.P. (1972). Deathday and birthday: An unexpected connection. In *Statistics: A Guide to the Unknown*, 2nd edition, edited by J. Tanur *et al.*, pp. 52-65. San Francisco: Holden-Day.

Whitaker's Almanac, 1977. London: Whitaker.

Table 71.1
Deaths of All Individuals
Classified by Full Calendar Months Elapsed
Since Last Birthday to Day of Death

	Full calendar months since last birthday												
	0	1	2	3	4	5	6	7	8	9	10	11	
Sample	No. of deaths												Total
1	41	28	35	40	40	30	34	25	31	27	39	30	400
2	24	34	33	33	28	29	33	52	31	33	35	35	400
3	30	32	39	23	34	36	33	28	34	39	32	39	399
4	4	5	7	10	4	8	7	9	13	5	4	6	82

Table 71.2
Deaths of Individuals Classified by Type of Notoriety, Month of Birth and Month of Death

Month of birth	Month of death: Jan	Feb	March	April	May	June	July	Aug	Sept	Oct	Nov	Dec	Total
						Sample 1							
Jan	2	6	7	4	3	4	1	7	1	2	3	3	43
Feb	3	4	1	3	4	3	1	1	2	5	1	6	34
March	2	4	7	3	3	1	4	4	1	1	4	3	37
April	0	3	0	2	4	2	2	5	0	2	1	2	23
May	1	3	3	2	1	2	1	5	5	2	1	1	27
June	4	1	0	2	7	1	1	2	1	3	4	0	26
July	5	2	5	1	2	5	2	3	1	2	0	3	31
Aug	4	3	4	3	4	3	1	4	2	2	3	5	38
Sept	1	4	4	3	1	0	3	3	7	0	3	2	31
Oct	2	5	7	2	1	2	5	1	1	3	2	2	33
Nov	3	6	2	4	4	2	4	3	1	1	5	2	37
Dec	1	5	3	4	6	2	3	2	5	2	5	2	40
Total	28	46	43	33	40	27	28	40	27	25	32	31	400
						Sample 2							
Jan	4	4	1	6	1	3	4	5	5	5	2	4	44
Feb	4	0	3	1	3	2	0	2	5	2	0	2	24
March	4	7	0	1	5	4	3	2	3	1	4	1	35
April	1	7	7	2	2	4	1	5	2	3	1	4	39
May	2	0	2	2	1	7	1	2	1	1	2	1	22
June	3	1	4	1	1	2	3	1	3	1	1	4	25
July	2	3	6	2	2	2	2	1	1	0	2	1	24
Aug	2	3	3	4	3	4	1	4	3	1	2	3	33
Sept	2	1	4	4	4	4	1	1	2	3	2	2	30
Oct	5	2	2	4	5	4	1	5	4	2	3	4	41
Nov	3	1	6	4	2	6	1	3	4	4	0	3	37
Dec	1	8	4	7	4	4	4	6	2	2	0	4	46
Total	33	37	42	38	33	46	22	37	35	25	19	33	400

Table 71.2 *(cont.)*

Month of birth	Month of death Jan	Feb	March	April	May	June	July	Aug	Sept	Oct	Nov	Dec	Total
						Sample 3							
Jan	2	4	4	4	4	2	5	3	2	4	5	6	45
Feb	5	3	2	3	2	0	3	2	0	3	0	3	26
March	5	3	2	3	7	4	2	4	5	0	5	4	44
April	3	1	1	3	3	5	0	1	1	1	5	5	29
May	3	2	5	2	1	2	6	1	4	2	5	2	35
June	5	3	1	4	4	2	3	1	4	5	2	1	35
July	3	5	0	0	1	4	3	2	1	7	1	2	29
Aug	1	1	4	3	3	2	0	0	3	4	3	2	26
Sept	4	7	3	2	4	7	0	4	3	3	2	3	42
Oct	8	0	2	1	1	4	5	2	4	2	1	1	31
Nov	1	1	6	1	3	2	2	2	2	4	3	1	28
Dec	4	4	1	0	4	2	1	1	3	2	5	2	29
Total	44	34	31	26	37	36	30	23	32	37	37	32	399
						Sample 4							
Jan	1	0	0	0	1	2	0	0	1	0	1	0	6
Feb	1	0	0	1	0	0	0	0	0	1	0	2	5
March	1	0	0	0	2	1	0	0	0	0	0	1	5
April	3	0	2	0	0	0	1	0	1	3	1	1	12
May	2	1	1	1	1	1	1	1	1	1	1	0	12
June	2	0	0	0	1	0	0	0	0	0	0	0	3
July	2	0	2	1	0	0	0	0	1	1	1	2	10
Aug	0	0	0	3	0	0	1	0	0	1	0	2	7
Sept	0	0	0	1	1	0	0	0	0	0	1	0	3
Oct	1	1	0	2	0	0	1	0	0	1	1	0	7
Nov	0	1	1	1	2	0	0	2	0	1	1	0	9
Dec	0	1	1	0	0	0	1	0	0	0	0	0	3
Total	13	4	7	10	8	4	5	3	4	9	7	8	82

Table 71.3
The Exact Date of Birth and Death for Sample 4, The Royal Family

Birth		Death		Birth		Death	
Month	Day	Month	Day	Month	Day	Month	Day
1	8	6	22	6	14	1	13
1	8	1	14	6	23	5	28
1	13	9	12	7	6	12	3
1	15	5	1	7	7	12	7
1	18	11	8	7	7	4	1
1	27	6	4	7	11	11	11
2	2	10	27	7	12	1	18
2	2	12	2	7	13	3	7
2	20	1	4	7	14	9	22
2	25	12	5	7	19	3	6
2	26	4	27	7	24	10	1
3	17	1	12	7	27	1	27
3	18	12	3	8	1	10	25
3	20	5	2	8	6	7	30
3	24	5	11	8	9	4	26
3	31	6	10	8	12	12	8
4	2	10	26	8	14	4	20
4	3	12	14	8	26	12	14
4	5	9	24	8	30	4	8
4	7	3	28	9	1	4	16
4	12	11	13	9	6	11	5
4	14	10	29	9	9	5	24
4	14	10	26	10	3	10	27
4	20	7	13	10	5	1	20
4	22	1	21	10	11	11	15
4	22	1	26	10	15	2	6
4	25	1	14	10	15	4	4
4	25	3	28	10	24	4	15
5	1	1	16	10	29	7	18
5	3	3	13	11	6	4	8
5	6	7	20	11	9	5	6
5	12	8	26	11	20	5	30
5	12	5	20	11	21	8	5
5	17	2	26	11	23	2	23
5	21	4	23	11	25	10	9
5	22	9	8	11	25	3	2
5	24	1	22	11	26	11	20
5	24	11	15	11	30	8	27
5	25	6	9	12	4	3	6
5	30	10	16	12	14	2	6
6	3	1	20	12	17	7	17

Index

The page numbers correspond to the first page of the chapter containing the item listed.

MIX
Papier aus verantwortungsvollen Quellen
Paper from responsible sources
FSC® C105338

If you have any concerns about our products,
you can contact us on
ProductSafety@springernature.com

In case Publisher is established outside the EU,
the EU authorized representative is:
Springer Nature Customer Service Center GmbH
Europaplatz 3, 69115 Heidelberg, Germany

Printed by Libri Plureos GmbH
in Hamburg, Germany